Electronics and Electrical Engineering

Electronics and Electrical Engineering

Edited by **Marko Silver**

CLANRYE INTERNATIONAL

New Jersey

Published by Clanrye International,
55 Van Reypen Street,
Jersey City, NJ 07306, USA
www.clanryeinternational.com

Electronics and Electrical Engineering
Edited by Marko Silver

International Standard Book Number: 978-1-63240-539-5 (Hardback)

Printed in China.

Contents

Preface

Electronics engineering is a discipline which deals with electrical devices of low intensity current such as mobiles, radio, television, etc. It is used to design and develop new and improved devices for daily usage. On the other hand electrical engineering is applied to devices that use high currents. Although these fields are different but mostly they are applied in combination in a variety of devices. This book traces the progress of these areas and highlights some of their key concepts and applications. The extensive content of this book provides the readers with a thorough understanding of the subjects. Students, researchers, experts and all associated with electronics and electrical engineering will benefit alike from this book.

This book is a comprehensive compilation of works of different researchers from varied parts of the world. It includes valuable experiences of the researchers with the sole objective of providing the readers (learners) with a proper knowledge of the concerned field. This book will be beneficial in evoking inspiration and enhancing the knowledge of the interested readers.

In the end, I would like to extend my heartiest thanks to the authors who worked with great determination on their chapters. I also appreciate the publisher's support in the course of the book. I would also like to deeply acknowledge my family who stood by me as a source of inspiration during the project.

Editor

1

Active Fault Tolerant Control Based on Bond Graph Approach

Manel Allous[1] and Nadia Zanzouri[1,2]

[1] *Department of Electrical Engineering, LACS Laboratory, National Engineering School of Tunis, Université de Tunis El Manar, P.O. Box 37, Le Belvedere, 1002 Tunis, Tunisia*
[2] *Preparatory Engineering Institute of Tunis 2, Université de Tunis, Rue Jawaher Lel Nahrou-Monfleury, 1089 Tunis, Tunisia*

Correspondence should be addressed to Manel Allous; allousmanel@yahoo.fr

Academic Editor: Gorazd Stumberger

This paper proposes a structural fault recoverability analysis using the bond graph (BG) approach. Indeed, this tool enables structural analysis for diagnosis and fault tolerant control (FTC). For the FTC, we propose an approach based on the inverse control using the inverse BG. The fault tolerant control method is also compared with another approach. Finally, simulation results are presented to show the performance of the proposed approach.

1. Introduction

Due to the growing complexity of the dynamical systems, there is an increasing demand for safe operation, fault diagnosis (FDI) (fault detection and isolation), and fault tolerant control (FTC) (strategies for control redesign). Different approaches have been developed for the designing and the implementation of FDI and FTC procedures [1]. These techniques are based on the knowledge of the system model (model-based methods) [2, 3] or its structure (data-based methods) [4, 5]. FTC is categorized into two different techniques: passive FTC (PFTC) [6, 7] and active FTC (AFTC) [8, 9]. In PFTC, controllers are fixed and designed to be robust against a class of presumed faults. The AFTC approach reacts to system component failures actively by reconfiguring control actions and acceptable performance of the entire system can be maintained.

This paper is focused on the design of a novel AFTC that integrates a reliable and robust fault diagnosis scheme with the design of a controller reconfiguration system. The FDI and FTC are fully integrated in dynamic systems design in several fields of engineering, such as robotic and automotive systems. Nevertheless, it must have tool that enables coupling the diagnosis results with fault tolerant control conditions. Therefore, the BG enables integrating both structural diagnosis results with control analysis. A BG model allows knowledge of a large amount of structural, functional, and behavioral information. This information enables computing appropriate control actions that compensate the faults.

The BG has proven to be a powerful tool not only for generating the direct model of a system but also for obtaining its inverse model. In [10], the authors have proposed an inverse control strategy based on BG model.

The innovative interest of the present paper is to combine the inverse control strategy and observer designs to generate the FDI and FTC algorithms from the BG model. The proposed approach takes into account the parameter uncertainties and considers the fault recoverability with respect to fault compensation, without complex calculations.

In the first part of the paper, we propose a methodology based on BG model for fault detection and fault tolerant control. In the second part, we have developed a method which compensates the faults in the absence of complex calculations. Finally, an illustrative example is developed and simulation results show the advantage of the proposed approach.

2. FDI and FTC Approaches Based on Bond Graph

The bond graph approach is proposed by [11] and then developed by [12, 13]. This tool allows the multidomain

FIGURE 1: Power and orientation symbol on BG.

TABLE 1: BG elements.

$\overset{e}{\underset{f}{\rightharpoonup}} R$	$S_e \overset{e}{\underset{f}{\rightharpoonup}}; S_f \overset{e}{\underset{f}{\rightharpoonup}}$
$\overset{e=dp/dt}{\underset{f}{\rightharpoonup}} I$	$\overset{e_1}{\underset{f_1}{\rightharpoonup}} \text{Gy} \overset{e_2}{\underset{f_2}{\rightharpoonup}}$
$\overset{e}{\underset{f=dq/dt}{\rightharpoonup}} C$	$\overset{e_1}{\underset{f_1}{\rightharpoonup}} \text{TF} \overset{e_2}{\underset{f_2}{\rightharpoonup}}$

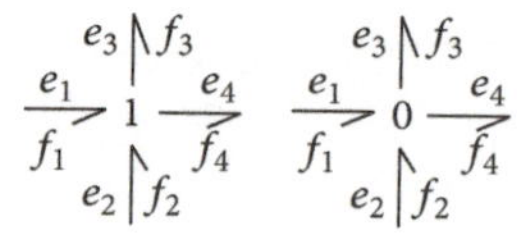

FIGURE 2: BG junctions.

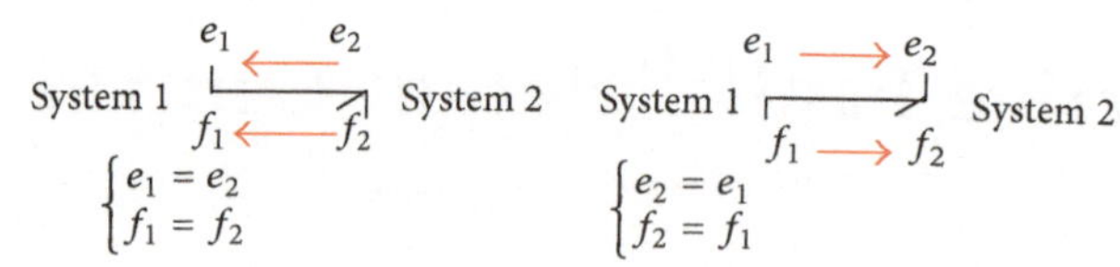

FIGURE 3: Bicausal bond graph concept.

systems (mechanical, electrical, thermal, etc.) to be described with the same components. Its causal structure was initially exploited to determine structural conditions of controllability, observability [14], and diagnosability [15, 16].

The bond graph is based on the graphical representation of the energy exchange within the system to be modeled. Table 1 represents the BG elements: resistor (R), compliance (C), and inertia (I) are passive elements. Effort source (S_e) and flow source (S_f) are the active elements.

Figure 1 indicates the power direction in the system.

There are only two types of junctions: the 1 and the 0 junctions (Figure 2). 1 junction has equality of flows and the efforts sum up to zero. 0 junctions have equality of efforts and the flows sum up to zero (Figure 2).

2.1. Luenberger Observer Based on BG. Bond graph approaches for observers design were developed in some works, such as Luenberger observers [17, 18], reduced order Luenberger observers [19], proportional integral observers [20], and nonlinear observers applied to electrical transformers [21].

The objective of this work is to design a Luenberger observer by bond graph for fault detection and isolation.

2.2. BG Modeling Bicausality Concept for System Inversion. The concept of "bicausal" introduced by Gawthrop [22] enlarges the possibilities of computation models that can be derived from a bond graph. The bicausal bond graph model is seen as half-strokes each associated with an effort and a flow variable that can be imposed independently at each end of the bond. Causal half-strokes indicate the fixed or known variables of the bond and so determine the right-hand side of the assignments form [23] (Figure 3).

The bicausality is used to get systems' inversion by imposing the output variable without modifying the energy structure. System inversion is an interesting analysis to know an input considering a given output. Therefore, in the next section, we use the bicausality property of the bond graphs.

Some conditions (structural invertibility) are proposed to present the bond graph-based procedure for system inversion [20].

Proposition 1. *A linear system modeled by bond graph is invertible if there is at least one causal path between the input variable and the output variable of the system.*

2.3. Control Law Design Directly on Inversion Bond Graph. The control strategy proposed by [24] computes the desired inputs based on the system objectives. Also, in [10], the authors have proposed an inverse control strategy from the BG model with parameter uncertainties estimated directly from the inverse BG-LFT (linear fractional transformation). The system inversion concept gives the basis to compute appropriate control actions that compensate the faults. Figure 4 shows the control design based on bond graph developed by [10].

In [10], the BG-LFT was used to estimate the faulty power. Then, to validate these structural results, a local adaptive compensation based on the inverse control strategy using the inverse BG-LFT was proposed. This strategy computes the desired inputs based on the system objectives and on the undesired power caused by the fault.

Limitations of this approach are as follows.

(i) The fault estimation is necessary for the control design. The inverted model of bond graph uses the estimate fault to compensate it.

(ii) The fault estimation with a BG-LFT causes FTC delay.

3. Proposed Approach

The principal of our proposed approach AFTC system is presented in Figure 5. There are basically two parts.

(i) The first part concerns the diagnosis by Luenberger observer using BG approach; in this part, the fault estimation and fault isolability are not necessary for system recoverability; just the residual is injected to the control loop.

(ii) The second part shows the control part determined by inverse BG for nominal system.

The following symbols have been used.

I.BG.N.S: Inverse BG for nominal system.

y_{ref}: Desired value.

y_{sys}: Measured output of system.

y_{obs}: Estimated output of observer.

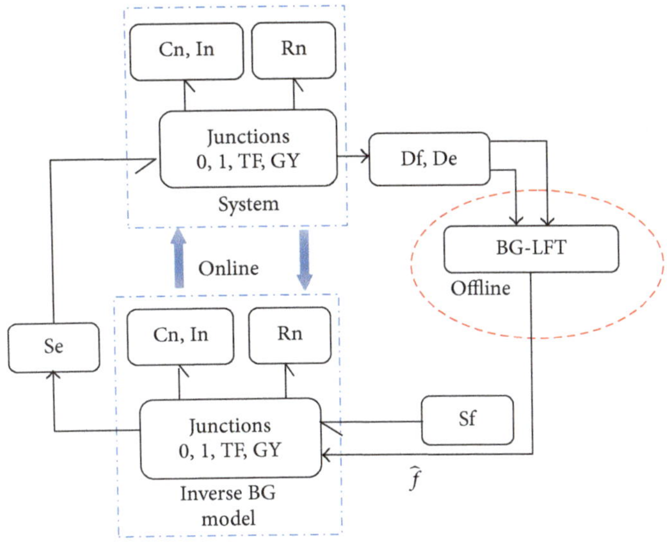

FIGURE 4: Control design on bond graph developed by [10].

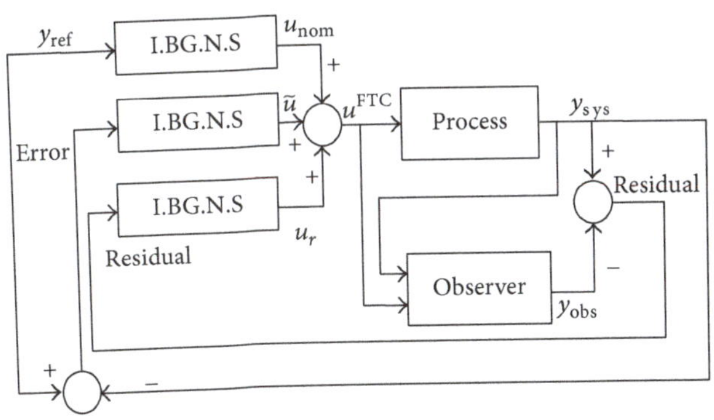

FIGURE 5: AFTC strategy based on bond graph.

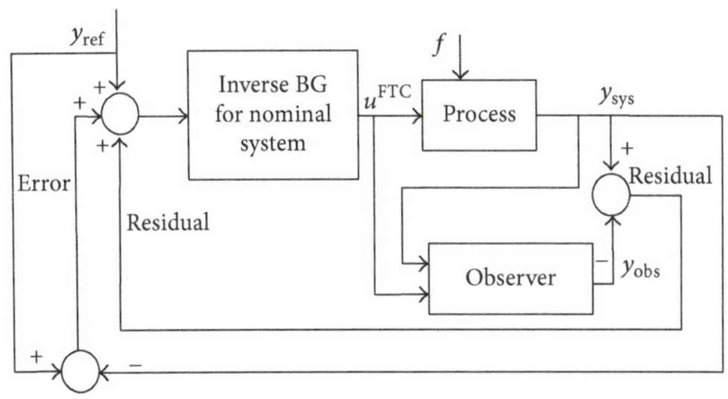

FIGURE 6: New control based on bond graph.

Computing the Control (u^{FTC}). Various methods have been proposed to recover as close as possible the system performance according to the considered fault representation.

Some extensions of the classical pseudoinverse method (PIM) have been proposed to guarantee both the performance and the stability of the faulty system. The authors in [25, 26] have synthesized a suitable feedback control K_{feedback}. In [27, 28], the authors have proposed to compute a reconfigurable forward gain K_{forward} controller in order to eliminate the steady-state tracking error in faulty case. Therefore, the control signal applied to the system is represented in

$$u^{\text{FTC}} = K_{\text{forward}} y_{\text{ref}} - K_{\text{feedback}} x. \tag{1}$$

A novel technique to adjust the command equation (1) is proposed by [29] in

$$u^{\text{FTC}} = K_{\text{forward}} y_{\text{ref}} + \tilde{u}. \tag{2}$$

$\tilde{u}$ given by

$$\tilde{u} = K_{\text{feedback}} \left(x - x^{n}\right). \tag{3}$$

According to the control law in (1) and (2), we propose a new control which uses the residual signal provided by the FDI based observer and error. So, the control law is expressed by

$$\begin{aligned} u^{\text{FTC}} &= K_{\text{forward}} y_{\text{ref}} + K_{\text{forward}} \Big(\underbrace{y_{\text{ref}} - y_{\text{sys}}}_{\text{error}} \Big) \\ &\quad + K_{\text{forward}} \Big(\underbrace{y_{\text{sys}} - y_{\text{obs}}}_{\text{residual}} \Big). \end{aligned} \tag{4}$$

Or K_{forward} is a gain of inverse BG for nominal system.

In Figure 5, the compensation term u_r (residual control) is useful to compensate the fault. Also $\tilde{u}$ (error control) is added to the nominal control (u_{nom}); this term ($\tilde{u}$) improves the compensation of the fault effect. So this additive control results' role is to reproduce the control signal (u^{FTC}: controlled input resulting from (4) for compensating for the effect of the fault every moment that the fault is detected).

To simplify the calculus of the control input represented in Figure 5, we propose to replace the three inverse BG models by a single inverse BG (Figure 6).

TABLE 2: Parameter values of the DC motor.

$R = 1\,\Omega$	Rotor resistance
$L = 5\,\text{mH}$	Rotor inductance
$b = 10^{-4}\,\text{Nm/rd}\cdot\text{s}^{-1}$	Coefficient of viscous
$J = 0.001\,\text{Kg}\,\text{m}^2$	Moment of inertia
$k_n = 0.2\,\text{Nm/A}$	Coefficient of the torque
$u_{\text{nom}} = 220\,\text{V}$	Motor voltage

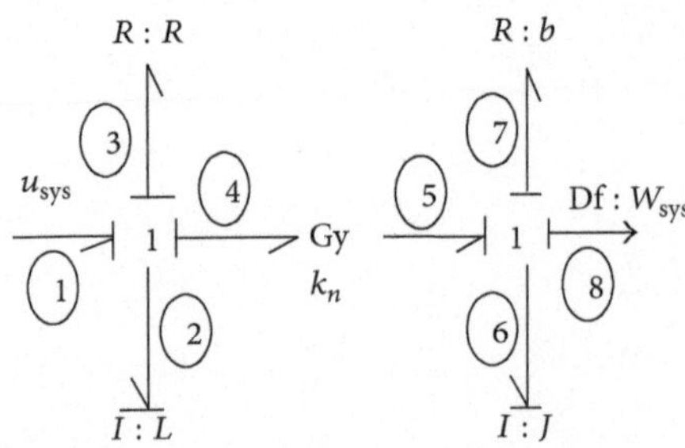

FIGURE 7: Bond graph model of DC motor.

So, the new control law is expressed by

$$u^{\text{FTC}} = K_{\text{forward}}\left(y_{\text{ref}} + \text{error} + \text{residual}\right). \quad (5)$$

4. Illustrative Example

An example of a DC motor is used to illustrate our new FTC technique. The BG model of the system is given in Figure 7, and the state-space equations are presented in (6);

$$\begin{bmatrix}\dot{x}_1\\ \dot{x}_2\end{bmatrix} = \begin{bmatrix}-\frac{b}{J} & \frac{k_n}{J}\\ -\frac{k_n}{L} & -\frac{R}{L}\end{bmatrix}\begin{bmatrix}x_1\\ x_2\end{bmatrix} + \begin{bmatrix}0\\ \frac{1}{L}\end{bmatrix}u,$$
$$y = \begin{bmatrix}1 & 0\\ 0 & 1\end{bmatrix}\begin{bmatrix}x_1\\ x_2\end{bmatrix} + \begin{bmatrix}0.1\\ 0\end{bmatrix}d, \quad (6)$$

with $x = (x_1, x_2) = (w, i)$ being the state vector, y the measured output variable, u the control input variable, and d the disturbance input variable.

The parameters of the DC motor are presented in Table 2.

Closed Loop System. Figure 8 shows that the AFTC strategy integrates the FDI module with an inverse BG for nominal system and the bond graph model is controllable and observable [14].

From the controlled input and output signals, the FDI module provides the residual which is injected to the control loop, in order to compensate the effect of fault.

The objective is to synthesize a controller so that the structure of the closed loop system is as close as possible to that of the desired reference model under the normal operation or in the presence of fault.

(i) Computing the Residual (r) in Normal Operating. By causal path, we deduce the structural equations from the BG of Figure 8. We compute the residual r_1 with the following equations:

(a) structural equations for system model:

$$\begin{aligned}
f_{1_s} &= f_{2_s} = f_{3_s} = f_{4_s},\\
e_{1_s} &= e_{2_s} + e_{3_s} + e_{4_s};\\
e_{4_s} &= k_n f_{5_s},\\
e_{5_s} &= k_n f_{4_s};\\
f_{5_s} &= f_{6_s} = f_{7_s} = f_{8_s},\\
e_{5_s} &= e_{6_s} + e_{7_s} + e_{8_s};\\
f_{5_s} &= \frac{e_{4_s}}{k_n},\\
f_{5_s} &= \frac{e_{1_s} - e_{2_s} - e_{3_s}}{k_n};
\end{aligned} \quad (7)$$

(b) structural equations for Luenberger observer:

$$\begin{aligned}
f_{1_{\text{ob}}} &= f_{2_{\text{ob}}} = f_{3_{\text{ob}}} = f_{4_{\text{ob}}},\\
e_{1_{\text{ob}}} &= e_{2_{\text{ob}}} + e_{3_{\text{ob}}} + e_{4_{\text{ob}}};\\
e_{4_{\text{ob}}} &= k_n f_{5_{\text{ob}}},\\
e_{5_{\text{ob}}} &= k_n f_{4_{\text{ob}}};\\
f_{5_{\text{ob}}} &= f_{6_{\text{ob}}} = f_{7_{\text{ob}}} = f_{8_{\text{ob}}},\\
e_{5_{\text{ob}}} &= e_{6_{\text{ob}}} + e_{7_{\text{ob}}} + e_{8_{\text{ob}}};\\
f_{5_{\text{ob}}} &= \frac{e_{4_{\text{ob}}}}{k_n},\\
f_{5_{\text{ob}}} &= \frac{e_{1_{\text{ob}}} - e_{2_{\text{ob}}} - e_{3_{\text{ob}}}}{k_n}.
\end{aligned} \quad (8)$$

From junctions equations (7) and (8), we generate the residual r_1:

$$\begin{aligned}
r_1 &= f_{8_s} - f_{8_{\text{ob}}} = f_{6_s} - f_{6_{\text{ob}}} = f_{5_s} - f_{5_{\text{ob}}} = \frac{e_{4_s}}{k_n} + \frac{e_{4_{\text{ob}}}}{k_n},\\
f_{5_s} &= \frac{1}{k_n}\left[U_s - L\frac{df_{2_s}}{dt} - Rf_{2_s}\right],\\
f_{5_{\text{ob}}} &= \frac{1}{k_n}\left[U_{\text{ob}} - L\frac{df_{2_{\text{obs}}}}{dt} - Rf_{2_{\text{ob}}} - r_1 K_{11}\right],\\
r_1 &= \frac{1}{k_n}\left[U_s - L\frac{df_{2_s}}{dt} - Rf_{2_s} - U_{\text{ob}} + L\frac{df_{2_{\text{ob}}}}{dt} + Rf_{2_{\text{ob}}}\right].
\end{aligned} \quad (9)$$

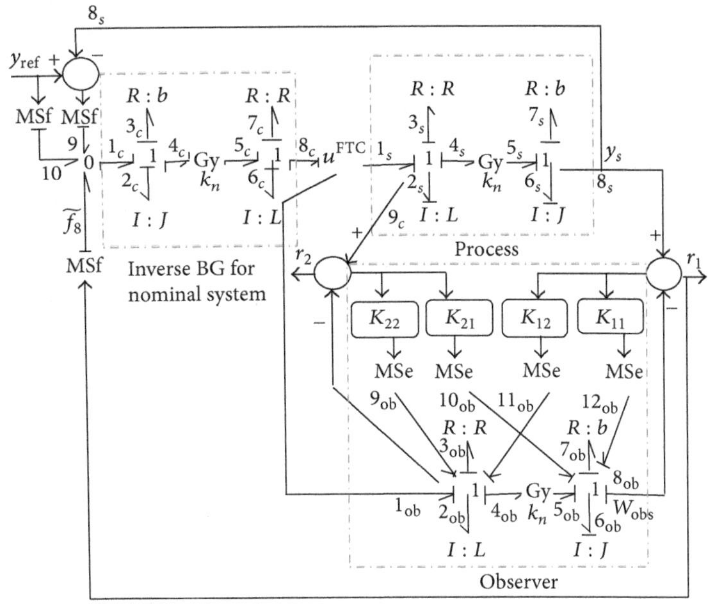

FIGURE 8: FTC based on BG model.

Or

$$
\begin{aligned}
U_s &= U_{\text{obs}}, \\
f_{2_s} &= \frac{e_{5_s}}{k_n} = \frac{1}{k_n}\left[J\frac{df_{6_s}}{dt} + bf_{6_s}\right], \\
f_{2_{\text{ob}}} &= \frac{e_{5_{\text{ob}}}}{k_n} = \frac{1}{k_n}\left[J\frac{df_{6_{\text{ob}}}}{dt} + bf_{6_s} - K_{11}r_1\right], \\
r_1 &= \frac{1}{k_n}\Bigg[U_s - L\frac{d(1/k_n)\left[J(df_{6_s}/dt) + bf_{6_s}\right]_2}{dt} \\
&\quad - R\frac{1}{k_n}\left[J\frac{df_{6_s}}{dt} + bf_{6_s}\right] - U_{\text{ob}} \\
&\quad + L\frac{d(1/k_n)\left[J\left(df_{6_{\text{ob}}}/dt\right) + bf_{6_s} - K_{11}r_1\right]}{dt}R \\
&\quad \times\frac{1}{k_n}\left[J\frac{df_{6_{\text{ob}}}}{dt} + bf_{6_s} - K_{11}r_1\right] - r_1K_{12}\Bigg].
\end{aligned}
\tag{10}
$$

The residual r_1 of the system is realized as

$$
\begin{aligned}
r_1 &= \frac{-JL}{k_n^2}\frac{d^2r_1}{dt} - \frac{dr_1}{dt}\left[\frac{RJ + Lb + K_{11}L}{k_n^2}\right] \\
&\quad - r_1\left[\frac{Rb + K_{11}R}{k_n^2} + \frac{K_{12}}{k_n}\right],
\end{aligned}
$$

$$
\begin{aligned}
&\left[1 + \frac{Rb + K_{11}R}{k_n^2} + \frac{K_{12}}{k_n}\right]r_1 + \left[\frac{RJ + Lb + K_{11}L}{k_n^2}\right]\frac{dr_1}{dt} \\
&\quad + \frac{JL}{k_n^2}\frac{d^2r_1}{dt} = 0.
\end{aligned}
\tag{11}
$$

The r_2 is deduced with similar method.

The inverse system enables computing appropriate control actions that compensate the faults.

(ii) Computing the Control (u^{FTC}). The control law can be designed directly from the BG model:

structural equations for inverse BG model:

$$
\begin{aligned}
f_{1_c} &= f_{2_c} = f_{3_c} = f_{4_c}, \\
e_{1_c} &= e_{2_c} + e_{3_s} + e_{4_s}; \\
e_{4_c} &= k_n f_{5_c}, \\
e_{5_c} &= k_n f_{4_c}; \\
f_{5_c} &= f_{6_c} = f_{7_c} = f_{8_c}, \\
e_{5_c} &= e_{6_c} + e_{7_c} + e_{8_c}.
\end{aligned}
\tag{12}
$$

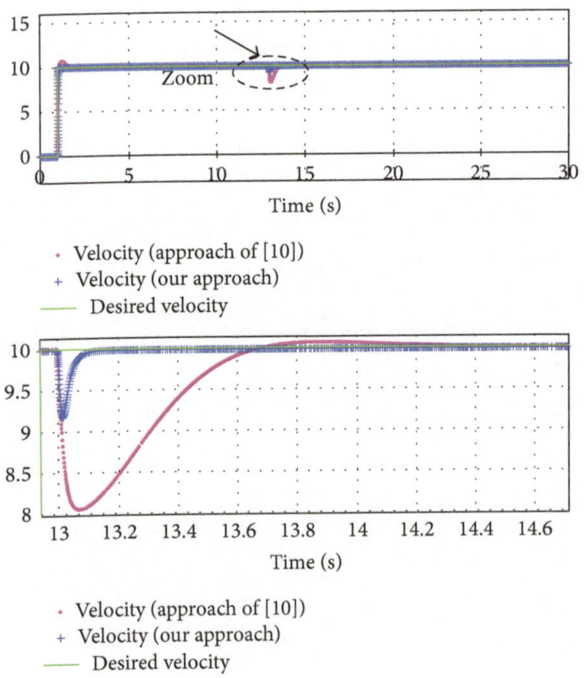

FIGURE 9: Simulation results when the fault is compensated.

The control law u^{FTC} of the system is shown as

$$
\begin{gathered}
u^{\mathrm{FTC}} = e_{8_c} = e_{5_c} + e_{6_c} + e_{7_c}, \\
e_{5_c} = k_n f_{1_c}, \\
f_{1_c} = f_{10} + \tilde{f}_8 + f_9, \\
f_{10} = w_{\mathrm{des}} = y_{\mathrm{ref}}, \qquad \tilde{f}_8 = f_{8_s} - f_{8_{\mathrm{ob}}} = r_1, \\
f_9 = w_{\mathrm{des}} - f_{8_s} = \varepsilon, \\
e_{6_c} = \frac{L d f_{6_c}}{dt}, \qquad e_{7_c} = R f_{7_c} = R f_{6_c}.
\end{gathered} \tag{13}
$$

Or r_1 = residual and ε = error.

So, the control law u^{FTC} is

$$
u^{\mathrm{FTC}} = R f_{6_c} + \frac{L d f_{6_c}}{dt} + k_n \left(\text{residual} + y_{\mathrm{ref}} + \text{error}\right). \tag{14}
$$

5. Simulation Results

Simulation results are carried out in the bond graph simulation software 20-sim [30] with parameter values described in Table 2.

Figure 9 shows the system output (velocity) evolution with a single fault (parameters R: b, $R_{\mathrm{fault}} = R + \delta$, and $\delta = 0.01$) introduced at the time 13 s.

We remark that the output decreases less than in the case of control considered in [10], and then it reaches the nominal values quicker at instant $t = 13.05$ s. So, the control law (FTC) is able to stabilize the system on the desired output and to compensate the fault in the system with a very short time delay.

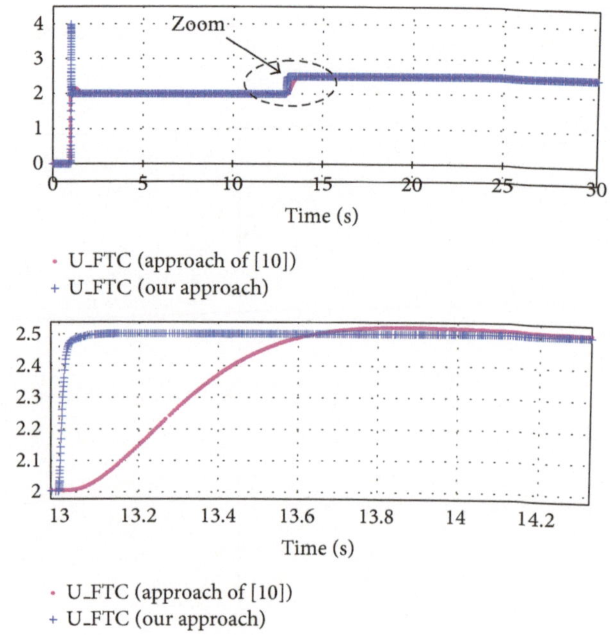

FIGURE 10: Control input.

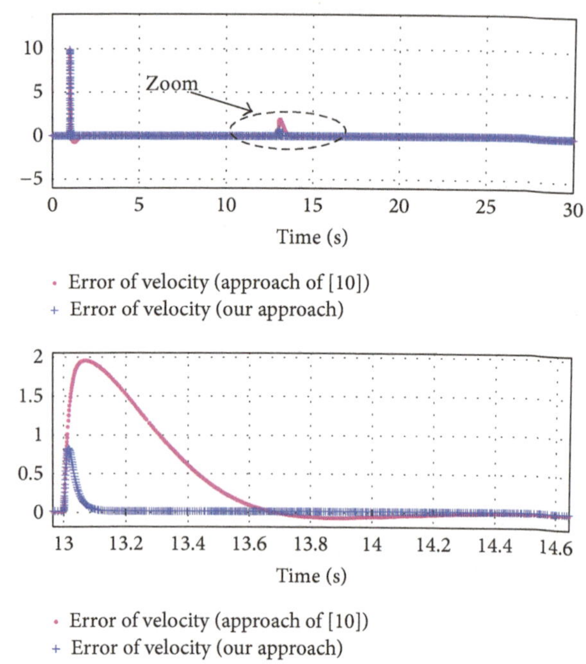

FIGURE 11: Velocity error.

From a control point of view, the reconfigurable control mechanism requires more energy to reach the target and to guarantee system performance, as shown in Figure 10.

These results can be confirmed by the control input U^{FTC} of Figure 10. In [10], the control input increases slowly trying to compensate for the fault affecting the system. In our approach, the control input increases quickly and enables rapid fault compensation on the controlled system output and allows compensating the convergence delay.

In Figure 11, the velocity error quickly converges to zero with the new approach.

6. Conclusion

In this paper, we propose an active FTC design based on BG approach. The novel strategy combines an observer based model and inverse BG model.

The proposed approach enables computing appropriate control actions for compensating the faults. The faults are detected by Luenberger observer technique based on BG modeling. Fault isolation and fault estimation are not necessary to the FTC. The comparison between the two approaches shows the efficiency of the proposed method. The application of a FTC approach to induction DC motor and simulation results illustrate the performance of the proposed FDI-FTC structure. Our future works concern the online implementation of the proposed techniques on a real process.

Conflict of Interests

The authors declare that there is no conflict of interests regarding the publication of this paper.

References

[1] D. Chilin, J. Liu, X. Chen, and P. D. Christofides, "Fault detection and isolation and fault tolerant control of a catalytic alkylation of benzene process," *Chemical Engineering Science*, vol. 78, pp. 155–166, 2012.

[2] P. M. Frank, "Fault diagnosis in dynamic systems using analytical and knowledge-based redundancy. A survey and some new results," *Automatica*, vol. 26, no. 3, pp. 459–474, 1990.

[3] J. Gertler, "Fault detection and isolation using parity relations," *Control Engineering Practice*, vol. 5, no. 5, pp. 653–661, 1997.

[4] S. Srinivas, "A probabilistic approach to hierarchical model-based diagnosis," in *Proceedings of the 10th Conference on Uncertainty in Artificial Intelligence (UAI '94)*, pp. 538–545, 1994.

[5] Y. Qian, L. Xu, X. Li, L. Lin, and A. Kraslawski, "LUBRES: an expert system development and implementation for real-time fault diagnosis of a lubricating oil refining process," *Expert Systems with Applications*, vol. 35, no. 3, pp. 1252–1266, 2008.

[6] F. Liao, J. L. Wang, and G.-H. Yang, "Reliable robust flight tracking control: an LMI approach," *IEEE Transactions on Control Systems Technology*, vol. 10, no. 1, pp. 76–89, 2002.

[7] H. Niemann and J. Stoustrup, "Passive fault tolerant control of a double inverted pendulum—a case study," *Control Engineering Practice*, vol. 13, no. 8, pp. 1047–1059, 2005.

[8] P. Mhaskar, "Robust model predictive control design for fault-tolerant control of process systems," *Industrial and Engineering Chemistry Research*, vol. 45, no. 25, pp. 8565–8574, 2006.

[9] Z. Zhang and W. Chen, "Adaptive output feedback control of nonlinear systems with actuator failures," *Information Sciences*, vol. 179, no. 24, pp. 4249–4260, 2009.

[10] R. Loureiro, R. Merzouki, and B. O. Bouamama, "Bond graph model based on structural diagnosability and recoverability analysis: Application to intelligent autonomous vehicles," *IEEE Transactions on Vehicular Technology*, vol. 61, no. 3, pp. 986–997, 2012.

[11] H. M. Paynter, *Analysis and Design of Engineering Systems*, The MIT Press, 1961.

[12] D. Karnopp, D. Margolis, and R. Rosenberg, *Systems Dynamics: A Unified Approach*, John Wiley and Sons, 1975.

[13] R. Rosenberg and D. C. Karnopp, *Introduction to Physical System Dynamics*, McGraw-Hill, 1983.

[14] C. Sueur and G. Dauphin-Tanguy, "Bond-graph approach for structural analysis of MIMO linear systems," *Journal of the Franklin Institute*, vol. 328, no. 1, pp. 55–70, 1991.

[15] B. Ould-Bouamama, A. K. Samantaray, M. Staroswiecki, and G. Dauphin-Tanguy, "Derivation of constraint relations from bond graph models for fault detection and isolation," in *Proceedings of the International Conference on Bond Graph Modeling and Simulation (ICBGM '03)*, pp. 104–109, Orlando, Fla, USA, January 2003.

[16] Y. Touati, R. Merzouki, and B. Ould Bouamama, "Robust diagnosis to measurement uncertainties using bond graph approach: application to intelligent autonomous vehicle," *Mechatronics*, vol. 22, no. 8, pp. 1148–1160, 2012.

[17] D. Karnopp, "Bond graphs in control: physical state variables and observers," *Journal of the Franklin Institute*, vol. 308, no. 3, pp. 219–234, 1979.

[18] P. J. Gawthrop and L. P. S. Smith, *Meta Modelling: Bond Graphs and Dynamic Systems*, Prentice Hall, 1995.

[19] C. Pichardo-Almarza, A. Rahmani, G. Dauphin-Tanguy, and M. Delgado, "Bond graph approach to build reduced order observers in linear time invariant systems," in *Proceedings of the 4th International Symposium on Mathematical Modelling*, 2003.

[20] C. Pichardo-Almarza, A. Rahmani, G. Dauphin-Tanguy, and M. Delgado, "Proportional-integral observer for systems modelled by bond graphs," *Simulation Modelling Practice and Theory*, vol. 13, no. 3, pp. 179–211, 2005.

[21] G. Gonzalez-A and I. Nuñez, "A nonlinear observer of an electrical transformer: a bond graph approach," *World Academy of Science, Engineering and Technology*, vol. 58, pp. 814–820, 2009.

[22] P. J. Gawthrop, "Bicausal bond graphs," in *Proceedings of the International Conference on Bond Graph Modelling and Simulation (ICBGM '95)*, pp. 83–88, 1995.

[23] R. F. Ngwompo, S. Scavarda, and D. Thomasset, "Inversion of linear time-invariant SISO systems modelled by bond graph," *Journal of the Franklin Institute B: Engineering and Applied Mathematics*, vol. 333, no. 2, pp. 157–174, 1996.

[24] N. Yadaiah and N. Venkata Ramana, "Linearisation of multi-machine power system: modeling and control—a survey," *International Journal of Electrical Power and Energy Systems*, vol. 29, no. 4, pp. 297–311, 2007.

[25] Z. Gao and P. J. Antsaklis, "Stability of the pseudo-inverse method for reconfigurable control systems," *International Journal of Control*, vol. 53, no. 3, pp. 717–729, 1991.

[26] M. Staroswiecki, "Fault tolerant control: the pseudo-inverse method revisited," in *Proceedings of the 16th Triennial World Congress of International Federation of Automatic Control (IFAC '05)*, pp. 418–423, July 2005.

[27] Y. M. Zhang and J. Jiang, "Active fault-tolerant control system against partial actuator failures," *IEE Proceedings on Control Theory and Applications*, vol. 149, no. 1, pp. 95–104, 2002.

[28] F. Guenab, D. Theilliol, P. Weber, J. C. Ponsart, and D. Sauter, "Fault tolerant control method based on cost and reliability analysis," in *Proceedings of the 16th Triennial World Congress of International Federation of Automatic Control (IFAC '05)*, pp. 490–495, Prague, Czech Republic, July 2005.

[29] Y. Zhang and J. Jiang, "Fault tolerant control system design with explicit consideration of performance degradation," *IEEE Transactions on Aerospace and Electronic Systems*, vol. 39, no. 3, pp. 838–848, 2003.

[30] 20 Sim Controllab Products B.V., http://www.20sim.com/.

Parameter Estimation of Turbo Code Encoder

Mehdi Teimouri[1] and Ahmadreza Hedayat[2]

[1] *Faculty of New Sciences and Technologies, University of Tehran, Tehran 14399, Iran*
[2] *Newracom, Irvine, CA 92618, USA*

Correspondence should be addressed to Mehdi Teimouri; mehditeimouri@ut.ac.ir

Academic Editor: Nikos D. Lagaros

The problem of reconstruction of a channel code consists of finding out its design parameters solely based on its output. This paper investigates the problem of reconstruction of parallel turbo codes. Reconstruction of a turbo code has been addressed in the literature assuming that some of the parameters of the turbo encoder, such as the number of input and output bits of the constituent encoders and puncturing pattern, are known. However in practical noncooperative situations, these parameters are unknown and should be estimated before applying reconstruction process. Considering such practical situations, this paper proposes a novel method to estimate the above-mentioned code parameters. The proposed algorithm increases the efficiency of the reconstruction process significantly by judiciously reducing the size of search space based on an analysis of the observed channel code output. Moreover, simulation results show that the proposed algorithm is highly robust against channel errors when it is fed with noisy observations.

1. Introduction

Channel code reconstruction problem is usually considered in noncooperative communication scenarios, where finding the structure and parameters of the channel code is a major goal of the receiver [1–14]. Since turbo codes are introduced in many standards and applications [15–22], blind reconstruction of turbo codes in practical noncooperative situations is a well motivated problem. This problem has been investigated in [1–4, 8, 11], where some of the code parameters are assumed to be known to the receiver, such as the number of input and output bits of each constituent convolutional code and the structure of puncturing matrix. Having the above parameters, the receiver becomes capable of separating the received coded stream into three sets [3]: systematic bits, parity bits of the first convolutional encoder, and parity bits of the second convolutional encoder. Now the turbo code reconstruction problem converts into three simpler reconstruction problems: two convolutional code reconstruction problems and one interleaver reconstruction problem.

However, in most practical noncooperative situations, the parameters of turbo code are unknown and thus should be estimated before applying reconstruction process or to test each possible set of parameters which is highly time consuming. To the best of the authors' knowledge, no algorithm has yet been proposed for estimating these parameters, and this paper is the first to propose a novel method to estimate above-mentioned parameters before applying the turbo code reconstruction process. The proposed method reduces the search space of the reconstruction process significantly compared to [3] and makes the reconstruction process faster and more efficient. The proposed method is first formulated for noiseless streams and then it is adjusted for noisy observations.

The rest of the paper is organised as follows. In Section 2, system model is described. The method for estimating the turbo code parameters is proposed in Section 3 and experimental results are presented in Section 4. Finally, the paper is concluded in Section 5.

2. System Model

We consider the system model of turbo code reconstruction as shown in Figure 1. In this model, input bits and their interleaved version are fed to the first and second convolutional encoders, respectively. The first encoder is

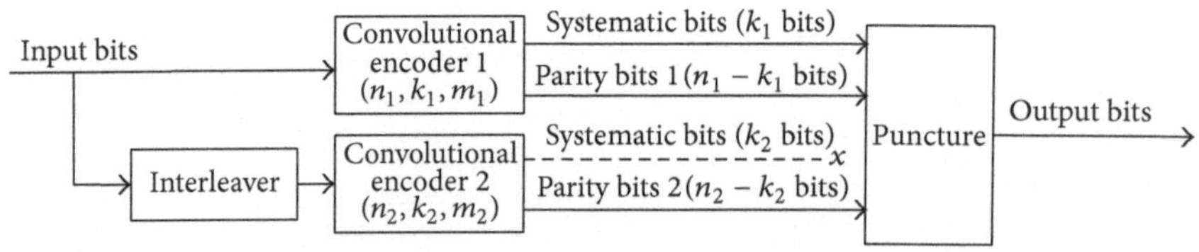

FIGURE 1: System model of turbo code reconstruction.

an (n_1, k_1, m_1) convolutional code, where n_1 is the number of output bits, k_1 is the number of input bits, and m_1 is the memory length of the longest shift register of the encoder. Since convolutional codes used for parallel turbo codes are systematic [15], k_1 bits out of n_1 output bits are systematic bits, and the remaining n_1-k_1 bits are parity bits. With such configuration, the value of any n_1-k_1 parity bits is a function of recent k_1 systematic output bits and at most $k_1 \times m_1$ values of shift registers content. In the system model of Figure 1, the second convolutional code is also assumed to be an (n_2, k_2, m_2) convolutional code, but unlike the first convolutional encoder, its systematic bits are removed and only the remaining n_2-k_2 parity bits are used.

Since parallel turbo codes have the same number of input bits per constituent convolutional code [15], we assume k_1 and k_2 are equal and denote both by k. Assume the length of input frames is N; then N is a multiple of k. Using this assumption, at each time step, k bits out of N input bits and k bits out of the interleaver are fed to the first and second convolutional encoders, respectively. These input bits result in $n_1 + n_2 - k$ output bits; thus we can say that N/k sets of input bits form an output frame with length $L = N/k \times (n_1 + n_2 - k)$.

As shown in Figure 1, a puncturing pattern can also be applied to the output frames to form smaller frame lengths and higher code rates. We use an $(n_1 + n_2 - k) \times p$ matrix $\mathbf{P} = [q_{i,j}]$ to represent puncturing pattern where p is the puncturing period. Every row of $\mathbf{P}$ corresponds to one output bit; that is, the first row corresponds to the first systematic bit, the second row corresponds to the second systematic bit (or the first parity bit of the first convolutional encoder), and so on. Note that $q_{i,j} \in \{0, 1\}$, where 0 implies that the corresponding bit is punctured. If n_0 elements out of $(n_1+n_2-k) \times p$ of $\mathbf{P}$ take the value of 1, then the output frame length is $L = (N \times n_0)/(k \times p)$. Note that we have implicitly assumed that N is a multiple of $k \times p$, since the input frame needs to be a multiple of the puncturing period. Since the turbo codes employed in many applications such as CDMA 2000 [16] and WCDMA [17] are systematic parallel turbo codes, in this paper we consider systematic parallel turbo codes. The term "systematic" is used for a turbo code when none of its systematic bits are punctured. In systematic turbo codes all the elements of the first k rows of $\mathbf{P}$ take the value of 1. Other elements of $\mathbf{P}$ can be either 0 or 1; however, it should be noted that none of the rows can be an all zeros vector.

All the previous works that address the turbo code reconstruction problem implicitly assume that the received coded stream is partitioned into three sets: systematic bits, parity bits of the first convolutional encoder, and parity bits of the second convolutional encoder. However, it should be noted that in order to do such partitioning we need to know the number of output bits of both encoders (i.e., n_1 and n_2) and the puncturing pattern (i.e., $\mathbf{P}$). But, in most practical noncooperative situations, these parameters are unknown and thus should be estimated before applying reconstruction process or to test each possible set of parameters which is highly time consuming. In this paper, a novel method is proposed to estimate the above-mentioned parameters before applying the turbo code reconstruction process. The proposed method reduces the search space of the reconstruction process significantly and makes the reconstruction process faster and more efficient.

3. Proposed Method

Assume that the number of elements with the value of 1 in puncturing matrix $\mathbf{P}_{(n_1+n_2-k)\times p}$ is equal to n_0. Using the same arguments as those detailed in [5], if the input data to the turbo encoder behaves as random data, we can say that the number of linearly independent blocks of received coded stream with length $n_0 \times i$, assuming i to be a positive integer, is at most equal to $(p \times k + p \times k) \times i + m_0$, where nonnegative integer m_0 takes at most the value of $m_1 + m_2$. This can be easily seen by noting that each block, out of i blocks with length n_0, is generated by linear combination of $p \times k + p \times k$ information bits and $m_1 + m_2$ initial memory bits. When the Hamming weight in the last $n_2 - k$ rows of $\mathbf{P}$ is smaller than $p \times k$, the value of the second $p \times k$ should be interchanged with this Hamming weight. Let us denote the minimum value between $p \times k$ and the Hamming weight in the last $n_2 - k$ rows of $\mathbf{P}$ by p_2. Thus, we can say that the number of linearly independent blocks of received coded stream with length $n_0 \times i$ is at most equal to $(p \times k + p_2) \times i + m_0$. When $p_2 < p \times k$ we can say that p_2 parity bits of the second convolutional encoder in each puncturing period act like random independent data, so it can be concluded that in this case m_0 can take at most the value of m_1. Since only the values of i that results in $((p \times k + p_2) \times i + m_0)/(n_0 \times i) < 1$ are acceptable, we have

$$i \geq \left\lfloor \frac{m_0}{n_0 - p_0} \right\rfloor + 1, \quad (1)$$

where $p_0 = p \times k + p_2$ and $\lfloor x \rfloor$ denotes the largest integer less than or equal to x.

Starting from the first bit we partition the received bit stream into blocks with length n and consider these blocks as the rows of a matrix named $\mathbf{Z}_n$ with size $a \times n$. Choosing the value of a is arbitrary, but it should be selected in order to be much bigger than n. Now, if we define function $R(n)$ for the received bit stream in the domain of positive integer as follows:

$$R(n) = \frac{\text{Rank}(\mathbf{Z}_n)}{n}, \quad (2)$$

where Rank($\mathbf{Z}_n$) denotes the rank of matrix $\mathbf{Z}_n$, according to the aforementioned discussion in the beginning of this section, it is obvious that this function should have local minimum points at $n = n_0 \times i$, where i is given by (1). Moreover, the function values at these local minimum points are given as follows:

$$R(i \times n_0) = \frac{p_0 \times i + m_0}{n_0 \times i}; \quad i \geq \left\lfloor \frac{m_0}{n_0 - p_0} \right\rfloor + 1. \quad (3)$$

By obtaining local minimum points and function values at these points for function $R(n)$ it is possible to estimate the values of m_0, p_0, and n_0 parameters of turbo code, which can facilitate the turbo code reconstruction process.

If in the construction process of $\mathbf{Z}_n$ (whilst $n = n_0 \times i$) we eliminate the first α bits of bit stream ($\alpha \in \{0, 1, 2, \ldots, n_0 - 1\}$), the obtained matrix denoted by $\mathbf{Z}_{n,\alpha}$ can help us to estimate the puncturing matrix. In fact, if the value of α is chosen such that the first considered bit in construction of $\mathbf{Z}_{n,\alpha}$ is a systematic bit in the first row of $\mathbf{P}$, it is obvious that Rank($\mathbf{Z}_{n,\alpha}$) = Rank($\mathbf{Z}_n$). On the other hand, if the value of α is chosen such that the first considered bit in construction of $\mathbf{Z}_{n,\alpha}$ is a systematic bit in other rows of $\mathbf{P}$ or a parity bit of the first convolutional code, we expect that Rank($\mathbf{Z}_{n,\alpha}$) > Rank($\mathbf{Z}_n$). As a matter of fact, in such situations degree of freedom is increased which results in larger rank values. Finally, if the value of α is chosen such that the first considered bit in construction of $\mathbf{Z}_{n,\alpha}$ is a parity bit of the second convolutional code, we expect that in (at most) p positions the equality Rank($\mathbf{Z}_{n,\alpha}$) = Rank($\mathbf{Z}_n$) holds, whereas, in the remaining positions, if any, the inequality Rank($\mathbf{Z}_{n,\alpha}$) > Rank($\mathbf{Z}_n$) holds.

Thus, we can define puncturing signature as a vector with length n_0 in the following way. For any $\alpha \in \{0, 1, 2, \ldots, n_0 - 1\}$, if Rank($\mathbf{Z}_{n_0 \times i,\alpha}$) > Rank($\mathbf{Z}_{n_0 \times i}$) the corresponding element in puncturing signature is set to 1; otherwise it is set to 0. Note that, by knowing the values of k, n_1, n_2, and p, construction of puncturing matrix $\mathbf{P}$ from puncturing signature is not unique, but definitely, the search space size for puncturing matrix is intensely reduced. Numerical results for implying this fact will be provided in the following section.

According to the above analysis, we are ready to present the proposed method for estimation of the number of input and output bits of both encoders (i.e., k, n_1, and n_2) and the puncturing pattern (i.e., $\mathbf{P}$).

(1) Plot $R(n)$ versus n for $n = n_{\min}, n_{\min}+1, \ldots, n_{\max}$ ($n_{\min}$ and $n_{\max}$ are two arbitrary integers). Denote the set of the points at which the local minimums occur by $\{\hat{n}_j\}$ and the set of the function local minimum values by $\{R_j\}$, where $j = 1, 2, 3, \ldots, j_{\max}$. Define the set of $\{\hat{k}_j\}$ such that $\hat{k}_j = R_j \times \hat{n}_j$ for any j.

(2) Estimate the value of n_0 as follows: $n_0 = \text{mode}(\{\hat{n}_{j+1} - \hat{n}_j; j = 1, 2, \ldots, j_{\max} - 1\})$, where mode$(X)$ denotes the most frequent value in set X.

(3) Estimate the value of p_0 as follows: $p_0 = \text{mode}(\{\hat{k}_{j+1} - \hat{k}_j; j = 1, 2, \ldots, j_{\max} - 1\})$.

(4) Estimate the value of m_0 as follows: $m_0 = \text{mode}(\{\hat{k}_j - (\hat{n}_j/n_0) \times p_0; j = 1, 2, \ldots, j_{\max}\})$.

(5) Plot function $R(\hat{n}_{j_{\max}}, \alpha) = \text{Rank}(Z_{\hat{n}_{j_{\max}},\alpha})/\hat{n}_{j_{\max}}$ versus α for $\alpha = 0, 1, \ldots, n_0 - 1$, and then obtain the puncturing signature as stated before.

(6) Using the estimated values of n_0, p_0, and puncturing signature we can consider feasible values of k, n_1, and n_2 to construct all possible puncturing matrices. Then, since the construction process of the turbo code can be called with any sets of these possible parameters, somewhere we expect that one of these calls successfully results in a turbo code structure. Estimated value of m_0 can as well help to make the size of the search space for constituent convolutional codes smaller.

Note that, compared to the current methods which should blindly pass any possible set of parameters to the turbo code construction process, the proposed method here tries to reduce the size of such large search space. Numerical results for comparison reasons are provided in the following section.

In formulation of the proposed method for estimation of turbo code parameters it is assumed that the observed turbo coded stream is noiseless. In such situation, the rows of matrix $\mathbf{Z}_n$ form a linear space denoted by Γ. Similarly, the rows of matrix $\mathbf{Z}_{n,\alpha}$ form a linear space denoted by Γ_α. So, if the received data is noisy it is obvious that we can use the proposed algorithm on noiseless rows of $\mathbf{Z}_n$ and $\mathbf{Z}_{n,\alpha}$, and yet the same turbo code parameters as the noiseless case will be obtained. However, we should choose the number of the rows of $\mathbf{Z}_n$ and $\mathbf{Z}_{n,\alpha}$ large enough in order that the number of the noiseless rows of these matrices becomes much bigger than n.

In order to find the noiseless rows of $\mathbf{Z}_n$, first we apply Valembois' algorithm [7] on (noisy) rows of $\mathbf{Z}_n$ to find some words in the dual space of Γ. Now, any rows of $\mathbf{Z}_n$ that is orthogonal to all obtained words in the dual space of Γ can be considered as a noiseless row. A similar method can be used to find noiseless rows of $\mathbf{Z}_{n,\alpha}$.

Numerical and simulation results provided in the next section show that this algorithm is quite efficient in space search size reduction, and moreover it is also robust when it is fed with noisy and erroneous input.

4. Experimental Results

For obtaining experimental results we consider three different turbo code structures. The characteristics of these codes are given in Table 1. The first code is the code applied in WCDMA standard and the two others are introduced in CDMA 2000 standard. In this table, $\mathbf{G}_1(D)$ and $\mathbf{G}_2(D)$ denote

Table 1: The structure of the simulated turbo code.

	Constituent convolutional codes $\mathbf{G}_1(D) = \mathbf{G}_2(D) = \mathbf{G}(D)$	Puncturing matrix	Overall code rate
Turbo code A	$\mathbf{G}(D) = \left[1, \dfrac{1+D+D^3}{1+D^2+D^3}\right]$	$\mathbf{P} = \begin{bmatrix}1\\1\\1\end{bmatrix}$	1/3
Turbo code B	$\mathbf{G}(D) = \left[1, \dfrac{1+D+D^3}{1+D^2+D^3}, \dfrac{1+D+D^2+D^3}{1+D^2+D^3}\right]$	$\mathbf{P} = \begin{bmatrix}1\\1\\1\\1\\1\end{bmatrix}$	1/5
Turbo code C	$\mathbf{G}(D) = \left[1, \dfrac{1+D+D^3}{1+D^2+D^3}, \dfrac{1+D+D^2+D^3}{1+D^2+D^3}\right]$	$\mathbf{P} = \begin{bmatrix}1&1\\1&1\\1&0\\0&1\\1&1\end{bmatrix}$	1/4

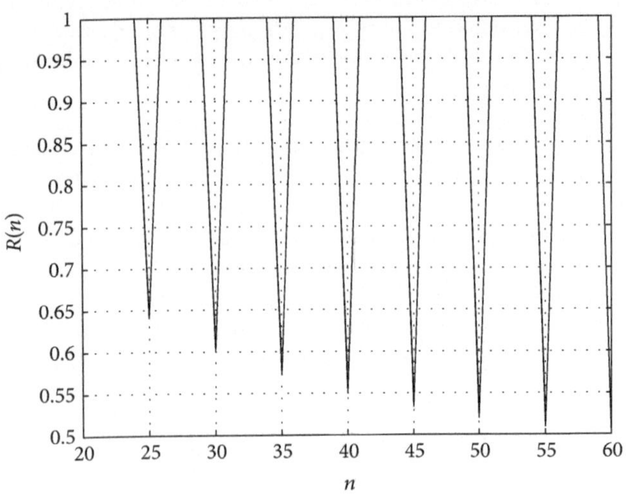

Figure 2: Plot of $R(n)$ for turbo code B.

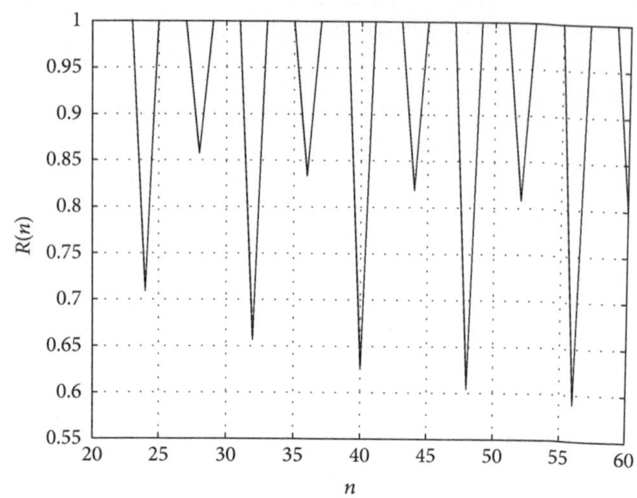

Figure 3: Plot of $R(n)$ for turbo code C.

polynomial representation of generator matrix of the first and the second convolutional codes, respectively.

For comparison reasons, we set the first dimension of matrix $\mathbf{Z}_n$ to $3 \times n$ and set the value of $n_{\min}$ and $n_{\max}$ to 21 and 60, respectively. In Figure 2, function $R(n)$ for turbo code B is plotted. For this code the set of the points at which the local minimums occur is $\{\hat{n}_j\} = \{25, 30, 35, \ldots, 55\}$. According to this set and as we expected, the distance between the consecutive points at which the local minimums occur is equal to $n_0 = 5$. For some turbo codes, more than one set of local minimums can be observed. For example, function $R(n)$ for turbo code C is shown in Figure 3. As it is apparent from this figure, two distinguishable sets of local minimum points are $\{\hat{n}_j\} = \{24, 32, 40, 48, 56\}$ and $\{\hat{n}_j\} = \{28, 36, 44, 52\}$. In such cases, the set with smallest local minimum values should be used to estimate turbo code parameters. The other local minimum set(s) in this case are formed due to the special structure of the constituent convolutional encoders and the puncturing pattern. For example, for code C, when the values of n are selected from the set $\{8 \times 3 + 4, 8 \times 4 + 4, 8 \times 5 + 4, \ldots\}$, linearly independent blocks of received coded stream would be less than n; however, the relative number of these linearly independent blocks, that is, $R(n)$, is higher compared to the situation in which the values of n are selected from the right set which is $\{8 \times 3, 8 \times 4, 8 \times 5, \ldots\}$.

In Table 2, estimated parameters of turbo codes defined in Table 1 for the case of noiseless coded stream are given. As we can see, the given values for parameters n_0, p_0, and m_0 match the expected value. The last column of Table 2 represents the number of the potential sets of $\{k, n_1, n_2, \mathbf{P}\}$ which can be considered according to the estimated values for n_0 and p_0. For turbo codes A and B, the only potential sets of $\{k, n_1, n_2, \mathbf{P}\}$ are those given in Table 1. But for turbo code C the number of the potential sets of $\{k, n_1, n_2, \mathbf{P}\}$ is equal to 56. Note that this value is much smaller compared to the number of the potential sets in blind reconstruction. For example, if the reconstruction process operates blindly in the case in which the values of k, n_1, n_2, and p are assumed to be at most

Table 2: Estimated parameters of turbo codes defined in Table 1 for noiseless case.

	n_0	p_0	m_0	Puncturing signature	Number of potential sets of $\{k, n_1, n_2, \mathbf{P}\}$
Turbo code A	3	2	3	$[0, 1, 0]$	1
Turbo code B	5	2	6	$[0, 1, 1, 0, 1]$	1
Turbo code C	8	4	5	$[0, 1, 1, 0, 0, 1, 0, 1]$	56

Table 3: Probability of success in parameters estimation of turbo codes defined in Table 1 for noisy channel.

	E_b/N_0 (dB)			
	5.2	4.2	3.7	3.2
Turbo code A	≈1	0.91	0.42	0.05
Turbo code B	≈1	0.52	0.05	≈0
Turbo code C	≈1	0.65	0.07	≈0

equal to 2, 3, 3, and 5, respectively, all possible puncturing matrices having these properties should be considered and the only constraint is that the number of 1 s in the examined puncturing matrices should be a factor of the output frames length which is assumed to be equal to 1200 in this case. It can be seen that in this case the search space size is equal to 461794 which is much larger than 56.

To evaluate the capability of the proposed method against channel noise, we consider that the turbo coded stream is first modulated by a binary phase shift keying (BPSK) modulator and then passed through an additive white Gaussian noise (AWGN) channel. Also, we define E_b/N_0 as the ratio of energy per transmitted bit (E_b) to the spectral noise density (N_0). In Table 3, success probability for turbo code parameter estimation is given under the assumption of different E_b/N_0 values, considering the turbo codes defined in Table 1. The number of tests for each turbo code and SNR pair is set to 1000. As it is apparent from this table, when the value of E_b/N_0 is more than 5 dB, the probability of success is near 1, but for E_b/N_0 values around 3 dB the probability of success reaches 0.

Since turbo codes are supposed to perform in noisy channels, the low noise assumption for code reconstruction may seem too optimistic. However, it should be noted that in practical noncooperative situations, recognition of a communication system is usually done in higher values of signal to noise ratio compared to cooperative situations. In fact, in order to achieve such signal quality, we have no choice other than using ultra low noise receivers. Finally, it should be noted that the channel code reconstruction process is done once based on many observations of the channel code outputs and decision on which of the estimated code parameters to be picked as the true code parameters can be easily made based on the multiplicity and inferred confidence of the estimated sets.

5. Conclusion

A novel method is proposed in this paper to estimate the parameters of turbo codes. These parameters are the number of input and output bits of both constituent convolutional encoders and the puncturing pattern while assuming that the target turbo code is systematic. It is assumed that receiver is only aware of the turbo coded bit stream. Numerical results confirm that the proposed method significantly reduces the size of the search space for reconstruction process of the turbo code. Moreover, simulation results indicate that when the proposed algorithm is fed with erroneous channel observations, the outcome is highly reliable as long as the E_b/N_0 value is more than 5 dB, demonstrating robustness of the proposed algorithm.

Conflict of Interests

The authors declare that there is no conflict of interests regarding the publication of this paper.

Acknowledgment

The authors would like to thank the anonymous reviewers for their helpful comments.

References

[1] J. Barbier, "Reconstruction of turbo-code encoders," in *Security and Defense, Space Communication Technologies Symposium*, Proceedings of SPIE, pp. 463–473, Orlando, Fla, USA, 2005.

[2] R. Gautier, M. Marazin, and G. Burel, "Blind recovery of the second convolutional encoder of turbo-code when its systematic outputs are punctured," *MTA Review*, vol. 19, pp. 213–232, 2009.

[3] M. Cluzeau, M. Finiasz, and J.-P. Tillich, "Methods for the reconstruction of parallel turbo codes," in *Proceedings of the IEEE International Symposium on Information Theory (ISIT '10)*, pp. 2008–2012, June 2010.

[4] M. Côte and N. Sendrier, "Reconstruction of a turbo-code interleaver from noisy observation," in *Proceedings of the IEEE International Symposium on Information Theory (ISIT '10)*, pp. 2003–2007, June 2010.

[5] J. Barbier, G. Sicot, and S. Houcke, "Algebraic approach for the reconstruction of linear and convolutional error correcting codes," *International Journal of Applied Mathematics and Computer Sciences*, vol. 2, pp. 113–118, 2006.

[6] M. Côte and N. Sendrier, "Reconstruction of convolutional codes from noisy observation," in *Proceedings of the IEEE International Symposium on Information Theory (ISIT '09)*, pp. 546–550, July 2009.

[7] A. Valembois, "Detection and recognition of a binary linear code," *Discrete Applied Mathematics*, vol. 111, no. 1-2, pp. 199–218, 2001.

[8] Y. G. Debessu, H.-C. Wu, and H. Jiang, "Novel blind encoder parameter estimation for turbo codes," *IEEE Communications Letters*, vol. 16, no. 12, pp. 1917–1920, 2012.

[9] M. Marazin, R. Gautier, and G. Burel, "Blind recovery of *k/n* rate convolutional encoders in a noisy environment," *EURASIP Journal on Wireless Communications and Networking*, vol. 2011, article 168, 2011.

[10] H. Lee, C.-S. Park, J.-H. Lee, and Y. J. Song, "Reconstruction of BCH codes using probability compensation," in *Proceedings of the 18th Asia-Pacific Conference on Communications (APCC '12)*, pp. 591–594, Jeju Island, Republoc of Korea, October 2012.

[11] A. Naseri, O. Azmoon, and S. Fazeli, "Blind recognition algorithm of turbo codes for communication intelligence systems," *International Journal of Computer Science Issues*, vol. 8, pp. 68–72, 2011.

[12] H. Xie, X.-M. Chai, F.-H. Wang, and Z.-T. Huang, "A method for blind identification of rate 1/2 convolutional code based on improved Euclidean algorithm," in *Proceedings of the 11th International Conference on Signal Processing (ICSP '12)*, pp. 1307–1310, Beijing, China, October 2012.

[13] A. Zahedi and G.-R. Mohammad-Khani, "Reconstruction of a non-binary block code from an intercepted sequence with application to reed-solomon codes," *IEICE Transactions on Fundamentals of Electronics, Communications and Computer*, vol. 95, pp. 1873–1880, 2012.

[14] M. Marazin, R. Gautier, and G. Burel, "Algebraic method for blind recovery of punctured convolutional encoders from an erroneous bitstream," *IET Signal Processing*, vol. 6, no. 2, pp. 122–131, 2012.

[15] S. Benedetto and G. Montorsi, "Design of parallel concatenated convolutional codes," *IEEE Transactions on Communications*, vol. 44, no. 5, pp. 591–600, 1996.

[16] D. J. Shah, V. K. Patel, and H. A. Patel, "Performance analysis of turbo code for CDMA 2000 with convolutional coded IS-95 system in wireless communication system," in *Proceedings of the International Conference on Electronic Computer Technology (ICECT '10)*, pp. 42–45, Kuala Lumpur, Malaysia, May 2010.

[17] K. Ramasamy, B. Balamuralithara, and M. U. Siddiqi, "Performance evaluation of proposed asymmetric turbo code in WCDMA HSDPA system," in *Proceedings of the 5th IEEE GCC Conference & Exhibition (GCC '09)*, pp. 1–5, March 2009.

[18] O. Bayat, "Intersymbol interference cancellation in CDMA 1xEVDO network," *International Journal of Communication Systems*, 2012.

[19] S. Y. Le Goff, "Bandwidth-efficient turbo coding over Rayleigh fading channels," *International Journal of Communication Systems*, vol. 15, no. 7, pp. 621–633, 2002.

[20] O. N. Ucan, K. Buyukatak, E. Gose, O. Osman, and N. Odabasioglu, "Performance of multilevel-turbo codes with blind/non-blind equalization over WSSUS multipath channels," *International Journal of Communication Systems*, vol. 19, no. 3, pp. 281–297, 2006.

[21] A. Ghrayeb and T. Abualrub, "Asymptotic performance comparison of concatenated (turbo) codes over *GF*(4)," *International Journal of Communication Systems*, vol. 17, no. 5, pp. 479–490, 2004.

[22] O. Osman, "Blind equalization of multilevel turbo coded-continuous phase frequency shift keying over MIMO channels," *International Journal of Communication Systems*, vol. 20, no. 1, pp. 103–119, 2007.

A Cancellation-Free Symbolic Sensitivity Technique Based on Network Determinant Expansion

Vladimir Filaretov, Konstantin Gorshkov, and Sergey Kurganov

Department of Electrical Engineering, Ulyanovsk State Technical University, 32 Severny Venets Street, Ulyanovsk 432027, Russia

Correspondence should be addressed to Vladimir Filaretov; vvfil@mail.ru, Konstantin Gorshkov; k.gorshkov@ulstu.ru and Sergey Kurganov; sakurganov@mail.ru

Academic Editor: Ramón M. Rodríguez-Dagnino

The generalization of Bode's sensitivity analysis technique for all types of the transfer functions and circuit elements is presented in the paper. The proposed formulae for first- and second-order symbolic sensitivity calculation provide the compact size of obtained expression and have the advantages of cancellation-free sum-of-product terms and matrix-free computation. This is achieved by means of the concept of high order summative cofactors and the generalized parameter extraction method. The proposed technique is implemented in symbolic circuit analysis program Cirsym. Illustrative example on symbolic sensitivity circuit analysis and comparison of the presented technique with the transimpedance method and the method based on the modified Coates flow-graph are given.

1. Introduction

The influence of network components on transfer function is expressed by sensitivities [1]. Sensitivity analysis is an important part of analog circuit design process. There are many methods of numerical calculation of circuit sensitivities, but only symbolic analysis of circuit sensitivity provides the way to get analytical sensitivity function of all interested circuit parameters. This is certainly advantageous in such applications as circuit parameters optimization and Monte Carlo simulation.

First researches on symbolic approach to sensitivity analysis of analog circuit were presented in [1–5], but those heuristic algorithms were not efficient enough and had limitations by the circuits scale.

Nowadays several more effective symbolic sensitivity analysis techniques have been developed. Some of them are based on sequence of expression computation such as two-port transimpedance method [6]. But the generated expressions are not compact enough, do not fully exploit sharing of subexpressions, cannot be easily manipulated, and complicate the circuit insight.

The others methods are based on implementation of graph theory for symbolic circuit analysis [7–10]. In particular, the work in [7] shows the technique based on the modified Coates flow-graph for performing sensitivity analysis. However, the proposed concept requires the nodal admittance matrix expansion, which may consist of the equal summands with opposite sign. It leads to appearing of the cancellation sum-of-product terms. In addition, the usage of graph-based methods does not provide the optimal size of obtaining expression.

Another widespread approach is implementation of the modified nodal matrix [10, 11]. But obtained results are not compact and include cancellations so the expressions often present only in semisymbolic form.

In this paper we propose a cancellation-free symbolic sensitivity technique for computation of compact expression. The first- and second-order symbolic sensitivities formulas for all types of the transfer functions and circuit elements are derived by means of the concept of higher order summative cofactors (HOSC) [12–15] and generalized parameter extraction method (GPEM) [16–21] to avoid drawbacks of the previous methods mentioned above.

The rest of this paper is organized as follows. In Section 2 we explain the basic idea of sensitivity analysis by means of GPEM and derive the formulae for symbolic sensitivity of

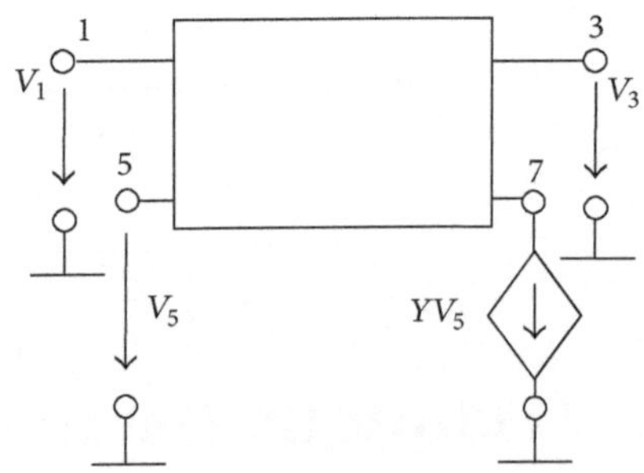

FIGURE 1: Test circuit.

different transfer functions with respect to controlled sources (CS) and two ports. The process of obtaining symbolic expressions can be easily automated. The example of the AC sensitivity analysis is presented in Section 3. In Section 4 the comparison of the presented method with the transimpedance method and the method based on the modified Coates flow-graph is performed. Section 5 concludes this paper.

2. Theoretical Basis of the Proposed Method

2.1. Symbolic Approach for Sensitivity Analysis. The sensitivity computation of the transfer function of $T(s)$ with respect to change of the parameter X requires finding the corresponding derivative:

$$S(H(s), X) = \frac{dH(s)}{H(s)}\frac{X}{dX}. \tag{1}$$

The study of network response by means of formula (1) is acceptable only when changes in its parameter X are small. When some element's parameters change significantly, this approximated result may be unsatisfied. Analytical sensitivity analysis would be much easier by using a symbolic approach than by a numerical approach where computation cost is a key concern.

If we consider the transfer function as $T(s) = N(s)/D(s)$ formula (1) will transform into expression proposed by Hoang [5]:

$$S(T(s), X) = X\left(\frac{dN(s)}{N(s)} - \frac{dN(s)}{D(s)}\right). \tag{2}$$

Calculation of the transfer function derivatives leads to appearance of multiple algebraic cofactors with different signs and thus they may contain many cancellations. The well-known technique proposed by Bode [1] presented the sensitivity function $T(s)$ in form of ratio of determinant and cofactors of nodal admittance or loop impedance matrix multiplication. For example, the sensitivity of $T(s) = V_3(s)/V_1(s)$ of the circuit in Figure 1 is expressed as follows:

$$S_X^T = -X\frac{\Delta_{15}\Delta_{73}}{\Delta\Delta_{13}}, \tag{3}$$

where X is a transfer impedance Z of current controlled voltage source (CCVS) or transfer admittance of voltage controlled current source (VCCS), Δ is a determinant of loop impedance matrix or $\Delta = \Delta_{11}$ for nodal admittance matrix, and Δ_{11}, Δ_{15}, Δ_{13}, and Δ_{73} are cofactors.

Expression (3) derived from (1) by means of Jacobi identity in form of

$$\Delta_{13,75}\Delta = \Delta_{13}\Delta_{75} - \Delta_{15}\Delta_{73}, \tag{4}$$

where $\Delta_{13,75}$ is a second-order cofactor. The rows and columns with numbers 1 and 3, 5, and 7 correspond to input and output variables (controlled and controlling variables of controlled source).

Sensitivity analysis by means of Bode's technique leads to appearance of cancellations because of multiple entries of elements parameters into the matrix. The sensitivity computation formulae with numbered ports of multiport circuit model presented in [5] are devoid of this drawback. But the tedious combinatorial search of the transfer function from one port to another is needed to use them.

We propose in this paper the generalization of Hoang formula and Bode's technique for all types of transfer functions and circuit elements, based on implementation of the concept of higher order summative cofactors (HOSC) [12–15] instead of ordinary algebraic cofactors and computation of HOSC by means of GPEM [16–21].

2.2. Higher Order Summative Cofactors and Generalized Parameter Extraction Method. The higher order cofactor is a cofactor of a cofactor. The nth order cofactor is represented by a symbol $\Delta_{r_1,k_1,r_2,k_2,\ldots,r_n,k_n}$, where $r_1, r_2, \ldots, r_n$ and $k_1, k_2, \ldots, k_n$ are the numbers of deleted rows and columns, respectively. The summative cofactors are cofactors with at least one deletion in a form $(a + d)$ or $(a - d)$, which means the following: add or subtract the row (column) a to or from the row (column) d. For example the summative cofactors of the same matrix can be expressed as follows [15]:

$$\Delta_{(a+b)(c+d)} = \Delta_{ac} - \Delta_{ad} - \Delta_{bc} + \Delta_{bd}, \tag{5}$$

where $\Delta_{ac} = \Delta_{(a+0)(c+0)}$, $\Delta_{ad} = \Delta_{(a+0)(d+0)}$, and so on.

In accordance with GPEM basis [16–21] we consider the sums of row (column) indexes as a result of nullator (norator) connection to the network nodes corresponding to the indexes.

Let us express the voltage gain function for circuit model in Figure 2(a) as follows:

$$T(s) = \frac{V_{\text{out}}(s)}{V_{\text{in}}(s)} = \frac{\Delta_{(1+2)(3+4)}}{\Delta_{(1+2)(1+2)}}, \tag{6}$$

where HOSC in the numerator is the determinant of the network, in which the independent source and arbitrary response are replaced by oriented nullor [16] and HOSC in the denominator is the determinant of the network, in which the independent excitation and the arbitrary response are zero.

The calculation procedure of the network determinants is based on the recursive usage of the parameter extraction formula [16]:

$$\Delta = \chi\Delta(\chi \longrightarrow \infty) + \Delta(\chi = 0), \tag{7}$$

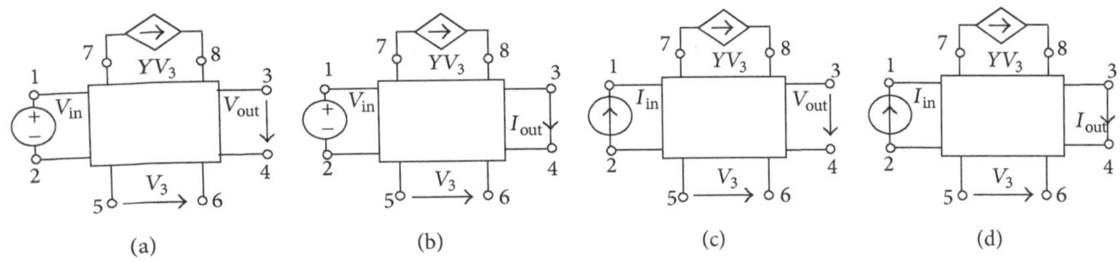

FIGURE 2: The circuit for sensitivity computation of transfer functions: voltage gain (a), transfer admittance (b), transfer impedance (c), and current gain (d).

where χ is a parameter of arbitrary linear circuit element and $\Delta(\chi \to \infty)$ and $\Delta(\chi = 0)$ correspond to the determinants of the circuit matrix in which the parameter of extracted elements $\chi \to \infty$ or $\chi = 0$, respectively.

The parameter extraction formula (7) is applicable for determinant calculation of the network consisting of any kind of linear active circuit models including the pathological elements (nullors and pathological mirrors). In case of nullor extraction the special formula was proposed in [16]

$$\Delta = \pm\Delta_n, \quad (8)$$

where χ is the determinant of the circuit matrix after the extraction procedure of the nullor number n. Equation (8) means that the nullor extraction will change the sign of the initial determinant. The choice of the sign depends on the orientation of the nullator and norator. If these elements have got the same orientation with respect to the basic node that the sign will be positive, in the opposite case the sign will be negative.

The procedure of nullor extraction can be formalized by the following steps.

(1) The choice of the supporting nodes: first of the supporting nodes should be connected to norator and the second to nullator. If there is a common node of a nullator and a norator it must be chosen as single supporting node. Note that supporting node may correspond to the ground node.

(2) The terminals of non-extracted norators and current sources connected to supporting node are moved to the opposite node of extracted norator. In that case the non-extracted nullators and controlling voltages keep connections to supporting node. Then in the same way, the terminals of non-extracted nullators and controlling voltages connected to supported node are moved to the opposite node of extracted nullator. In that case non-extracted norators and current sources keep connections to supporting node.

(3) A norator and a nullator of the extracted nullor are deleted from circuit. In case of two supporting nodes they must combine.

(4) If extracted norator and nullator have got the same orientation with respect to the supporting node then the determinant sign will be positive. Otherwise the sign will be negative. In case of two supporting nodes the inverted rule is needed.

Formula (8) can be implemented not only for the nullor extraction from circuit model, but also for Coates flow-graph modification as shown in Section 4.

The generalized parameter extraction method does not have any limitation by circuit size. The GPEM has been realized by Filaretov in the program Cirsym (circuit simulator) as a part of the software tool SCAD. The program can be downloaded from http://intersyn.narod.ru/. The symbolic expressions obtained by means of Cirsym are comparable by the number of arithmetical operations with computation results by means of the professional mathematic programs as Maple and Matlab [22].

As can be seen from the comparison of symbolic expressions of notch filter transfer function [23] presented in "Symbolic circuit analysis: Library of benchmark circuits" (http://rodanski.net/ben/work/symbolic/index.htm) the result of program Cirsym named by Filaretov 2 is second ranked. Also this result was replicated in [24] as example of compact size expression.

2.3. Derivation of Symbolic Sensitivity Computation Formulae. Let us express the transfer function (6) by means of parameter extraction formula as

$$T(s) = \frac{Y\Delta_{(1+2)(3+4),(7+8)(5+6)} + \Delta^{0}_{(1+2)(3+4)}}{Y\Delta_{(1+2)(1+2),(7+8)(5+6)} + \Delta^{0}_{(1+2)(1+2)}}, \quad (9)$$

and then expressed the Hoang formula (2) as follows:

$$S_Y^T = Y \cdot \left(\frac{\Delta^{\infty}_{(1+2)(1+2)}}{\Delta_{(1+2)(1+2)}} - \frac{\Delta^{\infty}}{\Delta} \right). \quad (10)$$

Note, that upper indexes $\langle\langle 0\rangle\rangle$ and $\langle\langle\infty\rangle\rangle$ in formulas (9) and (10) mean that those HOSC are derived with $Y = 0$ and $Y \to \infty$ correspondingly. In case of controlled source the operation $\chi \to \infty$ leads to substitution of CS by the nullor.

Formula (10) is applicable for sensitivity computation of the arbitrary circuit elements. Just replace the $\langle\langle Y\rangle\rangle$ by $\langle\langle\chi\rangle\rangle$.

Now let us consider the transformation of Bode formula (1) by substitution of expression (9) and cancellation by $\Delta_{(1+2)(1+2)}$:

$$S_Y^T = Y\frac{\Delta_{(1+2)(3+4),(7+8)(5+6)}\Delta_{(1+2)(1+2)}}{\Delta_{(1+2)(1+2)}\Delta_{(1+2)(3+4)}} - Y\frac{\Delta_{(1+2)(3+4)}\Delta_{(1+2)(1+2),(7+8)(5+6)}}{\Delta_{(1+2)(1+2)}\Delta_{(1+2)(3+4)}}. \quad (11)$$

Table 1: Sensitivity symbolic calculation formulae.

Number	Sensitivity type	Formula
1	S_{χ}^{T}	$-\dfrac{\chi\Delta_{(1+2)(5+6)}\Delta_{(1+2)(1+2),(7+8)(3+4)}}{\Delta_{(1+2)(1+2)}\Delta_{(1+2)(3+4)}}$
2	S_{χ}^{Y}	
3	S_{χ}^{B}	$-\dfrac{\chi\Delta_{(1+2)(5+6)}\Delta_{(7+8)(3+4)}}{\Delta\Delta_{(1+2)(3+4)}}$
4	S_{χ}^{Z}	

We obtain the transfer function sensitivity expression with respect to a VCCS parameter. It is possible to minimize the numerator by means of Jacobi identity. We need to use the HOSC $\Delta_{(1+2)(1+2)}$ instead of determinant Δ:

$$S_{Y}^{T} = -Y\frac{\Delta_{(1+2)(5+6)}\Delta_{(1+2)(1+2),(7+8)(3+4)}}{\Delta_{(1+2)(1+2)}\Delta_{(1+2)(3+4)}}. \quad (12)$$

As can be seen, $\Delta_{(1+2)(1+2)}$ is a numerator of transfer function from circuit input to controlling voltage of VCCS, $\Delta_{(1+2)(1+2),(7+8)(3+4)}$ is a numerator of transfer function from controlled current source to circuit output, and $\Delta_{(1+2)(1+2)}\Delta_{(7+8)(3+4)}$ is a product of numerator and denominator of voltage gain.

The process of sensitivity computation with respect to other elements and/or for different transfer function types is the same. In Table 1 the generalized sensitivity formulae are proposed.

Note that the formula in first line of Table 1 is appropriate not only for voltage gain, but also for transfer impedance function. Let us show it for circuit in Figure 2(b):

$$Y(s) = \frac{I_{\text{out}}(s)}{V_{\text{in}}(s)} = \frac{\Delta_{(1+2)(3+4)}}{\Delta_{(1+2)(1+2)}}. \quad (13)$$

Even if output nodes of transfer impedance function in Figure 2(b) are connected in contrast to Figure 2(a), it is obvious that expression (13) is equal to (6). The formulae for transfer impedance (Figure 2(c)) and current gain (Figure 2(d)) are similar like expressions (6) and (13):

$$Z = \frac{V_{\text{out}}}{I_{\text{in}}} = \frac{\Delta_{(1+2)(3+4)}}{\Delta},$$
$$B = \frac{I_{\text{out}}}{I_{\text{in}}} = \frac{\Delta_{(1+2)(3+4)}}{\Delta}. \quad (14)$$

The denominator of any sensitivity formulae in Table 1 is a product of initial transfer function numerator and denominator and the numerator of the sensitivity formulae is a product of numerators of transfer functions from circuit input to controlled branch and from controlling source to circuit output.

The proposed formulae are easy to use for sensitivity computation of two ports. For example, if we change indexes 7 and 8 to 5 and 6 correspondingly, we will obtain the sensitivity function for admittance element:

$$S_{Y}^{T} = -Y\frac{\Delta_{(1+2)(5+6)}\Delta_{(1+2)(1+2),(5+6)(3+4)}}{\Delta_{(1+2)(1+2)}\Delta_{(1+2)(3+4)}}. \quad (15)$$

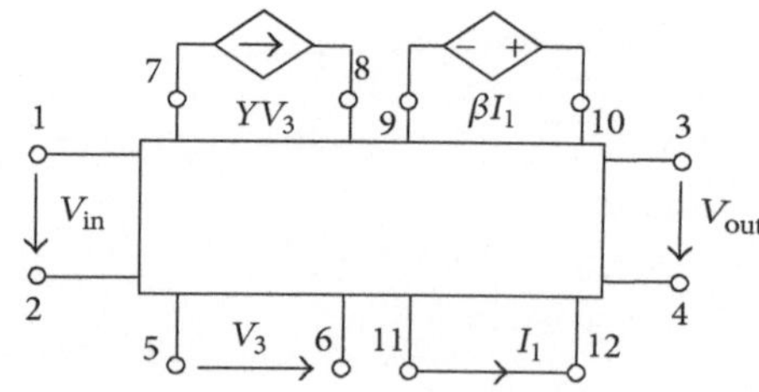

Figure 3: The test circuit for second-order transfer impedance sensitivity computation.

The formula for transfer function sensitivity with respect to impedance parameter ($\chi = Z$) can be derived in the same way. We just have to replace the number of index 8 by 6.

For sensitivity of the input impedance or input admittance functions derivation it is necessary to replace indexes 3 and 4 by 1 and 2 correspondingly.

2.4. The Special Topological Cases of Sensitivity Formulae Simplification. If the voltage source of CS is connected in parallel with a voltage output port or if the current source of CS is connected in series with a current output port, we can simplify the sensitivity formulae like that shown in Table 2. As it can be seen, the HOSC indexes 3 and 4 correspond to 7 and 8. Taking into account the identity $\chi\Delta_{(1+2)(5+6)} = \Delta_{(1+2)(7+8)}$ we obtain the expressions consisting of just two HOSC.

In the same way, we can easily simplify the sensitivity formulae for other special cases if the input excitation transfers directly to the controlling branch of CS, if the controlled voltage source is connected to the input voltage port in series or the controlled current source is connected in parallel with input current source, and if the controlling voltage of CS has the same port as output circuit voltage or the controlling current of CS corresponds to output current port.

2.5. The Second-Order Sensitivity Computation. Let us consider the process of computation of second-order sensitivity function. The repeated differentiation of first-order formulae from Table 1 is needed in this case. So the second-order transfer impedance sensitivity of circuit in Figure 3 with respect to the VCCS and CCVS parameters can be expressed as follows:

$$\begin{aligned} S_{\beta Y}^{Z} = \big(\beta Y \big[&\Delta_{(1+2)(5+6)}\Delta_{(7+8)(9+10),(1+2)(1+2)} \\ &\times \Delta_{(11+12)(3+4),(1+2)(1+2)} + \Delta_{(1+2)(9+10)} \\ &\times \Delta_{(11+12)(5+6),(1+2)(1+2)}\Delta_{(7+8)(3+4),(1+2)(1+2)}\big]\big) \\ &\times \left(\Delta^{2}\Delta_{(1+2)(3+4)}\right)^{-1}. \end{aligned} \quad (16)$$

We can see the product of squared determinant and the transfer function numerator in formula denominator and two terms in numerator. The first term here consists of the numerators of transfer functions, from circuit input to controlling voltage of VCVS, from voltage source of VCVS

Table 2: The simplified sensitivity formulae in case of controlled source connecting with circuit output.

Number	Sensitivity type	Circuit	Circuit	Sensitivity type	Formula
1	S_{μ}^{T}	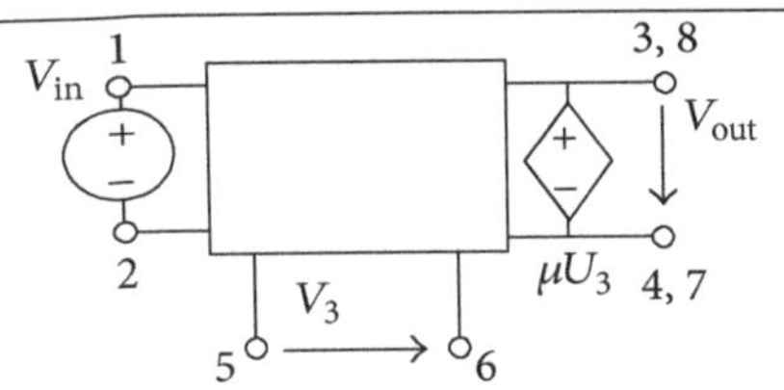	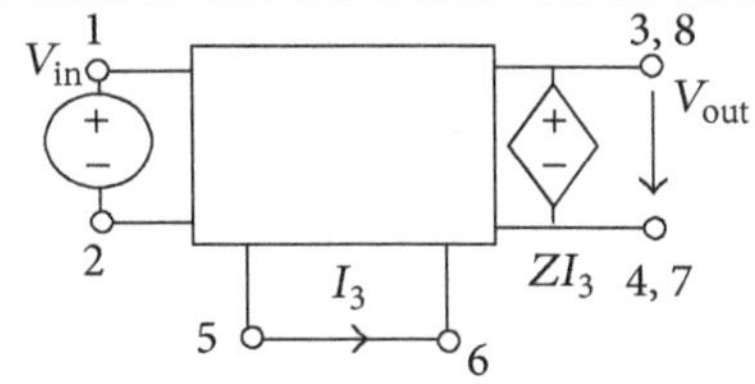	S_{Z}^{K}	$\frac{\Delta_{(1+2)(1+2),(3+4)(3+4)}}{\Delta_{(1+2)(1+2)}}$
2	S_{μ}^{Z}	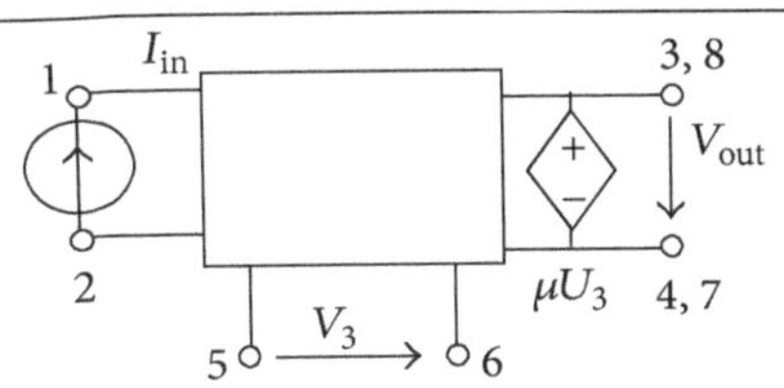	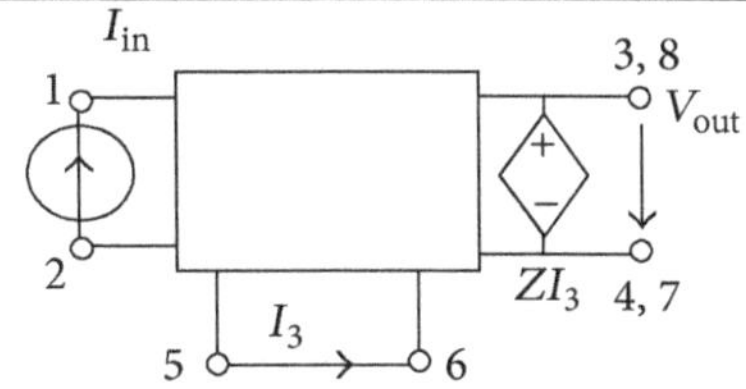	S_{Z}^{Z}	$\frac{\Delta_{(3+4)(3+4)}}{\Delta}$
3	S_{Y}^{Y}	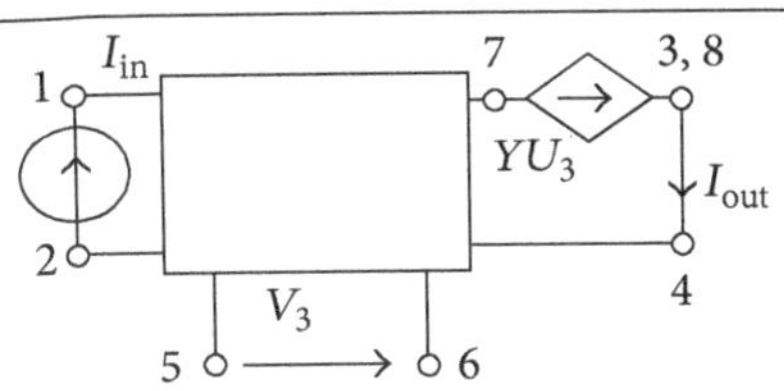	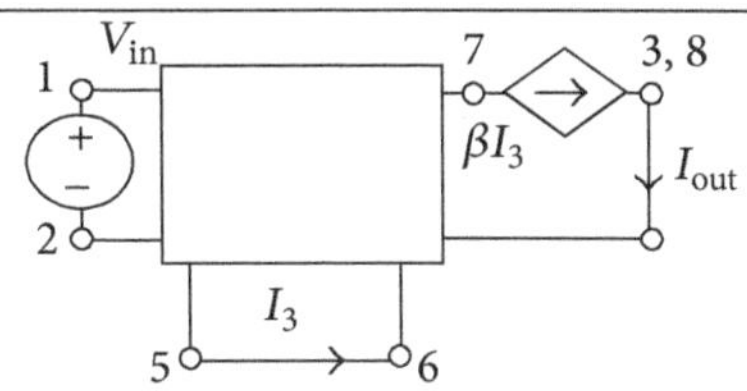	S_{β}^{Y}	$\frac{\Delta^{0}_{(1+2)(1+2)}}{\Delta_{(1+2)(1+2)}}$
4	S_{Y}^{B}	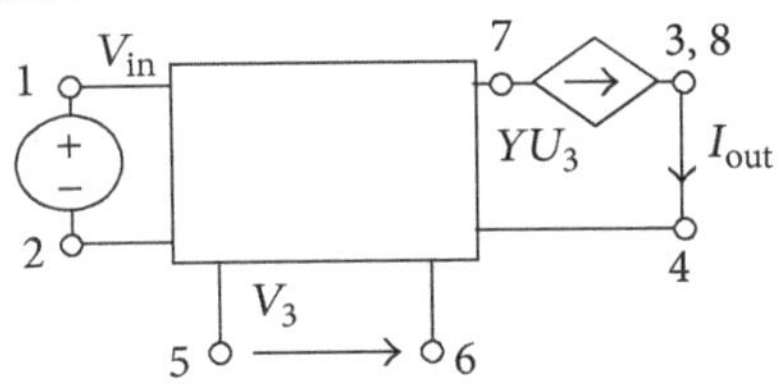	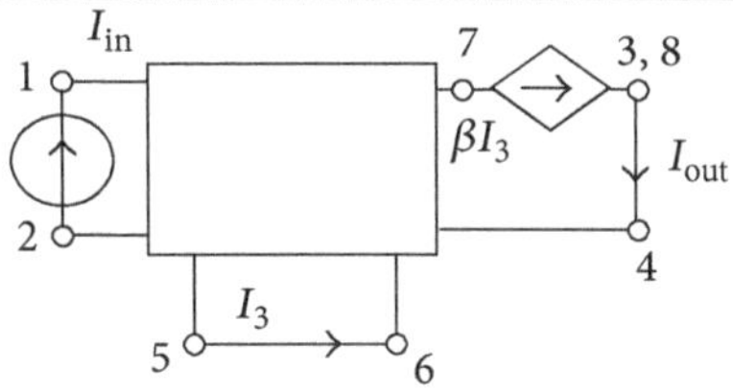	S_{β}^{B}	$\frac{\Delta^{0}}{\Delta}$

to controlling current of CCCS, and from current source of CCCS to circuit output. For the second term we take into account that succession of excitation transfer from input to output of circuit is different. In that case the signal transfers firstly to CCCS and then to VCVS.

For the purpose of obtaining the second-order sensitivity function in Hoang form we just need to differentiate and represent by means of HOSC formula (2). So for the circuit presented in Figure 3 the sensitivity expression for voltage transfer function with respect to parameters of CS is obtained as follows:

$$\begin{aligned} S_{\beta Y}^{T} = \Big(\beta Y \Big[& \Delta^{\infty}_{(1+2)(3+4)}\Delta^{2} - \Delta^{Y\to\infty}\Delta^{\beta\to\infty}_{(1+2)(3+4)}\Delta \\ & - \Delta^{Y\to\infty}_{(1+2)(3+4)}\Delta^{\beta\to\infty}\Delta - \Delta_{(1+2)(3+4)}\Delta^{\infty}\Delta \\ & + 2\Delta_{(1+2)(3+4)}\Delta^{\beta\to\infty}\Delta^{Y\to\infty}\Big]\Big) \\ \times & \left(\Delta^{2}\Delta_{(1+2)(3+4)}\right)^{-1}, \end{aligned} \tag{17}$$

where upper indexes ⟨⟨$Y \to \infty$⟩⟩ and ⟨⟨$\beta \to \infty$⟩⟩ mean that those HOSC are derived with $Y \to \infty$ and $\beta \to \infty$ correspondingly. The index ⟨⟨∞⟩⟩ means that this HOSC is derived with both parameters tending to infinity.

3. Symbolic Sensitivity Analysis Examples

Let us consider the voltage transfer function symbolic AC sensitivity analysis of second-order low-pass filter shown in Figure 4 [25]. The values of parameters elements are as follows: C_1 = 18,7571 nF; C_2 = 9,37857 nF; $G_1 = G_2 = 1$ mS; $\mu = 1$.

As can be seen, the circuit topology makes it possible to simplify sensitivity computation with respect to VCVS parameter by means of formula from the first line of Table 2. Taking into account the nodes numeration of filter circuit in Figure 4 the sensitivity function can be expressed as follows:

$$S_{\mu}^{T} = \frac{\Delta_{(1+0)(1+0),(2+0)(2+0)}}{\Delta_{(1+0)(1+0)}} = \frac{\Delta_{11,22}}{\Delta_{11}}. \tag{18}$$

The HOSC expansion is performed by means of GPEM. In accordance with (7) the VCVS parameter is extracted from circuit as shown in Figure 7.

The second summand of expression (Figure 7) is correspond to the HOSC $\Delta_{11,22}$, the numerator of formula (18). It is desirable to obtain the polynomial symbolic expression so the circuit expansion is performed by extracting capacitors admittances.

Note, that admittances extraction leads to appearance of two subcircuits with short circuit and deleting of the element correspondingly. The determinant of the first circuit

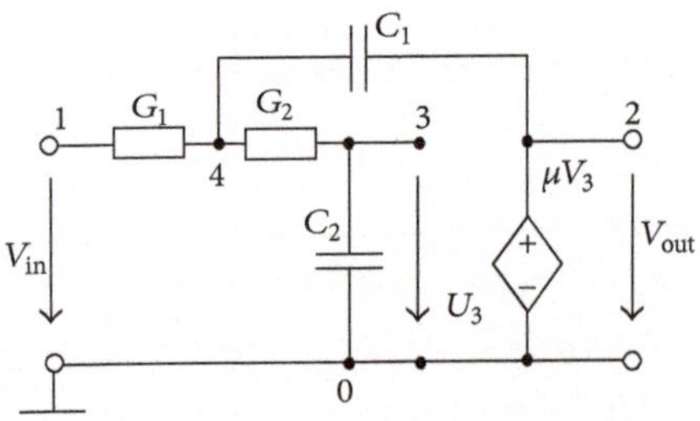

FIGURE 4: Second-order low-pass filter.

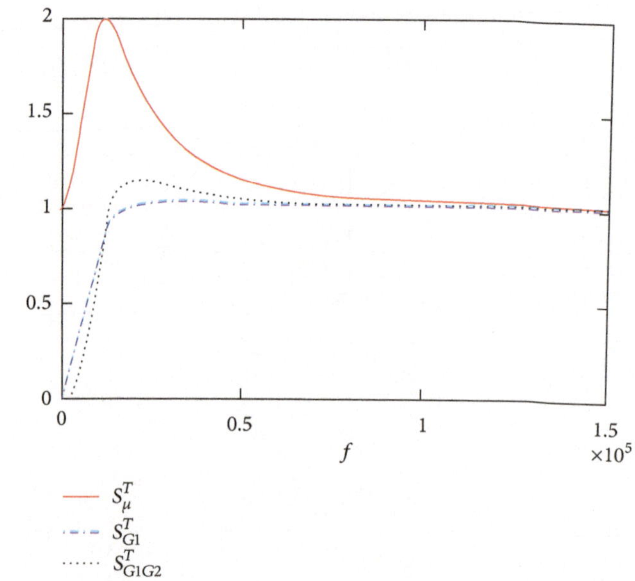

FIGURE 5: First- and second-order symbolic sensitivities versus frequency f.

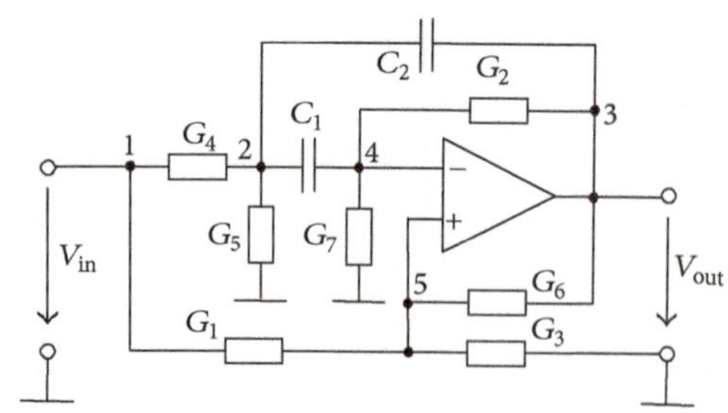

FIGURE 6: The STAR network.

in (Figure 8) is equal to unity because both admittances connected in short circuit. After expansion of element G_2 from second circuit in (Figure 8) we obtain the determinant of short-circuited admittance G_1 in the same way. In the third circuit with parallel admittances the determinant is equal to the elements sum. And the last circuit determinant is a product of G_2 and G_1.

The symbolic determinant expression for the first summand in (Figure 7) is shown in Figure 9

For computation of expression (Figure 9) the following procedures were done: deletion of capacities admittance sC_2 connecting in parallel with nullator, parameter extraction of admittance sC_1 connecting in series with nullator, followed by deletion of the admittance G_1 connecting in parallel with nullator. As a result we obtain norator-nullator pair connected in opposite direction and determinant of that kind of circuit is equal to "−1."

Now we need to substitute the symbolic expression from (Figures 8 and 9) in (18) to obtain the voltage gain sensitivity function:

$$S_{\mu}^{T} = \frac{s^2C_1C_2 + s\left[C_1G_2 + C_2\left(G_1 + G_2\right)\right] + G_1G_2}{s^2C_1C_2 + s\left[\left(1-\mu\right)C_1G_2 + C_2\left(G_1 + G_2\right)\right] + G_1G_2}. \tag{19}$$

The derivation process of sensitivity functions with respect to other parameters is similar. For example we derive the first- and second-order sensitivities $S_{G_1}^{T}$ and $S_{G_1G_2}^{T}$:

$$S_{G_1}^{T} = -\frac{s^2C_1C_2 + s\left[C_1G_2 + C_1G_2\left(1-\mu\right)\right]}{s^2C_1C_2 + s\left[\left(1-\mu\right)C_1G_2 + C_2\left(G_1 + G_2\right)\right] + G_1G_2},$$

$$\begin{aligned} S_{G_1G_2}^{T} = {} & (G_1G_2 [\Delta_{(1+0)(1+4)}\Delta_{(1+4)(4+3),(1+0)(1+0)} \\ & \times \Delta_{(4+3)(2+0),(1+0)(1+0)} + \Delta_{(1+0)(4+3)} \\ & \times \Delta_{(4+3)(1+4),(1+0)(1+0)}\Delta_{(1+4)(2+0),(1+0)(1+0)}]) \\ & \times \left(\Delta^2\Delta_{(1+0)(2+0)}\right)^{-1}, \end{aligned} \tag{20}$$

where

$$\begin{aligned} \Delta_{(1+0)(2+0)} &= \mu G_1G_2, \\ \Delta &= sC_2\left(sC_1 + G_2\left(1-\mu\right)\right) + sC_1\left(G_2 + G_1\right) + G_1G_2, \\ \Delta_{(1+0)(1+4)} &= s_2\left(C_1C_2\right) + s\left(C_1G_2\left(1-\mu\right) + C_2G_2\right), \\ \Delta_{(1+4)(4+3),(1+0)(1+0)} &= -sC_2, \\ \Delta_{(4+3)(2+0),(1+0)(1+0)} &= s\left(-\mu C_1\right) - \mu G_1, \\ \Delta_{(1+0)(4+3)} &= sC_2G_1, \\ \Delta_{(4+3)(1+4),(1+0)(1+0)} &= s\left(C_1\mu - C_2\right), \\ \Delta_{(1+4)(2+0),(1+0)(1+0)} &= -\mu G_2 0. \end{aligned} \tag{21}$$

Using Mathcad software symbolic sensitivity simulation results versus frequency f are shown in Figure 5.

As it is seen, the symbolic sensitivity analysis technique based on the concept of high order summative cofactors and GPEM rules can be implemented in circuit design process and for circuit characterization, successfully.

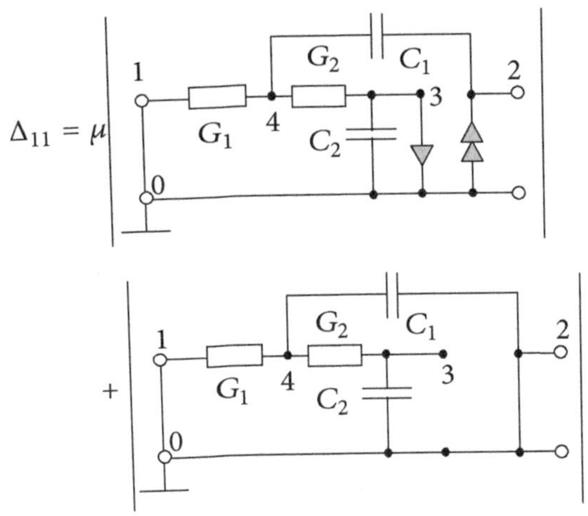

FIGURE 7

4. Comparison Test

4.1. The Comparison Conditions. In this section we present the comparison of sensitivity expressions obtained by means of three different approaches: the two-port transimpedance method [6], the method based on the modified Coates flow-graph [7], and the proposed technique. The test circuit is shown in Figure 6 [7].

Let us suppose that voltage transfer function $T(s)$ and the first- and second-order symbolic sensitivities with respect to parameters G_4 and C_2 are under consideration. We compare the formulae by following criteria in order to estimate the methods calculation efficiency: the number of arithmetical operations and occurrence of cancellations.

4.2. Symbolic Expressions Obtained by means of the GPEM-Based Technique. The symbolic expressions $T(s)$, $S^T_{G_4}$, $S^T_{sC_2}$, and $S^T_{G_4sC_2}$ shown below are calculated by means of the symbolic circuit analysis program Cirsym. The sensitivity functions are presented here in two forms: Hoang form (10) and Bode form (Table 1).

The transfer function

$$T(s) = \frac{\Delta_{(1+0)(3+0)}}{\Delta}, \quad (22)$$

where

$$y1 = s * C1; \qquad y2 = s * C2;$$

$$\Delta_{(1+0)(3+0)} = y1 * ((G7 + G2 + G5 + y2) * (G1) - G4 * (G6 + G3)) + (G7 + G2) * G1 * (G5 + y2 + G4);$$

$$\Delta = y1 * ((G7 + G5 + G4) * (-G6) + (G2 + y2) * (G3 + G1)) + ((G5 + G4 + y2)) * ((G3 + G1) * (G2) - G6 * G7). \quad (23)$$

The sensitivity functions in Hoang form

$$S^T_{G_4} = G_4 \cdot \left(\frac{\Delta^{\infty}_{(1+0)(3+0)}}{\Delta_{(1+0)(3+0)}} - \frac{\Delta^{\infty}}{\Delta} \right), \quad (24)$$

where

$$y1 = s * C1; \qquad y2 = s * C2;$$

$$\Delta^{\infty}_{(1+0)(3+0)} = y1 * (-(G3 + G6)) + G1 * (G7 + G2); \quad (25)$$

$$\Delta^{\infty} = (G3 + G1) * (G2) - G6 * (G7 + y1).$$

$$S^T_{sC_2} = sC_2 \cdot \left(\frac{\Delta^{\infty}_{(1+0)(3+0)}}{\Delta_{(1+0)(3+0)}} - \frac{\Delta^{\infty}}{\Delta} \right), \quad (26)$$

where

$$y1 = s * C1; \qquad y2 = s * C2;$$

$$\Delta^{\infty}_{(1+0)(3+0)} = G1 * (G7 + G2 + y1); \quad (27)$$

$$\Delta^{\infty} = (G2 + y1) * ((G3 + G1)) - G6 * G7,$$

$$S^T_{G_4sC_2} = G4 * y2 * ((\Delta^{\infty}_{(1+0)(3+0)} * \Delta * \Delta - \Delta^{sC_2 \to \infty} * \Delta^{G_4 \to \infty}_{(1+0)(3+0)} * \Delta - \Delta^{sC_2 \to \infty}_{(1+0)(3+0)} * \Delta^{G_4 \to \infty} * \Delta - \Delta_{(1+0)(3+0)} * \Delta^{\infty} * \Delta + 2 * \Delta_{(1+0)(3+0)} * \Delta^{G_4 \to \infty} * \Delta^{sC_2 \to \infty}) \times (\Delta_{(1+0)(3+0)} * \Delta * \Delta)^{-1}), \quad (28)$$

where

$$y1 = s * C1; \qquad y2 = s * C2;$$

$$\Delta^{G_4 \to \infty}_{(1+0)(3+0)} = y1 * (-(G6 + G3)) + G1 * (G7 + G2);$$

$$\Delta^{sC_2 \to \infty}_{(1+0)(3+0)} = G1 * (G7 + G2 + y1);$$

$$\Delta^{\infty}_{(1+0)(3+0)} = 0; \quad (29)$$

$$\Delta^{G_4 \to \infty} = (G3 + G1) * (G2) - G6 * (G7 + y1);$$

$$\Delta^{sC_2 \to \infty} = (G2 + y1) * ((G3 + G1)) - G6 * G7;$$

$$\Delta^{\infty} = 0.$$

The sensitivity functions in Bode form

$$S^T_{G_4} = \frac{-G_4 \Delta_{(1+0)(1+2)} \Delta_{(1+2)(3+0),(1+0)(1+0)}}{\Delta \Delta_{(1+0)(3+0)}}, \quad (30)$$

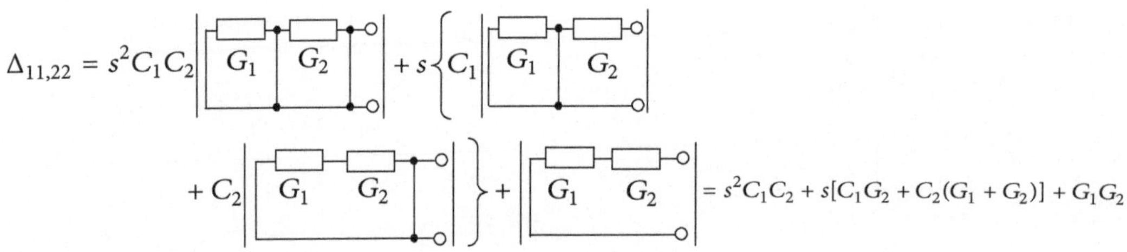

Figure 8

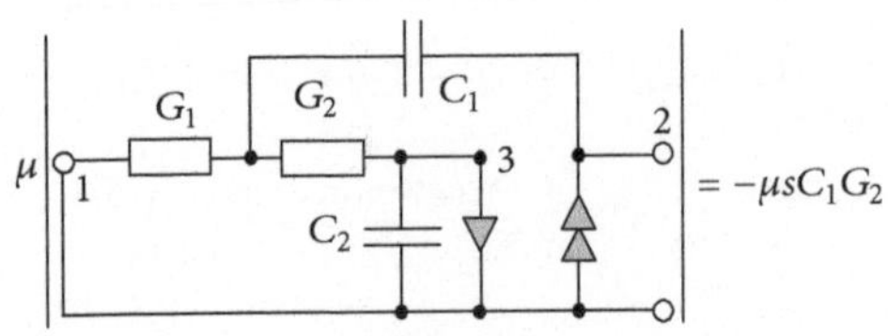

Figure 9

where

$$
\begin{aligned}
y1 &= s * C1; \qquad y2 = s * C2;\\
\Delta_{(1+0)(1+2)} &= G1 * (y2 * (-G7) + G2 * G5)\\
&\quad + y1 * ((G2 + y2) * (G3) - G6 * (G7 + G5))\\
&\quad + ((G5 + y2)) * (G2 * (G3) - G6 * G7);\\
\Delta_{(1+2)(3+0),(1+0)(1+0)} &= y1 * (G6 + G3 + G1),
\end{aligned}
\tag{31}
$$

$$
S^K_{sC_2} = \frac{-y_2\Delta_{(1+0)(2+3)}\Delta_{(2+3)(3+0),(1+0)(1+0)}}{\Delta\Delta_{(1+0)(3+0)}}, \tag{32}
$$

where

$$
\begin{aligned}
y1 &= s * C1; \qquad y2 = s * C2;\\
\Delta_{(1+0)(2+3)} &= y1 * ((G7 + G5) * (-G1) + G3 * G4)\\
&\quad + G4 * (G1 * (-G7) + G2 * (G3) - G6 * G7)\\
&\quad - G1 * G5 * (G7 + G2);\\
\Delta_{(2+3)(3+0),(1+0)(1+0)} &= (-y1 * (G6 + G3 + G1)).
\end{aligned}
\tag{33}
$$

$$
\begin{aligned}
S^T_{G_4sC_2} &= (G_4y_2[\Delta_{(1+0)(1+2)}\Delta_{(1+2)(2+3),(1+0)(1+0)}\\
&\quad \times \Delta_{(2+3)(3+0),(1+0)(1+0)} + \Delta_{(1+0)(2+3)}\\
&\quad \times \Delta_{(2+3)(1+2),(1+0)(1+0)}\Delta_{(1+2)(3+0),(1+0)(1+0)}])\\
&\quad \times (\Delta^2\Delta_{(1+0)(2+0)})^{-1},
\end{aligned}
\tag{34}
$$

where

$$
\begin{aligned}
y1 &= s * C1; \qquad y2 = s * C2;\\
\Delta_{(1+0)(1+2)} &= G1 * (y2 * (-G7) + G2 * G5)\\
&\quad + y1 * ((G2 + y2) * (G3) - G6 * (G7 + G5))\\
&\quad + ((G5 + y2)) * (G2 * (G3) - G6 * G7);\\
\Delta_{(1+2)(2+3),(1+0)(1+0)} &= y1 * (-(G3 + G1))\\
&\quad - ((G3 + G1) * (G2) - G6 * G7);\\
\Delta_{(2+3)(3+0),(1+0)(1+0)} &= -y1 * (G6 + G3 + G1);\\
\Delta_{(1+0)(2+3)} &= y1 * ((G7 + G5) * (-G1) + G3 * G4)\\
&\quad + G4 * (G1 * (-G7) + G2 * (G3) - G6 * G7)\\
&\quad - G1 * G5 * (G7 + G2);\\
\Delta_{(2+3)(1+2),(1+0)(1+0)} &= y1 * (G6)\\
&\quad - ((G3 + G1) * (G2) - G6 * G7);\\
\Delta_{(1+2)(3+0),(1+0)(1+0)} &= y1 * (G6 + G3 + G1).
\end{aligned}
\tag{35}
$$

4.3. Symbolic Expressions Obtained by means of Modified Coates Flow-Graph Concept. The Coates flow-graph is useful and often used in the network theory. The modification of Coates flow-graph provides the symbolic sensitivity analysis of nullor-based equivalent circuits [7]. We will show here that modification of Coates flow-graph is corresponding to implementation of the nullor extraction formula (8).

Let us consider the flow-graph presented in Figure 8(a) (see publication [7]) for the passive part of the equivalent circuit of STAR network shown in Figure 6. The oriented norator connected between nodes 3 and 0. The oriented nullator connected between nodes 5 and 4.

First, we choose the node 3 as supporting and make the norator short-circuited:

(1) the ends of all incoming edges ($-sC_2$ and $-G_2$) into node 3 are moved to node 0;

(2) the self-loop $Y33$ transforms into the incoming edge into node 0;

(3) the edges connected to node 0 are not shown on graph.

Second, we choose the node 5 as supporting and make the nullator short-circuited:

(1) the end of incoming edge $-G_6$ into node 5 is moved to node 4;

(2) the self-loop $Y55$ transformed into the incoming edge into node 4.

Third, we combine the two supporting nodes 3 and 5:

(1) the edge $-G_6$ between nodes 3 and 5 transforms into the self-loop in the new node named by 3;

(2) the parallel codirectional edges $-G_6$ and $Y55$ are combined into the edge $Y55$.

Consequently we obtained the modified Coates flow-graph shown in Figure 8(b) (see publication [7]). Note that in case of single supporting node the procedure of nodes combining is not needed.

The estimation process of the determinant sign of modified Coates flow-graph is not formalized in [7]. It leads to complications of method usage and high error probability.

Implementation of the oriented nullor concept [16] provides the simple rule for estimation of the sign of Coates graph determinant: the sign is positive (negative), if norator and nullator have the same (different) orientations against the supporting nodes. In case of two supporting nodes the inverted rule is needed.

The symbolic formulae $T(s)$, $S_{G_4}^T$, $S_{sC_2}^T$, and $S_{G_4 sC_2}^T$ presented here were replicated from [7] with correction of typing error in subexpression $Y55$ and sign of determinants:

$$T(s) = (-Y22 * G1 * Y44 + y1 * y1 * G1 + y1 * Y55 * G4) \times (Y22 * G6 * Y44 - Y22 * G2 * Y55 - y1 * y1 * G6 - y1 * y2 * Y55)^{-1}, \tag{36}$$

where

$$\begin{aligned} Y22 &= (G4 + G5 + y1 + y2); \\ Y44 &= (G2 + G7 + y1); \\ Y55 &= (G1 + G3 + G6); \\ y1 &= s * C1; \qquad y_2 = s * C2. \end{aligned} \tag{37}$$

Let us substitute the subexpressions $Y22$, $Y44$, and $Y55$ into the transfer function (36):

$$\begin{aligned} T(s) = {} & ((-(G4 + G5 + y1 + y2) * G1 \\ & * (G2 + G7 + y1) + y1 * y1 * G1 \\ & + y1 * (G1 + G3 + G6) * G4)) \\ & \times (((G4 + G5 + y1 + y2) * G6 \\ & * (G2 + G7 + y1) - (G4 + G5 + y1 + y2) \\ & * G2 * (G1 + G3 + G6) - y1 * y1 * G6 \\ & - y1 * y2 * (G1 + G3 + G6)))^{-1}. \end{aligned} \tag{38}$$

It is easy to verify that the expanded expressions of the numerator and denominator will consist of the 16 and 28 summands correspondingly. After simplification only 12 terms in numerator and 16 terms in denominator will remain. So, the transfer function (36) includes the 8 pairs of cancellation sum-of-product terms:

$$\begin{aligned} S_{G_4}^T = {} & \frac{G4}{(-Y22 * G1 * Y44 + y1 * y1 * G1 + y1 * Y55 * G4)} \\ & * \left(1 - \frac{(Y44 * (G4 * G6 - G1 * y2) - G2 * G4 * Y55 - G2 * G1 * y1)}{(Y22 * G6 * Y44 - Y22 * G2 * Y55 - Y1 * y1 * G6 - y1 * y2 * Y55)}\right) * y1 * Y55, \end{aligned} \tag{39}$$

$$\begin{aligned} S_{sC_2}^T = {} & \left(\frac{y2}{(-Y22 * G1 * Y44 + y1 * y1 * G1 + y1 * Y55 * G4)}\right. \\ & \left.\times (Y22 * G6 * Y44 - Y22 * G2 * Y55 - y1 * y1 * G6 - y1 * y2 * Y55)^{-1}\right) \\ & * (-Y44 * (G4 * G6 - G1 * y2) + G2 * G4 * Y55 + G2 * G1 * y1 - Y22 * G1 * Y44 + y1 * y1 \\ & * G1 + y1 * Y55 * G4) * y1 * Y55, \end{aligned} \tag{40}$$

$$S^{T}_{G_4 s C_2} = \left(\frac{-G4 * y2}{(-Y22 * G1 * Y44 + y1 * y1 * G1 + y1 * Y55 * G4)} \times \left(Y22 * G6 * Y44 - Y22 * G2 * Y55 - y1^2 * G6 - y1 * y2 * Y55\right)^{-1} \right) * \left(-(-Y44 * G6 + Y55 * G2) * y1 * Y55 * \frac{(-Y44 * (G4 * G6 - G1 * y2) + G2 * G4 * Y55 + G2 * G1 * y1 - Y22 * G1 * Y44 + y1 * y1 * G1 + y1 * Y55 * G4)}{(Y22 * G6 * Y44 - Y22 * G2 * Y55 - y1 * y1 * G6 - y1 * y2 * Y55)} - (-Y44 * G6 + Y55 * G2 + y1 * Y55) * y1 * Y55 * \left(1 - \frac{(Y44 * (G4 * G6 - G1 * y2) - G2 * G4 * Y55 - G2 * G1 * y1)}{(Y22 * G6 * Y44 - Y22 * G2 * Y55 - y1 * y1 * G6 - y1 * y2 * Y55)}\right)\right). \quad (41)$$

4.4. Symbolic Expressions Obtained by means of Two-Port Transimpedance Method. The sequences of expressions, derived by symbolic analysis program STAINS [6], are listed below. The calculation of second-order sensitivity is not support in the current version of the program. So we present here only the transfer function and two first-order symbolic sensitivities with respect to parameters G_4 and C_2:

$$T(s) = \frac{Zoi}{Zii}, \quad (42)$$

where

$$y1 = s * C1; \quad y2 = s * C2;$$

$$x1 = -G1; \quad x2 = -G2 - G6;$$

$$x3 = G1 + G3 + G6; \quad d1 = \frac{G2}{(y1)};$$

$$x4 = G4 * d1;$$

$$x5 = G2 + G6 + y2 + y2 * d1;$$

$$x6 = -y2 - (G4 + G5 + y1 + y2) * d1;$$

$$d2 = -\frac{(G2 + G7 + y1)}{(y1)};$$

$$x7 = x1 + G4 * d2; \quad x8 = x2 + y2 * d2;$$

$$x9 = -y1 - (G4 + G5 + y1 + y2) * d2;$$

$$d3 = -\frac{G1}{(x3)}; \quad x10 = G1 + G4 - x7 * d3;$$

$$x11 = -x8 * d3; \quad x12 = -G4 - x9 * d3;$$

$$d4 = -\frac{G6}{(x3)}; \quad x13 = x4 - x7 * d4;$$

$$x14 = x5 - x8 * d4; \quad x15 = x6 - x9 * d4;$$

$$Z21 = 1 * x12; \quad Zoi = Z21;$$

$$Z11 = -1 * x15; \quad Zii = Z11. \quad (43)$$

After substitution of the subexpressions into (42) we obtain the transfer function expression as follows:

$$T(s) = \left(-1 * \left(G4 * (G1 + G3 + G6) - G1 * \left((-s * C1) - (G4 + G5 + s * (C1 + C2)) * \left(-\frac{(G2 + G7 + s * C1)}{(s * C1)}\right)\right)\right)\right) \times \left(-1 * \left(G6 * \left((-s * C1) - (G4 + G5 + s * (C1 + C2)) * \left(-\frac{(G2 + G7 + s * C1)}{(s * C1)}\right)\right) + \left(-s * C2 - (G4 + G5 + s * (C1 + C2)) * \left(\frac{G2}{(s * C1)}\right)\right) * (G1 + G3 + G6)\right)\right)^{-1}. \quad (44)$$

The expansion of the transfer function numerator and denominator leads to obtaining the expressions with 16 and 28 terms correspondingly. Just as formula (36) the transfer function (42) includes the 8 pairs of cancellation sum-of-product terms:

$$S^T_{G_4} = G4 * Zki * \frac{(Zok * Zii - Zik * Zoi)}{(Zoi * D00 * Zii)}, \quad (45)$$

where

$$y1 = s * C1; \quad y2 = s * C2;$$
$$x1 = -G1; \quad x2 = -G2 - G6;$$
$$x3 = G1 + G3 + G6; \quad d1 = -\frac{G1}{(x3)};$$
$$x4 = G1 + G4 - x1 * d1; \quad x5 = -x2 * d1;$$
$$x6 = y1 * d1 - G4; \quad x7 = -(G2 + G7 + y1) * d1;$$
$$d2 = -\frac{G6}{(x3)}; \quad x8 = -x1 * d2;$$
$$x9 = G2 + G6 + y2 - x2 * d2;$$
$$x10 = y1 * d2 - y2;$$
$$x11 = -G2 - (G2 + G7 + y1) * d2;$$
$$Z21 = 1 * (x6 * y1 + x7 * (G4 + G5 + y1 + y2));$$
$$Zoi = Z21;$$
$$Z11 = -1 * (x10 * y1 + x11 * (G4 + G5 + y1 + y2));$$
$$Zii = Z11; \quad Z13 = 1 * (x8 * y1 - x11 * G4);$$
$$Zik = Z11 - Z13; \quad Z23 = -1 * (x4 * y1 - x7 * G4);$$
$$Zok = Z21 - Z23; \quad Z31 = 1 * (x6 * x11 - x7 * x10);$$
$$Zki = Z11 - Z31;$$
$$D00 = x4 * Z11 + x8 * Z21 - G4 * Z31. \quad (46)$$

$$S^T_{sC_2} = y2 * Zki * \frac{(Zik * Zoi - Zok * Zii)}{(Zoi * D00 * Zii)}, \quad (47)$$

where

$$y1 = s * C1; \quad y2 = s * C2;$$
$$x1 = -G1; \quad x2 = -G2 - G6;$$
$$x3 = G1 + G3 + G6; \quad d1 = -\frac{G1}{(x3)};$$
$$x4 = G1 + G4 - x1 * d1; \quad x5 = -x2 * d1;$$
$$x6 = y1 * d1 - G4; \quad x7 = -(G2 + G7 + y1) * d1;$$
$$d2 = -\frac{G6}{(x3)}; \quad x8 = -x1 * d2;$$
$$x9 = G2 + G6 + y2 - x2 * d2;$$
$$x10 = y1 * d2 - y2;$$
$$x11 = -G2 - (G2 + G7 + y1) * d2;$$
$$Z21 = 1 * (x6 * y1 + x7 * (G4 + G5 + y1 + y2));$$
$$Zoi = Z21;$$
$$Z11 = -1 * (x10 * y1 + x11 * (G4 + G5 + y1 + y2));$$
$$Zii = Z11; \quad Z12 = -0;$$
$$Z13 = 1 * (x8 * y1 - x11 * G4);$$
$$Zik = Z12 - Z13; \quad Z22 = 0;$$
$$Z23 = -1 * (x4 * y1 - x7 * G4);$$
$$Zok = Z22 - Z23;$$
$$Z31 = 1 * (x6 * x11 - x7 * x10);$$
$$Zki = Z21 - Z31;$$
$$D00 = x4 * Z11 + x8 * Z21 - G4 * Z31. \quad (48)$$

Table 3: Comparison of symbolic analysis methods efficiency.

Symbolic function	Symbolic analysis method	M/D	A/S
$T(s)$	GPEM (22)	14	20
	MCFG (36)	17	13
	TI (42)	29	34
$S^T_{G_4}$	GPEM (Hoang form) (24)	20	28
	GPEM (Bode form) (30)	27	32
	MCFG (39)	28	17
	TI (45)	37	42
$S^T_{sC_2}$	GPEM (Hoang form) (26)	19	26
	GPEM (Bode form) (32)	27	33
	MCFG (40)	34	20
	TI (47)	37	43
$S^T_{G_4sC_2}$	GPEM (Hoang form) (28)	50	72
	GPEM (Bode form) (34)	54	63
	MCFG (41)	68	38
	TI	—	—

4.5. Efficiency Comparison. In Table 3 we present the comparison of symbolic analysis methods efficiency by the number of arithmetical operations. The multiplication and division are denoted as M/D and the addition and subtraction are denoted as A/S. The method based on modified Coates flow-graph is denoted as MCFG and the transimpedance method is denoted as TI.

As can be seen, the symbolic expressions obtained by transimpedance method consist of the largest arithmetical operations amount. The results of MCFG-computation are comparable to expressions presented in Bode form obtained by GPEM. The symbolic sensitivity functions presented in

Hoang form are superior by number of arithmetical operations.

Both Bode and Hoang forms implement the four minor calculations. But the presentation of sensitivity function in Hoang form provides the optimal compact size of symbolic expressions. On the other hand the formulae in Bode form for special topological cases proposed in this paper provide the significant simplification of circuit analysis in many cases.

5. Conclusions

In this paper we have presented a method of computation the circuit functions sensitivities in symbolic form. The method requires neither matrix nor ordinary graph description of the circuit. The main advantages of the proposed technique are that it is cancellation-free and provided the compact size of obtaining sensitivity expression as distinct from other methods like the two-port transimpedance method and the method based on the modified Coates flow-graph. It is shown that presentation of sensitivity function in Hoang form provides the optimal compact size of symbolic expressions, and the formulae in Bode form provide the significant simplification of active circuit symbolic analysis. The process of calculation of first- and second-order and multiparameter symbolic sensitivity is automated by computer program Cirsym and allows obtaining all symbolic sensitivities simultaneously.

Conflict of Interests

The authors declare that there is no conflict of interests regarding the publication of this paper.

References

[1] H. W. Bode, *Network Analysis and Feedback Amplifier Design*, D. Van Nostrand, New York, NY, USA, 1945.

[2] S. R. Parker, E. Peskin, and P. Chirlian, "Application of a bilinear theorem to network sensitivity," *IEEE Transactions on Circuit Theory*, vol. 12, no. 3, pp. 448–450, 1965.

[3] E. V. Sorensen, "General relations governing the exact sensitivity of linear networks," *Proceedings of the Institution of Electrical Engineers*, vol. 114, no. 9, pp. 1209–1212, 1967.

[4] S.-D. Shieu and S.-P. Chan, "Topological formulation of symbolic network functions and sensitivity analysis of active networks," *IEEE Transactions on Circuits and Systems*, vol. 21, no. 1, pp. 39–45, 1974.

[5] S. Hoang, "Direct topological method in network sensitivity analysis," *Archiwum Elektrotechniki*, vol. 21, no. 4, pp. 767–784, 1975.

[6] F. Balik and B. Rodanski, "Calculation of symbolic sensitivities for large-scale circuits in the sequence of expressions form via the transimpedance method," *Analog Integrated Circuits and Signal Processing*, vol. 40, no. 3, pp. 265–276, 2004.

[7] I. Asenova and F. Balik, "Multiparameter symbolic sensitivity analysis of active circuits by using nullor model and modified Coates flow graph," in *Proceedings of the 9th International Conference on ELEKTRO*, pp. 401–406, Rajeck Teplice, Slovakia, May 2012.

[8] H. Yang, M. Ranjan, W. Verhaegen, M. Ding, R. Vemuri, and G. Gielen, "Efficient symbolic sensitivity analysis of analog circuits using element coefficient diagrams," in *Proceedings of the Asia South-Pacific Design Automation Conference (ASPDAC '05)*, pp. 230–235, Yokohama, Japan, 2005.

[9] G. Shi and X. Meng, "Variational analog integrated circuit design via symbolic sensitivity analysis," in *Proceedings of the IEEE International Symposium on Circuits and Systems (ISCAS '09)*, pp. 3002–3005, Taipei, Taiwan, May 2009.

[10] S. Rodriguez-Chavez, A. A. Palma-Rodriguez, E. Tlelo-Cuautle, and S. X.-D. Tan, "Graph-based symbolic and symbolic sensitivity analysis of analog integrated circuits," in *Analog/RF and Mixed-Signal Circuit Systematic Design*, vol. 233 of *Lecture Notes in Electrical Engineering*, pp. 101–122, Springer, 2013.

[11] L. Mandache, M. Iordache, L. Dumitriu, I. Gabriela Sirbu, and D. Niculae, "Sensitivity analysis of analog circuits based on a modified nodal approach," in *Proceedings of the International Conference on Optimization of Electrical and Electronic Equipment (OPTIM '14)*, pp. 83–88, Bran, Romania, May 2014.

[12] G. E. Alderson and P. M. Lin, "Computer generation of symbolic network functions—a new theory and implementation," *IEEE Transactions on Circuit Theory*, vol. 20, no. 1, pp. 48–56, 1973.

[13] P. M. Lin, *Symbolic Network Analysis*, Elsevier, Amsterdam, The Netherlands, 1991.

[14] S. Lasota, "Parameter decision diagrams in the analysis and structural synthesis: part I, III," in *Proceedings of the 10th International Workshop on Symbolic and Numerical Methods, Modeling and Application to Circuit Design (SMACD '08)*, pp. 149–157, 180–187, Erfurt, Germany, 2008.

[15] S. Lasota, "Multilevel hierarchical always cancellation-free symbolic analysis method for large electric networks," *Elektronika: Konstrukcje, Technologie, Zastosowania*, vol. 54, no. 2, pp. 51–57, 2013.

[16] V. V. Filaretov and A. S. Korotkov, "Generalized parameter extraction method in symbolic network analysis," in *Proceedings of the European Conference on Circuits Theory and Design (ECCTD '03)*, vol. 2, pp. 406–409, Kraków, Poland, September 2003.

[17] V. V. Filaretov and A. S. Korotkov, "Generalized parameter extraction method in case of multiple excitation," in *Proceedings of the 8th International Workshop on Symbolic Methods and Applications to Circuit Design (SMACD '04)*, pp. 8–11, Wrocław, Poland, September 2004.

[18] V. Filaretov and K. Gorshkov, "The generalization of the extra element theorem for symbolic circuit tolerance analysis," *Journal of Electrical and Computer Engineering*, vol. 2011, Article ID 652706, 5 pages, 2011.

[19] V. Filaretov and K. Gorshkov, "Transconductance realization of block-diagrams of electronic networks," in *Proceedings of the International Conference on Signals and Electronic Systems (ICSES '08)*, pp. 261–264, Krakow, Poland, September 2008.

[20] V. Filaretov, K. Gorshkov, and A. Mikheenko, "A circuit synthesis technique based on network determinant expansion," in *Proceedings of the International Conference on Synthesis, Modeling, Analysis and Simulation Methods and Applications to Circuit Design (SMACD '12)*, pp. 293–296, Seville Sevilla, Spain, September 2012.

[21] V. V. Filaretov and K. S. Gorshkov, "Topological analysis of active networks containing pathological mirror elements," in *Proceedings of the IEEE 33rd International Scientific Conference Electronics and Nanotechnology (ELNANO '13)*, pp. 293–296, Kiev, Ukrain, April 2013.

[22] R. Dmytryshyn and A. Kubaszek, "Sequence of expressions generation for the repetitive analysis acceleration," in *Proceedings of the International Workshop on Symbolic Methods, Modeling and Application in Circuit Design (SMACD '98)*, pp. 154–159, Kaiserslautern, Germany, 1998.

[23] J. A. Starzyk and A. Konczykowska, "Flowgraph analysis of large electronic networks," *IEEE Transactions on Circuits and Systems*, vol. 33, no. 3, pp. 302–315, 1986.

[24] B. Rodanski, "Computational efficiency of symbolic sequential formulae," in *Proceedings of the 6th International Workshop on Symbolic Methods and Applications to Circuit Design (SMACD '00)*, pp. 45–50, Lisbon, Portugal, October 2000.

[25] J. Vlach and K. Singhal, *Computer Methods for Circuit Analysis and Design*, Van Nostrand Reinhold Company, New York, NY, USA, 1983.

4

Fast Transforms in Image Processing: Compression, Restoration, and Resampling

Leonid P. Yaroslavsky

Department of Physical Electronics, School of Electrical Engineering, Tel Aviv University, 69978 Tel Aviv, Israel

Correspondence should be addressed to Leonid P. Yaroslavsky; yaro@eng.tau.ac.il

Academic Editor: George E. Tsekouras

Transform image processing methods are methods that work in domains of image transforms, such as Discrete Fourier, Discrete Cosine, Wavelet, and alike. They proved to be very efficient in image compression, in image restoration, in image resampling, and in geometrical transformations and can be traced back to early 1970s. The paper reviews these methods, with emphasis on their comparison and relationships, from the very first steps of transform image compression methods to adaptive and local adaptive filters for image restoration and up to "compressive sensing" methods that gained popularity in last few years. References are made to both first publications of the corresponding results and more recent and more easily available ones. The review has a tutorial character and purpose.

1. Introduction: Why Transforms? Which Transforms?

It will not be an exaggeration to assert that digital image processing came into being with introduction, in 1965 by Cooley and Tukey, of the *Fast Fourier Transform* algorithm (FFT, [1]) for computing the *Discrete Fourier Transform* (DFT). This publication immediately resulted in impetuous growth of activity in all branches of digital signal and image processing and their applications.

The second wave in this process was inspired by the introduction into communication engineering and digital image processing, in the 1970s, of Walsh-Hadamard transform and Haar transform [2] and the development of a large family of fast transforms with FFT-type algorithms [3–5]. Whereas Walsh-Hadamard and Haar transforms have already been known in mathematics, other transforms, for instance, quite popular at the time Slant Transform [6], were being invented "from scratch." This development was mainly driven by the needs of data compression, though the usefulness of transform domain processing for image restoration and enhancement was also recognized very soon [3]. This period ended up with the introduction of the Discrete Cosine Transform (DCT, [7, 8]), which was soon widely recognized as the best choice among all available at the time transforms and resulted in JPEG and MPEG standards for image, audio, and video compression.

The third large wave of activities in transforms for signal and image processing was caused by the introduction, in the 1980s, of a family of transforms that was coined the name "*wavelet transform*" [9]. The main motivation was achieving a better local representation of signals and images in contrast to the "global" representation that is characteristic to Discrete Fourier, DCT, Walsh-Hadamard, and other fast transforms available at the time. During 1980s–1990s a large variety of discrete wavelet transforms were suggested [10] for solving various tasks in signal and in image processing.

Presently, fast transforms with FFT-type fast algorithms and wavelet transforms constitute the basic instrumentation tool of digital image processing.

The main distinctive feature of transforms that makes them so efficient in digital image processing is their *energy compaction capability*. In regular image representations, in form of sets of ordered pixels, some pixels, for instance, those that belong to object borders, are more important than the others and there are always some pixels in each particular

TABLE 1: Register of relevant fast transforms, their main characteristic features, and application areas.

	Relevance to imaging optics	Main characteristic features	Main application areas
Discrete Fourier Transforms	Represents optical Fourier Transform	Cyclic shift invariance vulnerable to boundary effects	(i) Analysis of periodicities. (ii) Fast convolution and correlation. (iii) Fast and accurate image resampling. (iv) Image compression. (v) Image denoising and deblurring. (vi) Numerical reconstruction of holograms.
Discrete Cosine Transform	Represents optical Fourier Transform	Cyclic shift invariance (with double cycle); virtually not sensitive to boundary effects	
Discrete Fresnel Transforms	Represent optical Fresnel Transform	Computable through DFT/DCT	Numerical reconstruction of holograms
Walsh-Hadamard Transform	No direct relevance	Binary basis functions. Provides piece-wise constant separable image band-limited approximations	(i) Image compression (marginal). (ii) Coded aperture imaging.
Haar Transform and other Discrete Wavelet Transforms	Subband decomposition	Binary basis functions. The fastest algorithm. Multiresolution. Shift invariance in each particular scale.	(i) Signal/image wideband noise denoising. (ii) Image compression.

image that are of no such importance and can be dropped out from image representation and restored from the remaining "important" pixels. But the problem is that one never knows in advance which pixels in the image are "important" and which are not.

The situation is totally different in image representation in transform domain. For orthogonal transforms that feature good energy compaction capability, a lion share of total image "energy" (sum of squared transform coefficients) is concentrated in a small fraction of transform coefficients, which indices are, as a rule, known in advance for the given type of images or can be easily detected. It is this feature of transforms that is called their *energy compaction capability*. It allows replacing images with their "band-limited," in terms of a specific transform, approximations, that is, approximations defined by a sufficiently small fraction of image transform coefficients.

The optimal transform with the best energy compaction capability can be defined for groups of images of the same type or ensembles of images that are subjects of processing. Customarily, the "band-limited" image approximation accuracy is evaluated in terms of the root mean square approximation error (RMSE) for the image ensemble. By virtue of this, the optimal transform that secures the least band limited approximation error is defined by the *ensemble correlation function*, that is, image autocorrelation function averaged over the image ensemble. For continuous (not sampled) signals, this transform is called the *Karhunen-Loeve Transform* [11, 12]. Its discrete analog for sampled signals is called the *Hotelling transform* [13] and the result of applying this transform to sampled signals is called signal *principal component decomposition*. Karhunen-Loeve and Hotelling transforms provide the theoretical lower bound for compact (in terms of the number of terms in signal decomposition) signal discrete representation for RMS criterion of signal approximation.

However, being optimal in terms of the energy compaction capability, Karhunen-Loeve and Hotelling transforms have, generally, high computational complexity: the per pixel number of operations required for their computation is of the order of image size. This is why for practical needs only fast transforms that feature fast transform algorithms with computational complexity of the order of the logarithm of image size are considered. A register of the most relevant fast transforms, their main characteristic features, and application areas is presented in Table 1. Their mathematical definitions are given in Table 2.

Different fast transforms have different energy compaction capability for different types of images. Figure 1 illustrates the energy compaction capabilities of Discrete Fourier, Discrete Cosine, Walsh, and Haar transforms on a particular test image. It shows that for this image DCT demonstrates the best energy compaction capability: 95% of the image energy is contained in only 9.5% of all DCT spectral coefficients, whereas for DFT, Walsh and Haar transforms these fractions are 11%, 15%, and 9.8%, correspondingly. Note also that, though for Haar Transform the fraction of the "meaningful" transform coefficients (9.8%) is close to that for DCT, quality degradations of the band-limited approximation to the image are noticeably more severe.

Experience shows that DCT is in most applications advantageous with respect to other transforms in terms of the energy compaction capability. This property of DCT has a simple and intuitive explanation.

TABLE 2: Selected fast transforms most relevant for digital image processing and their computational complexity (in 1D denotations).

Name	Definition	Computational complexity (operations per signal sample) "Global": applied to entire signal of N samples	"Local": applied in mowing window of WSz pixels
Discrete Cosine Transform (DFT)	$\alpha_r = \frac{1}{\sqrt{N}} \sum_{k=0}^{N-1} a_k \cos\left(2\pi \frac{k+1/2}{N} r\right)$	Real number operations $C_{\mathrm{DCT}}^{G} \log N$	Real number operations $C_{\mathrm{DCT}}^{L} (< WSz)$
Discrete Fourier Transforms			
Canonic DFT	$\alpha_r = \frac{1}{\sqrt{N}} \sum_{k=0}^{N-1} a_k \exp\left(i2\pi \frac{kr}{N}\right)$	Complex number operations $C_{\mathrm{DFT}}^{G} \log N$	Complex number operations $C_{\mathrm{DFT}}^{L} (< WSz)$
General Shifted Scaled DFT	$\alpha_r = \frac{1}{\sqrt{N}} \sum_{k=0}^{N-1} a_k \exp\left[i2\pi \frac{(k+u)(r+v)}{\sigma N}\right];$ u, v—shift parameters that indicate shifts $u\Delta x$, $v\Delta f$ of sampling lattices in signal and spectrum domains; σ—sampling scale parameter ($\Delta x \Delta f = 1/\sigma N$)		
Discrete Fresnel Transforms (DFrT)			
Canonic DFrT	$\alpha_r = \frac{1}{\sqrt{N}} \sum_{k=0}^{N-1} a_k \exp\left[i2\pi \left(\frac{k\mu - r/\mu}{N}\right)\right];\quad \mu \geq 1$ $\mu = \sqrt{\lambda Z / N \Delta f^2}$	Complex number operations $C_{\mathrm{DFrT}}^{G} \log N$	n/a
Convolutional DFrT	$\alpha_r = \frac{1}{\sqrt{N}} \sum_{k=0}^{N-1} a_k \exp\left[i2\pi \frac{(k-r)^2}{\mu^2 N}\right];\quad \mu \leq 1$		
Walsh-Hadamard Transform			
Hadamard Transform	$\alpha_r = \frac{1}{\sqrt{N}} \sum_{k=0}^{N-1} a_k \mathrm{had}_k(r);$ $\mathrm{had}_k(r) = (-1)^{\sum_{m=0}^{n-1} k_m r_m}$ $k = \sum_{m=0}^{n-1} k_m 2^m;\ r = \sum_{m=0}^{n-1} r_m 2^m;\ N = 2^n$	Addition operations only $C_{\mathrm{WHT}}^{G} \log N$	n/a
Walsh Transform	$\alpha_r = \frac{1}{\sqrt{N}} \sum_{k=0}^{N-1} a_k \mathrm{wal}_k(r);$ $\mathrm{wal}_k(r) = (-1)^{\sum_{m=0}^{n-1} k_m r_{n-m-1}^{GrayCode}}$ $r_{n-m-1}^{GrayCode}$—binary digits of Gray code of r taken in bit reversal order		
Haar Transform (the simplest discrete wavelet transform)	$\alpha_r = \frac{1}{\sqrt{N}} \sum_{k=0}^{N-1} a_k \mathrm{har}_k(r);$ $\mathrm{har}_k(r) = 2^{\mathrm{msb}} (-1)^{k_{n-1-\mathrm{msb}}} \delta\left(r_{\bmod 2^{\mathrm{msb}}} - \lfloor 2^{1-\mathrm{msb}} k \rfloor\right)$ Msb—index of the most significant nonzero bit of binary code of msb; $\lfloor \cdot \rfloor$—residual of $(\cdot)$	Addition operations only C_{Haar}^{G}	n/a

DCT of a discrete signal is essentially, to the accuracy of an unimportant exponential shift factor, DFT for the same signal extended outside its border to a double length by means of its mirrored, in the order of samples, copy [15, 16]. This way of signal extension has a profound implication on the speed of decaying to zero of signal DCT spectra and is a paramount and fundamental reason of the good energy compaction capability of DCT. DFT implies periodical extension of signals, which may cause severe discontinuities at signal borders and thus may require intensive high frequency components to reproduce them. DCT also implies periodical extension, however extension of not the signal itself but of its above-described extended copy. Thanks to such an "even" extension, the extended signal has no discontinuities at its borders, as well as at borders of the initial signal. Therefore such an extended signal is "smooth" and its DFT spectrum decays to zero faster than DFT spectrum of the initial, not extended signal.

It is noteworthy to mention that introducing DCT in [7] was motivated not by the above reasoning but by the finding that basis functions of DCT provide a good approximation to the eigenvectors of the class of *Toeplitz matrices* with entries $\{\mu_{k,l} = \rho^{|k-l|}\}$; $\rho = 0.9$; $k, l = 0, 1, \ldots$; that is, DCT can be considered as a good approximation to the Karhunen-Loeve

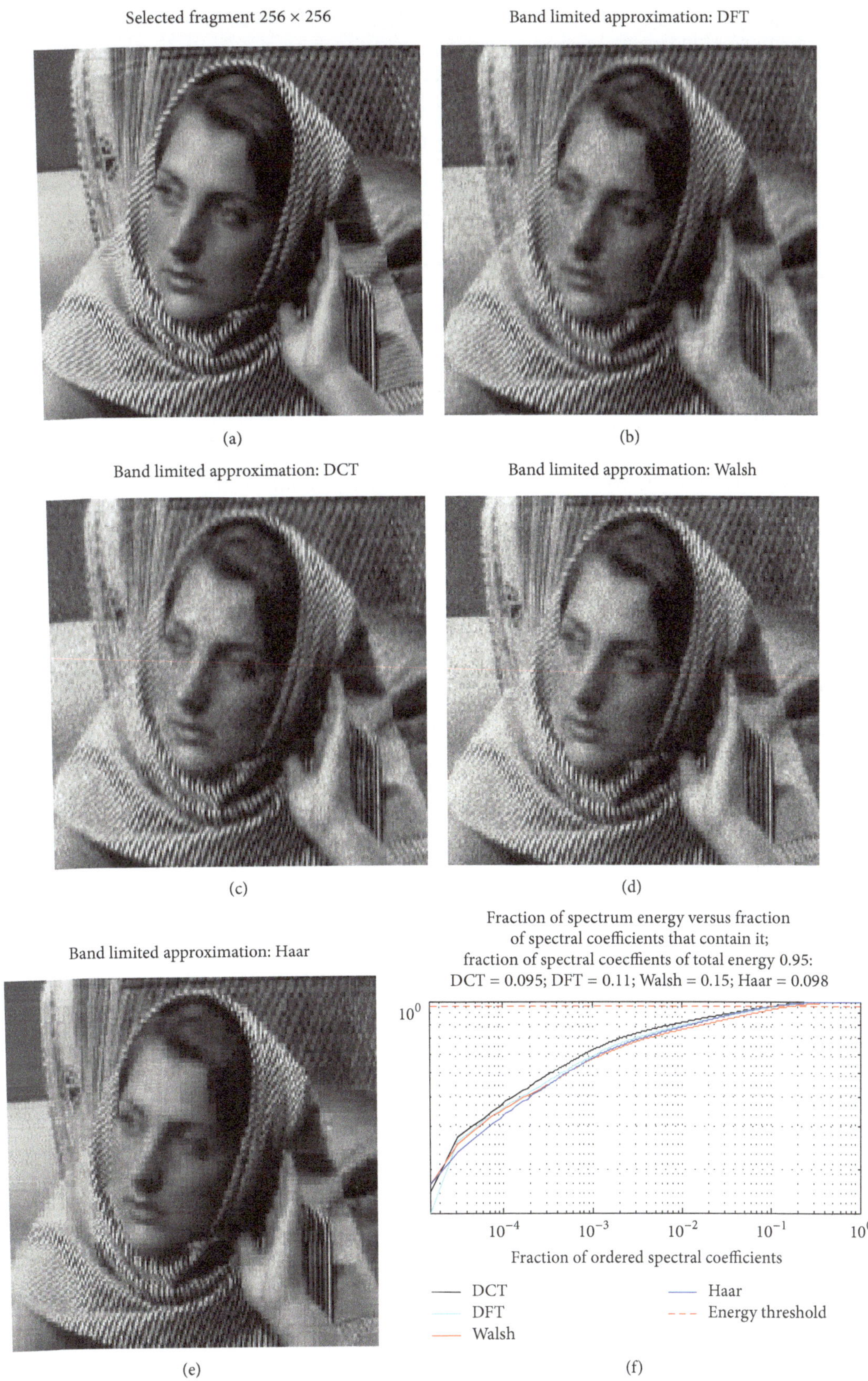

Figure 1: Illustration of the energy compaction capability of Discrete Fourier, Discrete Cosine, Walsh, and Haar transforms for a test image shown in the upper left corner. The rest of images show band-limited approximations of the test image in the domain of corresponding transforms for the approximation error variance equal to 5% of image variance. Graphs on the plot in the bottom right corner present energy contained in fractions of transform coefficients for different transforms.

Transform for signals with the Toeplitz correlation matrix. In fact, the above-mentioned fundamental reason for good energy compaction capability of DCT is hidden in the even symmetry of the Toeplitz matrix.

In the family of orthogonal transforms, DFT and DCT occupy a special place. DFT and DCT are two discrete representations of the Integral Fourier Transform, which plays a fundamental role in signal and image processing as a mathematical model of signal transformations and of wave propagation in imaging systems [17, 18]. Thanks to this, DFT and DCT are applicable in much wider range of applications than the other fast transforms (see Table 1).

In conclusion of this section, come back to the common feature of fast transform, to the availability for all of these transform of fast algorithms of FFT type. Initially, fast transforms were developed only for signals with the number of samples equal to an integer power of two. These are the fastest algorithms. At present, this limitation is overcome, and fast transform algorithms exist for arbitrary number of signal samples. Note that 2D and 3D transforms for image and video applications are built as separable; that is, they work separably in each dimension. Note that the transform separability is the core idea for fast transform algorithms. All fast transform algorithms are based on representation of signal and transform spectrum sample indices as multidimensional, that is represented by multiple digits, numbers.

The data on computational complexity of the transform fast algorithms are collected in Table 2. These data are more or less commonly known. Not as widely known is the existence of so-called *pruned algorithms* for the cases, when transform input data contain substantial fraction of zero samples and/or one does not need to compute all transform coefficients [15–24]. These algorithms are useful, for instance, in numerical reconstruction of holograms and in interpolation of sampled data by means of *transform zero padding* [17, 25, 26].

In some applications it is advisable to apply transforms locally in running window rather than globally to entire image frames. For such application, Discrete Cosine and Discrete Fourier Transforms have an additional very useful feature. Computing these transforms in running window can be carried out recursively: signal transform coefficients at each window position can be found by quite simple modification of the coefficients found at the previous window position with per pixel computational complexity proportional to the window size rather than to the product of the window size and its logarithm, which is the computational complexity of the fast transforms [17, 26–28].

The rest of the paper is arranged as follows. In Section 2, applications of fast transforms for image data compression are reviewed. In Section 3, we proceed to transform methods for image restoration and enhancement. In Section 4, we show how DFT and DCT can be applied for image perfect, that is, error less, resampling. And finally in Section 5, we address the usage of fast transforms for solving a specific task of image resampling, the task of image recovery from sparse and irregularly taken samples, and, in particular, compressive sensing approach to this problem.

2. Image Compression

2.1. Dilemma: Compressive Discretization or Compression. Ideally, image digitization, that is, converting continuous signals from image sensors into digital signals, should be carried out in such a way as to provide as compact image digital representation as possible provided that quality of image reproduction satisfies given requirements. Due to technical limitation, image digitization is most frequently performed in two steps: discretization, that is, converting image sensor signal into a set of real numbers, and scalar quantization, that is rounding off, of those numbers to a set of fixed quantized values [15, 18].

Image discretization can, in general, be treated as obtaining coefficients of image signal expansion over a set of discretization basis functions. In order to make the set of these representation coefficients as compact as possible, one should choose discretization basis functions that secure the least number of the signal expansion coefficients sufficient for image reconstruction with a given required quality. One can call this general method of signal discretization "*Compressive discretization*" because it secures the most compact discrete representation of signals, which cannot be further compressed. Note that this term should not be confused with terms "*Compressive sensing*" and "*Compressive sampling*," that gained large popularity in recent years [29–36]. We will discuss the "compressive sensing" approach later in Section 5.

There are at least two examples of practical implementation of the principle of general discretization by means of measuring transform domain image coefficients: *Coded Aperture Imaging* and *Magnetic Resonance Imaging* (MRI). In coded aperture imaging, images are sensed through binary masks that implement binary basis functions. In MRI imaging, coefficients of Fourier series expansion of sensor data are measured.

However, by virtue of historical reasons and of the technical tradition, image discretization is most frequently in imaging engineering is implemented as image sampling at nodes of a uniform rectangular sampling lattice using *sampling basis functions*, which are formed from one *"mother" function* by its shifts by multiple of fixed intervals called *"sampling" intervals.*

The theoretical foundation of image sampling is the *sampling theorem.* The traditional formulation of the sampling theorem states that signals with Fourier spectrum limited with bandwidth B can be perfectly restored from their samples taken with sampling interval $\Delta x = 1/B$, commonly called the *Nyquist sampling interval* [37–39].

In reality no continuous signal is band-limited, and the image sampling interval is defined not through specifying, in one or another way, of the image bandwidth, but directly by the requirement to reproduce smallest objects and borders of larger objects present in images sufficiently well. The selected in this way image sampling interval Δx specifies image *base band* $B = 1/\Delta x$.

Since small objects and object borders usually occupy relatively small fraction of the image area, vast portions of images are oversampled, that is, sampled with redundantly small sampling intervals. Hence, substantial compression of

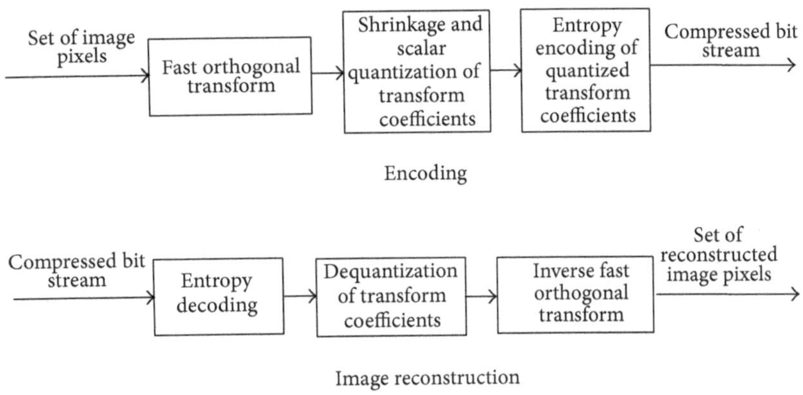

FIGURE 2: Flow diagrams of image transform coding and reconstruction.

image sampled representation is possible. It can be implemented by means of applying to the image primary and redundant sampled representation a discrete analog of the general compressive discretization, that is, by means of image expansion over a properly chosen set of discrete optimal bases functions and limitation of the amount of the expansion coefficients. This is exactly what is done in all transform methods of image compression.

2.2. Transform Methods of Image Compression. Needs of image compression were the primary motivations of digital image processing. Being started in 1950s from various predictive coding methods (a set of good representative publications can be found in [40]), by the beginning of 1970s image compression research began concentrating mostly on what we call now "transform coding" [41, 42], which literally implements the compressed discretization principle.

The principle of image transform coding is illustrated by the flow diagrams sketched in Figure 2.

According to these diagrams, set of image pixels is first subjected to a fast orthogonal transform. Then low intensity transform coefficients are discarded, which substantially reduces the volume of data. This is the main source of image compression. Note that discarding certain image transform coefficients means replacement of images by their band-limited, in terms of the selected transform, approximations. The remaining coefficients are subjected, one by one, to optimal nonuniform scalar quantization that minimizes the average number of transform coefficient quantization levels. Finally, quantized transform coefficients are entropy encoded to minimize the average number of bits per coefficient. For image reconstruction from the compressed bit stream, it must be subjected to corresponding inverse transformations.

In 1970s, main activities of researches were aimed at inventing, in addition to known at the time Discrete Fourier, Walsh-Hadamard, and Haar Transforms, new transforms that have guaranteed fast transform algorithms and improved energy compaction capability [3–5].

This transform invention activity gradually faded after introduction, in a short note, of the Discrete Cosine transform [7], which, as it was already mentioned, was finally recognized as the most appropriate transform for image compression due to its superior energy compaction capability and the availability of the fast algorithm. But the true breakthrough happened, when it was realized that much higher image compression efficiency can be achieved, if transform coding is applied not to entire image frames but blockwise [41, 42].

Images usually contain many objects, and their global spectra are a mixture of object spectra, whereas spectra of individual image fragments or blocks are much more specific and this enables much easier separation of "important" (most intense) from "unimportant" (least intense) spectral components. This is vividly illustrated in Figure 3, where global and block wise DCT power spectra of a test image are presented for comparison. Although, as one can see in the figure, spectral coefficients of DCT global spectrum are more sparse (spectrum sparsity of this image, i.e., fraction of spectral coefficients that contain 95% of the total spectrum energy, is 0.095) than, on average, those of image of 16×16 pixel fragments (average sparsity of spectra fragments is 0.14), spectra of individual fragments are certainly more specific and their components responsible for small objects and object borders have much higher energy than in the global spectrum and will not be lost when 95% of the most intense components are selected.

Blockwise image DCT transform compression proved to be so successful, that, by 1992, it was put in the base of the image and video coding standards "JPEG" and "MPEG" [43–46], which eventually resulted in the revolution in the industry of photographic and video cameras and led to the emergence of digital television as well [47].

Though the standard does not fix the size of blocks, blocks of 8×8 pixels are used in most cases. Obviously, the compression may, in principle, be more efficient, if the block size is adaptively selected in different areas of images; hence the use of variable size of blocks was discussed in number of publications (see, for instance, [48]).

3. Transform Domain Filters for Image Restoration and Enhancement

3.1. Transform Domain Scalar Empirical Wiener Filters. Very soon after the image transform compression methods

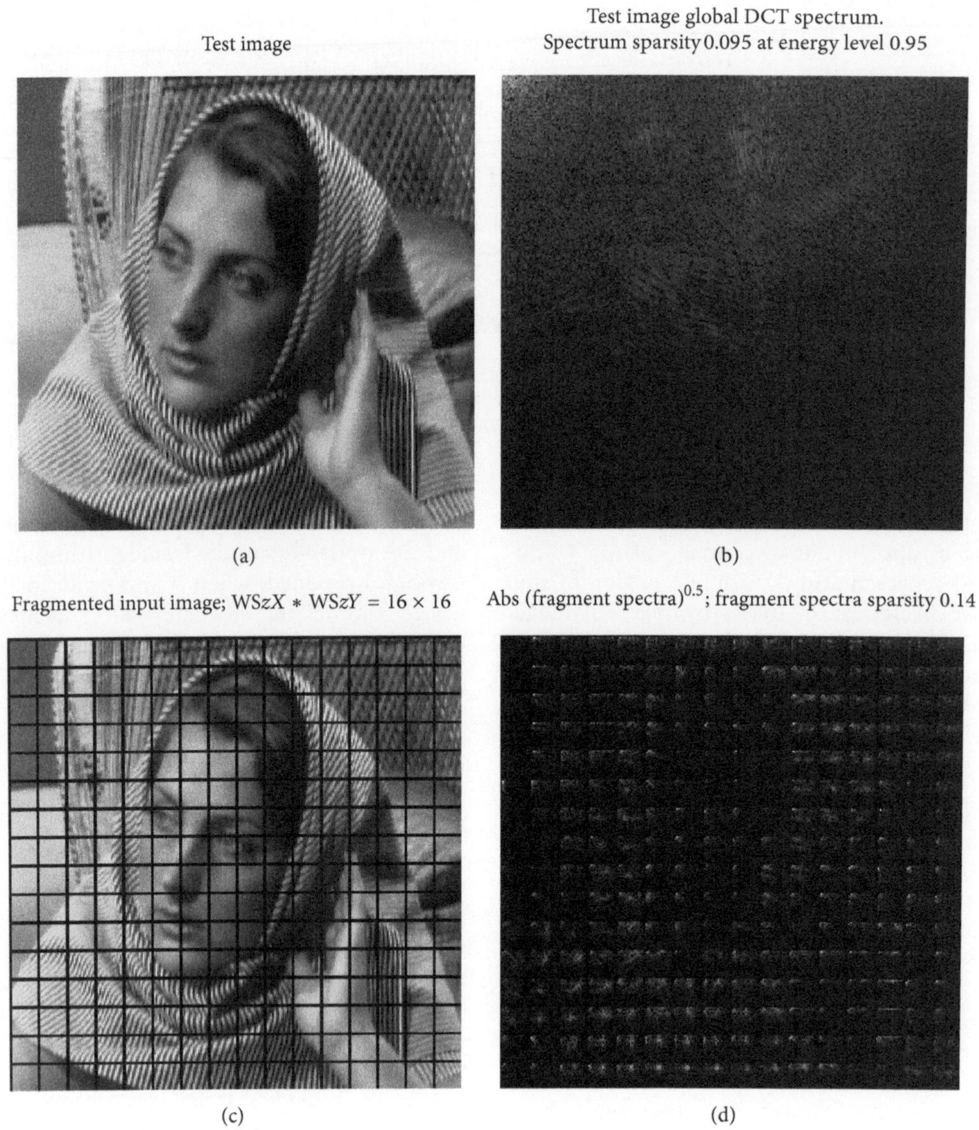

FIGURE 3: Comparison of global and local (blockwise) image DCT spectra.

FIGURE 4: Flow diagram of transform domain filtering.

emerged, it was recognized that transforms represent a very useful tool for image restoration from distortions of image signals in imaging system and for image enhancement [49, 50]. The principle is very simple: for image improvement, image transform coefficients are modified in a certain way and then the image is reconstructed by the inverse transform of modified coefficients (Figure 4). This way of processing is called *transform domain filtering*.

For the implementation of modifications of image transform coefficients, two options are usually considered.

(i) Modification of absolute values of transform coefficients by a nonlinear transfer function; usually it is a function that compresses the dynamic range of transform coefficients, which redistributes the coefficients' intensities in favor of less intensive coefficients and results in contrast enhancement of image small details and edges.

(ii) Multiplication of transform coefficients by scalar weight coefficients; this processing is called transform domain *"scalar filtering"* [49].

For defining optimal scalar filter coefficients, a Wiener-Kolmogorov [51, 52] approach of minimization of mean squared filtering error is used and thus filters implement empirical Wiener filtering principles [17, 18, 26].

Three modifications of the filters based on this principle are (i) proper *Empirical Wiener Filters*, (ii) *Signal Spectrum Preservation Filters,* and (iii) *Rejecting Filters* [18]. For image

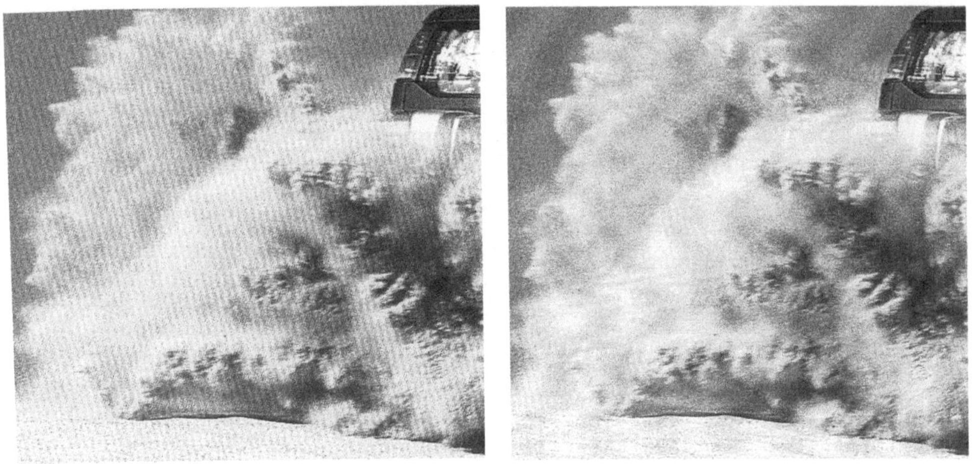

Before

(a)

After

(b)

FIGURE 5: Image cleaning from moire noise.

denoising from additive signal independent noise and image deblurring, they modify input image Fourier spectra Sp_{inp} at each frequency component according to the equations correspondingly.

Empirical Wiener Filter

$$Sp_{\text{out}} = \frac{1}{\text{ISFR}} \max\left(0, \frac{\left|Sp_{\text{inp}}\right|^2 - Sp_{\text{noise}}}{\left|Sp_{\text{inp}}\right|^2}\right) Sp_{\text{inp}}; \quad (1)$$

Signal Spectrum Preservation Filter

$$Sp_{\text{out}} = \frac{1}{\text{ISFR}} \max\left(0, \frac{\left|Sp_{\text{inp}}\right|^2 - Sp_{\text{noise}}}{\left|Sp_{\text{inp}}\right|^2}\right)^{1/2} Sp_{\text{inp}}; \quad (2)$$

Rejecting Filter

$$Sp_{\text{out}} = \begin{cases} \dfrac{Sp_{\text{inp}}}{\text{ISFR}}, & \text{if } \left|Sp_{\text{inp}}\right|^2 > Sp_{\text{noise}}, \\ 0, & \text{otherwise,} \end{cases} \quad (3)$$

where Sp_{out} is Fourier spectrum of output image, Sp_{noise} is power spectrum of additive noise, assumed to be known or to be empirically evaluated from the input noisy image [15–18], and ISFR is the imaging system frequency response assumed to be known, for instance, from the imaging system certificate.

As one can see in these equations, all these filters eliminate image spectral components that are less intensive than those of noise and the remaining components are corrected by the "*inverse*" *filter* with frequency response of 1/ISFR. Division of image spectra by ISFR compensates image blur in the imaging system due to imperfect ISFR; that is, it performs image deblurring. It is applicable in processing specifically in domains of DFT or DCT, which are, as it was already mentioned, two complementing discrete representations of the integral Fourier Transform.

In addition, empirical Wiener filter modifies image spectrum through deamplification of image spectral components according to the level of noise in them; spectrum preservation filter modifies magnitude of the image spectrum as well by making it equal to a square root of image power spectrum empirical estimate as a difference between power spectrum of noisy image and that of noise. Rejecting filter does not modify remaining, that is, not rejected, spectral components at all. In some publications, image denoising using the spectrum preservation filter is called "*soft thresholding*" and rejecting filtering is called "*hard thresholding*" [53, 54].

Versions of these filters are filters that combine image denoising-deblurring and image enhancement by means of P-law nonlinear modifications of the corrected signal spectral components

$$Sp_{\text{out}} = \begin{cases} \dfrac{1}{\text{ISFR}\left|Sp_{\text{inp}}\right|^{(1-P)}} Sp_{\text{inp}}, & \text{if } \left|Sp_{\text{inp}}\right|^2 > Sp_{\text{noise}}, \\ 0, & \text{otherwise,} \end{cases} \quad (4)$$

where $P \leq 1$ is a parameter that controls the degree of spectrum dynamic range compression and, by this, the degree of enhancement of low intensity image spectral components [18].

The described filters proved to be very efficient in denoising so-called *narrow band noises*, that is, noises, such as "*moirè noise*" or "*banding noise*," that contain only few nonzero components in its spectrum in the selected transform. Some illustrative examples are given in Figures 5 and 6.

However for image cleaning from wideband or "white" noise, above transform domain filtering applied globally to entire image frames is not efficient. In fact it can even worsen

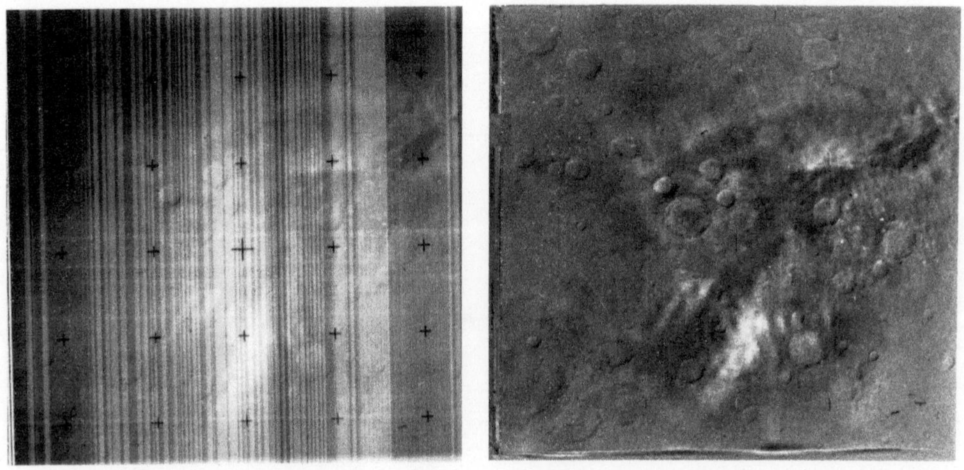

Before (a) After (b)

FIGURE 6: Cleaning a Mars satellite image from banding noise by means of separable (row-wise and column-wise) high pass rejecting filtering. Fiducial marks seen in the left image were extracted from the image before the filtering [14].

the image visual quality. As one can see from an illustrative example shown in Figure 7, filtered image loses its sharpness and contains residual correlated noise, which is visually more annoying than the initial white noise.

The reason for this inefficiency of global transform domain filtering is the same as for the inefficiency of the global transform compression, which was discussed in Section 2.2: in global image spectra low intensity image components are hidden behind wideband noise spectral components. And the solution of this problem is the same: replace global filtering by local fragment wise filtering. This idea is implemented in *local adaptive linear filters*. Being applied locally, filters are becoming locally adaptive, as their frequency responses are determined, according to (1)–(4), by local spectra of image fragments within the filter window. Local adaptive linear filters date back to mid-1980s and were refined in subsequent years [16, 18, 26, 55]. As a theoretical base for the design of local adaptive filters, local criteria of processing quality were suggested [56].

It is instructive to note that this is not an accident that the evolution of human vision came up with a similar solution. It is well known that when viewing image, human eye's optical axis permanently hops chaotically over the field of view and that the human visual acuity is very nonuniform over the field of view. The field of view of a man is about 30°. Resolving power of man's vision is about $1'$. However such a relatively high resolving power is concentrated only within a small fraction of the field of view that has size of about 2° (see, for instance, [57]); that is, the area of the acute vision is about 1/15th of the field of view. For images of 512×512 pixels this means window of roughly 35×35 pixels.

The most straightforward way to implement local filtering is to do it in a hopping window, just as human vision does, and this is exactly the way of processing implemented in the transform image coding methods. However "hopping window" processing, being very attractive from the computational complexity point of view, suffers from "*blocking effects*," artificial discontinuities that may appear, in the result of processing, at the edges of the hopping window. Blocking effects are artifacts characteristic for transform image coding. The desire to avoid them motivated the advancement of "Lapped" transforms [58], which are applied blockwise with the half of the block size overlap. Obviously, the ultimate solution of the "blocking effects" problem would be pixel by pixel processing in sliding window that scans image row-wise/column/wise.

Thus, local adaptive linear filters work in a transform domain in a sliding window and, at each position of the window, modify, according to the type of the filter defined by (1)–(4), transform coefficients of the image fragment within the window, and then compute an estimate of the window central pixel by means of the inverse transform of the modified transform coefficients. As an option, accumulation of estimates of the all window pixels overlapping in the process of image scanning by the filter window is possible as well [59]. As for the transform for local adaptive filtering, DCT proved to be the best choice in most cases. Figures 8 and 9 give examples of local adaptive filtering for image deblurring and enhancement.

Local adaptive filters can work with color and multicomponent images and videos as well. In this case, filtering is carried out in the corresponding multidimensional transform domains, for instance, domains of 3D (two spatial and one color component coordinates) spectra of color images or 3D spectra of a sequence of video frames (two spatial and time coordinates). In the latter case filter 3D window scans video frame sequence in both spatial and time coordinates. As one can see from Figures 10 and 11, the availability of the additional third dimension substantially adds to the filter efficiency [18, 55, 60].

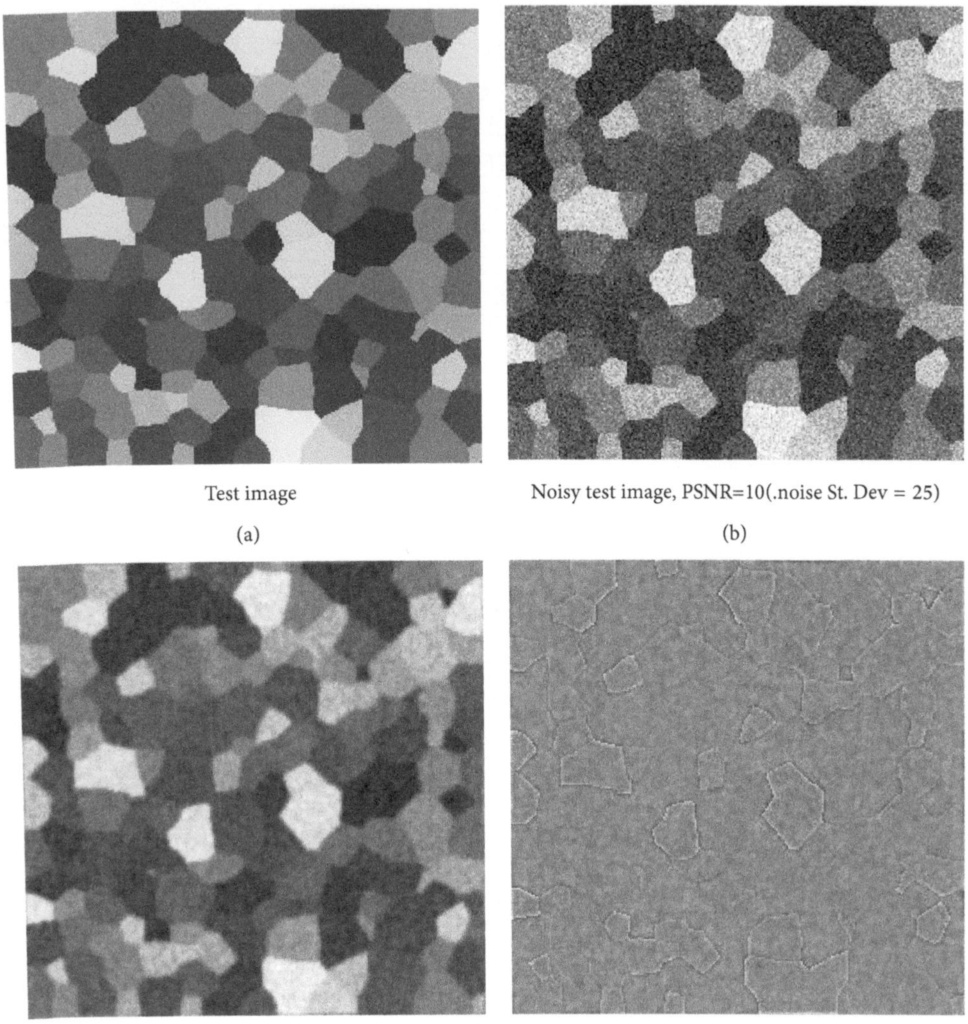

Test image

(a)

Noisy test image, PSNR=10(.noise St. Dev = 25)

(b)

Filtered image

(c)

Difference between original test and filtered images. Standard deviation 10.8

(d)

FIGURE 7: Denoising of a piece-wise test image using empirical Wiener filter applied globally to the entire image.

Certain further improvement of image denoising capability can be achieved if, for each image fragment in a given position of the sliding window, similar, according to some criterion, image fragments over the whole image frame are found and included in the stack of fragments, to which the 3D transform domain filtering is applied. This approach in its simplest form was coined the name "*nonlocal means*" filtering [61–63]. In fact, similar method has been much earlier known as the *correlational averaging* (see, for instance, [64]). It assumes finding similar image fragments by their cross-correlation coefficient with a template object and averaging of similar fragments, that is, leaving only the dc component of the 3D spectra in this dimension. In [65] the simple averaging was replaced by full scale 3D transform domain filtering. One should, however, take into account that the correlational averaging is prone to produce artifacts due to false image fragments that may be erroneously taken to the stack just because of the presence of noise that has to be cleaned [66].

Obviously, image restoration efficiency of the sliding window transform domain filtering will be higher, if the window size is selected adaptively at each window position. To this goal, filtering can be, for instance, carried out in windows of multiple sizes and, at each window position, the best filtering result should be taken as the signal estimate at this position using methods of statistical tests, for instance, "*intersection of confidence intervals method*" described in [67]. Another option in multiple window processing is combining, in a certain way (say, by weighted summation or by taking a median or in some other way), filtering results obtained for different windows. All this, however, increases correspondingly the computational complexity of the filtering, which may become prohibitive, especially in video processing.

In conclusion of this section note that DFT and DCT spectra of image fragments in sliding window form the so called "*time-frequency representation*" of signals, which can be traced back to 1930s-1940s to Gabor and Wigner

Before

(a)

After

(b)

Figure 8: Denoising and deblurring of a satellite image by means of local adaptive filtering. Top row: raw image and its fragments magnified for better viewing (marked by arrows); bottom row: resulting image and its corresponding magnified fragments.

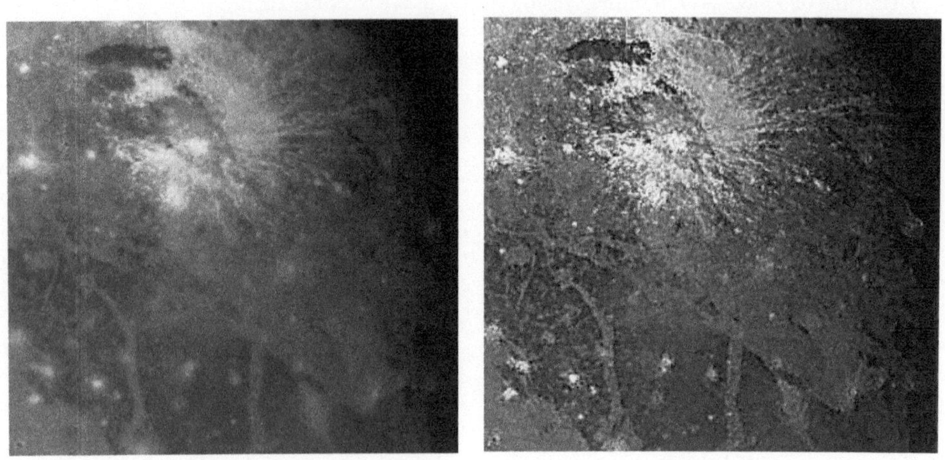

Before

(a)

After

(b)

Figure 9: Enhancement of an astronomical image by means of Pth law modification of its spectrum.

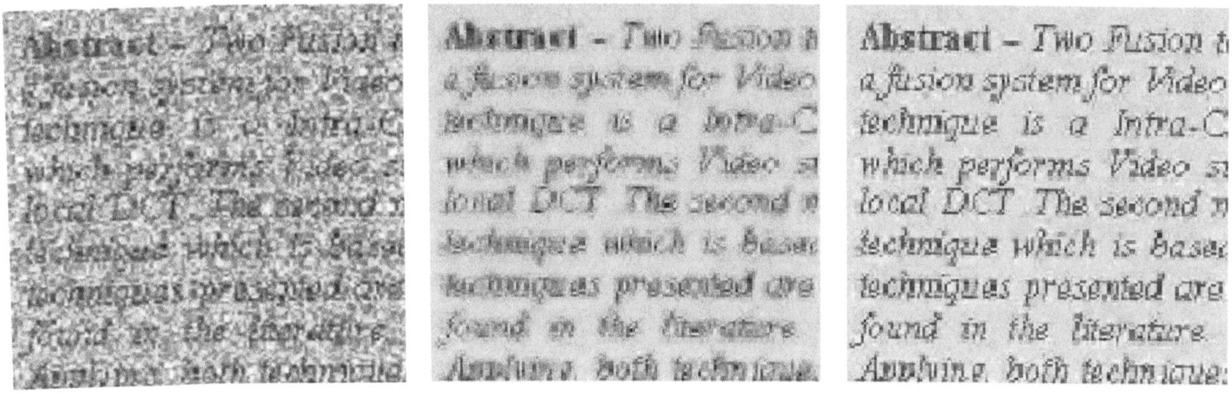

FIGURE 10: 2D and 3D local adaptive filtering of a simulated video sequence: (a) one of noisy frames of a test video sequence (image size 256×256 pixels); (b) a result of 2D local adaptive empirical Wiener filtering (filter window 5×5 pixels); (c) a result of 3D local adaptive empirical Wiener filtering (filter window 5×5 pixels $\times$ 5 frames).

FIGURE 11: 3D local adaptive empirical Wiener filtering for denoising and deblurring of a real thermal video sequence: (a) a frame of the initial video sequence; (b) a frame of a restored video sequence. Filter window is (5×5 pixels) $\times$ 5 frames; image size is 512×512 pixels.

and works on "*visible speech*" [68–70]. For 1D signals, the dimensionality of the representation is 2D; for images, it is 3D. Such representation is very redundant and this opens an interesting option of applying to it image (for 1D signals) and video (for 2D signals) processing methods for signal denoising and, in general, signal separation.

3.2. Wavelet Shrinkage Filters for Image Denoising. In 1990s, a specific family of transform domain denoising filters, the so-called *wavelet shrinkage* filters, gained popularity after publications [53, 54, 71, 72]. The filters work in the domain of one of wavelet transforms and implement, for image denoising, the above-mentioned Signal Spectrum Preservation Filter (2) and Rejecting Filter (3), except that they do not include the "inverse filtering" component (1/ISFR) that corrects distortions of image spectra in imaging systems. As it was already mentioned, the filters were coined in these publications the names "*soft thresholding*" and "*hard thresholding*" filters, correspondingly.

Basis functions of wavelet transforms (wavelets) are formed by means of a combination of two methods of building transform basis functions from one *"mother" function*, shifting and scaling. Thanks to this, wavelets combine local sensitivity of shift basis functions and global properties of *"scaled" basis functions* and feature such attractive properties as multiresolution and good localization in both signal and transform domain. Thus, the wavelet shrinkage filters, being applied to entire image frames, possess both global adaptivity and local adaptivity in different scales and do not require for the latter specification of the window size, which is necessary for the sliding window filters.

These filters did show a good denoising capability. However, a comprehensive comparison of their denoising capability with that of sliding window DCT domain filters reported in [59] showed that even when for particular test images the best wavelet transform from a transform set [73] was selected, sliding window DCT domain filters demonstrated in most cases better image denoising results.

The capability of wavelets to represent images in different scales can be exploited for improving the image denoising performance of both families of filters and for overcoming the above-mentioned main drawback of sliding window DCT domain filters, the fixed window size that might not be optimal for different image fragments. This can be achieved by means of incorporating sliding window DCT domain (SWTD) filters into the wavelet filtering structure through replacing soft/hard thresholding of image wavelet decomposition components in different scales by their filtering with SWTD filters working in the window of the smallest size 3×3 pixels. This idea was implemented in hybrid wavelet/SWTD filters and proved to be fruitful in ([74], see also [55]).

3.3. Local Adaptive Filtering and Wavelet Shrinkage Filtering as Processing of Image Subband Decompositions. Obviously, sliding window transform domain and wavelet processing are just different implementation of the scalar linear filtering in transform domain. This motivates their unified interpretation in order to gain a deeper insight into their similarities and dissimilarities. The common base for such unified interpretation is the notion of signal subband decomposition [75]. One can show that both filter families can be treated as special cases of signal subband decomposition by band-pass filters with point spreads functions defined by the corresponding basis functions of the transform [26, 55]. From the point of view of signal subband decomposition, the main difference between sliding window transform domain and wavelet signal analysis is arrangement of bands in the signal frequency range. While for sliding window transform domain filtering subbands are uniformly arranged with the signal base band, subbands of the wavelet filters are arranged in a logarithmic scale. Hybrid wavelet/SWDCT filtering combines these two types of subband arrangements: "coarse" subband decomposition in logarithmic scale, provided by wavelet decomposition, is complemented with "fine" uniformly arranged sub-subbands within each wavelet subband, provided by the sliding window DCT filtering of the wavelet subbands.

It is curious to note that this "logarithmic coarse—uniform fine" subband arrangement resembles very much the arrangements of tones and semitones in music. In Bach's equal tempered scale, octaves are arranged in a logarithmic scale and 12 semitones are equally spaced within octaves [76].

4. Image Resampling and Building "Continuous" Image Models

As it was indicated in the introductory section, Discrete Fourier and Discrete Cosine Transforms occupy the unique position among other orthogonal transforms. They are two versions of discrete representation of the integral Fourier Transform, the fundamental transform for describing the physical reality. Among applications specific for DFT and DCT are signal and image spectral analysis and analysis of periodicities, fast signal and image convolution and correlation, and image resampling and building "continuous" image models [17, 18, 26]. The latter is associated with the efficiency, with which DFT and DCT can be utilized for fast signal convolution.

Image resampling is a key operation in solving many digital image processing tasks. It assumes reconstruction of an approximation of the original nonsampled image by means of interpolation of available image samples to obtain samples "in-between" the available ones. The most feasible is interpolation by means of digital filtering implemented as digital convolution. A number of convolutional interpolation methods are known, beginning from the simplest and the least accurate nearest neighbor and linear (bilinear, for 2D case) interpolations to more accurate cubic (bicubic, for 2D case) and higher order spline methods [77, 78]. In some applications, for instance, in computer graphics and print art, the simplest nearest neighbor or linear (bilinear) interpolation methods can provide satisfactory results. In applications that are more demanding in terms of the interpolation accuracy, higher order spline interpolation methods gained popularity. The interpolation accuracy of spline interpolation methods grows with the spline order. However their computational complexity grows as well.

There exists a discrete signal interpolation method that is capable, given finite set of signal samples, to secure virtually error free interpolation of sampled signals, that is, the method that does not add to the signal any distortions additional to the distortions caused by the signal sampling. This method is the *discrete sinc-interpolation* [18, 25]. Interpolation kernel for the discrete sinc-interpolation of sampled signals of N samples is the *discrete sinc-function* $\text{sincd}x = \sin x/N \sin(x/N)$. This function replaces, for sampled signals with a finite number of samples, the continuous *sinc-function* $\text{sinc}\, x = \sin x/x$, which, according to the sampling theory, is the ideal interpolation kernel for reconstruction of continuous signals from their samples provided that the number of samples is infinitely large. Note that the discrete sinc-function should not be confused with a windowed or truncated sinc-function with a finite number of samples, which is very commonly believed to be a good replacement in numerical interpolation of the continuous sinc-function. The windowed sinc-function does not secure interpolation error free signal numerical interpolation, whereas the discrete sinc-interpolation does.

Fast Fourier transform and fast Discrete Cosine transform are the ideal computational means for implementing the convolutional interpolation by the discrete sinc-function. Of these two transforms, FFT has a substantial drawback: it implements the cyclic convolution with a period equal to the number N of signal samples, which is potentially prone to causing heavy oscillations at signal borders because of a discontinuity which is very probable between signal values at its opposite borders. This drawback is to a very high degree surmounted when signal convolution is implemented through DCT [18, 26]. As it was already indicated, convolution through DCT is also a cyclic convolution but with the period of $2N$ samples and of signals that are evenly extended to double length by their inversed, in the order of samples, copies. This completely removes the danger of appearance of heavy oscillations due to discontinuities at signal borders that are characteristic for discrete sinc-interpolation through processing in DFT domain.

Several methods of implementation of DFT/DCT based discrete sinc-interpolation. The most straightforward one is DFT/DCT spectrum zero padding. Interpolated signal is generated by applying inverse DFT (or, correspondingly, DCT) transform to its zero pad spectrum. Padding DFT/DCT spectra of signals of N samples with $L(N-1)$ zeroes produces LN samples of the discrete sinc interpolated initial signal with subsampling rate L, that is, with intersample distance $1/L$ of the initial sampling interval. The computational complexity of this method is $O(\log LN)$ per output sample.

The same discrete sinc interpolated L times subsampled signal can be obtained more efficiently computation-wise by applying to the signal L times the signal *perfect shifting filter* [25, 79] with shifts by k/L at each kth application and by subsequent combining the shifted signal copies. The computational complexity of this method is $O(\log N)$ per output sample.

The method is based on the property of *Shifted DFT* (SDFT, [18, 26, 80]): if one computes SDFT of a signal with some value of the shift parameter in signal domain and then computes inverse SDFT with another value of the shift parameter, the result will be a discrete sinc interpolated copy of the signal shifted by the difference between those values of the shift parameter. In order to avoid objectionable oscillations at signal borders, the DCT based version of this filter is recommended [81–83].

Image subsampling using the perfect shifting filter can be employed for creating "continuous" image models for subsequent image arbitrary resampling with any given accuracy [18, 84]. It can also be used for computing image correlations and image spectral analysis with a subpixel accuracy [18, 26]. Particular examples of using created in this way continuous signal models for converting image spectra from polar coordinate system to Cartesian coordinate system in the direct Fourier method for image reconstruction from projections and for converting (rebinning) of image fan beam projections into parallel projections one can find in [18].

Except for creating "continuous" image models, the perfect shifting filter is very well suited for image sheering in the three-step method for fast image rotation by means of three subsequent (horizontal/vertical/horizontal) image sheerings [85].

Perfect interpolation capability of the discrete sinc-interpolation was demonstrated in a comprehensive comparison of different interpolation methods in experiments with multiple 360° rotations reported in [25]. The results of experiments illustrated in Figure 12 clearly evidence that, in contrast to other methods, including high order spline interpolation ones [78, 86, 87], discrete sinc-interpolation does not introduce any appreciable distortions into the rotated image.

In some applications, "elastic" or space variant image resampling is required, when shifts of pixel positions are specified individually for each image pixel. In these cases, the perfect shifting filter can be applied to image fragments in sliding window for evaluating interpolated value of the window central pixel at each window position. Typical application examples are imitation of image retrieval through turbulent media (for illustration see [88]) and stabilization and superresolution of turbulent videos [89–91].

Being working in DCT transform domain, "elastic" resampling in sliding window can be easily combined with above-described local adaptive denoising and deblurring [25]. Yet another additional option is making resampling in sliding window adaptive to contents of individual image fragments by means of switching between the perfect shifting and another interpolation method better suited for specific image fragments. This might be useful in application to images that contain very heterogeneous fragments such as photographs and binary drawings and text [25].

A certain limitation of the perfect shifting filter in creating "continuous" image models is its capability of subsampling images only with a rate expressed by an integer or a rational number. In some cases, this might be inconvenient, for instance, when the required resampling rate is a value between one and two, say 1.1, 1.2, or alike. For such cases, there exists the third method of signal resampling with discrete sinc-interpolation. It is based on the general Shifted Scaled DFT, which includes arbitrary analog shift and scale parameters. Using Shifted Scaled (ShSc) DFT, one can apply to the image DFT spectrum inverse ShScDFT with the desired scale parameter and obtain a correspondingly scaled discrete sinc interpolated image [82, 83]. Being the discrete sinc-interpolation method, this method also does not cause any image blurring, which other interpolation methods are prone to do. Figure 13 illustrates this on an example of multiple iterative image zooming-in and zooming-out.

The Shifted Scaled DFT can be presented as a convolution and, as such, can be computed using Fast Fourier or Fast Cosine Transforms [18, 26]. Thus, the computational complexity of the method is $O(\sigma \log N)$ per output signal sample, where σ is the scale parameter. As in all other cases, the use of DCT guaranties the absence of severe boundary effects. This method is especially well suited for numerical reconstruction of digitally recorded color holograms, when it is required to rescale images reconstructed from holograms recorded with different wave lengths of coherent illumination [82].

Among the image processing tasks, which involve "continuous" image models, are also signal differentiation and integration, the fundamental tasks of numerical mathematics that date back to such classics of mathematics as Newton and Leibnitz. One can find standard numerical differentiation and integration recipes in numerous reference books, for instance, [92, 93]. All of them are based on signal approximation through Taylor expansion. However, according to the sampling theory, approximation of sampled signal by Taylor expansion is not optimal, and the direct use of these methods for sampled signals may cause substantial errors in cases, when signals contain substantial high frequency components. In order to secure sufficiently accurate signal differentiation and integration by means of those standard algorithms, one must substantially oversample such signals.

The exact solution for the discrete representation of signal differentiation and integration provided by the sampling theory tells that, given the signal sampling interval

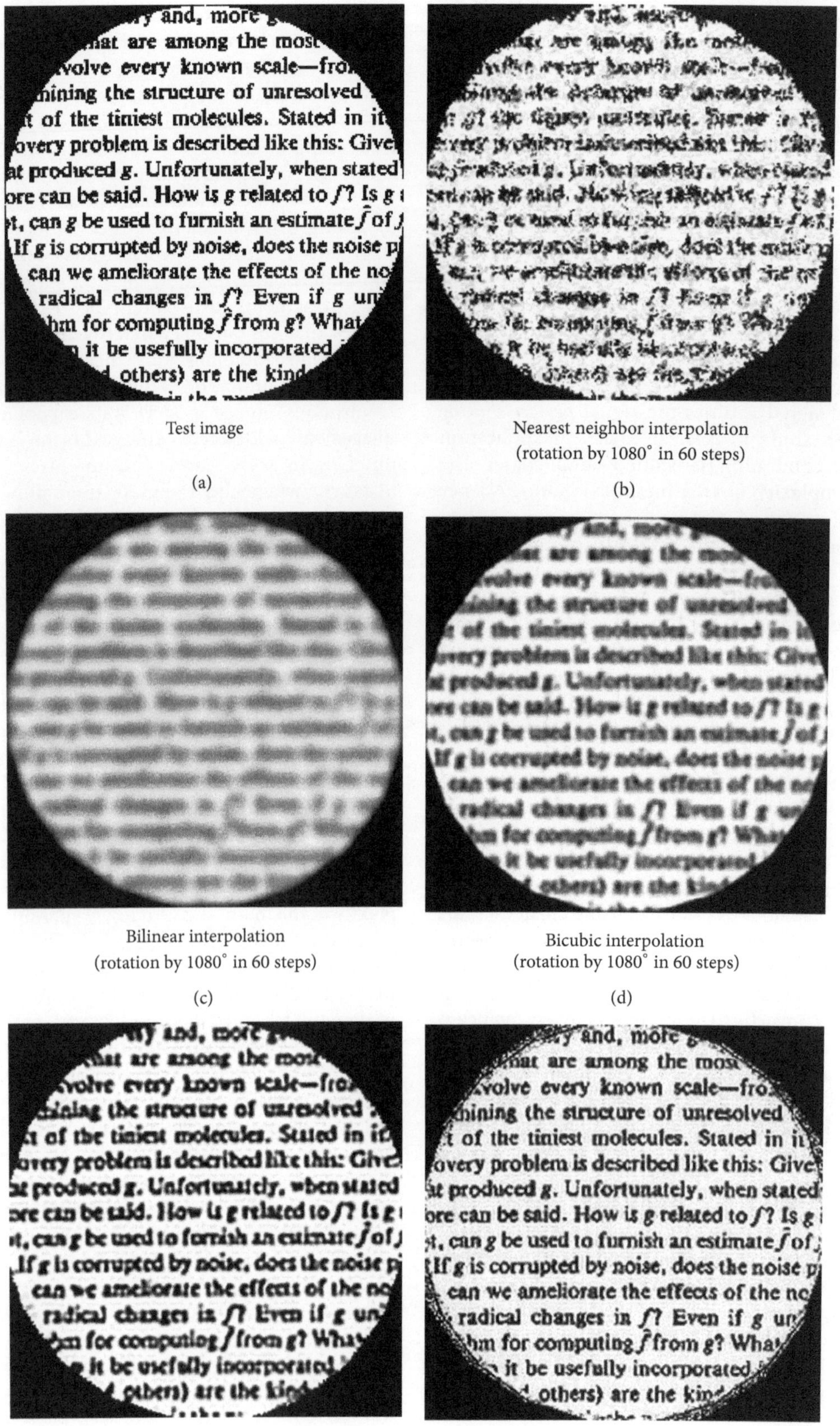

Test image

(a)

Nearest neighbor interpolation
(rotation by 1080° in 60 steps)

(b)

Bilinear interpolation
(rotation by 1080° in 60 steps)

(c)

Bicubic interpolation
(rotation by 1080° in 60 steps)

(d)

Spline interpolation Mems531 ([78], [87])
(rotation by 18000° in 1000 steps)

(e)

Discrete sinc-interpolation
(rotation by 18000° in 1000 steps)

(f)

FIGURE 12: Discrete sinc-interpolation versus other interpolation methods: results of multiple image rotations.

[original image, bilinear intrp; bicubic intrp; discr-sinc intrp]; Nit = 75

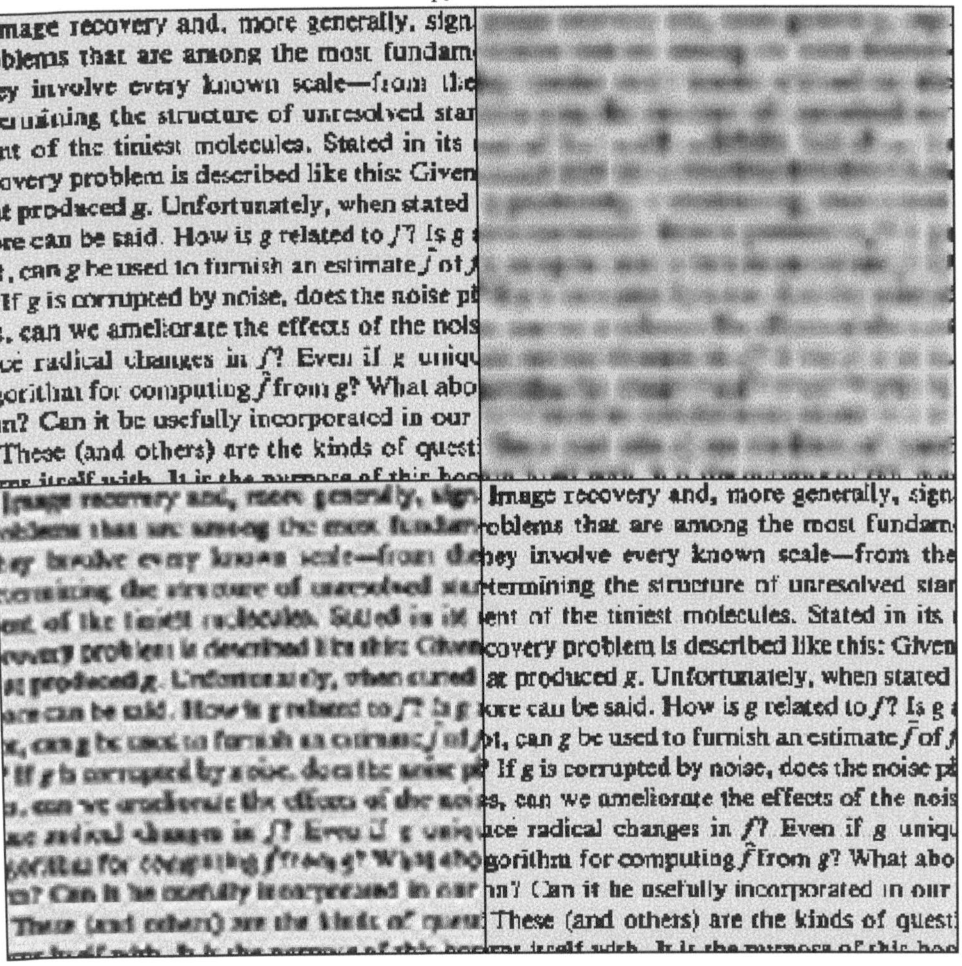

FIGURE 13: Discrete sinc-interpolation versus bilinear and bicubic interpolations in image iterative zoom-in/zoom-out with the scale parameter $\sqrt{2}$.

and signal sampling and reconstruction devices, discrete frequency responses (in DFT domain) of digital filters for perfect differentiation and integrations should be, correspondingly, proportional and inversely proportional to the frequency index [18, 26]. This result directly leads to fast differentiation and integration algorithms that work in DFT or DCT domains using corresponding fast transforms with computational complexity $O(\log N)$ operation per signal sample for signals of N samples. As in all cases, realization of the integration and, especially, differentiation filters in DCT domain is preferable because of much lower, for DCT, border effect artifacts associated with the finite number of signal samples.

The comprehensive comparison of the accuracy of standard numerical differentiation and integration algorithms with perfect DCT-based differentiation and integration reported in [25, 94] shows that the standard numerical algorithms in reality work not with original signals but with their blurred to a certain degree copies. This conclusion is illustrated in Figure 14 on an example of multiple alternative differentiations and integrations of a test rectangular impulse.

Computational efficiency of the DFT/DCT based interpolation error free discrete sinc-interpolation algorithms is rooted in the use of fast Fourier and Fast DCT transforms. Perhaps, the best concluding remark for this discussion of the discrete sinc-interpolation DFT/DCT domain methods would be mentioning that a version of what we call now Fast Fourier Transform algorithm was invented more than 200 hundred years ago by Gauss just for the purpose of facilitating numerical interpolation of sampled data of astronomical observation [95].

5. Image Recovery from Sparse and Irregularly Sampled Data

5.1. Discrete Sampling Theorem Based Method. As we mentioned at the beginning of Section 2, image discretization is usually carried out by image sampling at nodes of a uniform rectangular sampling lattice. In this section we address using fast transforms for solving the problem closely associated with the general compressive discretization: how and with

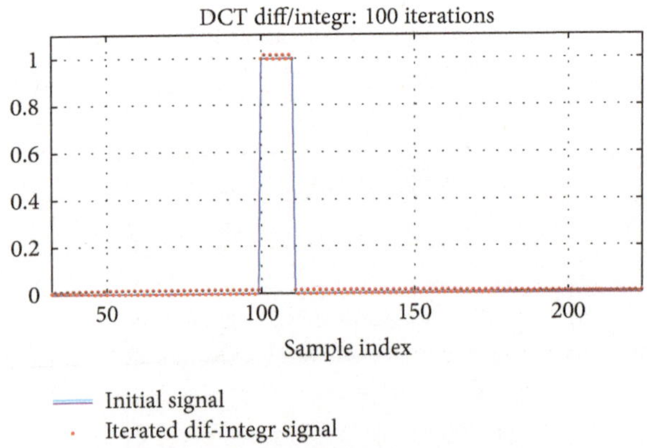

(a)

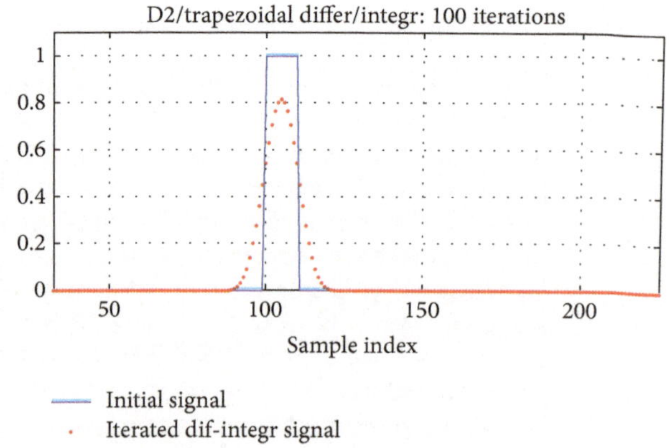

(b)

FIGURE 14: Comparison of results of iterative alternated 100 differentiations and integrations of a test rectangular signal using the DCT-based algorithms (left) and standard numerical algorithms (right, differentiation filter with point spread function [−0.5, 0, 0.5] and trapezoidal rule integration algorithm).

what accuracy can one recover an image from its sparse and irregularly taken samples.

There are many applications, where, contrary to the common practice of uniform sampling, sampled data are collected in irregular fashion. Because image display devices as well as computer software for processing sampling data assume using regular uniform rectangular sampling lattice, one needs in all these cases to convert irregularly sampled images to regularly sampled ones.

Generally, the corresponding regular sampling grid may contain more samples than it is available, because coordinates of positions of available samples might be known with a "subpixel" accuracy, that is, with the accuracy (in units of image size) better than $1/K$, where K is the number of available pixels. Therefore one can regard available K samples as being sparsely placed at nodes of a denser sampling lattice with the total amount of nodes $N > K$.

The general framework for recovery of discrete signals from a given set of their arbitrarily taken samples can be formulated as an approximation task in the assumption that continuous signals are represented in computers by their $K < N$ irregularly taken samples and it is believed that if all N samples in a regular sampling lattice were known, they would be sufficient for representing those continuous signals [18, 96]. The goal of the processing is generating, out of an incomplete set of K samples, a complete set of N signal samples in such a way as to secure the most accurate, in terms of the reconstruction mean square error (MSE), approximation of the discrete signal, which would be obtained if the continuous signal it is intended to represent were densely sampled at all N positions.

Above-described discrete sinc-interpolation methods provide band-limited, in terms of signal Fourier spectra, approximation of regularly sampled signals. One can also think of signal band-limited approximation in terms of their spectra in other transforms. This approach is based on the *Discrete Sampling Theorem* [18, 96, 97].

The meaning of the Discrete Sampling Theorem is very simple. Given K samples of a signal one can, in principle, compute K certain coefficients of a certain signal transform and then reconstruct N samples of a band-limited, in this transform, approximation of the signal by means of inverse transform of those K coefficients supplemented with N-K zero coefficients. If the transform has the highest, for this particular type of signals, energy compaction capability and selected were those nonzero transform coefficients that concentrate the highest, for this particular transform, signal energy, the obtained signal approximation will have the least mean square approximation error. If the signal is known to be band-limited in the selected transform and the computed nonzero coefficients corresponded to this band limitation, signal will be reconstructed precisely.

The discrete sampling theorem is applicable to signals of any dimensionality, though the formulation of the signal band-limitedness depends on the signal dimensionality. For 2D images and such transforms as Discrete Fourier, Discrete Cosine, and Walsh Transforms, the most simple is compact "low-pass" band-limitedness by a rectangle or by a circle sector.

It was shown in ([96], see also [18, 97]) that, for 1D signal and such transforms as DFT and DCT, signal recovery from sparse samples is possible for arbitrary positions of sparse samples. For images signals, the same is true, if band-limitation conditions are separable over the image dimensions. For nonseparable band-limitations, such as circle or circle sector, this may not be true and certain redundancy in the number of available samples might be required to secure exact recovery of band-limited images.

As it was indicated, the choice of the transform must be made on the base of the transform energy compaction capability for each particular class of images, to which the image to be recovered is believed to belong. The type of the band-limitation also must be based on a priori knowledge regarding the class of images at hand. The number of samples to be recovered is a matter of a priori belief of how many samples of a regular uniform sampling lattice would be sufficient to represent the images for the end user.

Implementation of signal recovery/approximation from sparse nonuniformly sampled data according to the discrete sampling theorem requires matrix inversion, which is, generally, a very computationally demanding procedure. In applications, in which one can be satisfied with image reconstruction with a certain limited accuracy, one can apply to the reconstruction a simple iterative reconstruction algorithm of the Gerchberg-Papoulis [98, 99] type, which, at each iteration step, alternatively applies band limitation in spectral domain and then restores available pixels in their given positions in the image obtained by the inverse transform.

Among reported applications, one can mention superresolution from multiple video frames of turbulent video and superresolution in computed tomography that makes use of redundancy of object slice images, in which usually a substantial part of their area is an empty space surrounding the object [96].

5.2. "Compressive Sensing": Promises and Limitations. The described discrete sampling theorem based methods for image recovery from sparse samples by means of their band-limited approximation in a certain transform domain require explicit formulation of the desired band limitation in the selected transform domain. While for 1D signal this formulation is quite simple and requires, for most frequently used low pass band limitation, specification of only one parameter, signal bandwidth, in 2D case formulation of the signal band limitation, requires specification of a 2D shape of signal band-limited spectrum. The simplest shapes, rectangle and circle sector ones, may only approximate, with a certain redundancy, real areas occupied by the most intensive image spectral coefficients for particular images. Figure 15 illustrates this on an example of spectral binary mask that corresponds to 9.5% of the DCT spectral coefficients of the test image shown in Figure 3 that contain 95% of the spectrum energy and a rectangular outline of this mask, which includes 47% of the spectral coefficients.

Spectrum mask; spectrum sparsity 0.095
Spectrum sparsity for the rectangle 0.47

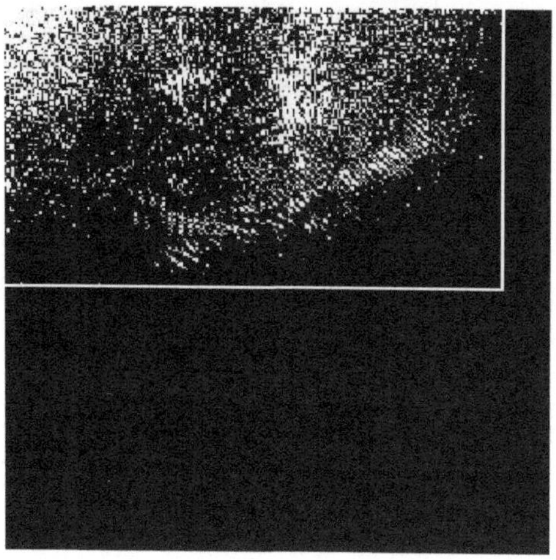

FIGURE 15: Spectral binary mask (shown white with the dc component in the left upper corner) that indicates components of DCT spectrum of the image in Figure 3, which contain 95% of the total spectrum energy and its rectangular outline.

In the cases, when exact character of spectrum band limitation is not known, image recovery from sparse samples can be achieved using the "compressive sensing" approach introduced in [29–32]. During last recent years, this approach to handling sparse spectral image representation has obtained considerable popularity [33–36].

The compressive sensing approach assumes obtaining a band-limited, in a certain selected transform domain, approximation of images as well, but it does not require explicit formulation of the image band-limitation and achieves image recovery from an incomplete set of samples by means of minimization of $L0$ norm of the image spectrum in the selected transform (i.e., of the number of signal nonzero transform coefficients), conditioned by preservation in the recovered image of its available pixels.

However, there is a price one should pay for the uncertainty regarding the band limitation: the number M of samples required for recovering N signal samples by this approach must be redundant with respect to the given number K nonzero spectral coefficients: $M = K \log N$ [30]. In real applications, this might be a serious limitation. For instance, for the test image of 256×256 pixels shown in Figure 3 spectrum sparsity (relative number of nonzero spectral components) on the energy level 95% is 0.095, whereas $\log N = \log(256 \times 256) = 16$, which means that the "compressive sensing" approach requires in this case more signal samples ($16 \cdot 9.5\% = 152\%$) than it is required to recover.

5.3. Discrete Signal Band-Limitedness and the Discrete Uncertainty Principle. Signal band-limitedness plays an important role in dealing with both continuous signals and discrete (sampled) signals representing them. It is well known that continuous signals cannot be both strictly band-limited and have strictly bounded support. In fact, continuous signals neither are band-limited nor have strictly bounded support. They can only be more or less densely concentrated in signal and spectral domains. This property is mathematically formulated in the form of the "*uncertainty principle*" [100]

$$X_{\varepsilon E} \times B_{\varepsilon E} > 1, \quad (5)$$

where $X_{\varepsilon E}$ and $B_{\varepsilon E}$ are intervals in signal and their Fourier spectral domains that contain a sufficiently large fraction εE of signal energy.

In distinction to that, sampled signals that represent continuous signals can be sharply bounded in both signal and spectral domains. This is quite obvious for some signal spectral presentation, such as Haar signal spectra: Haar basis functions are examples of sampled functions sharply bounded in both signal and Haar spectral domains. But it turns out that there exist signals that are sharply bounded in both signal and their spectral domains of DFT

(a)

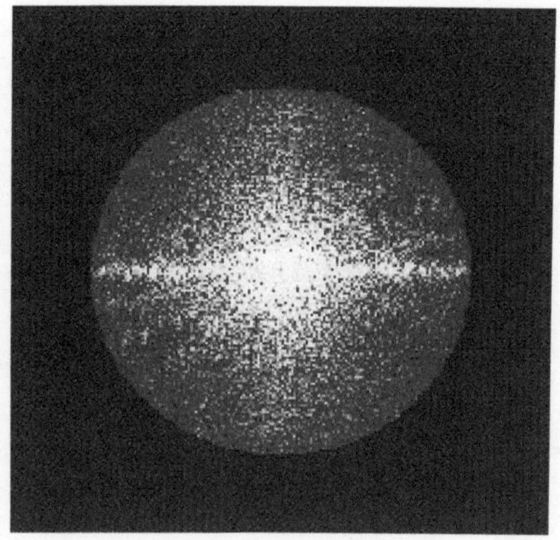

(b)

FIGURE 16: Space-limited image "C. Shannon" (left) and its band-limited DFT spectrum (right; shown centered at the signal dc component).

or DCT, which are discrete representations of the Fourier integral transform. An example of a space-limited/band-limited image is shown in Figure 16. Such images can be generated using the above-mentioned iterative Gershberg-Papoulis type algorithm which, at each iteration, applies to images alternatively space limitation in image domain and band limitation in transform domain. Such images are very useful as test images for testing image processing algorithms.

In a similar way one can generate space-limited and band-limited analogs of the discrete sinc-function, i.e. functions, which form band-limited shift bases in the given space limits. In [97] such functions were coined a name "*sinc-lets*." Figure 17 shows an example of a sinc-let in its three positions within a limited interval and, for comparison, the discrete sinc-function for the same band limitation.

The following relationship between signal bounds in signal and DFT spectral domains can be derived:

$$N_{\text{sign}} \times N_{\text{spect}} < N, \qquad (6)$$

where N_{sign} is the number of signal nonzero samples, N_{spect} is the number of its nonzero spectral samples, and N is the number of samples in the signal sampling lattice. This is what one can call the *discrete uncertainty principle.*

6. Conclusion

Two fundamental features of fast transforms, the energy compaction capability and fast algorithms for transform computation, make them perfect tool for image compression, restoration, reconstruction, and resampling. Of all fast transforms, Discrete Fourier Transform and Discrete Cosine Transform are the most important as they are complementing each other discrete representations of the integral Fourier transform, one of the most fundamental mathematical models for describing physical reality, and, in addition, they enable fast digital convolution. From these two transforms, DCT is preferable in most applications thanks to its ability to very substantially reduce processing artifacts associated with image discontinuities at the borders.

Modern tendency in the imaging engineering is computational imaging. Computer processing of sensor data enables substantial price reduction and sometimes even complete removal of imaging optics and similar imaging hardware. It also gives birth to numerous new imaging techniques in astrophysics, biology, industrial engineering, remote sensing, and other applications. No doubts, this area promises many new achievements in the coming years. And it is certain that fast transforms will remain to be the indispensable tool in this process.

Nomenclature

a_k: Image samples
α_r: Image spectral coefficients
$k = 0, \ldots, N-1$;
$r = 0, \ldots, N-1$: pixel and, correspondingly, spectra coefficient vertical and horizontal indices
$C_{(\cdot)}$: Various constants
Δx: Image sampling intervals
Δf: Sampling interval in the transform domain
N: Total number of image samples
λ: Wavelength of coherent radiation
Z: Distance between object plane and hologram.

Abbreviations

DCT: Discrete Cosine Transform
DFT: Discrete Fourier Transform
DFrT: Discrete Fresnel Transform
PSNR: Peak Signal-to-Noise Ratio: ratio of signal dynamic range to noise standard deviation

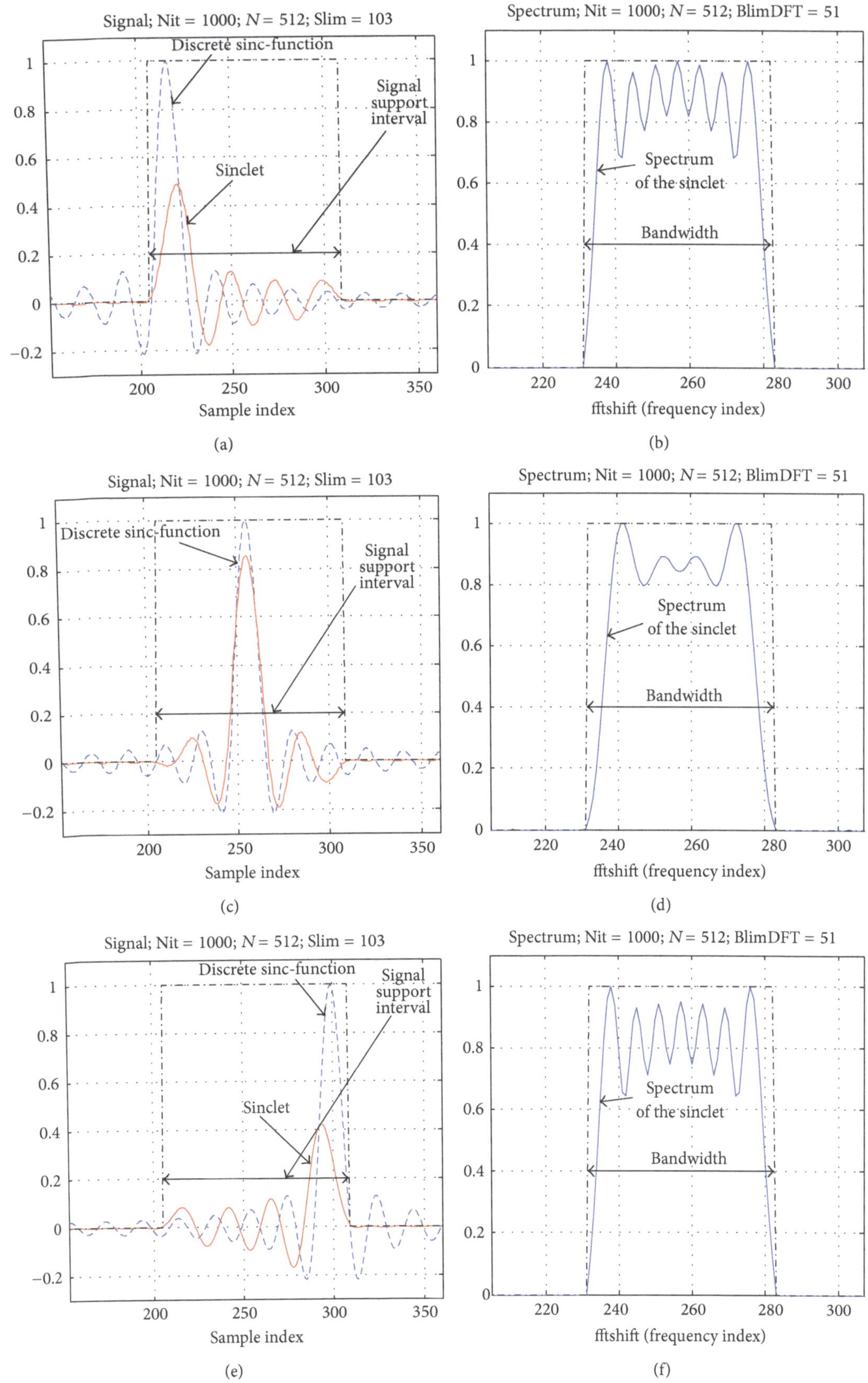

FIGURE 17: Examples of a "sinc-let" (red plots) and, for comparison, of the discrete sinc-function for the same band limitation (blue plots) in their three different positions within the interval of 103 samples of 512 samples (boxes a–c). Plots of their DFT spectra are shown in boxes (d–f).

RMSE: root mean square error
SNR: Signal-to-Noise Ratio (ratio of signal and noise variances).

Conflict of Interests

The author declares that there is no conflict of interests regarding the publication of this paper.

References

[1] J. W. Cooley and J. W. Tukey, "An Algorithm for the Machine Calculation of Complex Fourier Series," *Mathematics of Computation*, vol. 19, no. 90, pp. 297–301, 1965.

[2] H. F. Harmuth, *Transmission of Information by Orthogonal Functions*, Springer, Berlin, Germany, 1970.

[3] H. C. Andrews, *Computer Techniques in Image Processing*, Academic Press, New York, NY, USA, 1970.

[4] H. C. Andrews, "Two-dimensional Transforms," in *Picture Processing and Digital Filtering*, T. S. Huang, Ed., Springer, New, York, NY, USA, 1975.

[5] N. Ahmed and K. R. Rao, *Orthogonal Transforms for Digital Signal Processing*, Springer, Berlin, Germany, 1975.

[6] W. K. Pratt, W.-H. Chen, and L. R. Welch, "Slant Transform Image Coding," *IEEE Transactions on Communications*, vol. 22, no. 8, pp. 1075–1093, 1974.

[7] N. Ahmed, N. Natarajan, and K. R. Rao, "Discrete cosine transform," *IEEE Transactions on Computers*, vol. 23, no. 1, pp. 90–93, 1974.

[8] K. R. Rao and P. Yip, *Discrete Cosine Transform: Algorithms, Advantages, Applications*, Academic Press, Boston, Mass, USA, 1990.

[9] I. Daubechies, "Where do wavelets come from? A personal point of view," *Proceedings of the IEEE*, vol. 84, no. 4, pp. 510–513, 1996.

[10] S. Mallat, *A Wavelet Tour of Signal Processing*, Academic Press, Boston, Mass, USA, 3rd edition, 2008.

[11] H. Karhunen, *Über lineare Methoden in der Wahrscheinlichkeitsrechnung*, Series A.I, Annales Academiæ~Scientiarum Fennicæ, 1947.

[12] M. Loeve, "Fonctions aleatoires de seconde ordre," in *Processes Stochastiques et Movement Brownien*, P. Levy, Ed., Hermann, Paris, France, 1948.

[13] H. Hotelling, "Analysis of a complex of statistical variables into principal components," *Journal of Educational Psychology*, vol. 24, no. 6, pp. 417–441, 1933.

[14] T. P. Belokova, M. A. Kronrod, P. A. Chochia, and L. P. Yaroslavakii, "Digital processing of martian surface photographs from mars 4 and mars 5," *Space Research*, vol. 13, no. 6, pp. 898–906, 1975.

[15] L. P. Yaroslavsky, *Digital Picture Processing: An Introduction*, Springer, Berlin, Germany, 1985.

[16] L. Yaroslavsky, *Digital Signal Processing in Optics and Holography*, Radio i Svyaz', Moscow, Russia, 1987 (Russian).

[17] L. Yaroslavsky, *Digital Holography and Digital Image Processing. Principles, Methods, Algorithms*, Kluwer Academic Publishers, Boston, Mass, USA, 2004.

[18] L. Yaroslavsky, *Theoretical Foundations of Digital Imaging Using Matlab*, Taylor & Francis, Boca Raton, Fla, USA, 2013.

[19] L. Yaroslavsky, "Introduction to Digital Holography," in *Digital Signal Processing in Experimental Research*, L. Yaroslavsky and J. Astola, Eds., vol. 1 of *Bentham Series of e-books*, Bentham, 2009, http://www.benthamscience.com/ebooks/9781608050796/index.htm.

[20] L. P. Yaroslavskii and N. S. Merzlyakov, *Methods of Digital Holography*, Consultance Bureau, New York, NY, USA, 1980.

[21] L. P. Jaroslavski, "Comments on FFT algorithm for both input and output pruning," *IEEE Transactions on Acoustics, Speech, and Signal Processing*, vol. 29, no. 3, pp. 448–449, 1981.

[22] J. Markel, "FFT pruning," *IEEE Trans Audio Electroacoust*, vol. 19, no. 4, pp. 305–311, 1971.

[23] H. V. Sorensen and C. S. Burrus, "Efficient computation of the DFT with only a subset of input or output points," *IEEE Transactions on Signal Processing*, vol. 41, no. 3, pp. 1184–1200, 1993.

[24] R. G. Alves, P. L. Osorio, and M. N. S. Swamy, "General FFT pruning algorithm," in *Proceedings of the 43rd IEEE Midwest Symposium on Circuits and Systems*, vol. 3, pp. 1192–1195, August 2000.

[25] L. Yaroslavsky, "Fast discrete sinc-interpolation: a gold standard for image resampling," in *Advances in Signal transforms: Theory and Applications*, J. Astola and L. Yaroslavsky, Eds., vol. 7 of *Eurasip Book Series on Signal Processing and Communications*, pp. 337–405, 2007.

[26] L. Yaroslavky, "Fast transforms in digital signal processing: theory, algorithms, applications," in *Digital Signal Processing in Experimental Research*, L. Yaroslavsky and J. Astola, Eds., vol. 2 of *Bentham E-book Series*, 2011, http://www.benthamscience.com/ebooks/9781608052301/index.htm.

[27] R. Yu. Vitkus and L. P. Yaroslavsky, "Recursive Algorithms for Local Adaptive Linear Filtration," in *Mathematical Research. Computer Analysis of Images and Patterns*, L. P. Yaroslavsky, A. Rosenfeld, and W. Wilhelmi, Eds., pp. 34–39, Academie Verlag, Berlin, Germany, 1987, Band 40.

[28] J. Xi and J. F. Chicharo, "Computing running DCT's and DST's based on their second-order shift properties," *IEEE Transactions on Circuits and Systems I: Fundamental Theory and Applications*, vol. 47, no. 5, pp. 779–783, 2000.

[29] E. Candès, "Compressive sampling," in *Proceedings of the International Congress of Mathematicians*, Madrid, Spain, 2006.

[30] D. L. Donoho, "Compressed sensing," *IEEE Transactions on Information Theory*, vol. 52, no. 4, pp. 1289–1306, 2006.

[31] E. J. Candès, J. K. Romberg, and T. Tao, "Stable signal recovery from incomplete and inaccurate measurements," *Communications on Pure and Applied Mathematics*, vol. 59, no. 8, pp. 1207–1223, 2006.

[32] E. J. Candes and M. B. Wakin, "An introduction to compressive sampling: A sensing/sampling paradigm that goes against the common knowledge in data acquisition," *IEEE Signal Processing Magazine*, vol. 25, no. 2, pp. 21–30, 2008.

[33] R. G. Baraniuk, "Compressive sensing," *IEEE Signal Processing Magazine*, vol. 24, no. 4, pp. 118–124, 2007.

[34] M. Elad, *Spatse and Redundant Representations: From Theory to Applications in Signal and Image Processing*, Springer, New York, NY, USA, 2010.

[35] R. M. Willet, R. F. Mareia, and J. M. Nichols, "Compressed sensing for practical optical imaging systems: a tutorial," *Optical Engineering*, vol. 50, no. 7, Article ID 072601, 2011.

[36] "Compressive optical imaging: architectures and algorithms," in *Optical and Digital Image Processing. Fundamentals and*

Applications, G. Cristobal, P. Schelkens, and H. Thienpont, Eds., Wiley-VCH, New York, NY, USA, 2011.

[37] J. M. Whittaker, *Interpolatory Function Theory*, Cambridge University Press, Cambridge, UK, 1935.

[38] V. A. Kotelnikov, "on the carrying capacity of the ether and wire in telecommunications," Material for the First All-Union Conference on Questions of Communication, Izd. Red. Upr. Svyazi RKKA, Moscow, Russia, 1933 (Russian).

[39] C. Shannon, "A mathematical theory of communication," *Bell System Technical Journal*, vol. 27, no. 3, pp. 379–423, 623–656, 1948.

[40] T. S. Huang and O. J. Tretiak, *Picture Bandwidth Compression*, Gordon and Breach, New York, NY, USA, 1972.

[41] A. Habibi and P. Wintz, "Image coding by linear transformation and block quantization," *IEEE Transactions on Communications*, vol. 19, no. 1, pp. 50–62, 1971.

[42] T. A. Wintz, "Transform picture coding," *Proceedings of the IEEE*, vol. 60, no. 7, pp. 809–823, 1972.

[43] W. B. Pennebaker and J. L. Mitchell, *JPEG Still Image Data Compression Standard*, Springer, Berlin, Germany, 3rd edition, 1993.

[44] I. E. G. Richardson, *H.264 and MPEG-4 Video Compression*, John Wiley & Sons, Chichester, UK, 2003.

[45] T. Wiegand, G. J. Sullivan, G. Bjøntegaard, and A. Luthra, "Overview of the H.264/AVC video coding standard," *IEEE Transactions on Circuits and Systems for Video Technology*, vol. 13, no. 7, pp. 560–576, 2003.

[46] P. Schelkens, A. Skodras, and T. Ebrahimi, *The JPEG, 2000 Suite*, John Wiley & Sons, Chichester, UK, 2009.

[47] I. Pitas, *Digital Video and Television*, Ioannis Pitas, 2013.

[48] D. Vaisey and A. Gersho, "Variable block-size image coding, acoustics, speech, and signal processing," in *Proceedings of the IEEE International Conference on (ICASSP '87)*, vol. 12, pp. 1051–1054, 1987.

[49] W. K. Pratt, "Generalized wiener filtering computation techniques," *IEEE Transactions on Computers*, vol. 21, pp. 636–641, 1972.

[50] H. C. Andrwes, "Digital computers and image processing," *Endeavour*, vol. 31, no. 113, 1972.

[51] N. Wiener, *The Interpolation, Extrapolation and Smoothing of Stationary Time Series*, Wiley, New York, NY, USA, 1949.

[52] A. N. Kolmogorov, "Sur l'interpolation et extrapolation des suites stationnaires," *Comptes Rendus de l'Académie des Sciences*, vol. 208, pp. 2043–2045, 1939.

[53] D. L. Donoho and J. M. Johnstone, "Ideal spatial adaptation by wavelet shrinkage," *Biometrika*, vol. 81, no. 3, pp. 425–455, 1994.

[54] D. L. Donoho, "De-noising by soft-thresholding," *IEEE Transactions on Information Theory*, vol. 41, no. 3, pp. 613–627, 1995.

[55] L. Yaroslavsky, "Space variant and adaptive transform domain image and video restoration methods," in *Advances in Signal Transforms: Theory and Applications*, J. Astola and L. Yaroslavsky, Eds., EURASIP Book Series on Signal Processing and Communications, 2007.

[56] L. Yaroslavsky, "Local criteria: a unified approach to local adaptive linear and rank filterson," in *Signal Recovery and Synthesis III*, Technical Digest Series 15, 1989.

[57] M. D. Levine, *Vision in Man and Machine*, McGraw-Hill, New York, NY, USA, 1985.

[58] H. S. Malvar and D. H. Staelin, "LOT: transform coding without blocking effects," *IEEE Transactions on Acoustics, Speech, and Signal Processing*, vol. v, pp. 553–559, 1992.

[59] R. Öktem, L. Yaroslavsky, K. Egiazarian, and J. Astola, "Transform domain approaches for image denoising," *Journal of Electronic Imaging*, vol. 11, no. 2, pp. 149–156, 2002.

[60] L. P. Yaroslavsky, B. Fishbain, A. Shteinman, and S. Gepshtein, "Processing and fusion of thermal and video sequences for terrestrial long range observation systems," in *Proceedings of the 7th International Conference on Information Fusion (FUSION '04)*, pp. 848–855, Stockholm, Sweden, July 2004.

[61] A. Buades, B. Coll, and J. M. Morel, "A review of image denoising algorithms, with a new one," *Multiscale Modeling and Simulation*, vol. 4, no. 2, pp. 490–530, 2005.

[62] A. Buades, B. Coll, and J.-M. Morel, "A non-local algorithm for image denoising," in *Proceedings of the IEEE Computer Society Conference on Computer Vision and Pattern Recognition (CVPR '05)*, pp. 60–65, June 2005.

[63] J. M. Morel and A. Buades Capo, *Non Local Image Processing*, http://www.farman.ens-cachan.fr/6_JeanMichel_Morel.pdf.

[64] S. E. Hecht and J. J. Vidal, "generation of ECG prototype waveforms by piecewise correlational averaging," *IEEE Transactions on Pattern Analysis and Machine Intelligence*, vol. 2, no. 5, pp. 415–420, 1980.

[65] K. Dabov, A. Foi, V. Katkovnik, and K. Egiazarian, "Image denoising by sparse 3-D transform-domain collaborative filtering," *IEEE Transactions on Image Processing*, vol. 16, no. 8, pp. 2080–2095, 2007.

[66] L. Yaroslavsky and M. Eden, "Correlational accumulation as a method for signal restoration," *Signal Processing*, vol. 39, no. 1-2, pp. 89–106, 1994.

[67] V. Katkovnik, K. Egiazarian, and J. Astola, "Adaptive varying window methods in signal and image processing," in *Advances in Signal Transforms. Theory and Applications*, J. Astola and L. Yaroslavsky, Eds., vol. 7 of *Eurasip Book Series on Signal Processing and Communications*, pp. 241–284, 2007.

[68] D. Gabor, "Theory of Communication," *Journal of IEEE*, vol. 93, pp. 429–457, 1946.

[69] E. Wigner, "On the quantum correction for thermodynamic equilibrium," *Physical Review*, vol. 40, no. 5, pp. 749–759, 1932.

[70] R. K. Potter, G. Koppand, and H. C. Green, *Visible Speech*, Van Nostrand, New York, NY, USA, 1947.

[71] D. L. Donoho, "Nonlinear wavelet methods for recovery of signals, densities, and spectra from indirect and noisy data," *Proceedings of Symposia in Applied Mathematics, American Mathematical Society*, vol. 47, pp. 173–205, 1993.

[72] R. R. Coifman and D. L. Donoho, "Translation- Invariant De-Noising," in *Wavelets and Statistics*, A. Antoniadis, Ed., vol. 103 of *Lecture Notes in Statistics*, pp. 125–150, Springer, Berlin, Germany, 1995.

[73] ftp://ftp.cis.upenn.edu/pub/eero/matlabPyrTools.tar.gz.

[74] B. Z. Shaick, L. Ridel, and L. Yaroslavsky, "A hybrid transform method for image denoising," in *Proceedings of the 10th European Signal Processing Conference (EUSIPCO '00)*, M. Gabbouj and P. Kuosmanen, Eds., Tampere, Finland, September 2000.

[75] M. Vetterli and J. Kovacevic, *Wavelets and Sub-Band Coding*, Prentice Hall, Englewood Cliffs, NJ, USA, 1995.

[76] J. Backus, *The Acoustical Foundations of Music*, Norton and Co., New York, NY, USA, 1969.

[77] M. Unser, A. Aldroubi, and M. Eden, "Polynomial spline signal approximations: Filter design and asymptotic equivalence with Shannon's sampling theorem," *IEEE Transactions on Information Theory*, vol. 38, no. 1, pp. 95–103, 1992.

[78] A. Gotchev, "Image interpolation by optimized spline-based kernels," in *Advances in Signal transforms: Theory and Applications*, J. Astola and L. Yaroslavsky, Eds., vol. 7 of *Eurasip Book Series on Signal Processing and Communications*, pp. 285–335, 2007.

[79] L. P. Yaroslavsky, "Efficient algorithm for discrete sine interpolation," *Applied Optics*, vol. 36, no. 2, pp. 460–463, 1997.

[80] L. P. Yaroslavsky, "Shifted discrete fourier transforms," in *Digital Signal Processing*, V. Cappellini and A. G. Constantinides, Eds., pp. 69–74, Avademic Press, London, UK, 1980.

[81] L. Yaroslavsky, "Boundary effect free and adaptive discrete signal sinc-interpolation algorithms for signal and image resampling," *Applied Optics*, vol. 42, no. 20, pp. 4166–4175, 2003.

[82] L. Bilevich and L. Yaroslavsky, "Fast DCT-based image convolution algorithms and application to image resampling and hologram reconstruction," in *Real-Time Image and Video Processing*, Proceedings of SPIE, Brussels, Belgium, April 2010.

[83] L. Bilevich and L. Yaroslavsky, "Fast DCT-based algorithm for signal and image accurate scaling," in *Image Processing: Algorithms and Systems XI*, vol. 8655 of *Proceedings of SPIE*, February 2013.

[84] L. P. Yaroslavsky, "Linking analog and digital image processing," in *Optical and Digital Image Processing. Fundamentals and Applications*, G. Cristobal, P. Schelkens, and H. Thienpont, Eds., pp. 397–418, Wiley-VCH, New York, NY, USA, 2011.

[85] M. Unser, P. Thevenaz, and L. Yaroslavsky, "Convolution-based interpolation for fast, high-quality rotation of images," *IEEE Transactions on Image Processing*, vol. 4, no. 10, pp. 1371–1381, 1995.

[86] P. Thévenaz, EPFL/STI/IOA/LIB, Bldg. +BM-Ecublens 4.137, Station 17, CH-1015 Lausanne VD, Switzerland, http://bigwww.epfl.ch/thevenaz/.

[87] A. Gotchev, "Spline and wavelet based techniques for signal and image processing," Tampere University of Technology, Publication 429, Tampere, Finland, TTY-Paino 2003.

[88] http://www.eng.tau.ac.il/~yaro/Etudes/PDF/MonaLisa-320x256CurvMirr.avi.

[89] L. Yaroslavsky, B. Fishbain, G. Shabat, and I. Ideses, "Superresolution in turbulent videos: making profit from damage," *Optics Letters*, vol. 32, no. 21, pp. 3038–3040, 2007.

[90] B. Fishbain, I. A. Ideses, G. Shabat, B. G. Salomon, and L. P. Yaroslavsky, "Superresolution in color videos acquired through turbulent media," *Optics Letters*, vol. 34, no. 5, pp. 587–589, 2009.

[91] B. Fishbain, L. P. Yaroslavsky, and I. A. Ideses, "Real-time stabilization of long range observation system turbulent video," *Journal of Real-Time Image Processing*, vol. 2, no. 1, pp. 11–22, 2007.

[92] W. H. Press, B. P. Flannery, S. A. Teukolsky, and W. T. Vetterling, *Numerical Recipes. the Art of Scientific Computing*, Cambridge University Press, Cambridge, UK, 1987.

[93] J. H. Mathew and K. D. Fink, *Numerical Methods Using MATLAB*, Prentice-Hall, Englewood Cliffs, NJ, USA, 1999.

[94] L. P. Yaroslavsky, A. Moreno, and J. Campos, "Frequency responses and resolving power of numerical integration of sampled data," *Optics Express*, vol. 13, no. 8, pp. 2892–2905, 2005.

[95] C. F. Gauss, "Nachclass: theoria interpolationis methodo nova tractata," *IEEE ASSP Magazine*, vol. 1, no. 4, pp. 14–81, 1984.

[96] L. F. Yaroslavsky, G. Shabat, B. G. Salomon, I. A. Ideses, and B. Fishbain, "Nonuniform sampling, image recovery from sparse data and the discrete sampling theorem," *Journal of the Optical Society of America*, vol. 26, no. 3, pp. 566–575, 2009.

[97] L. P. Yaroslavsky, "Image recovery from sparse samples, discrete sampling theorem, and sharply bounded band-limited discrete signals," in *Advances in Imaging and Electron Physics*, P. Hawkes, Ed., vol. 167, chapter 5, pp. 295–331, Academic Press, 2011.

[98] R. W. Gerchberg and W. O. Saxton, "Practical algorithm for the determination of phase from image and diffraction plane pictures," *Optik*, vol. 35, no. 2, pp. 237–250, 1972.

[99] A. Papoulis, "New algorithm in spectral analysis and band-limited extrapolation," *IEEE Trans Circuits Syst*, vol. 22, no. 9, pp. 735–742, 1975.

[100] D. Gabor, "Theory of communication," *Journal of IEEE*, vol. 93, pp. 429–457, 1946.

5

The Design of Frequency Filters of Iterative Feedback Tuning Using Particle Swarm Optimization

Arman Sharifi

Department of Electrical Engineering, Islamic Azad University, Kermanshah Branch, Kermanshah 6718997551, Iran

Correspondence should be addressed to Arman Sharifi; arman85sh@gmail.com

Academic Editor: Ana Alejos

Iterative feedback tuning (IFT) is a data-based tuning approach that minimizes a quadratic performance index using some closed-loop experimental data. A control weighting coefficient, known as lambda, and two frequency filters are the most important parameters which can significantly improve the performance of the method. One of the major problems in IFT is tuning these parameters. This paper presents a new approach to tune frequency filters using particle swarm optimization (PSO). At the end, the performance of the proposed method is evaluated by two case study simulations.

1. Introduction

Iterative feedback tuning (IFT) is a data-based method for the tuning of controllers with restricted complexity which was proposed by Hjalmarsson et al. [1] in 1994. In this method, the problem of model bias could be avoided by replacing the information carried by the model with information achieved directly from the system itself. This leads to an iterative method where the controller parameters are successively updated using information from closed-loop experiments with the most recent controller in the loop. IFT has proved to be very effective in practice and is now greatly used in process control [2]. Since 1994 many experiences have been achieved by IFT algorithm and lots of improvements have been observed [3–13].

In order to fully take advantage of IFT, a new scheme needs to be introduced to tune the efficient parameters. In earlier studies, the importance of IFT parameters was confirmed and a few tuning key points were presented [8], but the tuning of these parameters has not been treated in much detail. However, some approaches have been proposed to tune similar parameters. Shridhar and Cooper [14] derived an analytical expression in which the suppression coefficients are calculated as a function of the plant model parameters. Kai et al. [15] proposed a min-max algorithm to design tuning parameters. Al-Ghazzawi et al. [16] presented an approach to tuning MPC on-line based on sensitivity equations derived from a step response model with linear constraints.

The IFT's parameters are mainly a scalar, named lambda, and two frequency filters. This paper provides a method to tune the frequency filters which can emphasize or suppress specific frequency bands of the output and control signals. These filets affect the IFT's performance greatly. Therefore, convenient design of these filters seems critical. In order to tune these filters, the particle swarm optimization (PSO) is proposed, which tunes frequency filters using the input-output data achieved from IFT algorithm. This method is proposed for the first order filters with two parameters.

The paper is organized as follows: a brief explanation of IFT algorithm is presented in Section 2. In Section 3, the particle swarm optimization method is briefly described and implemented to tune the frequency filters. Consequently, we shed some light on the performance of the proposed method for tuning the frequency filters, by applying the improved IFT to some case study systems in Section 4. Finally, the paper is concluded in Section 5 with a summary of key points and results.

2. Iterative Feedback Tuning Approach

Consider the SISO closed-loop system as presented in Figure 1. From the figure, it can be followed that

$$y(t) = Gu(t) + v(t),$$
$$u(t) = C_r(\rho) r_t - C_y(\rho) y_t, \quad (1)$$

where G is an unknown single input-single output (SISO) operator which is linear time invariant (LTI), $y(t)$ is the process output, $u(t)$ is the corresponding process input, and $v(t)$ is an immeasurable disturbance which is assumed to be stochastic. r_t is an external deterministic reference signal which is independent of $\{v(t)\}$. $C = \{C_r(\rho), C_y(\rho)\}$ are linear time invariant transfer functions parameterized by parameter vector ρ.

Assuming y^d as a desired output response to the reference signal r, for closed-loop system, the error between the achieved and desired response is

$$\tilde{y}(\rho) \triangleq y(\rho) - y^d. \quad (2)$$

The control design objective can be formulated as

$$J = \frac{1}{2N} E\left(\sum_{t=1}^{N} \left(L_y \tilde{y}_t(\rho)\right)^2 + \lambda \sum_{t=1}^{N} \left(L_u u_t(\rho)\right)^2\right), \quad (3)$$

where $E[\cdot]$ denotes expectation with respect to the disturbance v. Scalar λ is the weighting coefficient expressing the relative importance of the penalty on the control effort, and the symbols L_y and L_u are some frequency filters which can emphasize or suppress specific frequency bands of the output and control signals. These filters can be used for some purposes such as

(i) emphasizing or suppressing specific frequency bands of the outputs and control signals, for instance, to prevent unwanted oscillations in these signals:

(ii) using as notch filters in the frequency bands where the measurement noise dominates;

(iii) meeting specific frequency domain performance specifications, such as constraints on the sensitivities.

For minimizing J, IFT estimates the gradient $\partial J/\partial \rho$ only using some special closed-loop experiments. The detailed procedure of estimating the gradient can be seen in [8]. After estimating the gradient $\partial J/\partial \rho$, the parameters of the controller are updated by the following iterative algorithm:

$$\rho_{i+1} = \rho_i - \gamma_i R_i^{-1} \frac{\partial J}{\partial \rho}(\rho). \quad (4)$$

Here, R_i is an appropriate positive definite matrix which is typically a Gauss-Newton approximation of the Hessian of J while γ_i is a real positive scalar which defines the step size of the algorithm.

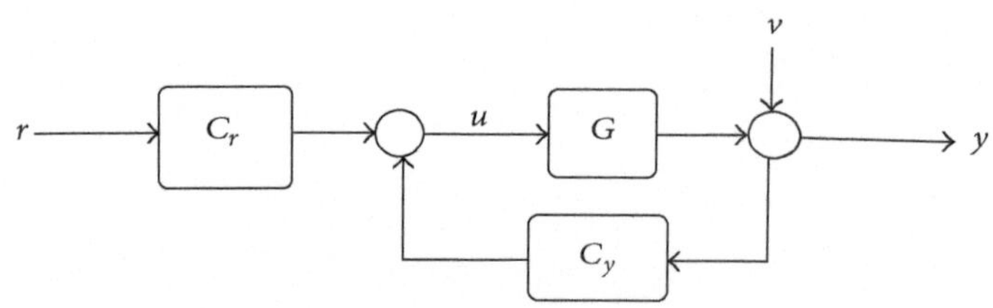

FIGURE 1: SISO closed loop system.

3. Particle Swarm Optimization

Particle swarm optimization (PSO) is a population-based optimization approach first proposed in 1995 [17, 18]. This method is urged by the observation of social interaction and animal behaviours such as fish schooling and bird flocking [19–21].

Similar to most optimization techniques, PSO requires a cost function evaluation function relevant to the particle's position. As L_y and L_u affect the output signal and the input signal, the proposed cost function consists of the integrated absolute error (IAE) defined as follows [22]:

$$\text{IAE} = \int_0^{\infty} \left|r(t) - y(t)\right| dt. \quad (5)$$

And the second part of the cost function is total variation (TV) of the manipulated input. This criterion is used to evaluate the required control effort which is a convenient measure of "smoothness" of control input and should be as small as possible [22]:

$$\text{TV} = \sum_{k=1}^{\infty} \left|u(k+1) - u(k)\right|. \quad (6)$$

Thus, the cost function can be defined as

$$\text{CF} = \text{IAE} + \text{TV}. \quad (7)$$

All solutions in PSO can be represented as particles in a swarm. Each particle has a position and velocity vector and each position coordinate represents a parameter value. The current position of ith particle of the swarm is denoted by x_i. x_i^{pbest} and x_i^{gbest} are the personal best (Pbest) position and global best (Gbest) position of the ith particle. Each particle is initialized with a random position and velocity. The velocity of each particle is accelerated toward the global best and its own personal best based on the following equation:

$$V_i(t+1) = wV_i(t) + c_1 r_1 \left(x_i^{\text{pbest}}(t) - x_i(t)\right) + c_2 r_2 \left(x_i^{\text{gbest}}(t) - x_i(t)\right). \quad (8)$$

Here r_1 and r_2 are two random numbers generated from a uniform distribution in the range $[0, 1]$; c_1 and c_2 are the acceleration constants and w is the inertia weight factor. Suitable selection of these parameters provides a balance between global and local explorations and helps the particles

TABLE 1: Controller setting for Example 1 (evaluating the proposed frequency filters effect).

Tuning method	K_1	T_1	K_2	T_2	S. T.	Os%	IAE	TV
Tuned filters	0.507	0.127	0.094	3.071	43.63	6.68	101.37	1.59
Random filters 1	0.71	0.14	0.06	3.22	41.89	29.8	114.65	3.24
Random filters 2	0.59	0.12	0.13	2.93	69.32	0	139.5	1.69
Random filters 3	0.53	0.15	0.1	2.14	125	0	2.08	2.05

S. T.: Settling time.

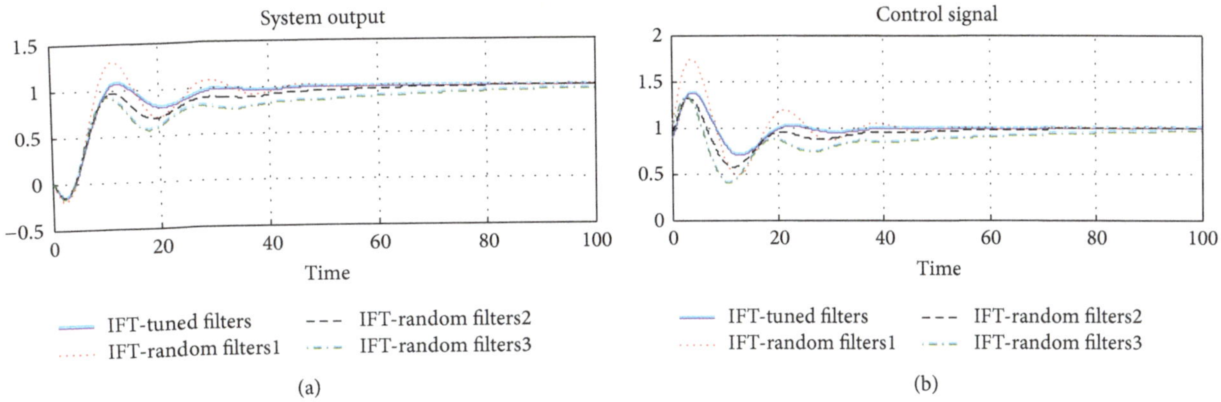

FIGURE 2: Simulation results of Example 1 for evaluating the effect of proposed filters.

converge to Gbest. In this paper, the constricted version of PSO [23] is used which sets the parameters as follows:

$$w = \frac{2}{\phi - 2 + \sqrt{\phi^2 - 4\phi}},$$
$$c_1 = w\phi_1,$$
$$c_2 = w\phi_2, \quad (9)$$

where $\phi_1 = \phi_2 = 2.05$, $\phi = \phi_1 + \phi_2$.

The positions are updated based on their movement over a discrete time interval (Δt) as follows:

$$x_i(t+1) = x_i(t) + V_i(t+1)\Delta t, \quad (10)$$

where Δt usually is set to 1. Then the cost function at each position is reevaluated to update Gbest and Pbest.

In this paper, the frequency filters are considered as the first order filter as follows:

$$L_y = \frac{K_1}{T_1 s + 1},$$
$$L_u = \frac{K_2}{T_2 s + 1}. \quad (11)$$

Therefore, the proposed PSO is employed to tune four parameters K_1, T_1, K_2, and T_2. In each iteration of PSO, the IFT algorithm obtains the optimal control parameter with the current filters parameters. Using the updated controller parameters, cost function (7) is changed and it leads to different parameters of frequency filters achieved in PSO algorithm.

4. Simulation Study

Example 1. A gas boiler system is considered where the gas valve position V_2 is used to control the temperature T of the water. The following transfer function is used to represent the relationship between the variations in V_2 and T when the flow of water is constant [24]:

$$P(s) = \frac{e^{-5s}}{(1+3s)(1+s)}. \quad (12)$$

This system is controlled by a PI controller tuned by IFT algorithm. The frequency filters tuned by the proposed PSO method are as follows:

$$L_y = \frac{0.507}{0.126s + 1},$$
$$L_u = \frac{0.094}{3.071s + 1}. \quad (13)$$

The simulation results in Figure 2 and Table 1 indicate that the IFT algorithm tuned by the proposed method gives an acceptable response which can substitute the trial-and-error tuning of the filters. As mentioned before, the frequency filters play a significant role in IFT performance. By many simulations it was observed that IFT is very sensitive to filters parameters. By a random selection of the filters parameters, it was observed that poorly tuned filters can lead to unstable responses. To show the filters effect on the responses, the filters parameters are obtained randomly around the optimal values by a 30% tolerance. From Figure 2 and Table 1, it can be seen that small variations of filters parameters can lead to large changes in output response.

Table 2: Controller setting for Example 2 (evaluating the proposed frequency filters effect).

Tuning method	K_1	T_1	K_2	T_2	S. T.	Os%	IAE	TV
Tuned filters	2.321	0.91	1.1	4.52	13.021	0.458	39.21	39.33
Random filters 1	2	0.22	1.7	2.1	33.71	98.81	96.43	5.19
Random filters 2	3.16	0.22	2.2	1.9	23.88	40.33	58.84	1.66
Random filters 3	3	0.15	1	0.83	24.75	13.098	86.81	0.3439

S. T.: Settling time.

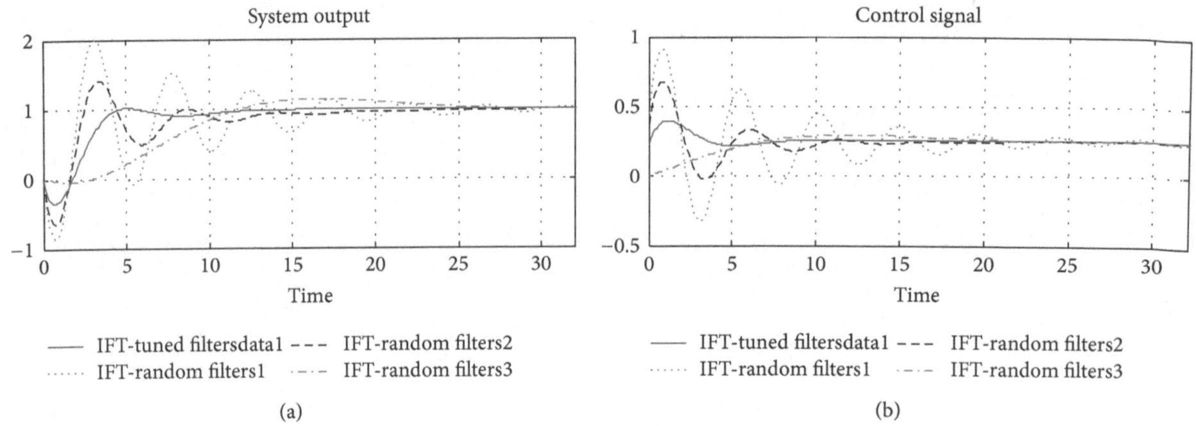

Figure 3: Simulation results of Example 2 for evaluating the effect of proposed filters.

Example 2. Consider the control system of a heated tank. The dynamics relating the control action and the temperature can be modelled with an FOPDT model as follows [24]:

$$P(s) = \frac{3.9e^{-2.1s}}{1.2s + 1}. \tag{14}$$

The output response and control signal of the system are depicted in Figure 3 in which the IFT tuned by the proposed method has been compared to IFT with the randomly tuned filters. Here, more freedom was assigned to select the random filters. Therefore, the effect of these filters has been shown better. The simulation results can be seen in Table 2.

5. Conclusion

The earlier attempts to tune the frequency filters were based on trial-and-error. In this study, these filters were designed using the PSO method. The proposed method was applied to a case study system. It was observed that the performance of IFT modified by the proposed methods was greatly improved. Therefore, IFT has become more capable with a better performance. The proposed method for designing frequency filters in this study is not merely limited to IFT algorithm; therefore, future studies are recommended to generalize the proposed approach to other methods with similar parameters. Also, a further study with focus on high order filters is suggested.

Conflict of Interests

The author declares that there is no conflict of interests regarding the publication of this paper.

References

[1] H. Hjalmarsson, S. Gunnarsson, and M. Gevers, “Convergent iterative restricted complexity control design scheme,” in *Proceedings of the 33rd IEEE Conference on Decision and Control*, pp. 1735–1740, IEEE, Orlando, Fla, USA, December 1994.

[2] H. Hjalmarsson, “Iterative feedback tuning—an overview,” *International Journal of Adaptive Control and Signal Processing*, vol. 16, no. 5, pp. 373–395, 2002.

[3] J. K. Huusom, N. K. Poulsen, and S. B. Jørgensen, “Improving convergence of iterative feedback tuning,” *Journal of Process Control*, vol. 19, no. 4, pp. 570–578, 2009.

[4] J. K. Huusom, H. Hjalmarsson, N. K. Poulsen, and S. B. Jorgensen, “A design algorithm using external perturbation to improve iterative feedback tuning convergence,” *Automatica*, vol. 47, no. 12, pp. 2665–2670, 2011.

[5] H. Hjalmarsson, “Efficient tuning of linear multivariable controllers using iterative feedback tuning,” *International Journal of Adaptive Control and Signal Processing*, vol. 13, no. 7, pp. 553–572, 1999.

[6] W. K. Ho, Y. Hong, A. Hansson, H. Hjalmarsson, and J. W. Deng, “Relay auto-tuning of PID controllers using iterative feedback tuning,” *Automatica*, vol. 39, no. 1, pp. 149–157, 2003.

[7] K. Tsang, A. B. Rad, and W. Chan, “Iterative feedback tuning for positive feedback time delay controller,” *International Journal of Control, Automation and Systems*, vol. 3, no. 4, pp. 640–645, 2005.

[8] H. Hjalmarsson, M. Gevers, S. Gunnarsson, and O. Lequin, “Iterative feedback tuning: theory and applications,” *IEEE Control Systems Magazine*, vol. 18, no. 4, pp. 26–41, 1998.

[9] A. E. Graham, A. J. Young, and S. Q. Xie, “Rapid tuning of controllers by IFT for profile cutting machines,” *Mechatronics*, vol. 17, no. 2-3, pp. 121–128, 2007.

[10] S. Kissling, P. Blanc, P. Myszkorowski, and I. Vaclavik, "Application of iterative feedback tuning (IFT) to speed and position control of a servo drive," *Control Engineering Practice*, vol. 17, no. 7, pp. 834–840, 2009.

[11] D. Rupp and L. Guzzella, "Iterative tuning of internal model controllers with application to air/fuel ratio control," *IEEE Transactions on Control Systems Technology*, vol. 18, no. 1, pp. 177–184, 2010.

[12] D. Liu, A. J. McDaid, K. C. Aw, and S. Q. Xie, "Position control of an Ionic Polymer Metal Composite actuated rotary joint using Iterative Feedback Tuning," *Mechatronics*, vol. 21, no. 1, pp. 315–328, 2011.

[13] A. J. McDaid, K. C. Aw, E. Haemmerle, and S. Q. Xie, "Control of IPMC actuators for microfluidics with adaptive "online" iterative feedback tuning," *IEEE/ASME Transactions on Mechatronics*, vol. 17, no. 4, pp. 789–797, 2012.

[14] R. Shridhar and D. J. Cooper, "A novel tuning strategy for multivariable model predictive control," *ISA Transactions*, vol. 36, no. 4, pp. 273–280, 1997.

[15] H. Kai, Z. Jun, and Q. Jixin, "A novel robust tuning strategy for model predictive control," in *Proceedings of the 6th World Congress on Intelligent Control and Automation (WCICA '06)*, pp. 6406–6410, Dalian, China, June 2006.

[16] A. Al-Ghazzawi, E. Ali, A. Nouh, and E. Zafiriou, "On-line tuning strategy for model predictive controllers," *Journal of Process Control*, vol. 11, no. 3, pp. 265–284, 2001.

[17] J. Kennedy and R. Eberhart, "Particle swarm optimization," in *Proceedings of the IEEE International Conference on Neural Networks*, pp. 1942–1948, IEEE Service Centre, Perth, Australia, December 1995.

[18] R. C. Eberhart and J. Kennedy, "A new optimizer using particles swarm theory," in *Proceedings of the 6th International Symposium on Micro Machine and Human Science*, pp. 39–43, IEEE Service Center, Nagoya, Japan, 1995.

[19] T. BartzBeielstein, K. E. Parsopoulos, and M. N. Vrahatis, "Analysis of particle swarm optimization using computational statistics," in *Proceedings of the International Conference on Numerical Analysis and Applied Mathematics (ICNAAM '04)*, pp. 34–37, 2004.

[20] J. Zhao, T. Li, and J. Qian, "Application of particle swarm optimization algorithm on robust PID controller tuning," in *Proceedings of the 1st International Conference on Natural Computation (ICNC '05)*, pp. 948–957, Springer, Berlin, Germany, August 2005.

[21] H. Hu, Q. Hu, Z. Lu, and D. Xu, "Optimal PID controller design in PMSM servo system via particle swarm optimization," in *Proceedings of the 31st Annual Conference of IEEE Industrial Electronics Society*, pp. 79–83, November 2005.

[22] D. Chen and D. E. Seborg, "PI/PID controller design based on direct synthesis and disturbance rejection," *Industrial and Engineering Chemistry Research*, vol. 41, no. 19, pp. 4807–4822, 2002.

[23] M. Clerc and J. Kennedy, "The particle swarm-explosion, stability, and convergence in a multidimensional complex space," *IEEE Transactions on Evolutionary Computation*, vol. 6, no. 1, pp. 58–73, 2002.

[24] J. E. Normey-Rico and E. F. Camacho, *Control of Dead-Time Processe*, Springer, New York, NY, USA, 2007.

6

Proton Irradiations on SJ HV Power MOSFETs to Realize Fast Diode Devices

Ignazio Bertuglia, Giuseppe Consentino, and Michele Laudani

Power Transistors Division, STMicroelectronics, Catania, Italy

Correspondence should be addressed to Michele Laudani; michele.laudani@st.com

Academic Editor: Changhwan Shin

This paper studies the effects of proton irradiations on Super Junction High Voltage power MOSFETs to realize transistors with fast diode. Experiments were performed on a sample of 600 V power MOSFETs and achieved results were compared to standard irradiated devices by electrons.

1. Introduction

In many modern applications and, in particular, when using full bridge converters, intrinsic diode of HV power MOSFET is even utilized to free wheel the current in the circuit without using any other external component [1–7]. Intrinsic diode of a power MOSFET is implemented by considering the body-drain junction of the same device. In fact, when the device is switched off and the transistor is polarized in reverse mode, the current flows from the source to the drain terminals working in third quadrant configuration. In such operating conditions, intrinsic diode needs to guarantee particular performances and features. In particular, diode needs to be fast in order to reduce switching losses when operating switching frequencies increase. A diode becomes fast when charges moved during rapid variation of V_{ds} against time are quite low as shown in Figure 1. In fact, in the example shown in Figure 1, MOSFET is rapidly switched off. When V_{ds} increases, drain current decreases down to negative values because charges stored in the body-drain junction need to be discharged. Standard devices (red curve) reach the lowest peak of drain current (I_{rm}) because the charge stored in the junction (Q_{std}) is higher than fast devices (green curve). That turns in a higher time to discharge (t_{rr}) and, thus, it makes the devices slower during commutation. To improve the performances of intrinsic diode, special processes need to be implemented. The typical action consists in the irradiation of the body-drain junction by energetic electrons in order to create suitable damage in the reticle. Such damage introduces deep energetic level traps in the silicon band-gap. Such traps capture the carriers lowering the quantity of charges moved during the fast transition from on to off states as described above. Typical irradiation doses are in the range of 5–50 MRad when considering power MOSFETs. Afterwards, a thermal process needs to be implemented in order to activate the traps. However, irradiation by considering electrons can bring issues related, for example, to the quality of gate oxide. In fact, interface traps states can be created increasing drain-source leakage current.

Recently, a new technique was implemented to realize devices with fast diodes based on proton irradiation. In fact, protons can be introduced in the body-drain junction of a power MOSFET's structure only concentrating the effect on the body-drain border junction differently from the electrons which are implanted on the entire structure.

2. Protons Irradiation, Experiment Setup, and Test Vehicles

A series of experiments were implemented by changing the dose of protons irradiation and acting on thermal process to evaluate the best solution and to compare the results with the standard electron irradiation process. The tests were performed by considering a tandem accelerator with protons at energies in the ranges of some MeV useful to

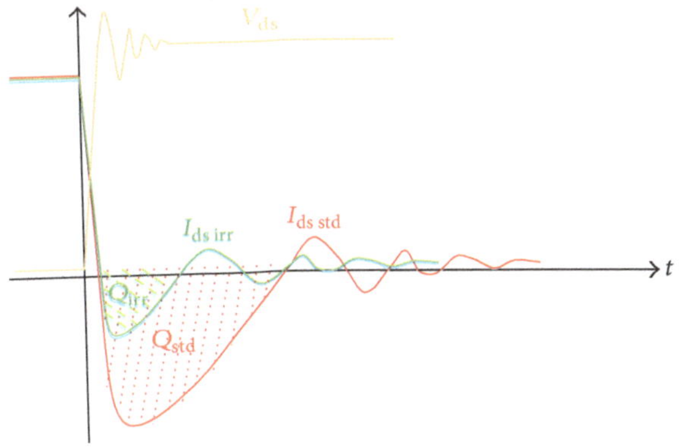

FIGURE 1: Example of commutation of the intrinsic diode and comparison with a fast diode.

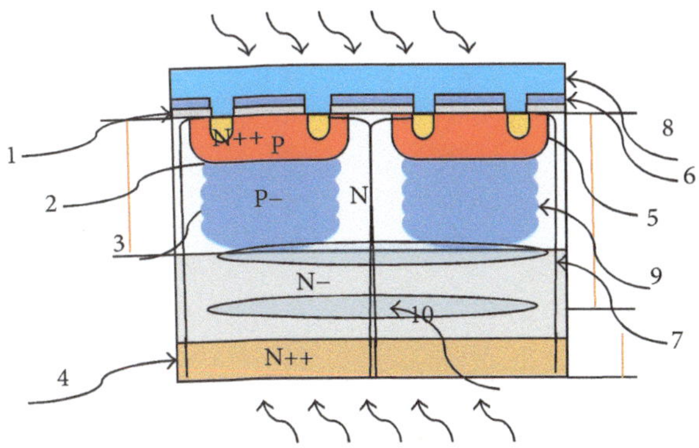

FIGURE 2: Example of cross section of a Super Junction HV power MOSFET. (1) Gate planar oxide; (2) body region; (3) EPY drift region; (4) back-drain contact; (5) source well-source contact; (6) polysilicon; (7) current flow; (8) metal; (9) drain column; and (10) region to be damaged by proton.

TABLE 1: Main characteristics of test vehicle.

Device type	SJ HV power MOSFETs used		
	BV_{dss} [V]	I_d continuous @ 25°C [A]	V_{th} [V]
n-ch	600	35	4

reach the depth inside of body-drain junction of the devices. The irradiation has been performed in air where at 1 cm of distance is located the wafer under test. Wafers were mounted on a frame connected to a precise step by step motor utilized to move it. To perform these experiments, a family of SJ HV power MOSFETs of 600 V was considered (see in Table 1 the peculiarities of the adopted test vehicle). Typical cross section of a Super Junction device is shown in Figure 2 where the depth of implant is even highlighted.

Irradiation of protons was implemented either in the front or in the back side of the wafer considered, reaching different depths in the region included between the body-drain junction and the same epitaxial layer under the pillars. Based on simulation activity, it was established that fluency of protons on the silicon needs to be in the range of 10^{11} cm^{-2} and $5 * 10^{12}$ cm^{-2} in order to compare with effects of electrons in silicon. It is necessary to observe that protons' irradiation

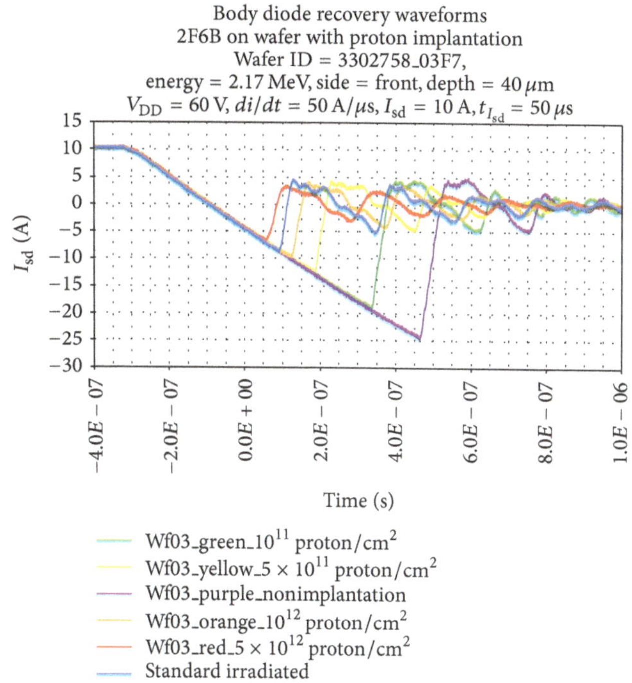

FIGURE 3: Example referred to data of Table 3.

implies a certain tail of damage along the initial path of the trajectory that could affect the electrical performances of the devices. For example, if the implantation is performed by the top of the wafer, the tail of damage involves essentially the first micrometers of silicon layers in the top of the die. Such vacancies could even affect the reliability of the gate oxide. These considerations will be more evident after studying the experimental results on the wafer irradiated on the front of the wafers. After the proton implantations, wafers under tests were thermally treated in furnaces and, afterwards, they finished the process (back end). Before packaging the dices, wafers were tested to evaluate the performances in terms of t_{rr} (diode recovery time), Q_{rr} (diode recovery charge), and I_{rm} (reverse peak current during the recovery time of diode). Figure 3 and Tables 2, 3, 4, and 5 report the data when a polarization which equals 60 V is applied.

As it is possible to see from the above data, by increasing proton dose I_{rm}, t_{rr}, and Q_{rr} decrease and the diode switching performances improve. However, results are different if compared to several trials and with standard irradiated devices with electrons. In fact, considering the devices irradiated around the body-drain junction with a concentration of 10^{12} cm^{-2}, Q_{rr} is comparable to standard irradiated transistors. Instead, considering devices with an implanted dose of $5*10^{12}$ cm^{-2}, Q_{rr} is lower than standard irradiated transistors. Considering devices irradiated in the epitaxial region with a concentration of $5*10^{12}$ cm^{-2}, Q_{rr} is quite similar to standard one. It is necessary to highlight that when the dose increases, a lowering of breakdown voltage is observed together with an increasing of drain-source leakage currents. Such results are more evident when dice implanted from the top are considered. This phenomenon can be explained taking into

TABLE 2: Main results on wafer bench—implantation on the front EPI layer.

Dose [cm^{-2}]	SJ HV power MOSFETs results			
	I_{rm} [A]	t_{rr} [ns]	Q_{rr} [nC]	BV_{dss} [V]
$5 * 10^{12}$	5.1	120	354	500
$1 * 10^{12}$	9.2	189	936	610
$5 * 10^{11}$	11.5	237	1466	650
$1 * 10^{11}$	21.1	400	4475	680
No irradiation	29.6	534	7582	680
Std. electrons	8.7	184	877	660

TABLE 3: Main results on wafer bench—implantation on the front body-drain junction.

Dose [cm^{-2}]	SJ HV power MOSFETs results			
	I_{rm} [A]	t_{rr} [ns]	Q_{rr} [nC]	BV_{dss} [V]
$5 * 10^{12}$	6.7	180	602	460
$1 * 10^{12}$	9.7	242	1176	570
$5 * 10^{11}$	12.5	308	1926	630
$1 * 10^{11}$	19.1	474	4524	670
No irradiation	29.6	534	7582	680
Std. electrons	8.7	184	877	660

TABLE 4: Main results on wafer bench—implantation on the back body-drain junction.

Dose [cm^{-2}]	SJ HV power MOSFETs results			
	I_{rm} [A]	t_{rr} [ns]	Q_{rr} [nC]	BV_{dss} [V]
$5 * 10^{12}$	7.7	194	747	490
$1 * 10^{12}$	10.9	271	1475	600
$5 * 10^{11}$	14.1	343	2421	640
$1 * 10^{11}$	17.3	427	3692	680
No irradiation	29.6	534	7582	680
Std. electrons	8.7	184	877	660

account that tails of the damage, located near the body-drain junction, change the charges balance inside the space charge region.

Comparable results can be achieved by considering packaged devices when current is switched off at around 27 A (different operating condition compared to the previous test). In Tables 6–9 and Figures 4, 5, 6, and 7 data regarding dynamic measurements of packaged devices are shown.

3. Conclusions

This paper has analyzed the effects on SJ HV power MOSFETs of proton irradiations to realize intrinsic "fast diode" components. Intrinsic diode needs to be fast in order to decrease switching losses and to increase operating switching frequencies without any issue related to dV/dt. In order to improve the performances of intrinsic diode, typical action consists in the irradiation of the body-drain junction by energetic electrons. Electronic irradiations can sometimes bring issues related, for example, to the quality of gate oxide

TABLE 5: Main results on wafer bench—implantation on the back EPI layer.

Dose [cm^{-2}]	SJ HV power MOSFETs results			
	I_{rm} [A]	t_{rr} [ns]	Q_{rr} [nC]	BV_{dss} [V]
$5 * 10^{12}$	9.5	232	1108	630
$1 * 10^{12}$	11.9	282	1678	660
$5 * 10^{11}$	13.5	335	2262	670
$1 * 10^{11}$	19.1	477	4550	680
No irradiation	29.6	534	7582	680
Std. electrons	8.7	184	877	660

TABLE 6: Main results on packaged power MOSFETs—implantation on the front EPY layer.

Dose [cm^{-2}]	SJ HV power MOSFETs results		
	I_{rm} [A]	t_{rr} [ns]	Q_{rr} [nC]
$5 * 10^{12}$	13	162	1180
$1 * 10^{12}$	18	203	1870
$5 * 10^{11}$	24	267	3460
$1 * 10^{11}$	36	448	8612
Std. electrons	18	180	1800

TABLE 7: Main results on packaged power MOSFETs—implantation on the front body-drain border.

Dose [cm^{-2}]	SJ HV power MOSFETs results		
	I_{rm} [A]	t_{rr} [ns]	Q_{rr} [nC]
$5 * 10^{12}$	10	116	630
$1 * 10^{12}$	19	233	2370
$5 * 10^{11}$	23	254	3133
$1 * 10^{11}$	35	414	7626
Std. electrons	18	180	1800

TABLE 8: Main results on packaged power MOSFETs—implantation on the back body-drain border.

Dose [cm^{-2}]	SJ HV power MOSFETs results		
	I_{rm} [A]	t_{rr} [ns]	Q_{rr} [nC]
$5 * 10^{12}$	18	190	1756
$1 * 10^{12}$	20	228	2410
$5 * 10^{11}$	26	286	4056
$1 * 10^{11}$	34	406	7400
Std. electrons	18	180	1800

TABLE 9: Main results on packaged power MOSFETs—implantation on the back EPY layer.

Dose [cm^{-2}]	SJ HV power MOSFETs results		
	I_{rm} [A]	t_{rr} [ns]	Q_{rr} [nC]
$5 * 10^{12}$	19	205	2058
$1 * 10^{12}$	23	253	3110
$5 * 10^{11}$	27	290	4200
$1 * 10^{11}$	35	431	8330
Std. electrons	18	180	1800

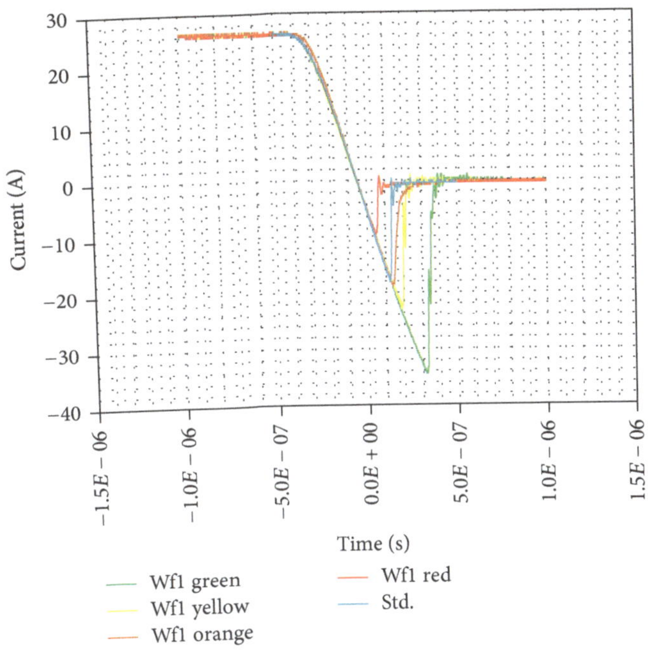

FIGURE 4: Example referred to data of Table 6.

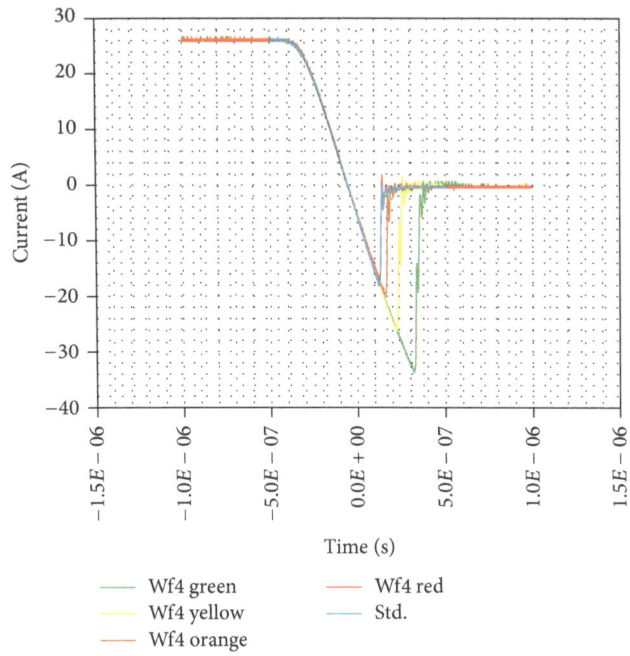

FIGURE 6: Example referred to data of Table 8.

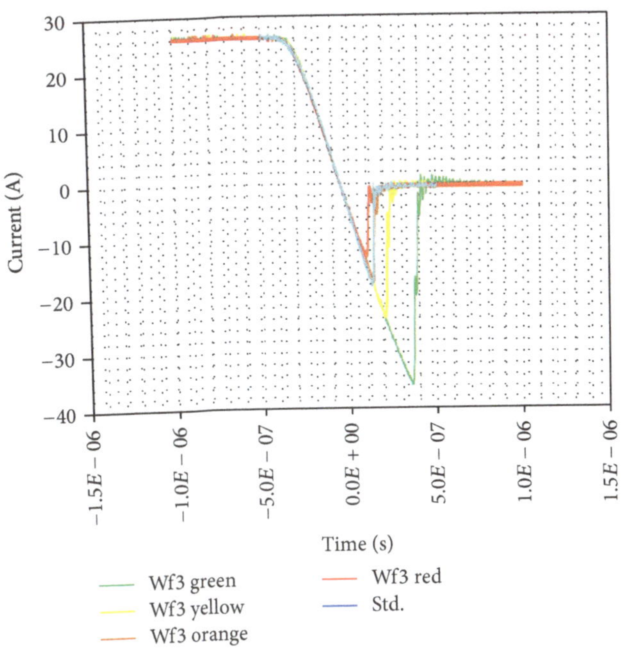

FIGURE 5: Example referred to data of Table 7.

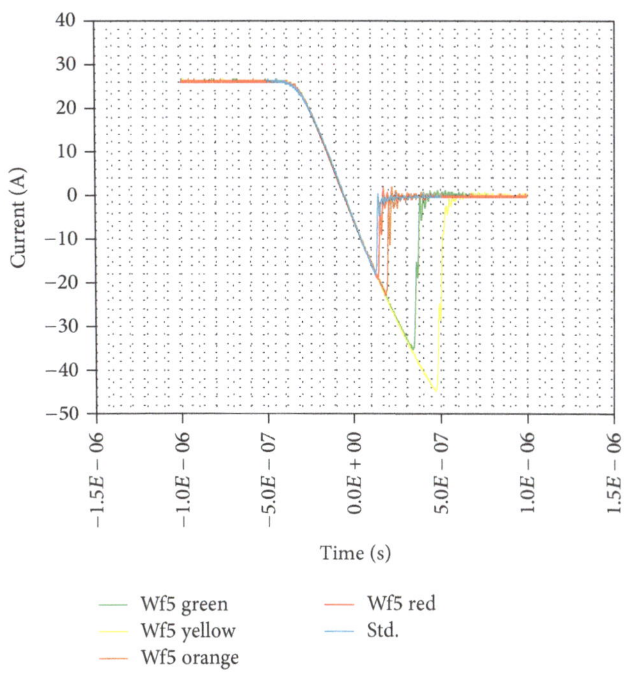

FIGURE 7: Example referred to data of Table 9.

because they are spread out in the entire wafer volume. Instead, protons are localized in a specific part of the die and, in particular, in the epitaxial layer slightly below the body-drain junction. A series of experiments was performed by considering a suitable proton irradiator on samples of Super Junction High Voltage power MOSFETs with a breakdown voltage of 600 V. Irradiations with protons were implemented either in the front or in the back side of the wafers with projected range in the region between the body-drain junction and the same epitaxial layer under the pillars. Based on simulations, it was established that fluency of protons on the silicon needs to be in the range of $10^{11}\,cm^{-2}$ and $5 * 10^{12}\,cm^{-2}$ to compare with effects of electrons irradiation on silicon. After the proton implantations, wafers under tests were thermally treated by considering dedicated process flow. Therefore, several measurements were performed either considering tested dice on wafers or considering static and dynamic characterizations after the packaging. Experimental results show that I_{rm}, t_{rr}, and Q_{rr} decrease by increasing proton doses. However, it is necessary to highlight that when the dose increases, a lowering of breakdown voltage is observed together with an increasing of drain-source leakage currents.

Therefore, it is possible to assume that the best trial, in terms of reverse recovery performance, was the one with protons implanted on the back side with a dose of $5 * 10^{12}$ cm^{-2}. Nevertheless, breakdown voltage degradation implies that a more optimized solution still has to be found by investigating different doses combined with thermal annealing process fine tuning, just to improve Q_{rr}/t_{rr} performances without BV degradation.

Conflict of Interests

The authors declare that there is no conflict of interests regarding the publication of this paper.

References

[1] P. Hazdra, J. Vobecký, and K. Brand, "Optimum lifetime structuring in silicon power diodes by means of various irradiation techniques," *Nuclear Instruments and Methods in Physics Research B: Beam Interactions with Materials and Atoms*, vol. 186, no. 1–4, pp. 414–418, 2002.

[2] J. Lutz, R. Siemieniec, F. J. Niedernostheide, and H. J. Schulze, "Characteristics of irradiation induced deep levels in silicon," in *Proceeding of the Solid State Chemistry and Modern Micro- and Nanotechnologies Conference*, Kislovodsk, Russia, 2005.

[3] M. Schmitt, H.-J. Schulze, A. Schlogl et al., "A Comparison of Electron, Proton and Helium Ion Irradiation for the Optimization of the CoolMOSTM Body Diode," Infineon Technologies AG, AI PS DD, Balanstr. 59, D-81541 Munich, Germany, 2011.

[4] V. Benda, "Design considerations for fast soft reverse recovery diodes," in *Proceedings of the 15th European Conference on Power Electronics and Applications*, Brighton, UK, September 1993.

[5] P. Lévêque, P. Pellegrino, A. Hallén, B. G. Svensson, and V. Privitera, "Hydrogen-related defect centers in float-zone and epitaxial n-type proton implanted silicon," *Nuclear Instruments and Methods in Physics Research, Section B: Beam Interactions with Materials and Atoms*, vol. 174, no. 3, pp. 297–303, 2001.

[6] T. Mizushima, M. Nemoto, H. Kuribayashi, T. Yoshimura, and H. Nakazawa, "Inhibiting effect of middle broad buffer layer diode using hydrogen-related shallow donor on reverse recovery oscillation," in *Proceedings of the 22nd International Symposium on Power Semiconductor Devices & IC's (ISPSD '10)*, pp. 115–118, June 2010.

[7] R. Siemieniec, F.-J. Niedernostheide, H.-J. Schulze, W. Südkamp, U. Kellner-Werdehausen, and J. Lutz, "Irradiation-induced deep levels in silicon for power device tailoring," *Journal of the Electrochemical Society*, vol. 153, no. 2, pp. G108–G118, 2006.

7

Multiobjective Genetic Algorithms Program for the Optimization of an OTA for Front-End Electronics

Abdelghani Dendouga,[1] Slimane Oussalah,[1] Damien Thienpont,[2] and Abdenour Lounis[3]

[1] *Division Microelectronique et Nanotechnologie, Centre de Développement des Technologies Avancées (CDTA), Cité é20 Août 1956, 16000 Algiers, Algeria*
[2] *OMEGA Micro, Microelectronics Design Center, Polytechnic School, LLR Aile 4, 91128 Palaiseau Cedex, France*
[3] *Laboratoire de l'Accélérateur Linéaire (LAL), Centre Scientifique d'Orsay, Université de Paris Sud XI, Bat 200, BP 34, 91898 Orsay Cedex, France*

Correspondence should be addressed to Abdelghani Dendouga; adendouga@cdta.dz

Academic Editor: Changhwan Shin

The design of an interface to a specific sensor induces costs and design time mainly related to the analog part. So to reduce these costs, it should have been standardized like digital electronics. The aim of the present work is the elaboration of a method based on multiobjectives genetic algorithms (MOGAs) to allow automated synthesis of analog and mixed systems. This proposed methodology is used to find the optimal dimensional transistor parameters (length and width) in order to obtain operational amplifier performances for analog and mixed CMOS-(complementary metal oxide semiconductor-) based circuit applications. Six performances are considered in this study, direct current (DC) gain, unity-gain bandwidth (GBW), phase margin (PM), power consumption (P), area (A), and slew rate (SR). We used the Matlab optimization toolbox to implement the program. Also, by using variables obtained from genetic algorithms, the operational transconductance amplifier (OTA) is simulated by using Cadence Virtuoso Spectre circuit simulator in standard TSMC (Taiwan Semiconductor Manufacturing Company) RF 0.18 μm CMOS technology. A good agreement is observed between the program optimization and electric simulation.

1. Introduction

Microelectronics industry is distinguished by the raising level of integration and complexity. It aims at decreasing exponentially the minimum feature sizes used to design integrated circuits [1]. The cost in time of design is a great problem to the continuation of this evolution. Senior designer's knowledge and skills are required to ensure a good analog integrated circuit design. To fulfill the given requirements, the designer must choose the suitable circuit architecture, although different tools which partially automated the topology synthesis appeared in the past [2, 3].

Therefore, the use of multiple-objective optimization algorithms is of a great importance to the automatic design of operational amplifier. Accuracy, ease of use, generality, robustness, and reasonable run-time are necessary for a circuit synthesis solution to gain acceptance by using optimization methods [4–9].

This method uses a program based on multiobjective optimization using a genetic algorithm to calculate the optimal transistors dimensions, length, and width of an operational amplifier (Figure 1) which is used as part of an electronic front-end for signal shaping stage. The method which handles a wide variety of specifications and constraints is extremely fast and results in globally optimal designs.

The aim of this work is to design and optimize an operational amplifier circuit in sight of a front-end electronics of the semiconductor tracker (SCT) detector in ATLAS (A Toroidal LHC Apparatus) experiment. ATLAS is a particle physics experiment at the Large Hadron Collider at CERN (the European Organization for Nuclear Research) in Switzerland.

This paper is organized as follows. The amplifier structure is analyzed in Sections 2 and 3. Section 4 describes the optimization approach proposed in this work. Section 5 presents

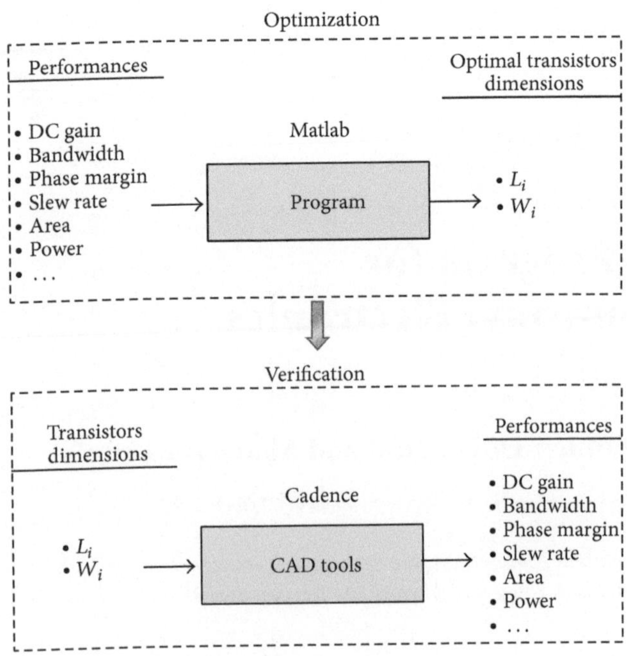

FIGURE 1: Operational amplifier design flow.

the obtained results, and there is a section for the comparison of our work with other optimization approaches. Finally some concluding remarks are provided after evaluating our study towards other works.

2. Design Methodology

Optimal design of analog circuits consists of finding a variable set $x = \{x_1, x_2, \ldots, x_n\}$ that optimizes performance functions, such as gain, offset, signal to noise ratio, and maximum operating frequency, while meeting imposed specifications and/or inherent constraints, for example, saturation conditions of transistors, technology limits, and impedance matching. Vector x may encompass biases, lengths (L), and widths (W) of MOS transistors, component values, and so forth [5].

3. Specifications

We concentrate on one operational amplifier topology and the two-stage operational amplifier shown in Figure 2. The main electrical parameters of the circuit are low frequency voltage gain (A_v), gain-bandwidth product (GBW), slew-rate (SR), dissipated power (P_{diss}), phase margin (PM), and area (A), among others. The design variables are the size of transistor (width and length), the value of the passive components (capacitors and resistors), and the value of bias currents and bias voltages. For this particular two-stage operational amplifier, there are fourteen design variables.

3.1. Open-Loop DC Gain. For the two-stage op-amp, the open-loop voltage gain is given by [3]

$$A_v = \frac{g_{m1}}{g_{ds2} + g_{ds4}} \cdot \frac{g_{m6}}{g_{ds7} + g_{ds6}}, \quad (1)$$

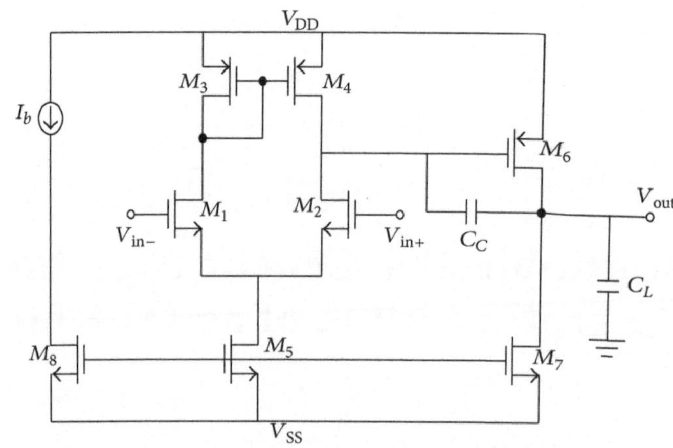

FIGURE 2: The two-stage operational amplifier architecture used in this study is composed of eight CMOS transistors.

where g_m (g_{m1} g_{m6}) is the transconductance of transistors (M_1 and M_6) and g_{ds} is the output conductance.

3.2. Unity-Gain Bandwidth. The unity-gain bandwidth is given by the expression [1]

$$\text{GBW} = \frac{g_{m1}}{C_C}, \quad (2)$$

where C_C is the compensation capacitance.

3.3. Phase Margin. The phase margin of operational amplifier depends on the sum of phase shifts, at the unity-gain frequency, contributed by the nondominant poles ($p1$ and $p2$) and zeros (z):

$$\text{PM} = \pm 180 - \tan^{-1}\left(\frac{\text{GBW}}{p1}\right) - \tan^{-1}\left(\frac{\text{GBW}}{p2}\right) - \tan^{-1}\left(\frac{\text{GBW}}{z}\right). \quad (3)$$

3.4. Slew Rate. For this operational amplifier, the slew rate is given by

$$\text{SR} = \frac{I_5}{C_C}, \quad (4)$$

where I_5 is the current that flows through transistor M_5.

3.5. Power Consumption. For the two-stage operational amplifier, the power consumption has the form [10]

$$P = (V_{DD} - V_{SS})(I_5 + 2I_7). \quad (5)$$

3.6. Area. The area A of the operational amplifier is given by the sum of transistors and capacitors areas:

$$\text{Area} = \sum_{i=1}^{k} W_i \cdot L_i. \quad (6)$$

4. Optimization

To make the system power level, it is obviously important to size the different constituent blocks. At this level, the performance of each unit becomes constraints to be respected. The performances are bound by a set of equations which depends on the considered characteristics (gain, SR, etc.). The set of equations is nonlinear, and there is no systematic analytical method to solve it. In addition, the solution is not unique. For this reason, the best way is to use an optimizer that will help automate the resolution of equations (synthesis of analog circuits). It is important to note that, in the design of analog circuits, compromises are made because there are many performance parameters used to describe them. Nonlinear relationships between them make them a more delicate design.

Optimal design of analog circuits is to find a set of variables $x = \{x_1, x_2, \ldots, x_n\}$ that optimizes performance, such as gain, offset, and signal to noise ratio, while respecting the imposed specifications and/or constraints [6].

In the program, every individual is presented by a binary code string. From Figure 2, we can see that there are 8 transistors and a biasing current to be adjusted. As a total, there are 10 parameters to be adjusted and each gene of the chromosome stands for one parameter. Thus, the parameter vector is compressed to [8] $[W_1, L_1, W_3, L_3, W_5, L_5, W_7, L_7, W_8, L_8]$.

Genetic algorithms start with an initial population of randomly generated individuals. Each individual in the population represents a possible solution to the problem of the study. Individuals evolve through successive iterations, called generations. In every generation, each individual in the population is evaluated using a measure of fitness. Then, the population of the next generation is created by genetic operators. The procedure continues until the stop condition is satisfied (Figure 3).

A weighted approach is used to optimize operational amplifiers. It uses adaptive weights along the optimization process to determine the overall fitness of an individual [7]:

$$F = \sum_{i=1}^{n} \omega_i \cdot f_i, \quad (7)$$

where ω_i is the weight coefficient of every subobjective, f_i is the overall fitness of every performance considered, and i is the number of the performances considered.

We used the Matlab optimization toolbox to implement optimization by MOGA. It starts by generating a random population of individuals. To pass from one generation k to generation $k + 1$, the following operations are performed. At first, the population is reproduced by good selection where individuals with the best evaluations tend to reproduce more often than those with bad evaluations. This population is applied to cross pairs of individuals (parents) of a certain proportion of the population (probability P_c, usually around 0.6) to produce new children. A mutation operator is applied to a certain proportion of the population (probability P_m, the P_c generally much lower). Finally, the new individuals are evaluated and incorporated into the population of the next generation and several stopping criteria of the algorithm are possible: the number of generations can be fixed a priori (time constant) or the algorithm can be stopped when the population does not evolve fast enough.

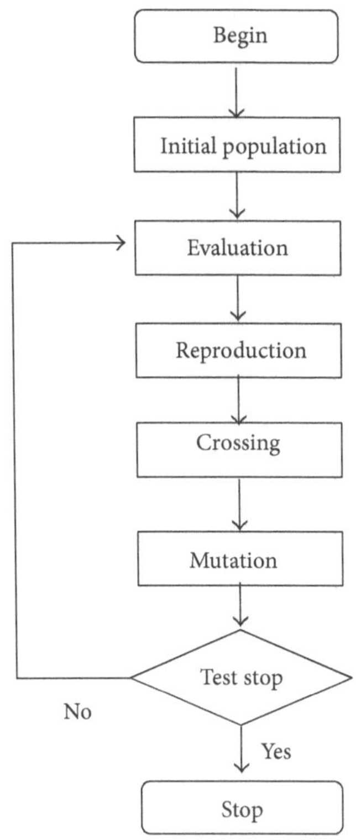

Figure 3: Basic procedures of genetic algorithms.

5. Results

Six performances are considered in this program. They are the DC gain, bandwidth of unity-gain, phase margin, power consumption, area, and slew rate. The optimization process optimizes the individual to improve its fitness score. This process will continue until the total number of generations is reached.

Also, by using variables obtained from GA, the OTA circuit is simulated by using Cadence Virtuoso Spectre in TSMC 0.18 μm CMOS process and simulation results are shown in Table 2 and Figure 4.

Table 2 shows the performance of the design obtained by Matlab optimization tools. The objective was to maximize the unity-gain bandwidth and minimize power consumption subject to the other given constraints. The simulation results confirm the efficiency of GA in determining the device sizes in an analog circuit.

According to the simulation results, the performance of the operational amplifier optimized by the proposed method represents a good method to optimize an analog circuit.

After introducing the transistors dimensions (Table 1) in Spectre and making the different simulations, we pass to the layout of the circuit which is represented in Figure 5 and the

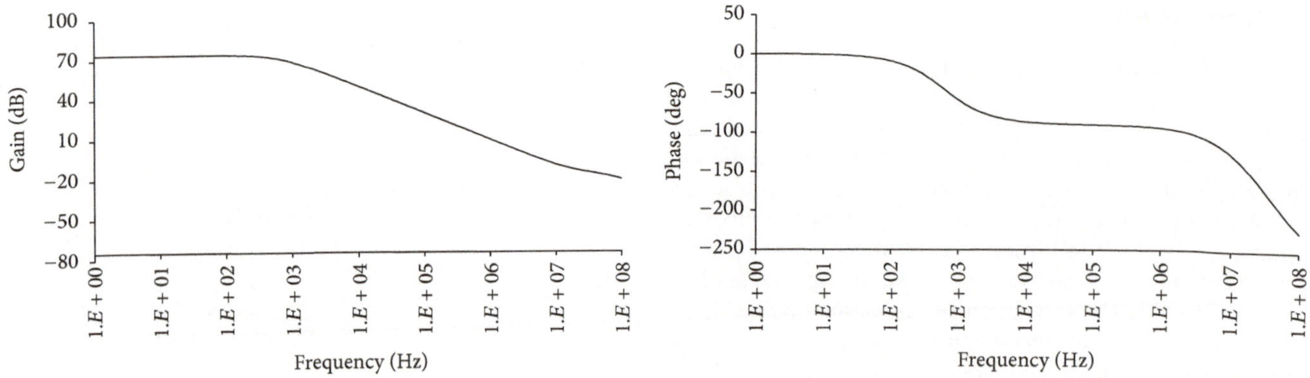

(a) (b)

FIGURE 4: Gain and phase simulation of the obtained transistors dimensions.

TABLE 1: Optimal transistors dimensions.

Variable	Value (μm)
$W_1 = W_2$	1.34
$L_1 = L_2$	1.5
$W_3 = W_4$	8.5
$L_3 = L_4$	1.11
W_5	1.4
L_5	0.44
W_6	48
L_6	0.92
W_7	6
L_7	0.8
W_8	23.5
L_8	0.68

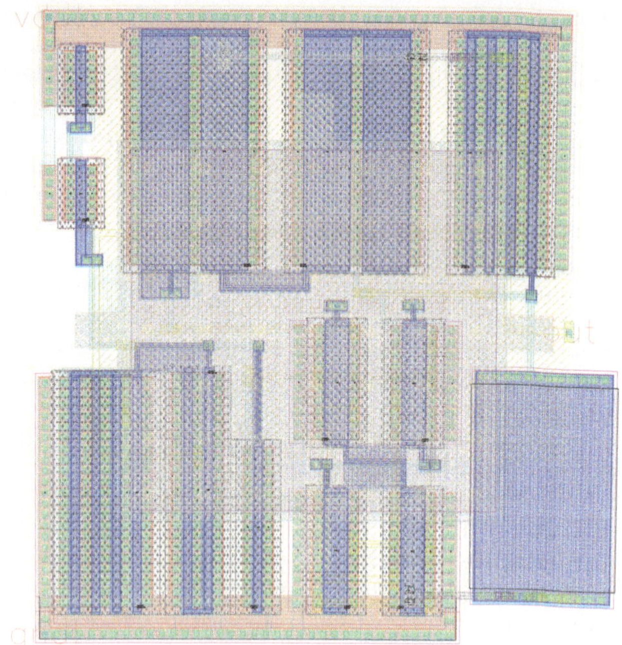

FIGURE 5: Layout of the operational amplifier.

TABLE 2: Simulation results in Matlab and Spectre.

Performances	Specifications	MOGA program	Spectre
DC gain (dB)	≥70	76	75
Unity gain (MHz)	Max	1.5	1.11
Phase margin (°)	≥60	70	64
Slew rate (V/μs)	Max	2.25	2.19
Area (μm^2)	Min	559	678
Power (mW)	≤0.5	0.047	0.051

postlayout simulations. Figure 4 represents the simulation of gain and phase of the circuit.

The results given in the two Tables 1 and 2, respectively, represent the dimensions and performance operational amplifiers obtained for different constraints and conditions. With constraints on the optimizer, the satisfactory results (GWB aspects gain and PM) are obtained. However, the program happens to minimize power consumption and layout area.

6. Comparison

The lack of detail necessary to compare the results (such as limits of design variables, the supply voltage, bias current, capacitive load time, and circuit optimization) makes the comparison of our work with other optimization approaches presented a difficult task [11].

Kubar and Jakovenko [11] present comparisons with works using Miller two-stage OTA design example. These works [10, 12] are using particle swarm optimization.

The differences between the work [9] and our case are that transistors of the current mirror ($M3$ and $M4$) do not have the same size. Table 3 presents results in comparison with our design.

Table 3

Variable	Lower bound	Upper bound	Our result	Result [9]
W_1 (μm)	5	100	1.34	87.1
L_1 (μm)	0.18	2	1.5	0.55
W_2 (μm)	5	100	1.34	87.1
L_2 (μm)	0.18	2	1.5	0.55
W_3 (μm)	3	200	8.5	400
L_3 (μm)	0.18	2	1.11	8.4
W_4 (μm)	7	100	8.5	400
L_4 (μm)	0.18	2	1.11	8.4
W_5 (μm)	3	60	1.4	62.5
L_5 (μm)	0.18	2	0.44	0.55
W_6 (μm)	3	60	48	331.8
L_6 (μm)	0.18	2	0.92	0.4
W_7 (μm)	3	60	6	62.5
L_7 (μm)	0.18	2	0.8	0.55
W_8 (μm)	3	60	23.5	62.5
L_8 (μm)	0.18	2	0.68	0.55

7. Conclusion

This work demonstrates the utility of an evolutionary algorithm for automating electronic design using algorithms called MOGAs, which have the ability to deal with a problem of multiobjective optimization with two or more goals and taking the constraints also into account.

In this paper, a program based on multiobjective genetic algorithm has been developed for analog integrated circuits design. The genetic algorithm and equation-based optimization are combined to produce an accurate tool in order to determine the device sizes in an analog circuit. A MOGAs-based approach is proposed to optimize the performances of two-stage OTAs.

The results prove the effectiveness of the approach in the analog design where the design space is too complicated to be done with the classical methods within a short time. It can be concluded that the proposed MOGAs-based approach is efficient and gives promising results for circuits design and optimization problems.

Conflict of Interests

The authors declare that there is no conflict of interests regarding the publication of this paper.

References

[1] H. D. Dammak, S. Bensalem, S. Zouari, and M. Loulou, "Design of folded cascode OTA in different regions of operation through gm/ID methodology," *International Journal of Electrical and Electronics Engineering*, pp. 178–183, 2008.

[2] M. G. R. Degrauwe, O. Nys, E. Dijkstra et al., "IDAC: an interactive design tool for analog CMOS circuits," *IEEE Journal of Solid-State Circuits*, vol. SC-22, no. 6, pp. 1106–1116, 1987.

[3] M. D. M. Hershenson, S. P. Boyd, and T. H. Lee, "GPCAD: a tool for CMOS operational amplifier synthesis," in *Proceedings of the IEEE/ACM International Conference on Computer-Aided Design (ICCAD '98)*, pp. 296–303, San Jose, Calif, USA, November 1998.

[4] J. Tao, X. Chen, and Y. Zhu, "Constraint multi-objective automated synthesis for CMOS operational amplifier," in *Life System Modeling and Intelligent Computing*, vol. 6329 of *Lecture Notes in Computer Science*, pp. 120–127, 2010.

[5] M. Takhti, A. Beirami, and H. Shamsi, "Multi-objective design automation of the folded-cascode OP-AMP using nsga-II strategy," in *Proceedings of the International Symposium on Signals, Circuits and Systems (ISSCS '09)*, pp. 1–4, Iași, Romania, July 2009.

[6] M. Köppen, G. Schaefer, and A. Abraham, *Intelligent Computational Optimization in Engineering*, Springer, 2011.

[7] P. Jianhai Yu and Z. Mao, "Automated design method for parameters optimization of CMOS analog circuits based on adaptive genetic algorithm," in *Proceedings of the 7th International Conference on ASIC (ASICON '07)*, pp. 1217–1220, Guilin, China, October 2007.

[8] S. Barra, A. Dendouga, S. Kouda, and N. Bouguechal, "Multi-Objective Genetic Algorithm optimization of CMOS operational amplifiers," in *Proceedings of the 24th International Conference on Microelectronics (ICM '12)*, pp. 1–4, Algiers, Algeria, December 2012.

[9] A. Dendouga, S. Oussalah, D. Thienpont, and A. Lounis, "Program for the optimization of an OTA for front end electronics based on multi objective genetic algorithms," in *Proceedings of the IEEE 29th International Conference on Microelectronics (MIEL '14)*, pp. 443–446, Belgrade, Serbia, May 2014.

[10] P. P. Kumar and K. Duraiswamy, "An optimized device sizing of analog circuits using particle swarm optimization," *Journal of Computer Science*, vol. 8, no. 6, pp. 930–935, 2012.

[11] M. Kubar and J. Jakovenko, "A powerful optimization tool for analog integrated circuits design," *Radioengineering*, vol. 22, no. 3, pp. 921–931, 2013.

[12] S. L. Sabat, K. S. Kumar, and S. K. Udgata, "Differential evolution and swarm intelligence techniques for analog circuit synthesis," in *Proceeding of the World Congress on Nature and Biologically Inspired Computing (NABIC '09)*, pp. 469–474, Coimbatore, India, December 2009.

8

Application of Firefly Algorithm in Voltage Stability Environment Incorporating Circuit Element Model of SSSC with Variable Susceptance Model of SVC

Luke Jebaraj,[1] Charles Christober Asir Rajan,[2] and Kumar Sriram[1]

[1] *Department of Electrical and Electronics Engineering, St.Anne's College of Engineering and Technology, Panruti, Tamil Nadu 607 110, India*

[2] *Department of Electrical and Electronics Engineering, Pondicherry Engineering College, Puducherry 605 014, India*

Correspondence should be addressed to Luke Jebaraj; jeba_meps@yahoo.co.in

Academic Editor: José R. C. Piqueira

This paper proposes an application of firefly algorithm (FA) based extended voltage stability margin and minimization of active (or) real power loss incorporating Series-Shunt flexible AC transmission system (FACTS) controller named as static synchronous series compensator (SSSC) combined with static var compensator (SVC). A circuit model of SSSC and variable susceptance model of SVC are utilized to control the line power flows and bus voltage magnitudes, respectively, for real power loss minimization and voltage stability limit improvement. The line quality proximity index (LQP) is used to assess the voltage stability of a power system. The values of voltage profile improvement, real power loss minimization, and the location and size of FACTS devices were optimized by FA. The results are obtained from the IEEE 14- and 30-bus test case systems under different operating conditions and compared with other leading evolutionary techniques such as shuffled frog leaping algorithm (SFLA), differential evolution (DE) and particle swarm optimization (PSO).

1. Introduction

Voltage stability is concerned with the ability of a power system to maintain acceptable voltage at all buses in the system under normal conditions and after being subjected to a disturbance [1]. The recent day power systems are undergoing numerous changes and becoming more complex from the standpoints of operation, control, and stability maintenance when they meet ever-increasing load demand [2]. A system enters a state of voltage instability when a disturbance, increase in load demand, or change in system condition causes a progressive and uncontrollable decline in voltage. The main factor causing voltage instability is the inability of the power system to meet the demand for reactive power [3, 4]. The authors [5, 6] discuss methods to assess voltage stability of a power system to find possible ways to improve the voltage stability.

Abnormal voltages and voltage collapse pose a primary threat to power system stability, security, and reliability. Moreover, with the fast development of restructuring, the problem of voltage stability has become a major concern in deregulated power systems. To maintain security of such systems, it is desirable to plan suitable measures to improve power system security and increase voltage stability margins [7]. Voltage instability is one of the phenomena which have resulted in major blackouts. Recently, several network blackouts have been related to voltage collapse [8]. The only way to counteract this problem is by reducing the reactive power load in the system or by adding new reactive power generation systems in the weakest points of the system, thereby, increasing the voltage at those points.

The flexible AC transmission system (FACTS) controllers are capable of supplying or absorbing reactive power at faster rates [9]. The introduction of FACTS controllers are increasingly used to provide voltage and power flow controls [10]. Insertion of FACTS devices is found to be highly effective in preventing voltage instability and minimize the active or real power loss on transmission lines [11, 12]. Series and shunt compensating devices are used to enhance the static voltage stability margin and reduce the real power loss appreciably [13, 14].

The generalized power injection model of SSSC needs modification of the Jacobian matrix and makes quite complex in coding. In the SSSC control parameters, voltage magnitude and angle of the series converters are presented as independent variables and their values are found through the traditional load flow iterative process [15]. In this case, the size of the Jacobian matrix increases to incorporate the additional independent variables. The new model of the SSSC changes only the bus admittance matrix and consequently reduces the coding of load flow problem incorporating SSSC simple. The SSSC control parameters, voltage magnitude and angle of the series converters, are presented as independent variables and their values are found through the traditional load flow iterative process. In this case, the size of the Jacobian matrix increases to incorporate the additional independent variables. Hence, a simple and easy way to implement SSSC model based on the circuit elements is used in this work [16].

Voltage stability assessment with appropriate representations of FACTS devices is investigated and compared under base case of study [17–19]. One of the shortcomings of those methods only considered the normal state of the system [20]. The author [21] presented the different computational techniques for voltage stability assessment and control. The author [22] focused on the contingency ranking obtained with respect to overloads in large power systems. Abdelaziz et al. [23] proposed the process of contingency ranking through fuzzy logic approach. However, voltage collapses are mostly initiated by a disturbance like line outages. Voltage stability limit improvement needs to be addressed during network contingencies. So, to locate FACTS devices, consideration of contingency conditions is more important than consideration of normal state of system and some approaches are proposed to locate FACTS devices with considerations of contingencies too [24, 25].

Line stability indices provide important information about the proximity of the system to voltage instability and can be used to identify the weakest bus as well as the critical line with respect to the bus of the system [26]. Different types of line stability indices are proposed to evaluate the proximity of the system to voltage instability [27, 28]. The line quality proximity index is used in this work for stability assessment [29, 30]. From the family of bioluminescence inspired computation, FA is used to solve the problem of real power loss minimization and voltage stability maximization of the system. In the firefly algorithm, the objective function of a given optimization problem is based on differences in light intensity. It helps the fireflies to move towards brighter and more attractive locations in order to obtain optimal solutions. All fireflies are characterized by their light intensity associated with the objective function [31–33].

Because of higher cost of the FACTS devices, the installation is not recommended to all possible line outages. Hence, line outage contingency screening and ranking were carried out to identify the most critical line during which outage FACTS controllers can be positioned and system can be operated under stable condition [34–36]. The prime objective of this work is to obtain the results from standard IEEE 14- and 30-bus test cases through FA incorporating circuit element model of SSSC combined with variable susceptance model of SVC and to compare the results of these series-shunt FACTS controllers with other important evolutionary techniques such as SFLA, DE, and PSO under normal, critical loading and single-line outage conditions.

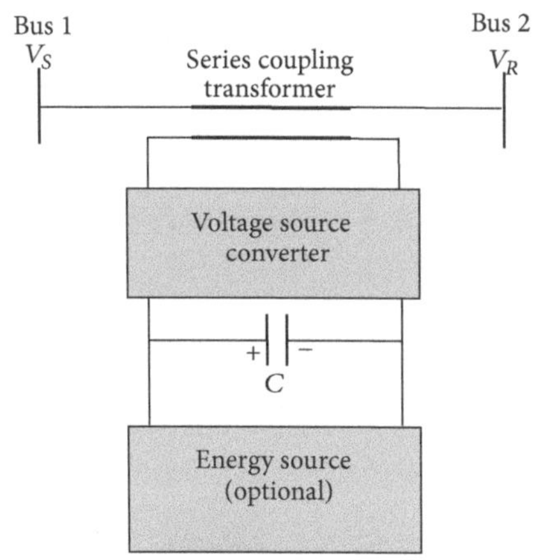

Figure 1: SSSC configuration.

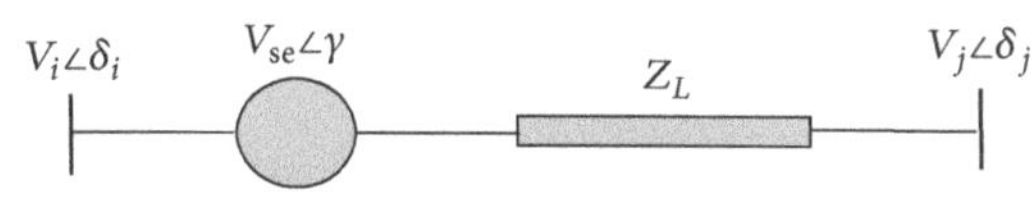

Figure 2: Circuit element model of SSSC.

2. Static Model of FACTS Devices

2.1. Circuit Element Model of SSSC. The SSSC can be operated without an external energy source as reactive power source and is fully controllable and independent of transmission line current for the purpose of increasing or decreasing the overall reactive voltage drop across the transmission line and thereby controlling the electric power flow. The widely used power injection model of SSSC requires modification of the Jacobian matrix and makes the Newton-Raphson load flow (NRLF) coding more complex. A new circuit elements based model of SSSC is utilized to control the line power flows and bus voltage magnitudes for voltage stability limit improvement. The new model of the SSSC changes only the bus admittance matrix and consequently reduces the coding of load flow problem incorporating SSSC simple. This converter performs the main function of injecting a controllable series voltage. The basic configuration of SSSC is depicted in Figure 1. The circuit element model of SSSC is also shown in Figure 2.

The real and reactive powers exchanged with the line by the series voltage inserted by SSSC are modeled as a negative resistance and reactance connected in parallel. The negative resistance represents injection of real power and the reactance may be either capacitive or inductive depending on whether reactive power is delivered or absorbed.

The complex power exchanged by the series converter with the line is expressed as

$$S_e = V_{se} I, \tag{1}$$

where V_{se} is the complex voltage injected by the converter and I the current through the line given by

$$I = \frac{V_i \angle \delta_i + V_{se} \angle \gamma - V_j \angle \delta_j}{Z_L}. \tag{2}$$

The active and reactive powers exchanged with the line are modeled as resistance and reactance associated as represented by

$$R = \frac{V_{se}^2}{P_{se}}, \qquad X = \frac{V_{se}^2}{Q_{se}}. \tag{3}$$

The elements R and X can be calculated by directly using the following equations:

$$R = \frac{V_{se} Z_L}{V_i \sin\left(\delta_i - \delta_j - \gamma\right) + V_j \sin\gamma}, \qquad X = \frac{V_{se} Z_L}{V_i \cos\left(\delta_i - \delta_j - \gamma\right) - V_j \cos\gamma + V_{se}}. \tag{4}$$

The resistance R and the reactance X representing the effect of series voltage are transformed into their equivalent series combination. This makes the line simple with only series-connected elements of the line (Z_L) and the R_{SSSC} and X_{SSSC} denoting the resistance and reactance of SSSC:

$$R_{SSSC} = \frac{RX^2}{R^2 + X^2}, \qquad X_{SSSC} = \frac{XR^2}{R^2 + X^2}. \tag{5}$$

2.2. Variable Susceptance Model of SVC. A variable susceptance B_{SVC} represents the fundamental frequency equivalent susceptance of all shunt modules making up the SVC. This model is an improved version of SVC models. Figure 3 shows the variable susceptance model of SVC which is used to derive its nonlinear power equations and the linearised equations required by Newton's load flow method.

In general, the transfer admittance equation for the variable shunt compensator is

$$I_{SVC} = jB_{SVC} V_j. \tag{6}$$

And the reactive power of SVC is

$$Q_{SVC} = -V_j^2 B_{SVC}. \tag{7}$$

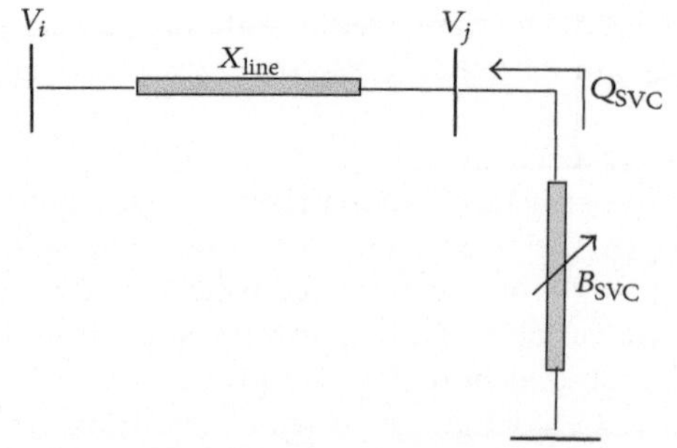

FIGURE 3: Variable susceptance model of SVC.

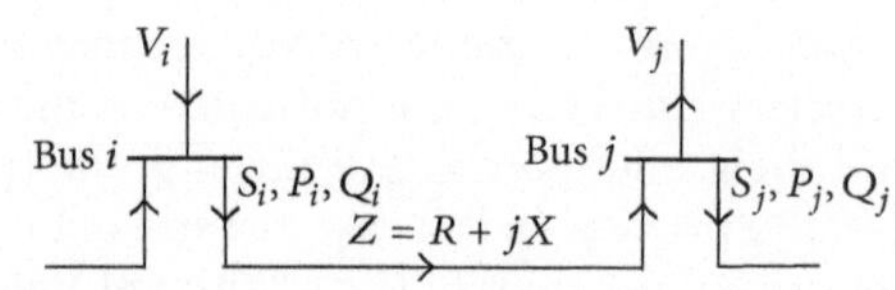

FIGURE 4: Single-line concept of power transmission.

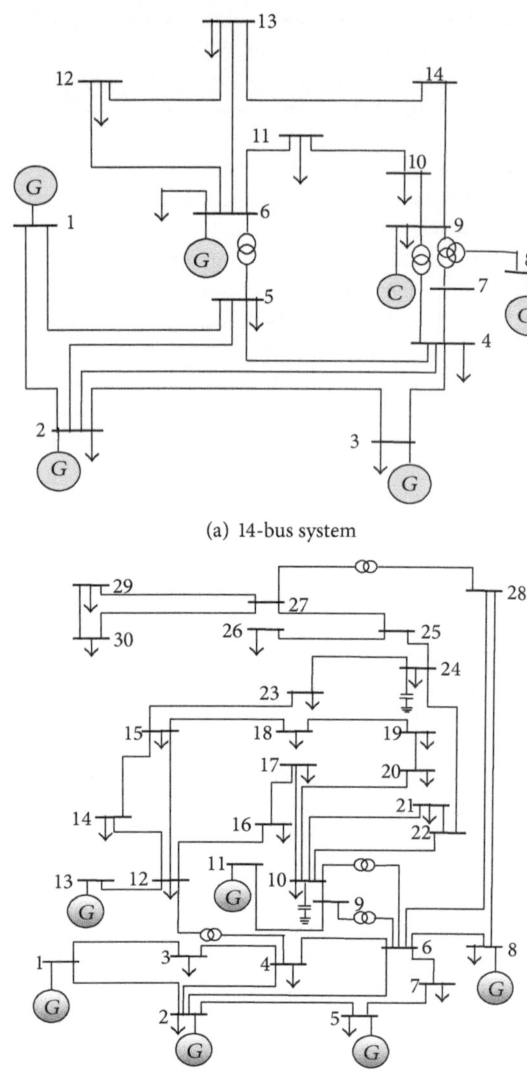

(a) 14-bus system

(b) 30-bus system

FIGURE 5: One-line diagram of IEEE test system.

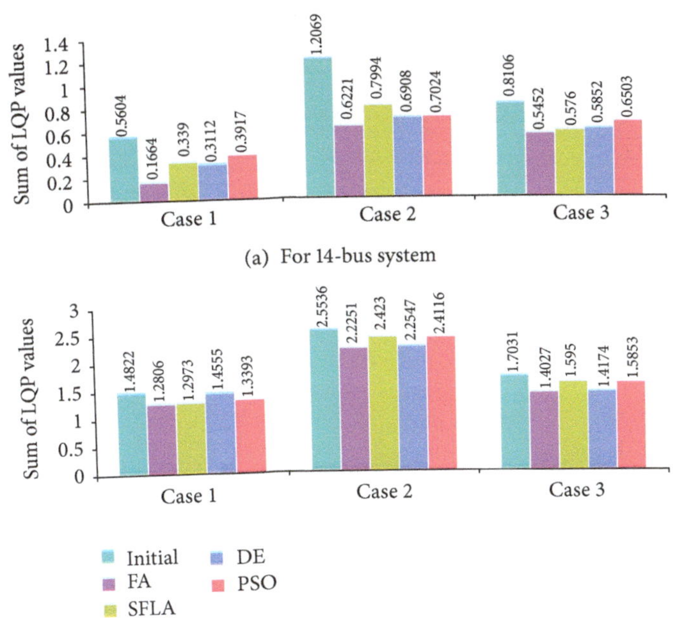

(a) For 14-bus system

(b) For 30-bus system

FIGURE 6: Sum of LQP index values comparison.

In SVC susceptance model, the total susceptance B_{SVC} is taken to be the state variable; therefore the linearised equation of the SVC is given by

$$\begin{bmatrix} \Delta P_j \\ \Delta Q_j \end{bmatrix} = \begin{bmatrix} 0 & 0 \\ 0 & \theta_j \end{bmatrix} \begin{bmatrix} \Delta \theta_j \\ \dfrac{\Delta B_{svc}}{B_{svc}} \end{bmatrix}, \qquad (8)$$

where ΔP_j, ΔQ_j, and $\Delta \theta_j$ are the change in real, reactive power, and voltage angle at jth bus.

At the end of iteration i, the variable shunt susceptance B_{SVC} is updated according to

$$B_{SVC}^{(i)} = B_{SVC}^{(i-1)} + \left(\frac{\Delta B_{SVC}}{B_{SVC}} \right)^{(i)} B_{SVC}^{(i-1)}. \qquad (9)$$

This changing susceptance value represents the total SVC susceptance which is necessary to maintain the nodal voltage magnitude at the specified value (1.0 p.u. in this paper).

3. Line Quality Proximity Index

Voltage stability can be assessed in a system by calculating the line based voltage stability index. The LQP index based on a power transmission concept is used in this paper. The value of line index shows the voltage stability of the system. The value close to unity indicates that the respective line is close to its stability limit and value much close to zero indicates light load in the line. The formulation begins with the power equation in a power system. Figure 4 illustrates a single line of a power transmission concept.

The power equation can be derived as

$$P_i = \sqrt{\frac{V_i^2}{X}\left(Q_i - Q_j\right) - Q_i^2}. \qquad (10)$$

The line stability factor isobtained by setting the discriminant of the reactive power roots at bus 1 to be greater than or equal to zero, thus defining the line stability factor, LQP, as

$$\text{LQP} = 4\left(\frac{X}{V_i^2}\right)\left(\frac{X}{V_i^2}P_i^2 + Q_j\right). \qquad (11)$$

4. Problem Formulation

The objective function of this work is to find the optimal rating and location of FACTS devices combination which minimizes the real power loss, minimizes voltage deviation, and maximizes the voltage stability limit. Hence, the objective function can be expressed as

$$F = \min \left\{ P_L + w\text{VD} + (1 - w)\,\text{LQP} \right\}, \qquad (12)$$

where w is the weighing factor for voltage deviation and is set to 10.

4.1. Minimization of Real Power Loss (P_L). The total real power of the system can be calculated as follows:

$$P_{\text{loss}} = \sum_{k=1}^{N_L} G_k \left[V_i^2 + V_j^2 - 2\left|V_i\right|\left|V_j\right| \cos\left(\delta_i - \delta_j\right) \right], \qquad (13)$$

where N_L is the total number of lines in the system; G_k is the conductance of the line "k"; V_i and V_j are the magnitudes of the sending end and receiving end voltages of the line; and δ_i and δ_j are angles of the end voltages.

4.2. Minimization of Load Bus Voltage Deviation (VD). Bus voltage magnitude should be maintained within the allowable range to ensure quality service. Voltage profile is improved by minimizing the deviation of the load bus voltage from the reference value (it is taken as 1.0 p.u. in this work):

$$\text{VD} = \sum_{k=1}^{N_{PQ}} \left| V_I - V_{\text{ref}} \right|, \qquad (14)$$

where V_i is the voltage at ith bus and V_{ref} is the reference voltage.

4.3. Minimization of Line Quality Proximity Index (LQP). Voltage stability limit of a power system is increased by minimizing voltage stability index value. The indicator takes values between 0 (no load) and 1 (full load). The line based stability index (LQP) is given as

$$\text{LQP} = \sum_{j=1}^{N_L} \text{LQP}_j. \qquad (15)$$

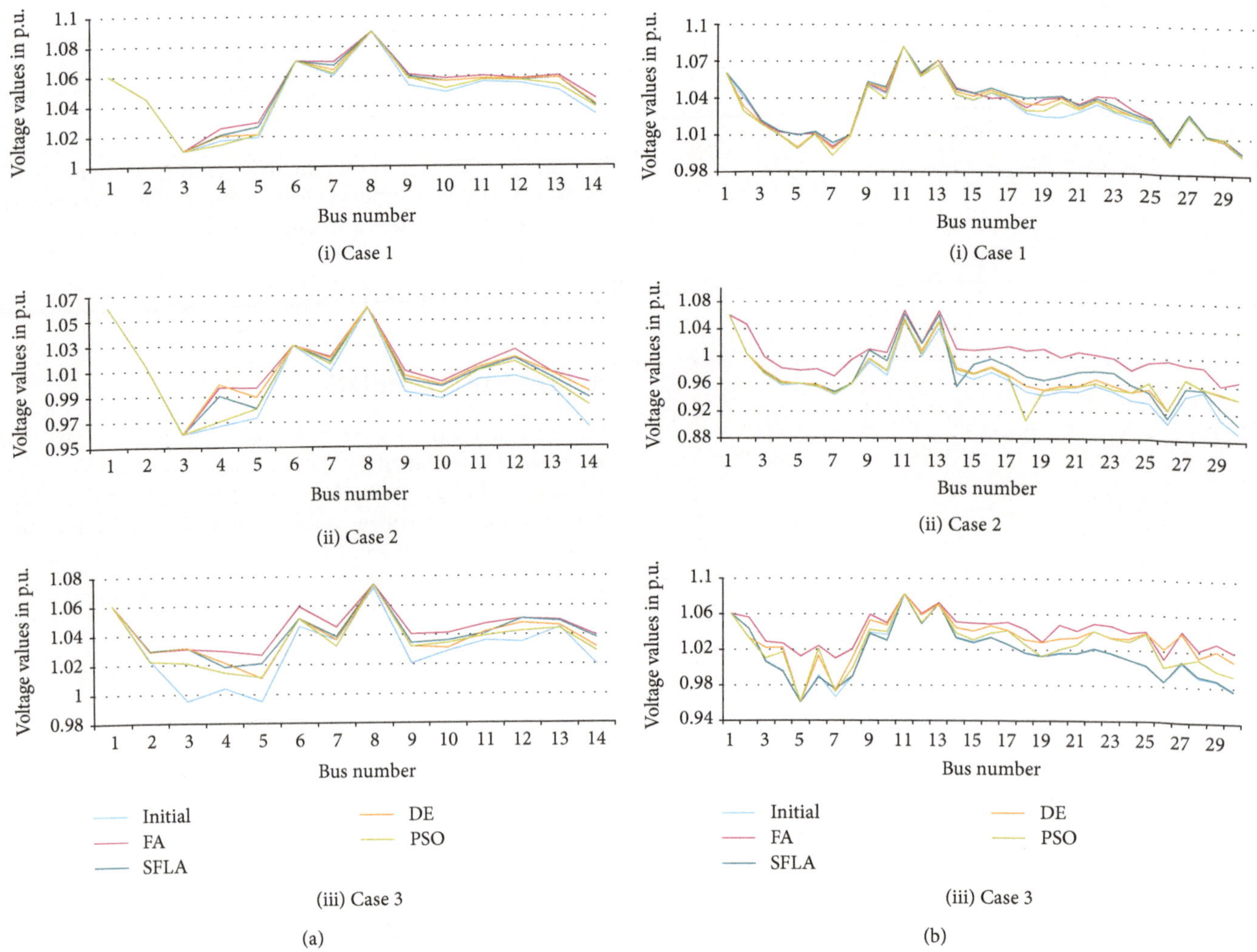

FIGURE 7: (a) Voltage profile values comparison for 14-bus system. (b) Voltage profile values comparison for 30-bus system.

4.4. Constraints. The minimization problem is subject to the following equality and inequality constraints.

4.4.1. Equality Constraints

Load Flow Constraints. Consider

$$P_{Gi} - P_{Di} - \sum_{j=1}^{N_B} V_i V_j Y_{ij} \cos\left(\delta_{ij} + \gamma_j - \gamma_i\right) = 0,$$
$$Q_{Gi} - Q_{Di} - \sum_{j=1}^{N_B} V_i V_j Y_{ij} \sin\left(\delta_{ij} + \gamma_j - \gamma_i\right) = 0, \quad (16)$$

where P_{Gi} *and* Q_{Gi} are the active and reactive powers of *i*th generator and P_{Di} and Q_{Di} are the active and reactive powers of *i*th load bus.

4.4.2. Inequality Constraints

Reactive Power Generation Limit of SVCs. By choosing small-sized compensator, capital cost involved can be minimum. The size limit of SVC has been taken from minimum of 5 MVAR to maximum of 20 MVAR:

$$Q_{SVCi}^{min} \le Q_{SVCi} \le Q_{SVCi}^{max}; \quad i \in N_{SVC}, \quad (17)$$

where Q_{SVCi}^{min} and Q_{SVCi}^{max} are the minimum and maximum VAR injection limits of *i*th shunt capacitor.

Voltage Constraints. The acceptable voltage limits in all load buses are taken from 0.95 p.u. to 1.05 p.u. to avoid voltage instability:

$$V_i^{min} \le V_i \le V_i^{max}; \quad i \in N_B, \quad (18)$$

where V_i^{min} and V_i^{max} are the minimum and maximum value voltage of bus "*i*".

Transmission Line Flow Limit. Consider

$$S_i \le S_i^{max}; \quad i \in N_l, \quad (19)$$

where S_i is the apparent power flow of *i*th branch and S_i^{max} is the maximum apparent power flow limit of *i*th branch.

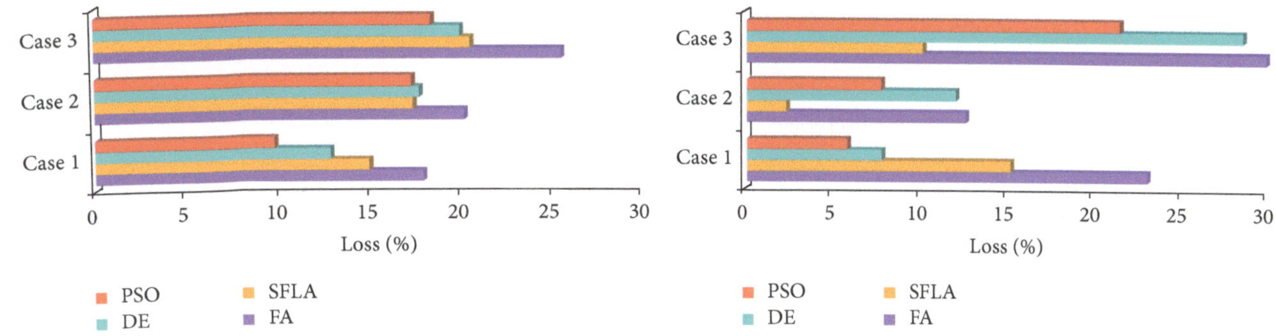

(a) For 14-bus system

(b) For 30-bus system

FIGURE 8: Percentage of real power loss reduction under all cases.

TABLE 1: Optimal values of FA parameters.

Parameters	Optimal Values
Number of fireflies	30
β_0	0.1
γ	0.2
α	0.25
m	4.0
d	2.0
Maximum iterations	100

5. Firefly Algorithm

5.1. Overview. The firefly algorithm (FA) is a metaheuristic, nature-inspired, optimization algorithm which is based on the social (flashing) behavior of fireflies, or lighting bugs, in the summer sky in the tropical temperature regions. It was developed by Dr. Xin-She Yang at Cambridge University in 2007, and it is based on the swarm behavior such as fish, insects, or bird schooling in nature. In particular, although the firefly algorithm has many similarities with other algorithms which are based on the so-called swarm intelligence, such as the famous particle swarm optimization (PSO), artificial bee colony optimization (ABC), and bacterial foraging algorithms (BFA), it is indeed much simpler in both concept and implementation. Furthermore, according to recent bibliography, the algorithm is very efficient and can outperform other conventional algorithms, such as genetic algorithms, for solving many optimization problems, a fact that has been justified in a recent research, where the statistical performance of the firefly algorithm was measured against other well-known optimization algorithms using various standard stochastic test functions. Its main advantage is the fact that it uses mainly real random numbers and it is based on the global communication among the swarming particles (i.e., the fireflies), and as a result, it seems more effective in multiobjective optimization such as the economic emissions load dispatch problem in our case.

The firefly algorithm has three particular idealized rules which are based on some of the major flashing characteristics of real fireflies. These rules are the following.

(i) All fireflies are unisex, and they will move towards more attractive and brighter ones regardless their sex.

(ii) The degree of attractiveness of a firefly is proportional to its brightness which decreases as the distance from the other firefly increases due to the fact that the air absorbs light. If there is no brighter or more attractive firefly than a particular one, it will then move randomly.

(iii) The brightness or light intensity of a firefly is determined by the value of the objective function of a given problem. For maximization problems, the light intensity is proportional to the value of the objective function.

5.2. Implementation

5.2.1. Attractiveness. In the firefly algorithm, the form of attractiveness functions of a firefly is the following monotonically decreasing function:

$$\beta(r) = \beta_0 * \exp(-\gamma r^m), \quad \text{with } m \geq 1, \tag{20}$$

where r is the distance between any two fireflies, β_0 is the initial attractiveness at $r = 0$, and γ is an absorption coefficient which controls the decrease of the light intensity.

5.2.2. Distance. The distance between any two fireflies i and j, at positions x_i and x_j, respectively, can be defined as a Cartesian or Euclidean distance as follows:

$$r_{ij} = \|x_i - x_j\| = \sqrt{\sum_{k=1}^{d} (x_{i,k} - x_{j,k})^2}, \tag{21}$$

where $x_{i,k}$ is the kth component of the spatial coordinate x_i of the ith firefly and d is the number of dimensions we have, for $d = 2$; we have

$$r_{ij} = \sqrt{(x_i - x_j)^2 + (y_i - y_j)^2}. \tag{22}$$

However, the calculation of distance r can also be defined using other distance metrics, based on the nature of the problem, such as Manhattan distance.

Table 2: Voltage stability index values of most stressed lines.

(a) For 14-bus system

Cases	Line number	Preinsertion of FACTS	Postinsertion of FACTS			
			FA	SFLA	DE	PSO
Case 1	14	0.1122	0.0714	0.0818	0.0807	0.0829
	20	0.0776	0.0582	0.0537	0.0544	0.0617
	1	0.0688	0.0447	0.0511	0.0491	0.0519
	17	0.0593	0.0310	0.0388	0.0408	0.0503
	9	0.0515	0.0291	0.0346	0.0307	0.0401
Case 2	10	0.2580	0.1106	0.1722	0.1651	0.1995
	14	0.1815	0.0994	0.0991	0.1012	0.1244
	20	0.1494	0.0912	0.0941	0.1109	0.1286
	9	0.1393	0.0833	0.0994	0.1081	0.1120
	1	0.0799	0.0517	0.0604	0.0597	0.0661
Case 3	10	0.1904	0.1014	0.0941	0.1057	0.1129
	15	0.1160	0.0809	0.0802	0.0914	0.0956
	6	0.0914	0.0529	0.0317	0.0441	0.0582
	4	0.0793	0.0447	0.0406	0.0533	0.0551
	9	0.0648	0.0361	0.0304	0.0369	0.0408

(b) For 30-bus system

Cases	Line number	Preinsertion of FACTS	Postinsertion of FACTS			
			FA	SFLA	DE	PSO
Case 1	12	0.3691	0.3664	0.3687	0.3700	0.3683
	5	0.1284	0.0695	0.1264	0.0987	0.1282
	13	0.1226	0.0722	0.0114	0.1169	0.1095
	15	0.0749	0.0713	0.0748	0.0752	0.0749
	14	0.0678	0.0667	0.0674	0.0676	0.0672
Case 2	12	0.3899	0.3861	0.3877	0.3874	0.3881
	5	0.3596	0.0884	0.3310	0.0793	0.3312
	13	0.2549	0.2115	0.2355	0.2351	0.2390
	16	0.1687	0.1241	0.1735	0.1767	0.1610
	15	0.1255	0.1167	0.1245	0.1244	0.1249
Case 3	12	0.3851	0.3688	0.3864	0.3686	0.3793
	13	0.1732	0.1159	0.1766	0.1152	0.1711
	16	0.0836	0.0445	0.0855	0.0475	0.0833
	1	0.0786	0.0540	0.0810	0.0568	0.0694
	15	0.0775	0.0741	0.0776	0.0749	0.0771

5.2.3. Movement. The movement of a firefly i which is attracted by a more attractive (i.e., brighter) firefly j is given by the following equation:

$$x_i = x_i + \beta_0 * \exp\left(-\gamma r_{ij}^2\right) * \left(x_i - x_j\right) + \alpha * \left(\text{rand} - \frac{1}{2}\right), \tag{23}$$

where the first term is the current position of a firefly, the second term is used for considering a firefly's attractiveness to light intensity seen by adjacent fireflies, and the third term issued for the random movement of a firefly in case there are no any brighter ones. The coefficient α is a randomization parameter determined by the problem of interest, while rand is a random number generator uniformly distributed in the space (0,1). As we will see in this implementation of the algorithm, we use $\beta_0 = 0.1$, $\alpha \in [0, 1]$, and the attractiveness or absorption coefficient $\gamma = 0.2$, which guarantees a quick convergence of the algorithm to the optimal solution. The optimal values of firefly parameters are in Table 1.

6. Results and Discussion

The proposed work is coded in MATLAB 7.6 platform using 2.8 GHz Intel Core 2 Duo processor based PC. The method is tested in the IEEE 14- and 30-bus test systems shown in Figure 5. The line data and bus data are taken from the standard power system test case archive. The 14-bus system has 5 generator buses, 9 load buses, and 20 transmission lines.

Table 3: Contingency ranking.

Rank	14-bus system		30-bus system	
	Line number	LQP values	Line number	LQP values
1	1	0.5795	5	0.9495
2	3	0.3826	9	0.6050
3	10	0.3449	2	0.4993
4	2	0.3157	4	0.4968
5	15	0.1984	7	0.4693

Table 4: Average load bus voltage values in p.u.

(a) For 14-bus system

Cases	Preinsertion of FACTS	Postinsertion of FACTS			
		FA	SFLA	DE	PSO
Case 1	1.0438	1.0517	1.0494	1.0483	1.0460
Case 2	0.9890	1.0082	1.0013	1.0051	0.9966
Case 3	1.0243	1.0409	1.0363	1.0331	1.0308

(b) For 30-bus system

Cases	Preinsertion of FACTS	Postinsertion of FACTS			
		FA	SFLA	DE	PSO
Case 1	1.0260	1.0296	1.0306	1.0288	1.0265
Case 2	0.9520	0.9961	0.9664	0.9626	0.9602
Case 3	1.0109	1.0380	1.0108	1.0298	1.0222

The 30-bus system has 6 generator buses, 24 load buses, and 41 transmission lines. System data and results are based on 100 MVA and bus number 1 is the reference bus. In order to verify the presented models and illustrate the impact of circuit element model of SSSC with variable susceptance model of SVC combination study, three different operating conditions are considered as mentioned below.

Case 1. The system with normal load in all the load buses is considered as normal condition and the Newton-Raphson load flow is carried out with loading factor value equal to 1.

Case 2. The system with 50% increased load in all the load buses is considered as a critical condition. Loading of the system went beyond this level and results in poor voltage profile in the load buses and unacceptable real power loss level.

Case 3. Contingency is imposed by considering the most critical line outage in the system. This is the most suitable condition for voltage stability analysis of a power system as voltage stability is usually triggered by line outages.

Newton-Raphson program is repeatedly run with the absence and presence of FACTS device combination. The voltage stability limit improvement is assessed by the value of LQP index. The voltage stability index values of top five stressed lines under all cases with preinsertion and postinsertion of FACTS devices are shown in Table 2. It is evident from the table that LQP values of the stressed lines are reduced after placement of FACTS devices in the system.

For quick assessment of voltage stability limit improvement of the system under the three different operating conditions, sum of the LQP index values of all the lines before and after the insertion of FACTS devices combination is compared in Figure 6. The reduction in the index value through the proposed method indicates the best voltage stability limit improvement.

The line outage is ranked according to the severity and the severity is taken on the basis of the line quality proximity index values and such values are arranged in descending order. The maximum value of index indicates most critical line for outage. Line outage contingency screening and ranking are carried out on the test systems and the results are shown in Table 3.

It is clear from Table 3 that outage of lines numbers 1 and 5 are the most critical line outages of 14- and 30-bus systems, respectively. This situation is considered for voltage stability improvement. Outage of other lines has not much impact on the system and therefore they are not given importance. Load flow is run on the system with the lines that have outages. Outage of these lines results in large real power loss and voltage profile reduction in most of the load buses. The system is under stressed conditions and needs to be relieved by some means. Installation of FACTS devices at suitable locations can relieve the system much from stressed conditions (reduced line losses).

FACTS devices help the system to maintain acceptable voltage profile in the load buses. Under normal operating conditions, most of the bus voltage magnitudes are within the normal value. During critical and contingency conditions, voltage magnitude of remote load buses are below 0.95 (lower

Table 5: Minimum-maximum load bus voltage values in p.u.

(a) For 14-bus system

Cases	V_{MIN} Preinsertion of FACTS	Bus number	Postinsertion of FACTS FA	SFLA	DE	PSO	V_{MAX} Preinsertion of FACTS	Bus number	Postinsertion of FACTS FA	SFLA	DE	PSO
Case 1	1.0173	4	1.0252	1.0211	1.0203	1.0143	1.0604	7	1.0696	1.0673	1.0642	1.0617
Case 2	0.9656	14	1.0004	0.9884	0.9933	0.9824	1.0101	7	1.0211	1.0176	1.0199	1.0155
Case 3	0.9913	5	1.0264	1.0208	1.0104	1.0111	1.0350	13	1.0496	1.0490	1.0460	1.0438

(b) For 30-bus system

Cases	V_{MIN} Preinsertion of FACTS	Bus number	Postinsertion of FACTS FA	SFLA	DE	PSO	V_{MAX} Preinsertion of FACTS	Bus number	Postinsertion of FACTS FA	SFLA	DE	PSO
Case 1	0.9953	30	0.9978	0.9976	0.9959	0.9954	1.0576	11	1.0820	1.0820	1.0820	1.0822
Case 2	0.8915	30	0.9662	0.9035	0.9415	0.9410	1.0004	11	1.0667	1.0620	1.0520	1.0550
Case 3	0.9661	5	1.0121	0.9600	0.9600	0.9607	1.0495	11	1.0820	1.0820	1.0820	1.0820

Table 6: Real power loss values comparison (in MW).

(a) For 14-bus system

Conditions	Preinsertion of FACTS (initial)	Postinsertion of FACTS FA	SFLA	DE	PSO
Case 1	13.401	11.014	11.409	11.693	12.106
Case 2	35.042	28.018	28.991	28.875	29.019
Case 3	24.183	18.046	19.255	19.389	19.771

(b) For 30-bus system

Conditions	Preinsertion of FACTS (Initial)	Postinsertion of FACTS FA	SFLA	DE	PSO
Case 1	17.514	13.467	14.852	16.157	16.504
Case 2	46.900	40.990	45.797	41.258	43.272
Case 3	32.569	22.801	29.272	23.234	25.561

bound of allowable value). These bus voltages are improved after the FACTS devices are installed. Voltage profile value comparison under all cases incorporating the series-shunt FACTS devices combination is depicted separately in Figures 7(a) and 7(b) under both test cases. It is obvious from the figure that voltage profile comparison of the system is improved better with the proposed method incorporating FACTS devices combination. The comparison of average value of load bus voltage also proves the enhancing behavior of voltage stability limit incorporating with FACTS devices of all cases shown in Table 4. Table 5 exposes the comparison of minimum and maximum load bus voltages in both the absence and presence of FACTS devices under all cases.

In real power loss minimization point of view through insertion of FACTS devices, the real power loss under Case 1 is decreased by 2.387 and 4.047 MW, respectively, for 14- and 30-bus systems. In Case 2, the reduction in real power loss rate is 7.024 and 5.910 MW, respectively. The reduction rate in real power loss is 6.137 and 9.768 MW, respectively, under Case 3 operation. The real power loss values of proposed method compared with other important evolutionary techniques under all cases are shown in Table 6.

The percentage of reduction in real power loss comparison under all cases including both combinations is depicted in Figure 8. Obviously from the results shown in Table 6, the proposed method is more efficient than other evolutionary techniques given here to reduce the real power loss minimization application. The much reduction in real power loss and increase in voltage magnitudes after the insertion of FACTS devices are highly efficient in relieving a power network from stressed condition and improving voltage stability limit. The most suitable size and location of combinations of FACTS devices to improve the voltage stability limit and real power loss minimization are also given in Table 7 under all cases with 14- and 30-bus systems, respectively. The optimized size of SVC inserted with both combinations is not in larger value which helps to reduce in cost.

Table 7: Best location and size of FACTS devices.

(a) For 14-bus system

Cases		Series		Shunt	
		Location (line number)	Size (degrees of compensation)	Location (bus number)	Size (MVar)
Case 1	FA	10	$0.0722 + j0.1464$	5	7.2015
	SFLA	4	$0.0377 + j0.0916$	12	8.6554
	DE	15	$0.0461 + j0.1540$	10	6.9988
	PSO	3	$0.0622 + j0.1067$	12	8.3037
Case 2	FA	12	$0.0823 + j0.0961$	13	11.211
	SFLA	1	$0.0536 + j0.1642$	11	10.6217
	DE	9	$0.0507 + j0.1145$	12	9.9654
	PSO	11	$0.0727 + j0.0836$	10	7.2510
Case 3	FA	12	$0.0769 + j0.0810$	13	9.6817
	SFLA	10	$0.0421 + j0.1162$	5	10.6528
	DE	2	$0.0489 + j0.0773$	11	8.9207
	PSO	7	$0.0673 + j0.1286$	10	11.4368

(b) For 30-bus system

Cases		Series		Shunt	
		Location (line number)	Size (degrees of compensation)	Location (bus number)	Size (MVar)
Case 1	FA	5	$0.0159 + j0.0924$	12	8.6962
	SFLA	1	$0.0086 + j0.0833$	19	6.9688
	DE	6	$0.0239 + j0.2664$	20	5.8131
	PSO	12	$0.0191 + j0.0977$	23	15.0806
Case 2	FA	6	$0.0137 + j0.0384$	17	9.0607
	SFLA	9	$0.0170 + j0.1243$	21	14.1058
	DE	5	$0.0289 + j0.0931$	30	6.2554
	PSO	8	$0.0281 + j0.0447$	18	19.0106
Case 3	FA	11	$0.0311 + j0.1417$	26	8.0115
	SFLA	9	$0.0115 + j0.1302$	28	6.7499
	DE	6	$0.0112 + j0.0478$	25	5.8559
	PSO	8	$0.0291 + j0.1189$	3	18.2640

7. Conclusion

In this paper, optimal location and size of series-shunt FACTS devices for voltage stability limit improvement and loss minimization through firefly algorithm are demonstrated. The voltage stability limit improvement and real power loss minimization are done under three operating cases such as normal, critical loading, and line outage contingency conditions. The LQP index is used for voltage stability assessment. The circuit element model of SSSC is considered to improve the voltage stability limit by controlling power flows and maintaining voltage profile. This model is easy to incorporate the effect of SSSC into Newton-Raphson load flow program coding. The variable susceptance model of SVC is used in a combined manner with SSSC. It is clear from the numerical results that voltage stability limit improvement and real power loss minimization are highly encouraging. The real power loss minimization and voltage stability limit improvement is remarkable by the combined action of power flow control of SSSC and reactive power compensation by SVC through firefly algorithm over other evolutionary optimization techniques given in this paper.

Conflict of Interests

The authors declare that there is no conflict of interests regarding the publication of this paper.

References

[1] T. V. Cutsem and C. Vournas, *Voltage Stability of Electric Power Systems*, Kluwer Academic, New York, NY, USA, 1998.

[2] P. Kundur, *Power System Stability and Control*, McGraw-Hill, New York, NY, USA, 1994.

[3] C. W. Taylor, *Power System Voltage Stability*, McGraw-Hill, New York, NY, USA, 1994.

[4] T. van Cutsem, "Voltage instability: phenomena, countermeasures, and analysis methods," *Proceedings of the IEEE*, vol. 88, no. 2, pp. 208–227, 2000.

[5] P. Kessel and H. Glavitsch, "Estimating the voltage stability of power system," *IEEE Transactions on Power Delivery*, vol. 1, no. 3, pp. 346–354, 1986.

[6] H. J. Wan, J. D. McCalley, and V. Vittal, "Risk based voltage security assessment," *IEEE Transactions on Power Systems*, vol. 15, no. 4, pp. 1247–1254, 2000.

[7] I. Dobson and H.-D. Chiang, "Towards a theory of voltage collapse in electric power systems," *Systems and Control Letters*, vol. 13, no. 3, pp. 253–262, 1989.

[8] "Technical Analysis of the August 14, 2003, Blackout: What Happened, Why, and What Did We Learn?" A report by the North American Electrical Reliability Council Steering Group, New Jersey, July 2004.

[9] N. G. Hingorani and L. Gyugyi, *Understanding FACTS: Concepts and Technology of Flexible AC Transmission Systems*, IEEE Press, 2000.

[10] K. K. Sen and M. L. Sen, *Introduction to FACTS Controllers Theory, Modeling, and Applications*, IEEE Press, 2009.

[11] G. Wu, A. Okoyama, J. He, and Y. Yu, "Allocation and control of FACTS devices for steady state stability enhancement of large scale power system," *International Conference on Power System Technology*, vol. 1, pp. 357–361, 1998.

[12] N. Yorino, E. E. El-Araby, H. Sasaki, and S. Harada, "A new formulation for FACTS allocation for security enhancement against voltage collapse," *IEEE Transactions on Power Systems*, vol. 18, no. 1, pp. 3–10, 2003.

[13] S. Biansoongnern, S. Chusanapiputt, and S. Phoomvuthisarn, "Optimal SVC and TCSC placement for minimization of transmission losses," in *Proceedings of the International Conference on Power System Technology*, pp. 1–5, October 2006.

[14] M. A. Abido, "Power system stability enhancement using FACTS controllers: a review," *The Arabian Journal For Science and Engineering*, vol. 34, no. 1, pp. 153–172, 2009.

[15] O. L. Bekri, "Optimal location of SVC and TCSC for voltage stability enhancement," in *Proceedings of the 4th International Power Engineering and Optimization Conference (PEOCO '10)*, pp. 7–12, June 2010.

[16] Y. Zhang and Y. Zhang, "A novel power injection model of embedded SSSC with multi-control modes for power flow analysis inclusive of practical constraints," *Electric Power Systems Research*, vol. 76, no. 5, pp. 374–381, 2006.

[17] A. Sode-Yome, N. Mithulananthan, and K. Y. Lee, "Static voltage stability margin enhancement using STATCOM, TCSC and SSSC," in *Proceedings of the IEEE/PES Transmission and Distribution Conference and Exhibition*, pp. 1–6, August 2005.

[18] C. A. Canizares and Z. T. Faur, "Analysis of SVC and TCSC controllers in voltage collapse," *IEEE Transactions on Power Systems*, vol. 14, no. 1, pp. 158–165, 1999.

[19] S. Musunuri and G. Dehnavi, "Comparison of STAT COM, SVC, TCSC, and SSSC performance in steady state voltage stability improvement," in *Proceedings of the North American Power Symposium (NAPS '10)*, pp. 1–7, September 2010.

[20] P. Kundur, K. Morison, and B. Gao, "Practical considerations in voltage stability assessment," *International Journal of Electrical Power and Energy Systems*, vol. 15, no. 4, pp. 205–215, 1993.

[21] V. Ajjarapu, *Computational Techniques for Voltage Stability Assessment and Control*, Springer Science, 2006.

[22] S. Fliscounakis, P. Panciatici, F. Capitanescu, and L. Wehenkel, "Contingency ranking with respect to overloads in large power systems taking into account uncertainity, preventive and corrective actions," *IEEE Transactions on Power Systems*, vol. 28, no. 4, pp. 4909–4917, 2013.

[23] A. Y. Abdelaziz, A. T. M. Taha, M. A. Mostafa, and A. M. Hassan, "Power systen contingency ranking using fuzzy logic based approach," *International Journal of Intelligent Systems and Applications*, no. 2, pp. 24–28, 2013.

[24] P. S. Venkataramu and T. Ananthapadmanabha, "Installation of unified power flow controller for voltage stability margin enhancement under line outage contingencies," *Iranian Journal of Electrical and Computer Engineering*, vol. 5, no. 2, pp. 90–96, 2006.

[25] M. Ja'fari and S. Afsharnia, "Voltage stability enhancement in contingency conditions using shunt FACTS devices," in *Proceedings of the EUROCON: The International Conference on Computer as a Tool*, pp. 1660–1665, September 2007.

[26] P.-A. Loef, G. Andersson, and D. J. Hill, "Voltage stability indices for stressed power systems," *IEEE Transactions on Power Systems*, vol. 8, no. 1, pp. 326–335, 1993.

[27] M. Moghavvemi and F. M. Omar, "Technique for contingency monitoring and voltage collapse prediction," *IEE Proceedings on Generation, Transmission and Distribution*, vol. 145, no. 6, pp. 634–640, 1998.

[28] M. Moghavvemi and O. Faruque, "Real time contingency evaluation and ranking technique," *IEE Proceedings on Generation, Transmission and Distribution*, vol. 145, no. 5, pp. 517–524, 1998.

[29] I. Musirin and T. K. A. Rahman, "Novel fast voltage stability index (FVSI) for voltage stability analysis in power transmission system," in *Proceedings of the Student Conference on Research and Development Proceedings*, pp. 265–268, 2002.

[30] C. Reis, A. Andrade, and F. P. Maciel, "Line stability indices for voltage collapse prediction," in *Proceedings of the 2nd IEEE International Conference on Power Engineering, Energy and Electrical Drives*, pp. 239–243, March 2009.

[31] X. S. Yang, *Nature-Inspired Metaheuristic Algorithms*, Luniver Press, 2008.

[32] X. S. Yang, "Firefly algorithms for multimodal optimization," in *Stochastic Algorithms: Foundations and Applications, SAGA*, vol. 5792 of *Lecture Notes in Computer Sciences*, pp. 169–178, 2009.

[33] S. Łukasik and S. Żak, "Firefly algorithm for continuous constrained optimization task, computational collective intelligence," in *Semantic Web, Social Networks and Multi-Agent Systems*, vol. 5796 of *Lecture Notes in Computer Science*, pp. 97–106, 2009.

[34] N. D. Reppen, R. R. Austria, J. A. Uhrin, M. C. Patel, and A. Galatic, "Performance of methods for ranking a evaluation of voltage collapse contingencies applied to a large-scale network," in *Proceedings of the IEEE Joint International Conference*, pp. 337–343, 1993.

[35] G. C. Ejebe, G. D. Irisarri, S. Mokhtari, O. Obadina, P. Ristanovic, and J. Tong, "Methods for contingency screening and ranking for voltage stability analysis of power systems," in *Proceedings of the IEEE Power Industry Computer Application Conference*, pp. 249–255, May 1995.

[36] E. Vaahedi, C. Fuchs, W. Xu et al., "Voltage stability contingency screening and ranking," *IEEE Transactions on Power Systems*, vol. 14, no. 1, pp. 256–265, 1999.

9

Filter Bank Multicarrier Modulation: A Waveform Candidate for 5G and Beyond

Behrouz Farhang-Boroujeny

ECE Department, University of Utah, Salt Lake City, UT 84112, USA

Correspondence should be addressed to Behrouz Farhang-Boroujeny; farhang@ece.utah.edu

Academic Editor: Gorazd Stumberger

Recent discussions on viable technologies for 5G emphasize on the need for waveforms with better spectral containment per subcarrier than the celebrated orthogonal frequency division multiplexing (OFDM). Filter bank multicarrier (FBMC) is an alternative technology that can serve this need. Subcarrier waveforms are built based on a prototype filter that is designed with this emphasis in mind. This paper presents a broad review of the research work done in the wireless laboratory of the University of Utah in the past 15 years. It also relates this research to the works done by other researchers. The theoretical basis based on which FBMC waveforms are constructed is discussed. Also, various methods of designing effective prototype filters are presented. For completeness, polyphase structures that are used for computationally efficient implementation of FBMC systems are introduced and their complexity is contrasted with that of OFDM. The problems of channel equalization as well as synchronization and tracking methods in FBMC systems are given a special consideration and a few outstanding research problems are identified. Moreover, this paper brings up a number of appealing features of FBMC waveforms that make them an ideal choice in the emerging areas of multiuser and massive MIMO networks.

1. Introduction

In the past, orthogonal frequency division multiplexing (OFDM) has enjoyed its dominance as the most popular signaling method in broadband wired [1, 2] and wireless [3, 4] channels. OFDM has been adopted in the broad class of DSL standards as well as in the majority of wireless standards, for example, variations of IEEE 802.11 and IEEE 802.16, 3GPP-LTE, and LTE-Advanced. OFDM is known to be a perfect choice for point-to-point communications, for example, from a base station to a mobile node and vice versa. It offers a minimum complexity and achieves very high bandwidth efficiency. However, it has been noted that OFDM has to face many challenges when considered for adoption in more complex networks. For instance, the use of OFDM in the uplink of multiuser networks, known as OFDMA (orthogonal frequency division multiple access), requires full synchronization of the users' signals at the base station input. Such synchronization was found to be very difficult to establish, especially in mobile environments where Doppler shifts of different users are hard to predict/track. Morelli et al. [5] have noted that carrier and timing synchronization represents the most challenging task in OFDMA systems. To combat the problem, some researchers have relaxed on the need for a close to perfect carrier synchronization among users and have proposed multiuser interference cancellation methods [6–9]. These methods are generally very complex to implement. Their implementation increases the receiver complexity by orders of magnitude [10]. Hence, one of the main advantages of OFDM, the low complexity, will be lost.

Another limitation of OFDM appears when attempt is made to transmit over a set of noncontiguous frequency bands, known as *carrier aggregation*. The poor response of the subcarrier filters in IFFT/FFT filter banks of OFDM introduces significant out-of-band *egress noise* to other users and also picks up significant *ingress noise* from them. The same problem appears if one attempts to adopt OFDM for filling in the spectrum holes in cognitive radios. Methods of reducing OFDM spectral leakage prove to be very limited in performance and may add significant complexity to the transmitter. For instance, the side lobe suppression techniques, like those proposed in [11–13], can achieve an out-of-band emission suppression of only 5 to 10 dB, while they may add significant complexity to the transmitter and they will incur some loss in bandwidth efficiency.

Filter bank multicarrier (FBMC) is an alternative transmission method that resolves the above problems by using high quality filters that avoid both ingress and egress noises. Also, because of the very low out-of-band emission of subcarrier filters, application of FBMC in the uplink of multiuser networks is trivial [14, 15]. It can be deployed without synchronization of mobile user nodes signals. In the application of cognitive radios, the filter bank that is used for multicarrier data transmission can also be used for spectrum sensing [16–20]. On the other hand, compared to OFDM, FBMC falls short in handing multiple-input multiple-output (MIMO) channels, although a few solutions to adopt FBMC in MIMO channels have been reported in the literature; for example, see [21–23]. Nevertheless, as our recent research study has shown (see Section 8, below), in the emerging area of *massive* MIMO, FBMC is found as powerful as OFDM and in some cases superior to OFDM.

In the past, many attempts have been made to adopt FBMC in various standards. Apparently, the earliest proposal to use FBMC for multicarrier communications is a contribution from Tzannes et al. of AWARE Inc., in one of the asymmetric digital subscriber lines (ADSL) standard meetings in 1993 [24]. The proposed method that was called discrete wavelet multitone (DWMT) was further studied in [25, 26]. Despite enthusiasm from the research community (see [27] and the cited references therein), DWMT was not adopted in the ADSL standard. This was partly because of the perceived complexity of this method as compared to its rival, DMT (discrete multitone, an equivalent name for OFDM in DSL literature). Indeed, the DWMT structure proposed by Tzannes et al. was significantly more complex than that of DMT. The major part of the complexity of DWMT came from the equalization method that was adopted. The detailed discussion presented in [26] assumed that one needs an equalizer that combines signals from each subcarrier band and its adjacent bands. Typical equalizer lengths suggested in [26] were 21 real-valued taps per subcarrier. It was later noted by the author of this paper that if each subcarrier is sufficiently narrow such that it can be approximated by a flat gain, two real taps per subcarrier would be sufficient for equalization [27]. This observation led to further study of DWMT. In power line communications (PLC) community, it has been named wavelet OFDM and was adopted in the IEEE P1901 standard [28]. The main motivation for use of DWMT in DSL and its adoption later in PLC was to deal with ingress and egress noises, since both DSL and PLC use unshielded copper lines that are subject to strong radio interference. Moreover, in 1999, an FBMC method with nonoverlapping subcarrier bands was proposed as a solution for filtering the narrow-band interferences in very high-speed DSL (VDSL) channels [29]. The proposed method was called filtered multitone (FMT). This proposal that was included as an annex in one of the initial draft documents of VDSL [30] was further developed by a number of researchers [31–34]. However, to avoid incompatibility with ADSL, FMT was not included in the final document of the VDSL standard [35]. Another unsuccessful story is an attempt by France Telecom [36] to introduce FBMC in the IEEE 802.22, a cognitive radio standard to access TV bands in wireless rural area networks (WRAN). Up to now, apparently, the only standard for radio transmission that uses FBMC is the TIAs Digital Radio Technical Standard [37].

Recent discussions on the fifth generation (5G) wireless communications have initiated a much stronger wave of interest in deviating from the main stream of OFDM systems. This shift of interest is clearly due to limitations of OFDM in the more dynamic and multiuser networks of future. A number of proposals have been made to adopt new waveforms with improved spectral containment. A good example of such activity is the 5GNOW project in Europe which challenges LTE and LTE-Advanced in coping with the dynamic needs of 5G. The 5GNOW has identified four alternative choices of waveforms to better serve 5G needs. These waveforms that are all built based on some sort of filtering may be thought as adoptions of FBMC method to suit different needs of various applications. We refer interested readers to the documents available at http://www.5gnow.eu for the details of the proposed waveforms by the 5GNOW group. Another major activity that has performed a broad study of FBMC is due to the PYHDYAS project (also in Europe). The PHYDYAS contributors have published heavily on various aspects of FBMC, including prototype filter design, equalization, synchronization, and application to MIMO channels. A summary of the major findings of the PHYDYAS contributors is presented in [21], and a complete list of their publications can be found at http://www.ict-phydyas.org.

The main thrust of this paper is to present a point of view of FBMC and its future applications as seen by the author. While we acknowledge the presence of a large body of works on FBMC, the paper details are geared towards the research outcomes of the author and his students in past 15 years. The paper emphasis is on the recent works of the author and his students. Many shortcomings of OFDM in dealing with the requirement of the next generation of wireless systems are discussed and it is shown how FBMC overcomes these problems straightforwardly. We present a derivation of FBMC systems that reveals the relationships among different forms of FBMC. A method of designing FBMC systems for a near-optimum performance in doubly dispersive channels is presented and its superior performance over OFDM is shown. The example considered is an underwater acoustic channel. Application of FBMC technique to massive MIMO communications is introduced and its advantages in this emerging technology are revealed. Last, but not the least, the problems of channel equalization and synchronization in FBMC systems are also given a special treatment and a number of outstanding research problems in this field, for future studies, are identified.

This paper begins with a historical overview of FBMC methods in Section 2. In order to keep the presentation in this paper a complement to those we have recently reported in [19, 38], the rest of this paper is organized as follows. A summary of the theoretical background based on which the various forms of FBMC are built are presented in Section 3. Methods of designing prototype filters for FBMC are presented in Section 4. Polyphase structures that are used for efficient implementation of FBMC systems are reviewed in Section 5. A few comments on complexity comparison of FBMC and OFDM are also made in this section. Channel equalization in FBMC systems is discussed in Section 6. Methods of carrier and symbol timing acquisition and tracking in FBMC systems are reviewed in Section 7. Here, a few particular features of FBMC systems that need special attention for their successful implementation are highlighted. Section 8 reminds the reader of a number of applications in the literature where FBMC has been found to be a good fit. This section also discusses the opportunities offered by FBMC in the emerging area of massive MIMO. The concluding remarks of the paper are made in Section 9.

Notations. The presentations in this paper follow a mix of continuous-time and discrete-time formulations, as appropriate. While $x(t)$ refers to a continuous function of time, t, $x[n]$ is used to refer to its discrete-time version, with n denoting the time index. The notation f is used as frequency variable in $X(f)$, the Fourier transform of the continuous-time signal $x(t)$, and also as normalized frequency in $X(e^{j2\pi f})$, the Fourier transform of the discrete-time signal $x[n]$. We use $\star$ to denote linear convolution, and the superscript $*$ to denote complex conjugate.

2. Review of FBMC Methods

FBMC communication techniques were first developed in the mid-1960s. Chang [39] presented the conditions required for signaling a parallel set of PAM symbol sequences through a bank of overlapping filters within a minimum bandwidth. To transmit PAM symbols in a bandwidth-efficient manner, Chang proposed vestigial sideband (VSB) signaling for subcarrier sequences. Saltzberg [40] extended the idea and showed how Chang's method could be modified for transmission of QAM symbols in a double-sideband- (DSB-) modulated format. In order to keep the bandwidth efficiency of this method similar to that of Chang's signaling, Saltzberg noted that the in-phase and quadrature components of each QAM symbol should be time staggered by half a symbol interval. Efficient digital implementation of Saltzberg's multicarrier system through polyphase structures was first introduced by Bellanger and Daguet [41] and later studied by Hirosaki [42, 43]. Another key development appeared in [44], where the authors noted that Chang's/Saltzberg's method could be adopted to match channel variations in doubly dispersive channels and, hence, minimize intersymbol interference (ISI) and intercarrier interference (ICI).

Saltzberg's method has received a broad attention in the literature and has been given different names. Most authors have used the name offset QAM (OQAM) to reflect the fact that the in-phase and quadrature components are transmitted with a time offset with respect to each other. Moreover, to emphasize the multicarrier feature of the method, the suffix OFDM has been added, hence, the name OQAM-OFDM. Others have chosen to call it staggered QAM (SQAM), equivalently SQAM-OFDM. In [38] we introduced the shorter name staggered multitone (SMT).

Chang's method [39], on the other hand, has received very limited attention. Those who have cited [39] have only acknowledged its existence without presenting much detail, for example [41, 45, 46]. In particular, Hirosaki who has extensively studied and developed digital structures for the implementation of Saltzberg's method [43, 47] has made a brief reference to Chang's method and noted that since it uses VSB modulation, and thus its implementation requires a Hilbert transformation, it is more complex to implement than Saltzberg's method. This statement is inaccurate, since as we have demonstrated in [38] Chang's and Saltzberg's methods are equivalent and, thus, with a minor modification, an implementation for one can be applied to the other. More on this is presented in Section 5. Also, as noted earlier, a vast literature in digital signal processing has studied a class of multicarrier systems that has been referred to as DWMT. It was later noted in [27, 48] that DWMT uses the same analysis and synthesis filter banks as the cosine modulated filter banks (CMFB) [49]. CMFB, on the other hand, may be viewed as a reinvention of Chang's method, with a very different application in mind [38]. In [38], we also introduced the shorter name cosine modulated multitone (CMT) to be replaced for DWMT and/or CMFB.

One more interesting observation is that another class of filter banks which were called modified DFT (MDFT) filter bank has appeared in the literature [50]. Careful study of MDFT reveals that this, although derived independently, is in effect a reformulation of Saltzberg's filter bank in discrete-time and with emphasis on compression/coding. The literature on MDFT begins with the pioneering works of Fliege [51] and later has been extended by others, for example [52–55].

Finally, before we proceed with the rest of our presentation, it should be reiterated that we identified three types of FBMC systems: (i) CMT: built based on the original idea of Chang [39]; (ii) SMT: built based on the extension made by Saltzberg [40]; and (iii) FMT: built based on the conventional method frequency division multiplexing (FDM) [34].

3. Theory

The theory of FBMC, particularly those of CMT and SMT, has evolved over the past five decades by many researchers who have studied them from different angles. Early studies by Chang [39] and Saltzberg [40] have presented their finding in terms of continuous-time signals. The more recent studies have presented the formulations and conditions for ISI and ICI cancellation in discrete-time, for example, [46, 50]. On the other hand, a couple of recent works [19, 38], from the author of this paper and his group, have revisited the more classical approach and presented the theory of CMT and SMT in continuous time. It is believed that this formulation greatly simplifies the essence of the theoretical concepts behind the theory of CMT and SMT and how these two waveforms are related. It also facilitates the design of prototype filters that are used for realization of CMT and SMT systems. Thus, here, also, we follow the continuous-time approach of [19, 38].

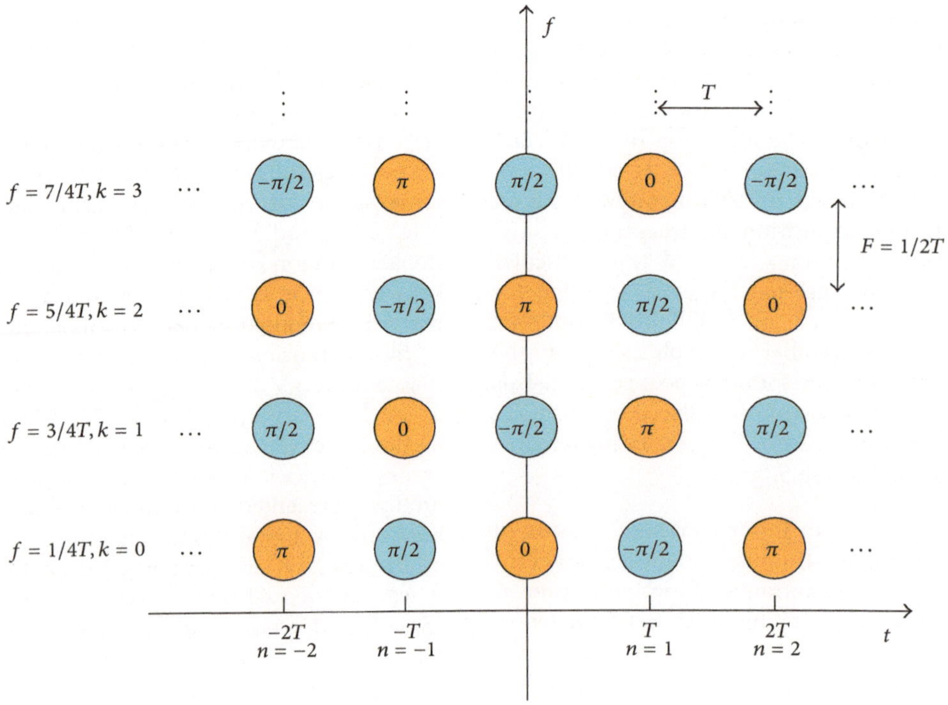

FIGURE 1: The CMT time-frequency phase-space lattice.

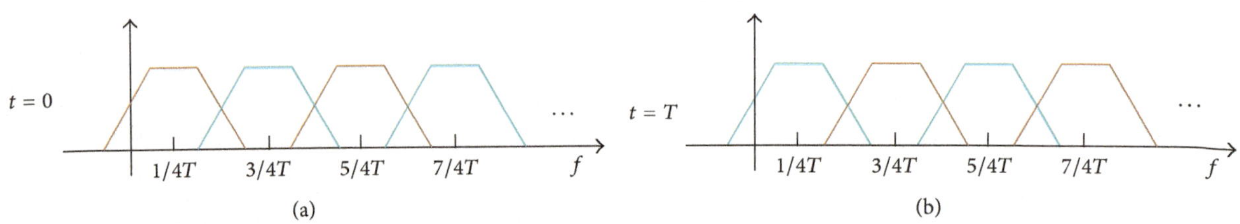

FIGURE 2: Magnitude responses of the CMT pulse-shaping filters at various subcarriers and time instants $t = 0$ and $t = T$.

Moreover, to give a complement presentation to those of [19, 38], an attempt is made to discuss the underlying theory mostly through the *time-frequency phase-space*, with minimum involvement in mathematical details. It is believed that this presentation also provides a new intuition into the relationship between SMT and CMT. Interested readers who wish to see the mathematical details are referred to [20, 38].

3.1. CMT. In CMT, data symbols are from a pulse amplitude modulated (PAM) alphabet and, hence, are real-valued. To establish a transmission with the maximum bandwidth efficiency, PAM symbols are distributed in a time-frequency phase-space lattice with a density of two symbols per unit area. This is equivalent to one complex symbol per unit area. Moreover, because of the reasons that are explained below a 90-degree phase shift is introduced to the respective carriers among the adjacent symbols. These concepts are presented in Figure 1. Vestigial side-band (VSB) modulation is applied to cope with the carrier spacing $F = 1/2T$. The pulse-shape used for this purpose at the transmitter as well as for matched filtering at the receiver is a square-root Nyquist waveform, $p(t)$, which has been designed such that $q(t) = p(t) \star p(-t)$ is a Nyquist pulse with regular zero crossings at $2T$ time intervals. Also, $p(t)$, by design, is a real-valued function of time and $p(t) = p(-t)$. These properties of $p(t)$, as demonstrated below, are instrumental for correct functionality of CMT.

Figure 2 presents a set of magnitude responses of the modulated versions of the pulse-shape $p(t)$ for the data symbols transmitted at $t = 0$ and $t = T$. The colors used for the plots follow those in Figure 1 to reflect the respective phase shifts.

Let $n = \ldots, -2, -1, 0, 1, 2, \ldots$, denote symbol time index, let $k = 0, 1, 2, \ldots$, denote symbol frequency index, let $s_k[n]$ denote the (n, k) data symbol in the time-frequency lattice, and let $\theta_k[n] = (k - n)(\pi/2)$ be the phase shift that is added to the carrier of $s_k[n]$. Accordingly, a CMT waveform that is constructed based on the pulse-shape/prototype filter $p(t)$ is expressed as

$$x(t) = \sum_n \sum_k s_k[n] \, a_{n,k}(t), \tag{1}$$

where

$$a_{n,k}(t) = e^{j\theta_k[n]} p_{n,k}(t), \qquad p_{n,k}(t) = p(t - nT) \, e^{j((2k+1)\pi/2T)t}. \tag{2}$$

The synthesis of $x(t)$ according to (1) has the following interpretations. The terms $a_{n,k}(t)$ may be thought as a set of basis functions that are used to modulate the data symbols $s_k[n]$. The data symbols $s_k[n]$ can be extracted from $x(t)$ straightforwardly if $a_{n,k}(t)$ are a set of orthogonal basis functions. The orthogonality for a pair of functions $v_1(t)$ and $v_2(t)$, in general, is defined as

$$\langle v_1(t), v_2(t)\rangle = \int_{-\infty}^{\infty} v_1(t)\, v_2^*(t)\, dt = 0. \tag{3}$$

For the case of interest here, where the data symbols $s_k[n]$ are real-valued, the orthogonality definition (3) can be replaced by the more relaxed definition:

$$\langle v_1(t), v_2(t)\rangle_R = \Re\left\{\int_{-\infty}^{\infty} v_1(t)\, v_2^*(t)\, dt\right\} = 0, \tag{4}$$

where $\Re\{\cdot\}$ indicates the real part. Definition (4) is referred to as *real orthogonality*.

It is not difficult to show that

$$\langle a_{n,k}(t), a_{m,l}(t)\rangle_R = \begin{cases} 1, & n = m,\ k = l \\ 0, & \text{otherwise} \end{cases} \tag{5}$$

and, hence, for any pair of n and k,

$$s_k[n] = \langle x(t), a_{n,k}(t)\rangle_R. \tag{6}$$

To develop an in-depth understanding of the CMT signaling, it is instructive to explore a detailed derivation of (5). To this end, we begin with the definition

$$\langle a_{n,k}(t), a_{m,l}(t)\rangle_R = \Re\left\{\int_{-\infty}^{\infty} e^{j\theta_k[n]} p_{n,k}(t)\, e^{-j\theta_l[m]} p_{m,l}^*(t)\, dt\right\} \tag{7}$$

and note that this can be rearranged as

$$\langle a_{n,k}(t), a_{m,l}(t)\rangle_R = \Re\left\{\int_{-\infty}^{\infty} e^{j(m-n+k-l)(\pi/2)} p(t-nT) \times p(t-mT)\, e^{j((k-l)\pi/T)t} dt\right\}. \tag{8}$$

When $m = n$ and $k = l$, after a change of variable t to $t+nT$, (8) reduces to

$$\langle a_{n,k}(t), a_{n,k}(t)\rangle_R = \int_{-\infty}^{\infty} p^2(t)\, dt = 1, \tag{9}$$

where the second equality follows from the fact that $p(t)$ is a real-valued square-root Nyquist pulse and $p(t) = p(-t)$.

When $k = l$, $m \neq n$, and $m-n = 2r$, where r is an integer,

$$\langle a_{n,k}(t), a_{n,k}(t)\rangle_R = (-1)^r \int_{-\infty}^{\infty} p(t-nT)\, p(t-mT)\, dt = 0, \tag{10}$$

where the second equality follows since $p(t)$ is a square-root Nyquist pulse, designed for a symbol spacing $2T$. On the other hand, when $k = l$, but $m-n = 2r+1$,

$$\langle a_{n,k}(t), a_{n,k}(t)\rangle_R = \Re\left\{j(-1)^r \int_{-\infty}^{\infty} p(t-nT)\, p(t-mT)\, dt\right\} = 0. \tag{11}$$

Next, consider the case where $k-l = 1$ and $m-n = 2r$. In that case, one finds that

$$\begin{aligned}\langle a_{n,k}(t), a_{m,l}(t)\rangle_R &= \Re\left\{j(-1)^r \int_{-\infty}^{\infty} p(t-nT)\, p(t-mT)\, e^{j(\pi/T)t} dt\right\} \\ &= -(-1)^r \int_{-\infty}^{\infty} p(t-nT)\, p(t-mT) \sin\left(\frac{\pi}{T}t\right) dt \\ &= 0,\end{aligned} \tag{12}$$

where the last identity follows, by applying the change of variable $t \to t + ((m+n)/2)T$ and noting that the expression under the integral will reduce to an odd function of t. Following similar procedures, it can be shown that the real orthogonality $\langle a_{n,k}(t), a_{m,l}(t)\rangle_R = 0$ also holds, when $k-l = 1$ and $m-n$ is an odd integer, and when $k-l = -1$ and $m-n$ is either an even or odd integer. Finally, for the cases where $|k-l| > 1$, the real orthogonality $\langle a_{n,k}(t), a_{m,l}(t)\rangle_R = 0$ is trivially confirmed by noting that the underlying basis functions correspond to filters that have no overlapping bands. The stop-band quality of the frequency response of the prototype filter $p(t)$ determines the accuracy of the equality $\langle a_{n,k}(t), a_{m,l}(t)\rangle_R = 0$ when $|k-l| > 1$.

It is also worth noting that some of the recent derivations of FBMC that are presented in discrete-time design the respective prototype filter $p[n]$ such that the respective real orthogonality is perfectly satisfied for the cases where $|k-l| > 1$ as well, for example, [56]. However, one may realize that, in practice, the presence of channel destroys the orthogonality of the basis functions. The orthogonality of the basis functions is commonly recovered at the receiver using per subcarrier equalizers; see Section 6, below. Such equalizers, unfortunately, will not be able to recover the orthogonality of basis functions that belong to nonadjacent subcarriers. Hence, the design of a $p[n]$ that satisfies the real orthogonality for $|k-l| > 1$ may be a waste.

To summarize, the above derivations revealed that the following settings of the CMT parameters lead to the real orthogonality of the basis functions $a_{n,k}(t)$ and allow symbol placement in the time-frequency lattice at the maximum density of two real symbols per unit area.

(1) The symbol spacing T along the time axis should be matched with the subcarrier spacing $F = 1/2T$ along the frequency axis.

(2) The pulse-shape/prototype filter $p(t)$ must be a real-valued square-root Nyquist filter for a symbol spacing

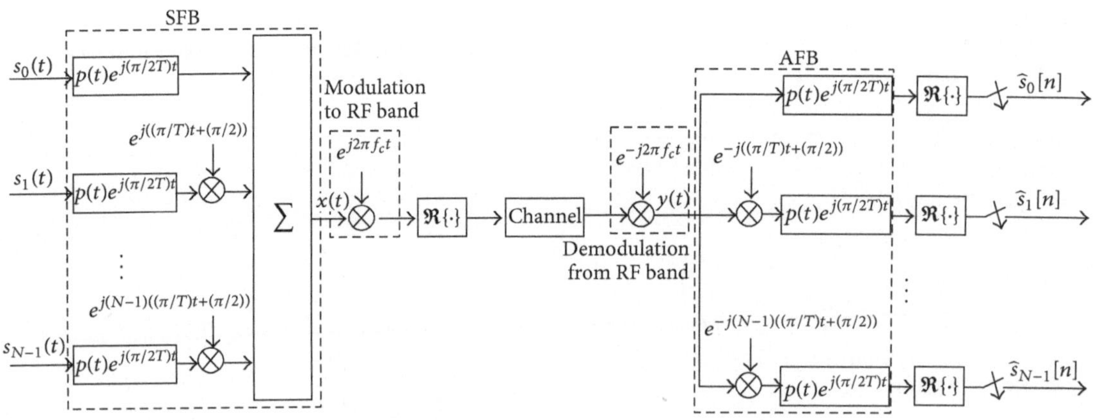

FIGURE 3: CMT transmitter and receiver blocks.

$2T$. It should be also a linear phase; hence, when viewed as a zero phase filter, it satisfies the condition $p(t) = p(-t)$. The latter is not a necessary condition but is satisfied in most of the designs; see [38] for a more relaxed condition.

(3) The above constraints are applied to assure orthogonality of the basis functions that are within the same subcarrier or adjacent subcarriers only. The orthogonality of basis functions that belong to the nonadjacent subcarriers is guaranteed by virtue of the fact that they correspond to filters with nonoverlapping bands. A more advanced design presented in Section 4.2 allows overlapping of the nonadjacent bands and yet satisfies the orthogonality condition.

(4) The phase shifts indicated in Figure 1 can be modified to other choices, as long as a phase difference $\pm\pi/2$ is preserved between each pair of adjacent points in the lattice.

(5) Although in CMT the position of the lattice points is fixed, these points can be moved in the time-frequency plane as long as their relative position and phase differences remain unchanged. We use this point below to arrive at the SMT waveform by applying a simple modification to the CMT waveform (1).

The above equations may be combined to arrive at the CMT transmitter and receiver structures that are presented in Figure 3. As shown a synthesis filter bank (SFB) is used to construct the transmit signal, and the received signal is passed through an analysis filter bank (AFB) to separate the data streams of different subcarrier bands. Here, it is assumed that there are N subcarrier streams and the data stream of the kth subcarrier at the input to the SFB is represented by the impulse train:

$$s_k(t) = \sum_n s_k[n]\,\delta(t - nT). \tag{13}$$

It is also worth noting that, in practice, the analyzed signals at the output of the AFB should be equalized. Here, to keep the presentation simple, we have not included the equalizers. Equalizer details are presented in Section 6.

3.2. SMT. SMT may be thought of as an alternative to CMT. Its time-frequency phase-space lattice is obtained from that of CMT (Figure 1) through a frequency shift of the lattice points down by $1/4T$, scaling the time axis by a factor $1/2$ and hence the frequency axis by a factor 2. Moreover, to remain consistent with the past literature of SMT (equivalently, OQAM-OFDM), some adjustments to the carrier phases of the lattice points have been made. This leads to the time-frequency phase-space lattice that is presented in Figure 4. The magnitude responses of the SMT pulse-shaping filters at various subcarriers and time instants $t = 0$ and $t = T/2$ are presented in Figure 5. Note that here the PAM symbols are spaced by $T/2$ and subcarriers are spaced by $1/T$. In SMT, each pair of adjacent symbols along time in each subcarrier is treated as real and imaginary parts of a QAM symbol. This leads to the transmitter and receiver structures that are presented as in Figure 6. Here, each data symbol $s_k[n]$ belongs to a QAM constellation and, thus, may be written in terms of its in-phase and quadrature components as $s_k[n] = s_k^I[n] + js_k^Q[n]$. Accordingly, the inputs to the SFB in Figure 6 are

$$s_k^I(t) = \sum_n s_k^I[n]\,\delta(t - nT),$$
$$s_k^Q(t) = \sum_n s_k^Q[n]\,\delta(t - nT). \tag{14}$$

3.3. FMT. FMT waveforms are synthesized following the conventional method of frequency division multiplexing (FDM). The subcarrier channels have no overlap, and thus ICI is resolved through use of well-designed filters with high stopband attenuation. ISI may be compensated for by adopting the conventional method of square-root Nyquist filtering that is used in single-carrier communications. For doubly dispersive channels, we have adopted a more advanced design [57] that compensates for both ICI and ISI through an effective method which is presented in Section 4.2.

For comparison with the CMT and SMT structures in Figures 3 and 6, respectively, and also as a basis for further

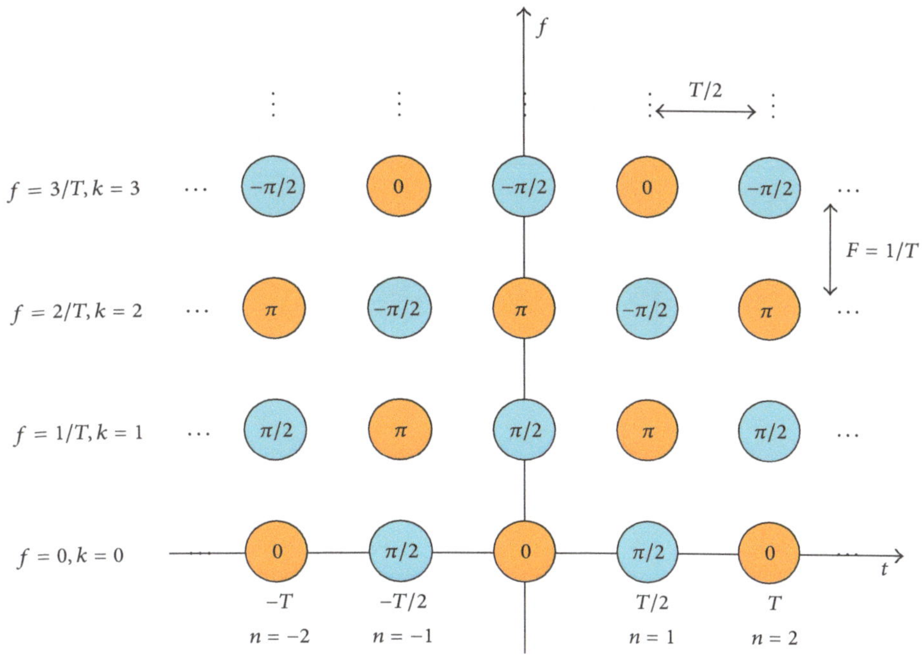

FIGURE 4: The SMT time-frequency phase-space lattice.

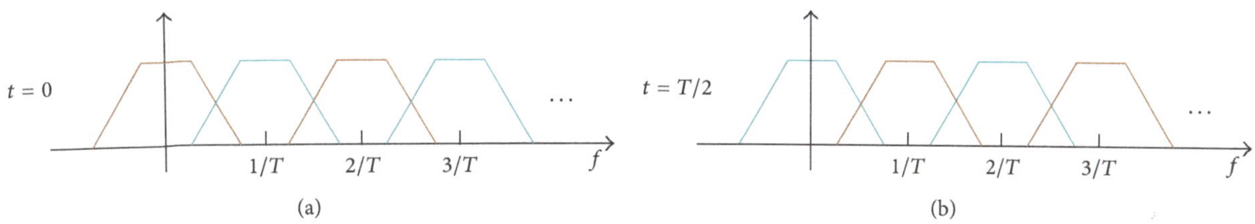

FIGURE 5: Magnitude responses of the SMT pulse-shaping filters at various subcarriers and time instants $t = 0$ and $t = T/2$.

development in later parts of this paper, Figure 7 presents the structures of transmitter and receiver blocks in an FMT communication system. Also, the magnitude responses of the FMT pulse-shaping filters at various subcarriers are presented in Figure 8. Note that here the subcarrier spacing is equal to $(1 + \alpha)/T$, where α is the roll-off factor of the pulse-shaping/prototype filter. Accordingly, FMT has a symbol density of $1/(1+\alpha)$ complex symbols per unit area in the time-frequency plane. Hence, FMT is less bandwidth efficient than CMT and SMT.

4. Prototype Filter Design

The results of the previous section suggest that linear phase square-root Nyquist filters that have been widely used for single carrier data transmission are the most trivial choice for prototype filter in FBMC systems. This indeed remains an accurate statement as long as the underlying channel is time-invariant or varies slowly. However, in cases where the channel is fast varying, a more general class of square-root Nyquist filters that satisfy the Nyquist condition both along the time and frequency axis should be adopted. In this section, we first review couple of classical Nyquist designs developed by us [58] and others [59], and then discuss an example of more advanced designs that also have been developed in our research group [57] and is more appropriate for time-varying channels.

4.1. Prototype Filters for Time-Invariant Channels. The classical prototype filter for FBMC systems that was suggested in [39, 40] was the square-root raised-cosine (SRRC) filter. More specifically, both [39, 40] suggested using a SRRC filter with the roll-off factor $\alpha = 1$. In practice, where FBMC systems are implemented in discrete-time, SRRC response should be sampled and truncated. The truncation of the SRRC response may result in a filter with poor frequency response and, hence, makes SRRC a poor choice. Advancements in digital filters design have led to very effective methods for designing square-root Nyquist (SR-Nyquist) filters for any specified finite length. These designs, two of which are presented here, aim at balancing the stopband attenuation and the Nyquist property of the designs.

Martin [59] has proposed a design method whose goal is to satisfy the Nyquist criterion approximately while achieving good attenuation in the stopband. A nice property of this method is that when the number of samples per symbol

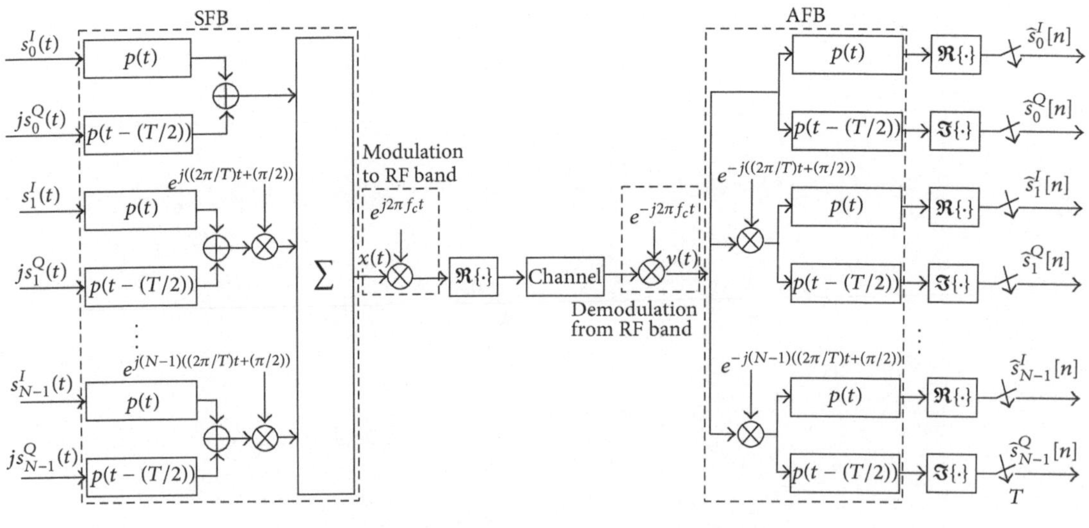

FIGURE 6: SMT transmitter and receiver blocks.

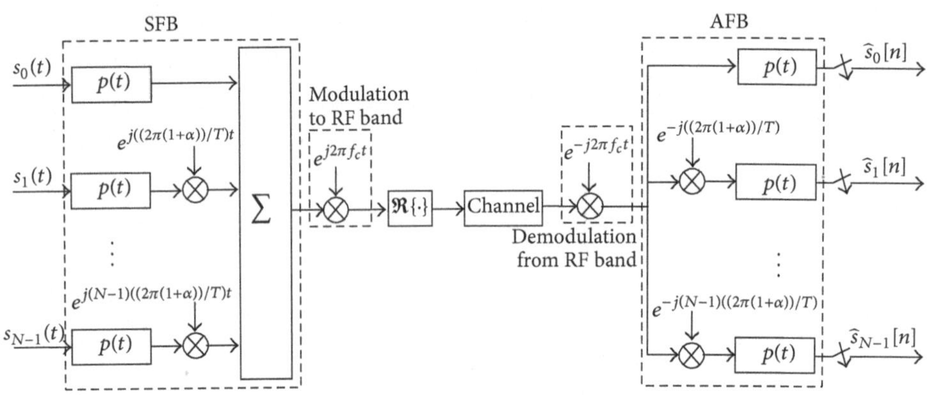

FIGURE 7: FMT transmitter and receiver blocks.

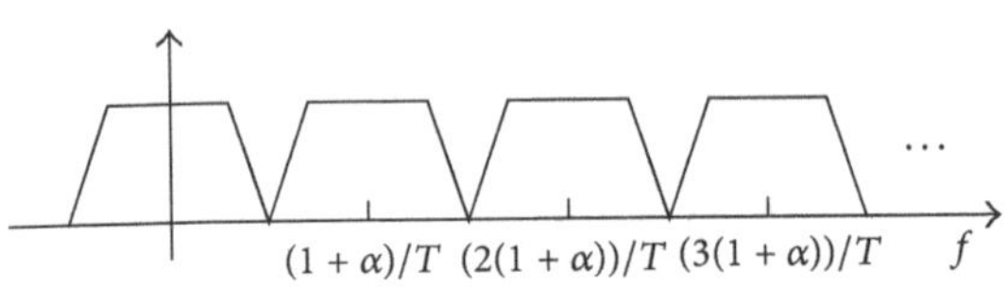

FIGURE 8: Magnitude responses of an FMT pulse-shaping filter at various subcarriers.

interval is N and the filter length is $\beta N + 1$, where β is an integer greater than 1, there are only β parameters that need to be found for the optimization of the design. More specifically, the prototype filter is constructed as

$$p[n] = \begin{cases} \frac{1}{\beta N + 1}\left(k_0 + 2\sum_{l=1}^{\beta-1} k_l \cos\left(\frac{2\pi l n}{\beta N + 1}\right)\right), & 0 \le n \le \beta N \\ 0, & \text{otherwise,} \end{cases} \quad (15)$$

where the coefficients k_0 through $k_{\beta-1}$ are to be optimized. The optimum choices of the coefficients k_0 through $k_{\beta-1}$, for different values of β, are tabulated in [59] and, also, in [60]. Beside being a simple and effective method, it is important to note that Martin's design for $\beta = 4$ has been adopted by the PHYDYAS project [21].

More recently, we have developed an algorithm for designing SR-Nyquist filters that can balance between the accuracy of the Nyquist constraints and the filter stopband attenuation [58]. To select the samples of the zero-phase impulse response $p[n] = p[-n]$ of the SR-Nyquist filter, this algorithm defines

$$q[n] = p[n] \star p[-n] \quad (16)$$

and minimizes the cost function

$$\xi = (q[0] - 1)^2 + \sum_{m=1}^{K/2} q^2[mN] + \gamma \int_{(1+\alpha)/2N}^{1-((1+\alpha)/2N)} |P(f)|^2 \, df \quad (17)$$

with respect to the elements $p[n]$. The first two terms on the right-hand side of (17) are to enforce Nyquist property on $q[n]$, and the third term is included to minimize the stopband response of $p[n]$. Also, the length of $p[n]$ is assumed to be

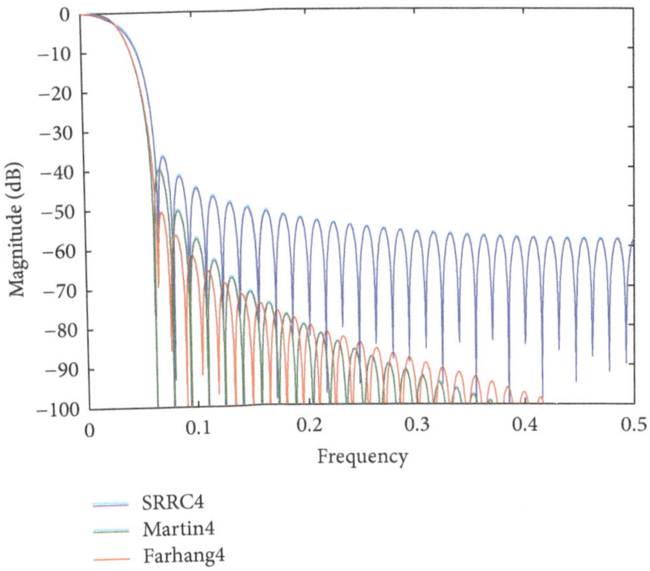

FIGURE 9: Magnitude responses of a sampled and truncated SRRC filter and two discrete-time designed SR-Nyquist filters.

equal to $\beta N + 1$. One may note that since ξ is a forth-order function of the parameters $p[n]$, it is a multimodal function and hence its minimization may not be straightforward. However, fortunately, if $p[n]$ could be initialized near its optimum choice, an iterative solution may be applied to search for the local minimum of ξ. The search algorithm proposed in [58] follows a similar procedure to the one originally suggested in [61].

Figure 9 presents the magnitude responses of three designs based on (i) a sampled and truncated version of the impulse response of a SRRC filter: SRRC4; (ii) a SR-Nyquist design obtained following [59]: Martin4; and (iii) a SR-Nyquist design obtained following [58]: Farhang4. All designs are based on the roll-off factor $\alpha = 1$ and are for an FBMC system with $N = 16$ subcarriers, and the suffix 4 on the names indicates that, for all designs, $\beta = 4$; that is, the designed filters have a length $\beta N + 1 = 4 \times 16 + 1 = 65$. In addition, for Farhang4, the parameter γ (defined in [58]) is set equal to 0.1.

As seen, and one would expect, SRRC design performs significantly inferior to both of the SR-Nyquist designs. From the two SR-Nyquist designs, the one based on [58] achieves a higher attenuation in the first few sidelobes of its stopband than the design based on [59]. The first sidelobe of the former design is 10 dB lower. Also, a study of the time domain responses of the designs reveals that none of them satisfies the Nyquist property perfectly. Hence, all are near Nyquist designs. However, the design of [59] is the closest to Nyquist, followed by the design delivered by [58] and, then, the SRRC design. Nevertheless, all the three designs are very close to Nyquist; hence, the choice of one against the others may be dominantly determined by its superior magnitude response.

4.2. Prototype Filters for Time-Varying Channels. When a channel is subject to dispersion both in time (due to multipath effects) and in frequency (due to variation of the channel in time), we say the channel is *doubly dispersive*. We also note that while the dominant effect of time dispersion in a channel is ISI, the dominant effect of frequency dispersion is ICI. Hence, in order to design prototype filters that address both time and frequency dispersions, it has been argued [44] that one should choose a pulse-shape $p(t)$ with similar behavior along time and frequency. In particular, if $p(t)$ is selected to be a SR-Nyquist along the time axis, it should be also made sure that $P(f)$ is a SR-Nyquist along the frequency axis. To this end, the pulse-shape, $p(t)$, with the following property may be adopted:

$$P(f) = p(\eta f), \quad \text{for a constant scaling factor } \eta, \tag{18}$$

that is, a function that has the same form in both time and frequency domains.

The parameter η in (18) is related to the symbol spacing in time, T, and frequency, F, and is given by

$$\eta = \frac{T}{F}. \tag{19}$$

Also, as one may understand intuitively, T and F are, respectively, chosen proportional to time dispersion, $\Delta\tau$, and frequency dispersion, $\Delta\nu$, of the channel; that is, $T/F = \Delta\tau/\Delta\nu$. Hence, the following identity also holds:

$$\eta = \frac{\Delta\tau}{\Delta\nu}. \tag{20}$$

The definitions for $\Delta\tau$ and $\Delta\nu$ are usually vague. The time dispersion $\Delta\tau$ may be thought of as a coarse estimate of the duration of the channel impulse response, equivalently the span of the multipaths of the channel. Similarly, the frequency dispersion $\Delta\nu$ may be thought of as a coarse estimate of the span of Doppler spread of the channel.

The design of the prototype filter for time-varying channels is closely tied to the ambiguity function [62–64]:

$$A_p(\tau, \nu) = \int_{-\infty}^{\infty} p\left(t + \frac{\tau}{2}\right) p^*\left(t - \frac{\tau}{2}\right) e^{-j2\pi\nu t} dt, \tag{21}$$

where τ is a time delay and ν is a frequency shift.

Let $p(t)$ be a prototype filter and let $p_{n,k}(t)$ be a time frequency translated version of it which is defined as

$$p_{n,k}(t) = p(t - nT) e^{j2\pi kFt}. \tag{22}$$

We note that

$$\begin{aligned} &\langle p_{m,k}(t), p_{n,l}(t) \rangle \\ &\quad = \int_{-\infty}^{\infty} p(t - mT) e^{j2\pi kFt} p(t - nT) e^{-j2\pi lFt} dt \\ &\quad \propto A_p((n-m)T, (l-k)F), \end{aligned} \tag{23}$$

where $\propto$ indicates proportionate to. The proportionate factor is a phase shift due to the delays of mT and nT whose value is irrelevant to our discussion here. Using (23), one may note that $p_{n,k}(t)$ of (22), for all choices of n and k, will form a set of complex-valued orthogonal basis functions, if $p(t)$ is chosen such that

$$A_p(nT, lF) = \begin{cases} 1, & n = l = 0 \\ 0, & \text{otherwise.} \end{cases} \tag{24}$$

We note that the case where $\nu = 0$ corresponds to the familiar (time) correlation function:

$$A_p(\tau, 0) = \int_{-\infty}^{\infty} p\left(t + \frac{\tau}{2}\right) p\left(t - \frac{\tau}{2}\right) dt \tag{25}$$

which reduces to the Nyquist constraints:

$$A_p(nT, 0) = \begin{cases} 1, & n = 0 \\ 0, & n \neq 0. \end{cases} \tag{26}$$

Also, one may note that the integral on the right-hand side of (25) is equal to the convolution of $p(t)$ and its matched pair $p(-t)$, evaluated at time $t = \tau$. Thus, $A_p(t, 0) = p(t) \star p(-t)$ and, hence, (26) implies that $p(t)$ is a SR-Nyquist pulse. Moreover, the ambiguity function $A_p(\tau, \nu)$ may be thought of as a generalization of the correlation function $A_p(\tau, 0)$ where correlation is found between $p(t)$ and its modulated version at the frequency $f = \nu$. Accordingly, we refer to the set of constraints (24) as the *generalized Nyquist constraints.*

Another key point that one should consider in the selection of $p(t)$, for time-varying channels, is minimization of its duration. Le Floch et al. [44] have noted that to maximize the density of the basis functions $p_{n,k}(t)$ in the time-frequency space and, hence, maximize the bandwidth efficiency of the transmission, one should choose a $p(t)$ that has maximum compactness in the time-frequency space. Maximum compactness, on the other hand, is quantified by the product $\sigma_t \sigma_f$, where σ_t^2 and σ_f^2 are, respectively, the second-order moments of the functions $p(t)$ and $P(f)$. Moreover, the Heisenberg-Gabor uncertainty principle states that [65]

$$\sigma_t \sigma_f \geq \frac{1}{4\pi}, \tag{27}$$

where the equality holds only when $p(t)$ is the Gaussian pulse $g(t) = e^{-\pi t^2}$. Also, the Gaussian pulse $g(t)$ has the interesting property that $G(f) = g(f)$; that is, it satisfies the desirable property (18), with $\eta = 1$. However, $p(t) = g(t)$ does not satisfy the orthogonality conditions (24).

Attempts to design filters that satisfy the orthogonality conditions (24) and at the same time approach the Heisenberg-Gabor uncertainty lower bound (27) as close as possible have been made and design methods have been developed [44, 66]. The design presented in [44] is called *isotropic* orthogonal transform algorithm (IOTA) filter. IOTA design/algorithm was first introduced in a patent by Alard [67]. The designs proposed in [66], on the other hand, are referred to as *Hermite pulses*, reflecting the fact that their construction is based on a linear combination of a set of Hermite functions. In the rest of this section, we limit our emphasis to the design of Hermite pulses and emphasize the flexibilities that these designs provide in adopting to doubly dispersive channels. More details on this topic can be found in [57] (also, see [68, 69]). Discussion on IOTA design can be found in [19, 20, 44, 70].

The design procedure proposed by Haas and Belfiore [66] constructs an isotropic filter according to the equation:

$$p(t) = \sum_{k=0}^{L} \alpha_k h_{4k}(t), \tag{28}$$

where $\{h_n(t)\}$ is the set of Hermite functions defined as

$$h_n(t) = \frac{1}{(2\pi)^{n/2}} e^{\pi t^2} \frac{d^n}{dt^n} e^{-2\pi t^2}. \tag{29}$$

Note that $h_0(t) = g(t)$ and, thus, it is an isotropic function with parameter $\eta = 1$. Moreover, it can be shown that the set of functions $h_n(t)$ for $n = 4k$, $k = 1, 2, \ldots$, are also isotropic, with the same parameter. This implies that the construction (28) for any set of coefficients α_k leads to an isotropic function. In [66], the coefficients α_k have been calculated to construct a filter $p(t)$ that satisfies the set of constraints (24), for parameters $T = F = \sqrt{2}$.

Haas and Belfiore's design [66] allows transmission of QAM symbols with a density of $1/TF = 0.5$ symbol per unit area in the time-frequency space. This design of $p(t)$ may also be used as the prototype filter in a CMT or SMT structure to increase the density to 2 PAM symbols per unit area, equivalent to one QAM symbol per unit area. Alternatively, one can aim for a design with larger density than 0.5 and stay with transmitting QAM symbols. Examples of both designs are presented later. In the rest of this section, we follow the approach of [57] to present a broad class of Hermite filter designs that includes the design presented in [66] as a special case.

The basic equations for the design of Hermite pulses are obtained by substituting (28) in (21). This gives

$$\begin{aligned} A_p(\tau, \nu) &= \int_{-\infty}^{\infty} \sum_{n=0}^{L} \sum_{l=0}^{L} \alpha_n \alpha_l h_{4n}\left(t + \frac{\tau}{2}\right) h_{4l}\left(t - \frac{\tau}{2}\right) e^{-j2\pi\nu t} dt \\ &= \sum_{n=0}^{L} \sum_{l=0}^{L} \alpha_n \alpha_l A_{n,l}(\tau, \nu), \end{aligned} \tag{30}$$

where

$$A_{n,l}(\tau, \nu) = \int_{-\infty}^{\infty} h_{4n}\left(t + \frac{\tau}{2}\right) h_{4l}\left(t - \frac{\tau}{2}\right) e^{-j2\pi\nu t} dt. \tag{31}$$

Defining

$$\boldsymbol{\alpha} = \begin{bmatrix} \alpha_0 & \alpha_1 & \cdots & \alpha_L \end{bmatrix}^T,$$

$$\mathbf{A}(\tau, \nu) = \begin{bmatrix} A_{0,0}(\tau, \nu) & A_{0,1}(\tau, \nu) & \cdots & A_{0,L}(\tau, \nu) \\ A_{1,0}(\tau, \nu) & A_{1,1}(\tau, \nu) & \cdots & A_{1,L}(\tau, \nu) \\ \vdots & \vdots & \ddots & \vdots \\ A_{L,0}(\tau, \nu) & A_{L,1}(\tau, \nu) & \cdots & A_{L,L}(\tau, \nu) \end{bmatrix}, \tag{32}$$

(30) may be rearranged as

$$A_p(\tau,\nu) = \boldsymbol{\alpha}^T \mathbf{A}(\tau,\nu)\,\boldsymbol{\alpha}. \tag{33}$$

(1) Haas and Belfiore Design. In [66], it was proposed that the coefficients α_0 through α_L can be determined by substituting (33) into (24) for $(n,l) = (0,0)$ and L other significant choices of (n,l) and solving the resulting system of equations. It was, numerically, demonstrated that this leads to very good designs. To further clarify this procedure and pave the way for additional developments in the sequel, we discuss the method of [66] in the context of a specific design. Figure 10 presents the grid of all choices of (n,l). Here, it is assumed that the constraints (24) are applied at the origin and the 12 nearest grid points to it. In this figure, three different sets of grid points are identified.

(1) The significant points are the ones at which the constraints (24) should be imposed. There are four such points and these are indicated by red disks.

(2) Once the desired constraints are imposed at the significant points defined in (1), it follows from the even symmetry and isotropic property of $p(t)$ that the same constraints will automatically be imposed at the rest of the points indicated by green disks.

(3) The remaining grid points, indicated by white disks, will satisfy the constraints (24), within a good approximation, since the designed pulse $p(t)$ decays exponentially/fast as $|t|$ increases.

We note that to satisfy the constraints (24) at the origin and the 12 nearest grid points to it, it is sufficient to apply constraints to only 3 of the latter points. A unique design is thus obtained if the design is made based on $h_0(t)$, $h_4(t)$, $h_8(t)$, and $h_{12}(t)$, that is, the choice of $L = 3$ in (30) through (33).

It should be also noted that since, here, the designed filter is isotropic, with the parameter $\eta = 1$, the identity $T = F$ should hold. Hence, given a desired density $D = 1/TF$, the design must be for the parameters $T = F = 1/\sqrt{D}$. Once $p(t)$ is obtained based on these parameters, applying a time scaling factor, one can set T to any desired value. This will set F equal to $1/TD$.

(2) Hexagonal Lattice. The spread of data symbols in Figure 10 and other lattice grids that have been presented in this paper, so far, follows an orientation that is referred to as *rectangular* lattice. In [71], it is noted that better designs may be obtained by adopting the *hexagonal* orientation. The reason behind this improvement in performance follows the fact that the hexagonal orientation allows maximum separation of points for a given symbol density. Interested readers are referred to the detailed discussions in [19, 20, 71].

To design prototype filters for an orientation that follows the hexagonal lattice, we choose the constrained points according to those depicted in Figure 11. It should be noted that we set $T = 2F$ and for this choice it is sufficient to enforce design constraints at the points indicated by red disks. As in

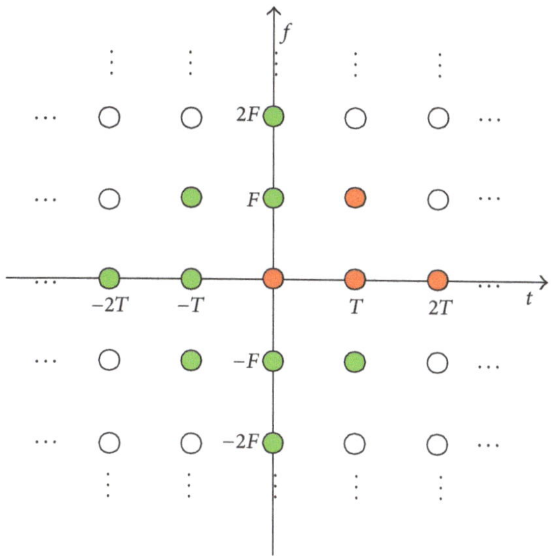

FIGURE 10: Grid of (n,k) points at which the constraints (24) should be imposed. The red disks indicate the points at which the constraints (24) are imposed. The green disks follow the red ones, thanks to symmetric and isotropic property of the design. The rest of the points (white disks) satisfy the constraints within a good approximation, since the construction is based on exponentially decaying pulses as $|t|$ increases.

the case of rectangular lattice, once the constraints are applied to these points, similar constraints will be automatically imposed to the points indicated by green disks, following the symmetric and isotropic property of the designs. Moreover, as in the case of rectangular lattice, by design, the remaining grid points, indicated by white disks, will satisfy the constraints (24), within a good approximation. Finally, a time scaling can be applied to $p(t)$ to set T, and accordingly F, to any desired value.

(3) Robust Design. Building on the above findings, we noted that, in a practical design, the time and frequency dispersion introduced by a channel smear the null points of the ambiguity function $A_p(\tau,\nu)$ [57, 68, 69]. Hence, as a result of the channel dispersion, the nulls will convert to shallow nulls. It is, thus, argued that instead of designing $p(t)$ to introduce perfect nulls in $A_p(\tau,\nu)$, a more robust design is obtained by aiming for a design that results in deep (but imperfect) null areas around the grid points (nT, kF). To this end, a robust Hermite pulse is designed by minimizing the cost function:

$$\begin{aligned}\zeta = {} & \gamma_0 \int_{\mathcal{A}_0} \left|\boldsymbol{\alpha}^T \mathbf{A}(\tau,\nu)\boldsymbol{\alpha} - 1\right|^2 d\tau\, d\nu \\ & + \sum_{k=1}^{L} \gamma_k \int_{\mathcal{A}_k} \left|\boldsymbol{\alpha}^T \mathbf{A}(\tau,\nu)\boldsymbol{\alpha}\right|^2 d\tau\, d\nu,\end{aligned} \tag{34}$$

where $\mathcal{A}_0$ is an area around the origin and $\mathcal{A}_1$ through $\mathcal{A}_L$ are the null areas that are aimed for.

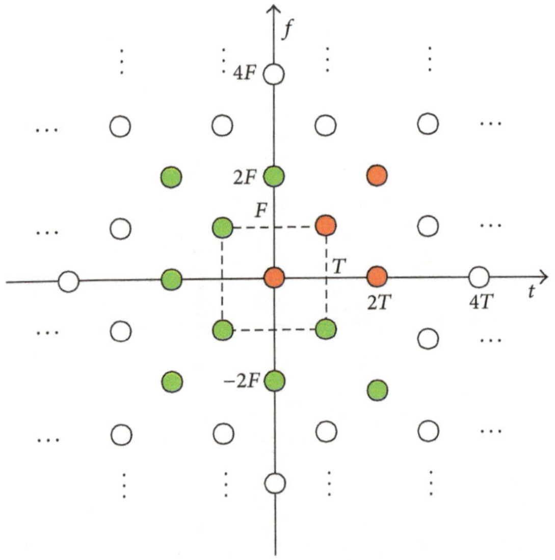

Figure 11: Grid of (n, k) points for hexagonal orientation. The red disks indicate the points at which the constraints (24) are imposed. The green disks follow the red ones, thanks to symmetric and isotropic property of the design. The rest of the points (white disks) satisfy the constraints within a good approximation, since the construction is based on exponentially decaying pulses as $|t|$ increases.

To develop a numerical method for the minimization of ζ, the ambiguity function is sampled along the time axis τ and the frequency axis ν. This converts (34) to

$$\zeta = T_s F_s \left(\sum_{k=0}^{L} \sum_{(mT_s, nF_s) \in \mathcal{A}_k} \gamma_k \left| \boldsymbol{\alpha}^T \mathbf{A}(mT_s, nF_s) \boldsymbol{\alpha} - u_k \right|^2 \right), \quad (35)$$

where T_s and F_s are the sampling periods along the time axis and frequency axis, respectively, and

$$u_k = \begin{cases} 1, & k = 0 \\ 0, & k \neq 0. \end{cases} \quad (36)$$

Removing the third term on the right-hand side of (17), one may realize that the cost functions ξ and ζ are similar. Both are fourth-order functions of the parameters that we seek to optimize. Hence, the iterative method developed in [58] can be readily adopted here as well. For this purpose, as in [58], an initial guess for the elements of the parameter vector $\boldsymbol{\alpha}$ should be made. It has been noted in [57] that a proper initial choice for $\boldsymbol{\alpha}$ is the vector whose first element is 1 and the rest of its elements are 0. That is, one should start with the Gaussian pulse as a first guess and add the higher order Hermite functions in the subsequent iterations.

(4) Numerical Examples. To conclude this section, we present the results arising from a few designs of IOTA and Hermite prototype filters. Also, we compare the results with those of the SR-Nyquist design of [58].

Figure 12 presents the time domain and the magnitude of frequency domain responses of (i) an IOTA design [67], (ii) a Hermite design [66], and (iii) a SR-Nyquist design [58]. All designs have the same filter length of $4T$ and designed to serve a CMT/SMT system with a maximum of 16 subcarrier bands.

Considering the results presented in Figure 12, one may make the following observations.

(1) Both IOTA and Hermite designs have time-domain responses that are more compact than the time-domain response of the SR-Nyquist filter.

(2) In the frequency domain, on the other hand, the SR-Nyquist design gives a more compact response.

(3) The compact responses of IOTA and Hermite designs in time will make them less prone to distortion introduced by channel variation with time.

(4) The broader responses of IOTA and Hermite design in frequency, on the other hand, make them more prone to the channel frequency selectivity.

(5) Considering (3) and (4), a balance has to be made between the choice of the symbol interval, T, and the carrier spacing, F. Obviously, by applying a proper time scaling to the prototype filter $p(t)$, such balance can be made.

(6) When channel is time-invariant or varies slowly, the choice of SR-Nyquist results in the maximum immunity to the channel frequency selectivity.

The robust design approach that was presented above was first introduced in [69] and was further studied in [57, 68]. Here, to emphasize on the significance of this design approach, we present Figure 13 from [57]. This figure presents a set of results that compare the signal-to-interference ratio (SIR) of the robust isotropic designs (named FMT-dd: FMT for doubly dispersive channels) with OFDM and an FMT design according to the SR-Nyquist design of [58] (named FMC-c: conventional FMT). To measure SIR, the channel noise is set equal to zero. Three sets of results, corresponding to density values $D = 1/TF = 1/2, 2/3$, and $4/5$ are presented. The robust isotropic designs are obtained for a channel model in which dispersion in time and frequency is uniform in the intervals $(-\delta\tau/2, \delta\tau/2)$ and $(-\delta\nu/2, \delta\nu/2)$, respectively, and $\Delta\tau = 0.2T$ and $\Delta\nu = 0.2F$. The design for each density is fixed and its performance is examined for varying values of $\Delta\tau\Delta\nu$ in the range of 0 to 0.1. For FMT-c, the prototype filter is designed in each of the three cases with the aim of achieving a stopband attenuation of 60 dB or better. Also, following the basic principle of FMT-c, the roll-off factors of the prototype filters for density values $D = 1/2, 2/3$, and $4/5$ are set equal to 1, 0.5, and 0.25, respectively. For OFDM, the density $D = 1/TF$ is set by adjusting the ratio of cyclic prefix length, T_{CP}, over the length of FFT, T_{FFT}. In particular, we note that since in OFDM $F = 1/T_{\text{FFT}}$ and $T = T_{\text{CP}} + T_{\text{FFT}}$, $D = T_{\text{FFT}}/(T_{\text{CP}} + T_{\text{FFT}})$ and, thus, $T_{\text{CP}}/T_{\text{FFT}} = 1/D - 1$.

The results presented in Figure 13 clearly show the expected superior performance of the robust designs over OFDM and the conventional FMT design. The more detailed results presented in [57] for an underwater acoustic (UWA) communication further confirm these results in an at-sea experimental setting.

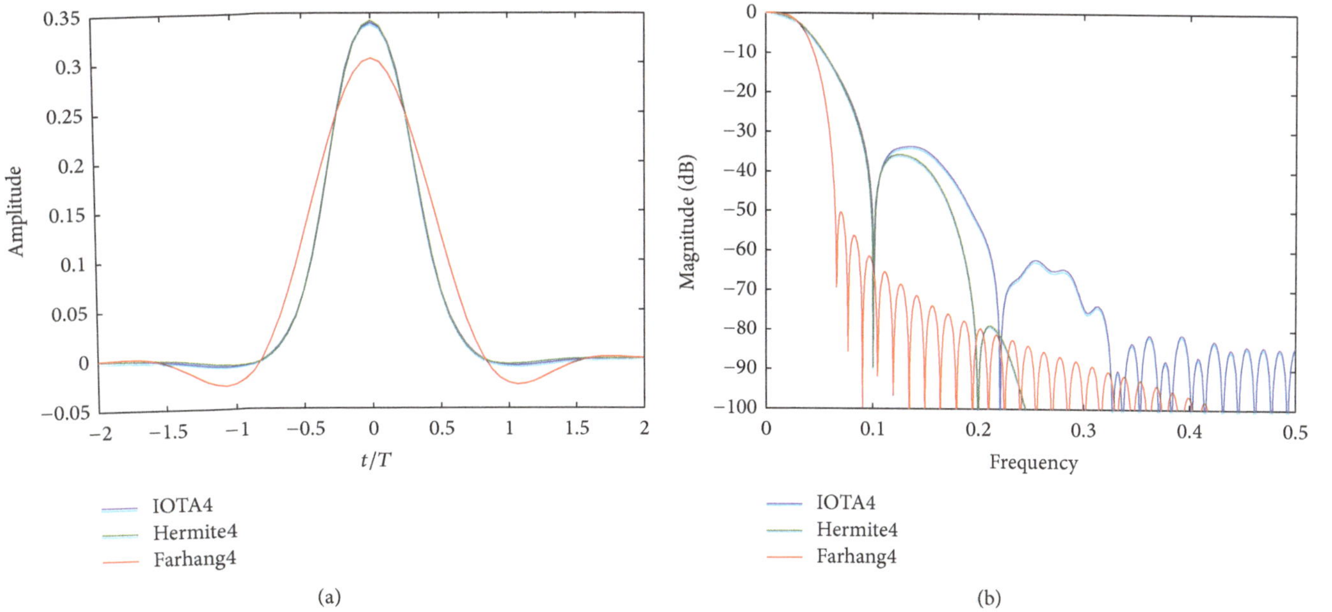

FIGURE 12: Responses of three designs of a prototype filter. (a) The time-domain responses. (b) The magnitude of the frequency-domain responses.

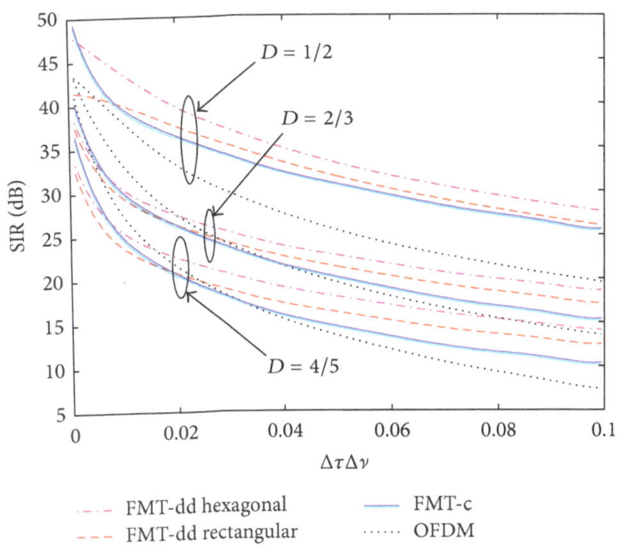

FIGURE 13: The impact of a doubly dispersive design and three designs of FMT. "FMT-c" is the case where a SR-Nyquist design is used. "FMT-dd" follows the robust design that was introduced above (the author obtained permission from IEEE to include this figure).

5. Polyphase Structures and Computational Complexity

Polyphase structures are commonly used to implement FBMC transmitter and receiver. Direct mimicking of the continuous-time structures of Figures 3 and 6 to implement CMT and SMT, respectively, may lead to structures whose complexity is not optimized. One polyphase structure per each set of real symbols, equivalent to two polyphase structures per each pair of real symbol sets, should be implemented. This section highlights the fact that the two polyphase structures at the transmitter side can be combined together and hence reduce the complexity to one half. At the receiver side, special attention has to be paid so that proper equalizers can be applied to the analyzed signal components. The presentation in this section, although results in structures with the same complexity to those already published, arrives at structures in more intuitive way and, thus, hopefully assists the reader to have a better grasp of the concepts.

Polyphase structures for FMT transmitter and receiver also need particular attention to take care of the fact that sampling rate changes in the AFB (at the transmitter side) and in the SFB (at the receiver side) is different from the size of IFFT/FFT block. A method that takes into account this change of sampling rate is presented in a later part of this section.

5.1. Polyphase Analysis and Synthesis Filter Banks. The basic polyphase SFB block that may be used to construct the SFB blocks in both CMT and SMT transmitter is presented in Figure 14. This block which is widely available and well developed in the literature, for example, see [49, 72], has the following characteristics.

(1) The inputs $\{s_k[n], k = 0, 1, \ldots, N-1\}$ are a set of data sequences with the nominal rate of unity.

(2) The SFB output has a rate of L or, equivalently, L times faster than the rate at the input.

(3) The synthesis is performed based on an IFFT of size $L > N$, with the inputs of index N and greater set equal to zero. This is to allocate some guard band between the generated baseband signal and its images

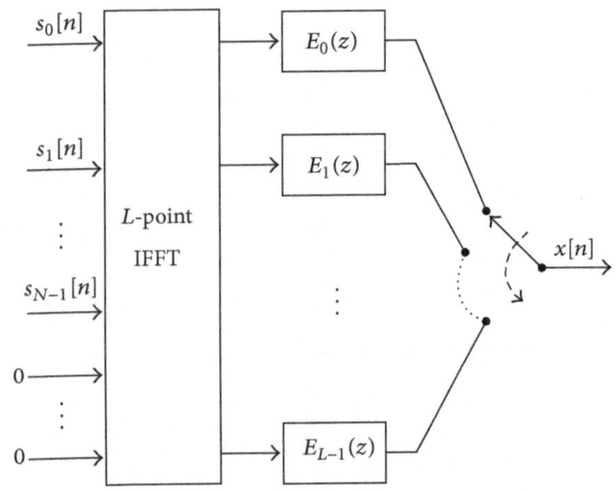

FIGURE 14: Basic synthesis polyphase filter bank.

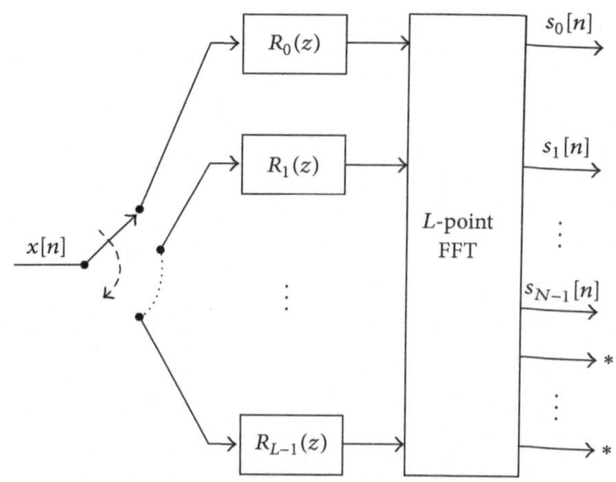

FIGURE 15: Basic analysis polyphase filter bank.

and, hence, facilitate additional filtering in the further stages of the transmitter.

(4) The filtering blocks $E_0(z)$ through $E_{L-1}(z)$ are the polyphase components of the prototype filter $p[n]$.

(5) This structure is an efficient implementation of a bank of filters with the respective inputs $s_0[n], s_1[n], \ldots, s_{N-1}[n]$ and the transfer functions $P(e^{j2\pi f})$, $P(e^{j2\pi(f-1/L)}), \ldots, P(e^{j2\pi(f-(N-1)/L)})$, where $P(e^{j2\pi f})$ is the Fourier transform of $p[n]$.

(6) The commutator at the output serializes the outputs of the polyphase component filters, after arrival of each set of data symbols at the input.

Figure 15 presents the basic polyphase AFB that matches with the SFB of Figure 14. Here, $R_k(z) = E_{L-k-1}(z)$. In the literature, $E_k(z)$ and $R_k(z)$ are distinguished by referring to them as type I and type II polyphase components [49]. When $x[n]$ is fed to its input, within the accuracy provided by the prototype filter $p[n]$, the data sequences $\{s_k[n], k = 0, 1, \ldots, N-1\}$ appear at its first N outputs. In other word, if $x[n]$ is passed through an ideal channel (free of distortion and noise), the SFB may be used to recover the transmitted data symbols $s_k[n]$. Obviously, channel introduces distortion and thus the AFB outputs should pass through a bank of equalizers to recover the data symbols $s_k[n]$.

5.2. Polyphase Structures for CMT and SMT. CMT and SMT systems, as was demonstrated in Section 3, are effectively the same modulations. For instance, a CMT waveform can be generated using an SMT waveform generator and subsequently apply a positive spectral shift of one half of the subcarrier spacing to the result. Alternatively, an SMT waveform may be generated using a CMT waveform generator and subsequently apply a negative spectral shift of one half of the subcarrier spacing to the result. It turns out that the development of structures for SMT waveform is more straightforward than those of the CMT. We thus continue this section with development of a pair of transmitter and receiver structures for SMT. The relevant CMT structures will

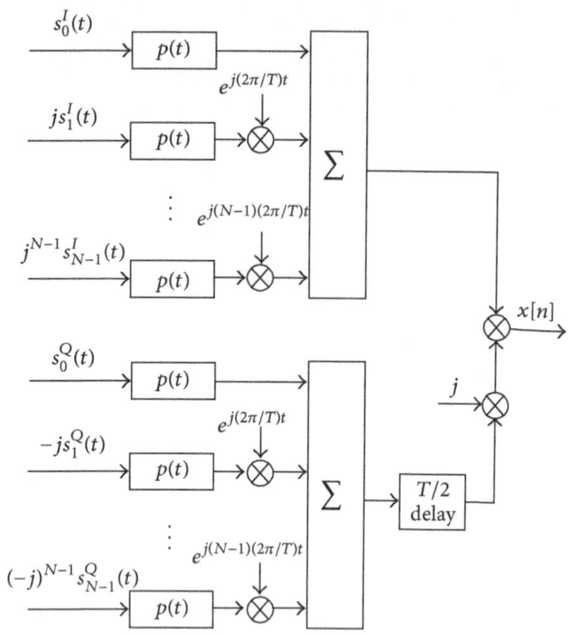

FIGURE 16: A rearrangement of the transmitter of SMT.

follow through simple modifications of the SMT structures following the details provided in Section 3.

(1) SMT Transmitter. To develop a computationally efficient polyphase structure for SMT transmitter, we begin with rearranging the transmitter part of Figure 6 as in Figure 16. This, clearly, is obtained by separating the phase and quadrature parts of the SFB, combining the phase shifts at different points in the structure and adding the combined results to the real data symbols, $s_k^I[n]$ and $s_k^Q[n]$, at the input. The delay of $T/2$ in the prototype filters in the quadrature paths has been shifted to the corresponding filter bank output.

The structure that is presented in Figure 16 consists of a pair for SFBs. Clearly, each of these filter banks may be implemented efficiently using the basic polyphase structure of Figure 14. This implementation is presented in Figure 17.

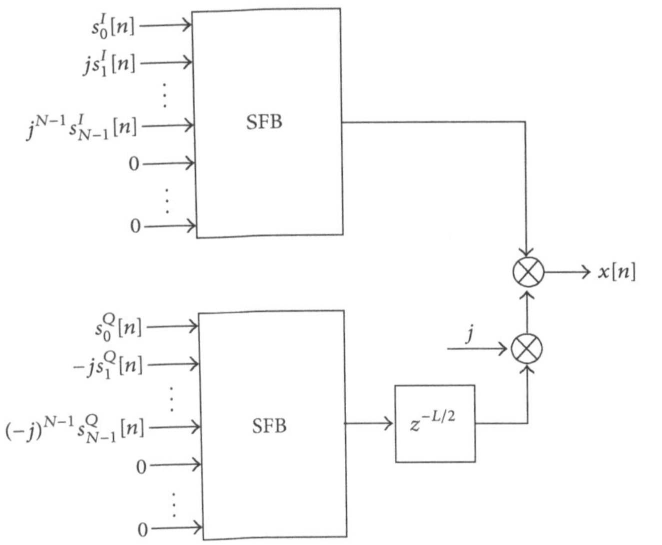

FIGURE 17: A rearrangement of the transmitter of SMT.

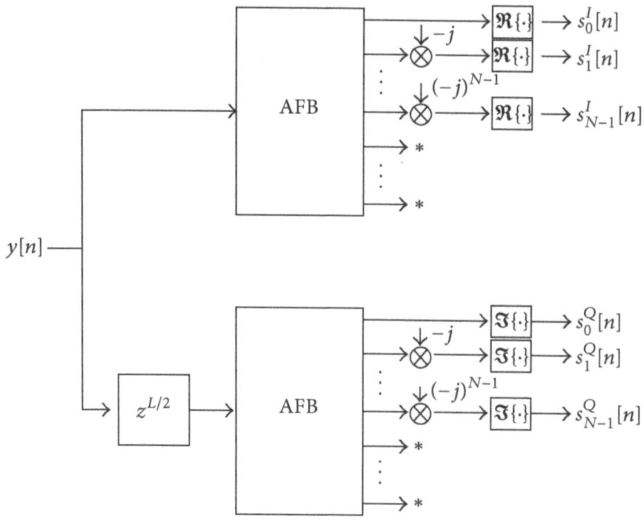

FIGURE 18: A rearrangement of the receiver of SMT.

It involves two separate polyphase structures. However, a closer look at the inputs to the polyphase structures reveals that they are modulated versions of two real vectors:

$$\mathbf{s}^I[n] = \left[s_0^I[n] \;\; s_1^I[n] \;\; \cdots \;\; s_{N-1}^I[n] \;\; 0 \;\; \cdots \;\; 0\right]^T,$$
$$\mathbf{s}^Q[n] = \left[s_0^Q[n] \;\; \cdots \;\; s_{N-1}^Q[n] \;\; 0 \;\; \cdots \;\; 0\right]^T. \qquad (37)$$

As demonstrated next, one may take advantage of this property to combine the two SFBs into one.

Looking back at the synthesis polyphase structure of Figure 14, it may be argued that one can start with the real vectors $\mathbf{s}^I[n]$ and $\mathbf{s}^Q[n]$ at the input to the IFFT block and apply the modulation effect at the output of the IFFT through a circular rotation of the results. In that case, one may start with the computation of IFFT of the real vectors $\mathbf{s}^I[n]$ and $\mathbf{s}^Q[n]$. It is well known that this pair of IFFTs can be implemented through a single IFFT with a computational complexity of $(L/2)\log_2 L$ complex multiplications. On the other hand, we note that since the coefficients of the prototype filter $p[n]$ are real-valued, the complexity of implementation of two sets of polyphase component filters in Figure 17 involves $2\beta L$ real by complex multiplications, equivalent to βL complex multiplications. Adding these results, the total complexity of the SMT transmitter is measured as $(L/2)\log_2 L + \beta L$ complex multiplications per each set of complex output symbols. Recall that the length of the prototype filter $p[n]$ was assumed to be equal $\beta L + 1$. Here, we have replaced this by βL for brevity of the expressions. Also, we are only counting the number of multiplications as the measure of complexity. This is because, in practice, each multiplication is usually followed by an addition. Hence, without involving ourselves with details, we argue that the number of additions in each implementation is about the same as the number of multiplications.

It is also worth noting that a number of other authors also have taken note of the same symmetry properties in SMT waveform and accordingly proposed methods of reducing the complexity [73–75]. All these works have reached the complexity numbers that are similar to those presented here.

(2) SMT Receiver. In the absence of channel, the analysis of an SMT signal at the receiver can be performed with the same complexity as its synthesis counterpart at the transmitter [74, 75]. However, the presence of the channel destroys the symmetry property of SMT waveform. Hence, the implementation approach used at the transmitter cannot be extended to the receiver. In particular, we note that although the final goal is to extract the real-valued data symbols, the analyzed signals are the preequalized ones and, hence, are complex-valued.

Following the same line of thoughts that led to Figure 17, the receiver side of Figure 6 can be rearranged as in Figure 18, where the AFB blocks follow the structure presented in Figure 15. This structure, yet, does not include the channel equalizers that should be added at each output branch. Also, as noted in the previous section, the equalizers may be single-tap or multitap. Furthermore, one may note that since the pair of AFBs in Figure 18 are the same, but separated by $L/2$ sample delay, they may be combined as one AFB. Moreover, the phase rotations $(-j)^k$ at the output side may be absorbed in the equalizers. Combining these points, we arrive at the SMT receiver structure presented in Figure 19.

The receiver structure presented in Figure 19 has a complexity of one AFB plus N equalizers for each set of real symbols, $s_k^I[n]$ or $s_k^Q[n]$. Assuming that each equalizer has M complex tap weights, and noting that at each equalizer output we need either the real or imaginary part of the result, the total complexity of the receiver structure of Figure 19 is obtained as $(L/2)\log_2 L + (MN + KL + 1)/2$ complex multiplications per each set of real output symbols.

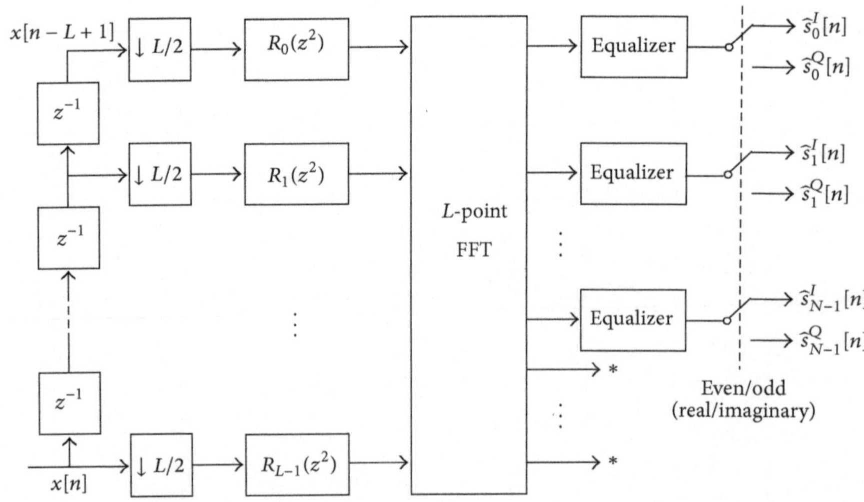

FIGURE 19: SMT receiver structure. The equalizers are single-tap or multitap. When they are multitap, they have a tap spacing of half a symbol interval.

Alternatively, we may say the total complexity of the receiver structure of Figure 19 is $L\log_2 L + MN + KL + 1$ complex multiplications per each set of complex output symbols.

5.3. Polyphase Structures for FMT. The polyphase SFB that is presented in Figure 14 implements a bank of synthesis filters whose center frequencies are at the normalized frequencies $0, 1/L, 2/L, \ldots$, equivalent to the unnormalized frequencies $0, 1/T, 2/T, \ldots$, respectively. Also, the sampling rate at the SFB output is $f_s = L/T$. Equivalently, the sampling interval at the SFB output is $T_s = T/L$. The same is true for the AFB that is presented in Figure 15.

In FMT, the situation is different. While the symbol rate remains equal to $1/T$, each subcarrier bandwidth and accordingly the FMT waveform bandwidth increase by a factor of $1 + \alpha$. Therefore, the sampling frequency of the synthesized waveform should be increased to $f_s = L(1+\alpha)/T$. This, in turn, implies that $T_s = T/(L(1+\alpha))$, and accordingly, in discrete-time, data symbols at the SFB input should be up-sampled $K = L(1+\alpha)$ fold (it is assumed that α is chosen such that K will be an integer number). Similarly, at the receiver side, the output of the AFB should be decimated K fold. Nevertheless, it should be noted that the center frequency of subcarriers remains unchanged; that is, it will be at 0, $1/L$, $2/L, \ldots$.

The fact that $K \neq L$ introduces some irregularity in the polyphase components which needs a special care. Most of the work in the literature that have addressed this issue have looked at the samples of underlying continuous time and accordingly have discussed how the corresponding polyphase filter bank should be implemented. This treatment, unfortunately, has led to a set of equations that are often hard to follow. Here, we present a solution that directly works with sampled signals/sequences [34]. This solution that has been presented for the first time in [72] is believed to be easier to follow. Here, we quote our original formulation from [72].

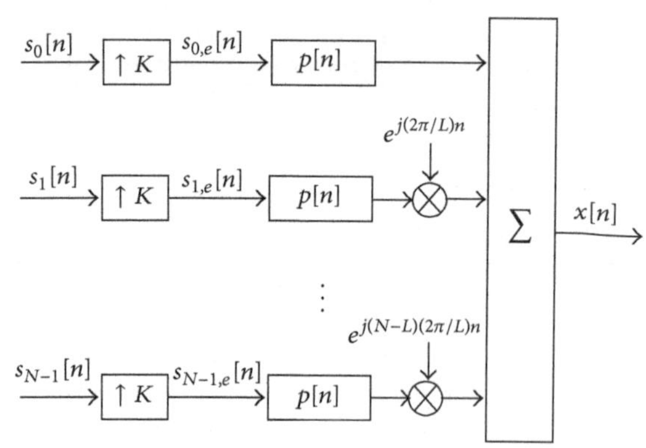

FIGURE 20: FMT transmitter in discrete-time.

(1) FMT Transmitter. Figure 20 presents a discrete-time equivalent of the SFB of Figure 7. Following this figure, one obtains

$$\begin{aligned} x[n] &= \sum_{k=0}^{N-1} \left(\sum_m s_k[m]\, p[n - mK] \right) e^{j(2\pi kn/L)} \\ &= \sum_m \left(\sum_{k=0}^{N-1} s_k[m]\, e^{j(2\pi kn/L)} \right) p[n - mK]. \end{aligned} \tag{38}$$

To proceed, we write the time index n as

$$n = \gamma L + \ell, \tag{39}$$

where γ is the integer part of n/L and $\ell = 0, 1, \ldots, L-1$, is the remainder of n/L. Similarly, n may be written as

$$n = \eta K + \kappa, \tag{40}$$

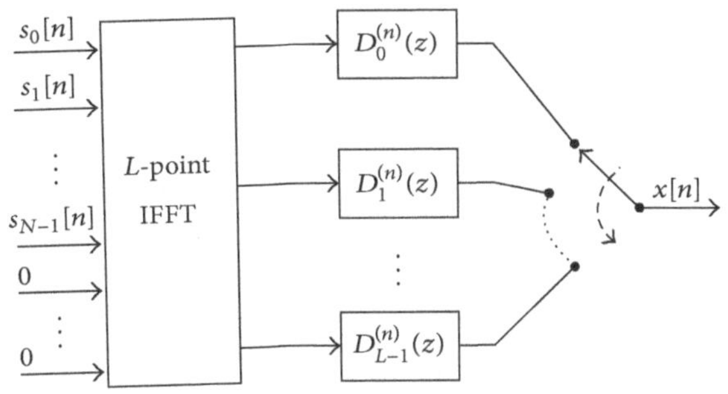

FIGURE 21: Polyphase implementation of an FMT transmitter.

where η is the integer part of n/K and $\kappa = 0, 1, \ldots, K-1$ is the remainder of n/K. Next, we use (39) to substitute for n in the exponential term in (38) to obtain

$$S_\ell[m] = \sum_{k=0}^{N-1} s_k[m] e^{j(2\pi k\ell/L)}, \quad (41)$$

where we have noted that $e^{j(2\pi kn/L)} = e^{j(2\pi k\ell/L)}$. We may also note that $S_\ell[m]$ is the L-point IFFT of input symbols $s_0[n], s_1[n], \ldots, s_{N-1}[n]$ for $n = m$. Using (41) and substituting (40) into (38), we obtain

$$x[\eta K + \kappa] = \sum_m S_\ell[m] p[(\eta - m)K + \kappa], \quad \kappa = 0, 1, \ldots, K-1. \quad (42)$$

Equation (42) may be interpreted and accordingly implemented as follows. For each set of input symbols $s_0[n], s_1[n], \ldots, s_{N-1}[n]$, K samples of $x[n]$ are calculated. This calculation begins with performing an L-point IFFT on the input symbols $s_0[n], s_1[n], \ldots, s_{N-1}[n]$. The output samples from the IFFT are passed through the polyphase components of the prototype filter $p[n]$, where the polyphase components are with respect to the decimation factor K. This process is depicted in a block diagram form in Figure 21. The commutator at the output of Figure 21 takes K steps after arrival of each set of input symbols. $D_0^{(n)}(z)$ through $D_{L-1}^{(n)}(z)$ are a set of time-varying filters that are picked up from the set of K polyphase components $E_0(z)$ through $E_{K-1}(z)$ of the prototype filter $P(z) = \sum_n p[n]z^{-n}$. For a given time n, $x[n]$ is taken from the output of $D_\ell^{(n)}(z)$ and $D_\ell^{(n)}(z) = E_\kappa(z)$, where ℓ and κ are the integers defined according to (39) and (40), respectively.

Following Figure 21, the complexity of implementation of FMT transmitter, for each set of data symbols, is obtained as $(L/2)\log_2 L + \beta L$ complex multiplications.

(2) FMT Receiver. Following the same line of arguments that led to Figure 20, the discrete-time implementation of an FMT receiver is obtained as in Figure 22. From this figure, it follows that, for $k = 0, 1, \ldots, N-1$,

$$\begin{aligned} y_k[n] &= \left(y[n] e^{-j(2\pi kn/L)}\right) \star p[n] \\ &= \sum_m y[m] e^{-j(2\pi km/L)} p[n-m]. \end{aligned} \quad (43)$$

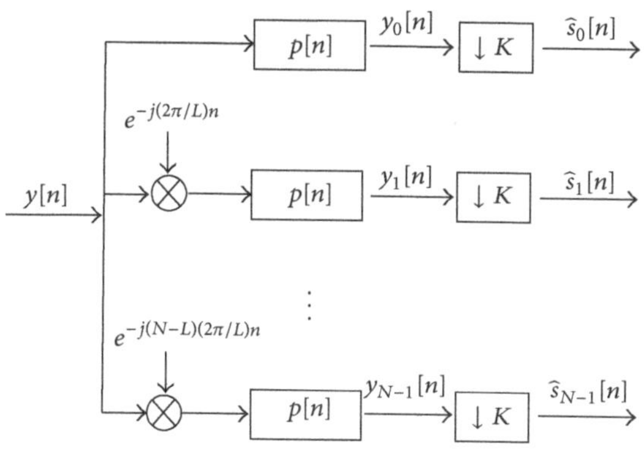

FIGURE 22: FMT receiver in discrete-time.

To proceed, we define

$$m = \gamma L + \ell, \quad (44)$$

where γ is the integer part of m/L and $\ell = 0, 1, \ldots, L-1$, is the remainder part of m/L. Using (44), (43) may be rearranged as

$$\begin{aligned} y_k[n] &= \sum_\gamma \sum_{\ell=0}^{L-1} y[\gamma L + \ell] e^{-j(2\pi k\ell/L)} p[n - \gamma L - \ell] \\ &= \sum_{\ell=0}^{L-1} \left(\sum_\gamma y[\gamma L + \ell] p[n - \gamma L - \ell] \right) e^{-j(2\pi k\ell/L)}. \end{aligned} \quad (45)$$

According to Figure 22, the recovered symbol $\hat{s}_k[n]$ is obtained after K-fold decimation of $y_k[n]$. Moreover, in practice, a timing recovery loop determines a timing phase at which $y_k[n]$ should be decimated. Equalization is also necessary to combat the distortion introduced by the channel. A single-tap or a multitap equalizer per subcarrier may be used. In the latter case, one may use a fractionally spaced equalizer to relax on the sensitivity of the FMT receiver to timing phase. In such cases, the decimation factor is smaller than K.

Here, we present a general polyphase structure for computation of the samples of $y_k[n]$, for an arbitrary value of n; hence, any arbitrary choice of decimation factor is made possible. To this end, we write

$$n - \ell = \eta L + \kappa, \quad (46)$$

where η is the integer part of $(n-\ell)/L$ and $\kappa = 0, 1, \ldots, L-1$, is the remainder of $n - \ell$ divided by L. Using (46) into (45), we get

$$y_k[n] = \sum_{\ell=0}^{L-1} \left(\sum_\gamma y[\gamma L + \ell] p[(\eta - \gamma)L + \kappa] \right) e^{-j(2\pi k\ell/L)}. \quad (47)$$

Following (47), one may propose the analysis polyphase structure that is presented in Figure 23. The commutator takes the successive samples of the received signal $y[n]$ and distributes them across the inputs of the time-varying filters

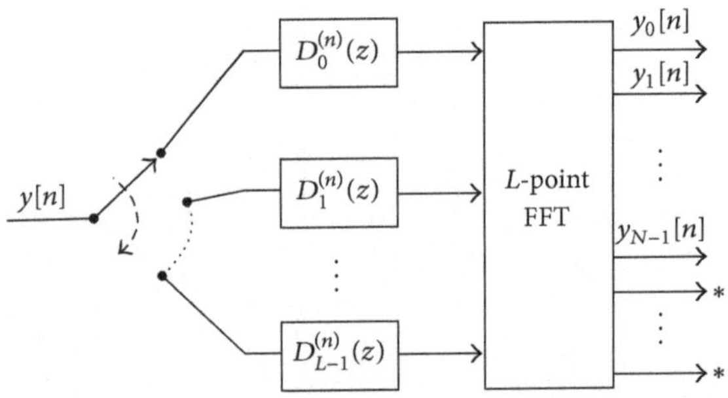

FIGURE 23: Polyphase implementation of an FMT receiver.

$D_0^{(n)}(z)$ through $D_{L-1}^{(n)}(z)$. The filters $D_0^{(n)}(z)$ through $D_{L-1}^{(n)}(z)$ are picked up from the set of L polyphase components $E_0(z)$ through $E_{L-1}(z)$ of the prototype filter $P(z) = \sum_n p[n]z^{-n}$. For a given time n, $D_\ell^{(n)}(z) = E_\kappa(z)$. The reader should be reminded that while in Figure 21 the time-varying filters $D_0^{(n)}(z)$ through $D_{L-1}^{(n)}(z)$ are picked up from the polyphase components of $P(z)$ with respect to the decimation factor K, in Figure 23 $D_0^{(n)}(z)$ through $D_{L-1}^{(n)}(z)$ are picked up from the polyphase components of $P(z)$ with respect to the decimation factor L.

Following Figure 23, the complexity of implementation of FMT receiver, for computation of each set of output samples, is obtained as $(L/2)\log_2 L + \beta L$ complex multiplications. In FMT, unlike CMT and SMT, the use of symbol spaced multitap equalizer is sufficient. If an M tap symbol spaced equalizer is used, one will find the complexity number $(L/2)\log_2 L + \beta L + MN$ complex multiplications per each set of detected data symbols.

5.4. Complexity Comparison with OFDM. Being an alternative to the widely adopted OFDM in various standards, it is always of interest and important to compare FBMC methods against OFDM. As noted in Section 1, the much higher quality of the prototype filter in FBMC makes it an ideal choice for many applications where OFDM faces challenges because of the poor frequency response of its prototype filter. Here, we discuss the additional complexity brought to FBMC (compared to OFDM) because of the use of a higher quality prototype filter.

We count the complexity of OFDM as one IFFT, $(L/2)\log_2 L$ complex multiplications, at the transmitter, and one FFT and N single-tap equalizers, $(L/2)\log_2 L+N$ complex multiplications, at the receiver. Thus, recalling the complexity numbers that were presented above, one finds that

$$\rho_{\text{SMT}}^{Tx} = \frac{\log_2 L + 2\beta}{\log_2 L}, \tag{48}$$

$$\rho_{\text{SMT}}^{Rx} = \frac{L\log_2 L + \beta L + MN}{(L/2)\log_2 L + N}, \tag{49}$$

where ρ_{SMT}^{Tx} and ρ_{SMT}^{Rx}, respectively, denote the relative complexity of SMT transmitter and receiver against OFDM. These numbers remain about the same in the case of CMT. To get a better feel of these complexity ratios in practical cases, consider a case where $L = 1024$, $N = 0.8L$ (i.e., 20% of subcarriers are left as guard band), $M = 1$, and $\beta = 4$. For this case, we find $\rho_{\text{SMT}}^{Tx} = 1.8$ and $\rho_{\text{SMT}}^{Rx} = 2.55$. For $M = 3$, the most likely value of M if a multitap equalizer is used, $\rho_{\text{SMT}}^{Rx} = 2.83$. Hence, one may say, CMT and SMT have a complexity that is roughly two to three times that of the OFDM.

Similarly, for FMT, we find that

$$\rho_{\text{FMT}}^{Tx} = \frac{\log_2 L + 2\beta}{\log_2 L}, \tag{50}$$

$$\rho_{\text{FMT}}^{Rx} = \frac{(L/2)\log_2 L + \beta L + MN}{(L/2)\log_2 L + N}. \tag{51}$$

Comparing (48) and (50), one will find that SMT/CMT and FMT transmitters have the same complexity. On the other hand, comparing (49) and (51), one will find that FMT receiver has a lower complexity than SMT/CMT. In any case, it still remains fair to say FBMC systems, in general, are two to three times more complex than OFDM. Nevertheless, it should be noted that this comparison is with a basic OFDM. It does not consider the cases where advanced signal processing methods may have to be adopted to compensate for other channel effects, for example, carrier frequency offset among different users in an OFDMA network. In such cases, one often finds that OFDM is more complex than FBMC [10].

6. Equalization

In FBMC receivers, equalization is performed at the output of the analysis filter banks. It is often assumed that each subcarrier has a small bandwidth; hence, the channel may be assumed to be flat over each subcarrier band. In that case, a single-tap equalizer per subcarrier would suffice. In cases where the flat gain approximation may be insufficient, a multitap equalizer per subcarrier band may be necessary. For CMT and SMT systems, it is necessary to use a fractionally spaced equalizer. A tap-spacing of half symbol interval is the most convenient choice and, hence, has been suggested in the literature, for example, [43]. FMT on the other hand allows the use of symbol spaced equalizers.

Although, in many scenarios it may appear that a single-tap equalizer per subcarrier is sufficient, in practice, where carrier and clock mismatch between the transmitter and receiver is inevitable, a multitap equalizer can be instrumental. In particular, the difference between the clock used at transmitter to pass the signal samples to a digital-to-analog converter (DAC) and the one used at the receiver to control the rate of samples taken by an analog-to-digital converter (ADC) at the receiver introduces a constant drift in the timing phase of sampled signal. This drift, as explained in the next section, can significantly deteriorate the receiver performance. The use of a multitap equalizer can resolve this problem to a great extent. This is demonstrated through a numerical example in the next section.

Another point that should be emphasized here is the following. Most researchers, who have published in the area of FBMC, have limited their results to the use of one single-tap equalizer per subcarrier band. This has been to emphasize on the simplicity of equalization in FBMC systems and also to stay in par with OFDM. Alternatively, one may look at the use of multitap equalizers in FBMC systems as a mechanism that brings with itself a number of advantages. Firstly, the use of multitap equalizers removes any small frequency selectivity over each subcarrier band and, hence, facilitates adoption of a large symbol constellations and this, in turn, increases the bandwidth efficiency of transmission. Secondly, it allows reducing the number of subcarriers which, in turn, reduces the system (i) complexity (possibly compensating for the added complexity resulting from the use of multitap equalizers), (ii) latency, (iii) sensitivity to carrier frequency offset, and (iv) peak-to-average power ratio (PAPR).

7. Synchronization and Tracking Methods

Synchronization methods are necessary in any receiver to compensate for any difference between the carrier frequency of the incoming signal and the local oscillator used to demodulate it as well as to compensate for any clock mismatch between the transmitter and receiver. Pilot aided and blind synchronization methods have been widely studied in the literature from the beginning of development of data modems [76].

In majority of standards, including those of OFDM and FBMC, the pilot aided approaches have been adopted. For instance, all the OFDM packet formats that have been proposed for WiFi and WiMAX as well as those suggested in LTE and LTE-Advanced start with a short-training periodic preamble, for coarse carrier acquisition. This follows by a long preamble, consisting of two or more similar OFDM symbols, that is used for fine tuning of the carrier, timing acquisition, and channel estimation; for example, see [72, 77]. In addition, to be able to track channel variations during the payload of each packet, it is often proposed to insert distributed pilots in the time-frequency space. Interpolation methods are then used to obtain the channel gain and accordingly the equalizers gain, at the data points [78–81].

A few researchers, who have looked at the packet design for FBMC, have also suggested a similar packet format to those of the OFDM [82–86]. In particular, a periodic training consisting of a few repetitions of a pilot FBMC symbol is suggested to be used along with a least squares or a maximum likelihood estimator to estimate CFO and symbol timing. This is equivalent to long preamble in OFDM. Moreover, in [82], it is noted that the use of a short preamble, similar to that of OFDM, may be useful for coarse estimation of carrier, so that fine tuning of carrier and timing phase based on the long preamble can be established more robustly. Distributed pilots to track channel variations in payload part of FBMC packets have also been proposed [87, 88]. In the rest of this section, we make comments/suggestions for future research in three directions related to synchronization and tracking methods in FBMC systems.

7.1. Preamble Design. Figure 24 presents the packet format that has been proposed in [82–86]. The short preamble

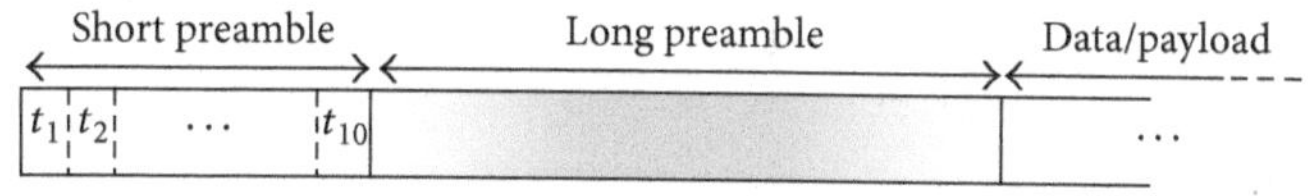

FIGURE 24: An FBMC data packet with short and long preambles.

consists of a few tones with wider separation than the subcarrier spacing to allow detection of large CFO values. It is used to obtain a coarse estimate of CFO, so that the residual CFO (after removal of the estimated CFO) will be within the range of half symbol spacing. The long preamble, then, may be used for fine tuning of the carrier. The works presented in [83–85] do not mention the short preamble, but they make assumption that CFO is within the range of half symbol spacing. Naturally, to satisfy this assumption a mechanism similar to short preamble may be necessary.

There is also some difference in the long preamble proposed in [83–85] and the one proposed in [82, 86]. In the context of SMT waveform, the long preamble proposed in [83–85] is constructed based on a set of complex-valued pilot symbols that occupy all subcarriers. The long preamble proposed in [82, 86], on the other hand, is real-valued and only occupies the alternate subcarriers; the other subcarriers are left empty. This modification/simplification of the long preamble as demonstrated through numerical results in [86] improves the accuracy of the CFO estimates significantly.

Here, to make suggestions to further improve the preamble design, we first note that the preamble is a packet component that constitutes a loss in the bandwidth efficiency of transmission. Clearly, an attempt to reduce the length of the preamble improves the bandwidth efficiency. The research so far in the above cited works has led to the following observations. (i) A preamble with a set of pilot symbols that are well separated in frequency is needed to assure detection and correction of large CFO values. (ii) A longer preamble that allows a more accurate estimate of CFO may be also necessary. On the other hand, one may realize that to obtain the CFO and symbol timing estimates, it is not necessary to send pilots on all the subcarriers. The number of parameters that characterize the channel response (determined by the length of the channel impulse response) is usually much smaller than the number of subcarriers. This means, a fraction of subcarriers, whose number is greater than or equal to the length of the channel impulse response, suffice for pilot subcarriers in the preamble. This, in turn, implies that one may propose to combine the short and long preambles into one preamble, and by taking such approach, obviously, a shorter preamble will result. We note that the preamble proposed in [82, 86], compared to those proposed in [83–85], may be viewed as a step forward in this direction. While the works in [83–85] use all subcarriers to transmit the preamble pilots, [82, 86] allocate pilots only to alternate (even or odd) subcarriers.

Another consideration that needs one's attention in designing an effective preamble is the choice of the prototype filter that is used for preamble. Assuming that the preamble consists of only a long preamble, the preamble length is determined by the length of the prototype filter and the

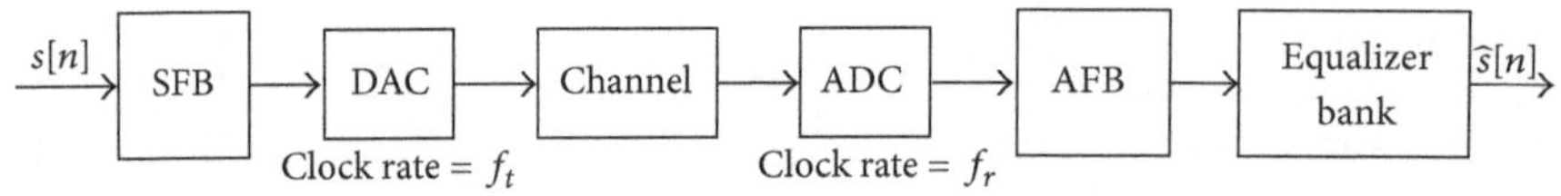

FIGURE 25: An FBMC transceiver with timing drift and the corresponding compensator block.

number of its repetitions. Accordingly, a shorter prototype filter leads to a shorter preamble and thus a more bandwidth efficient packet. The above suggestion to space out the pilot subcarriers in the preamble can be taken advantage of to shorten the length of the preamble. This is because wider subcarrier allows one to increase the bandwidth of each subcarrier and accordingly reduce the symbol period and thus the length of the corresponding prototype filter. Nevertheless, it should be also noted that a broader band of each subcarrier can make it prone to some level of frequency selectivity that should be taken into account while identifying the channel parameters.

7.2. Channel Tracking. Tracking of channel variation (including variations due to CFO and clock timing drift), during the payload transmission, is usually made through distributed pilots that are transmitted along with the data symbols. This topic, as noted above, has been well studied for OFDM. However, it has received very limited attention in the FBMC literature. Stitz et al. [88] have studied the problem of pilot-aided synchronization and tracking methods in CMT/SMT-type systems. They have shown that by applying specific pilots designed for CMT/SMT, the carrier frequency offset (CFO), fractional time delay (FTD), and channel response can be accurately estimated. The work presented in [89] has taken a different approach. It takes note of the fact that each signal component at the output of the analysis filter bank, before performing equalization and taking its real (or imaginary) part, has the form of $h_k[n](s_k[n] + j\psi_k[n])$ (or $h_k[n](js_k[n] + \psi_k[n])$), where $\psi_k[n]$ arises from ICI and ISI. It is thus argued that if the same (known) symbol is transmitted over two instants of times at the same subcarrier or allocated to the real and imaginary parts of an OQAM symbol in a SMT waveform, the analyzed signal samples give a pair of equations that can be solved to obtain an estimate of $h_k[n]$.

In [27] (also see [90]), it has been noted that the use of PAM symbols in CMT (and similarly in SMT, if the real and imaginary parts of each OQAM symbol are viewed as a pair of PAM symbols) allows development of a blind equalization algorithm. This algorithm may also be used for channel tracking. Hence, in CMT and SMT channel tracking is possible without any use of pilot symbols. To elaborate more on this possibility, we note that the convergence behavior of the blind equalizer that is developed in [27] was thoroughly analyzed in [91] (also see [92, 93]). It was found that the CMT/SMT blind equalizer has a well-behaved performance surface around its optimal setting. This implies that if the equalizer gain is set near its optimal value and channel subsequently varies slowly, the blind equalization algorithm of [27] can easily track the channel variation. This observation that yet to be thoroughly studied will add an additional advantage to CMT and SMT over OFDM in terms of bandwidth efficiency.

7.3. Timing Tracking. As noted above, any difference between the clock frequency at the transmitter and its counterpart at the receiver introduces a timing drift. The common method of compensating for this timing drift is to add a timing recovery loop to the system. In a single carrier receiver, the addition of a timing recovery loop as part of the receiver processing is a relatively straightforward task and is well understood [72, 94]. In an OFDM system, with a not very long payload, the timing drift may be absorbed by the CP, provided that not the full length of the CP has been consumed by the channel transient response. Drift of timing phase introduces a linear phase across frequency and that can be easily compensated by the single-tap equalizers at the FFT outputs. In FBMC, on the other hand, the absence of CP/guard interval makes an FBMC receiver more sensitive to timing drift. However, fortunately, this problem can be resolved trivially through the use of a set of multitap fractionally spaced equalizers, one per subcarrier.

To demonstrate the impact of per subcarrier fractionally spaced equalizers on resolving the problem of timing drift, we consider a CMT/SMT transceiver system which we simulate following the schematic diagram presented in Figure 25. The samples of the transmit signal are converted from digital to analog using a DAC that operated at a clock rate of f_t. At the receiver, the received signal is sampled through an ADC with a clock that has a rate of f_r. The clock frequencies f_t and f_r are different but typically are close. In practice, their difference may be as large as 50 ppm (part per million). This corresponds to the case where each clock has a precision of 25 ppm.

For the purpose of demonstration here, we consider the case where channel is absent. In that case, the effect of using a pair of DAC and ADC with different clock rates is like resampling the generated waveform with a resampling ratio f_r/f_t. This concept can be easily applied in a MATLAB simulation environment that we use, using the function "resample." The results of simulating a CMT system with 64 active subcarriers and a clock drift of 50 ppm are presented in Figure 26. The results show the signal-to-interference ratio (SIR) averaged over all subcarriers, as one proceed over successive CMT symbols. Timing phase is set to zero (i.e., perfect timing) at the beginning. The drift of timing phase introduces some ISI that cannot be resolved when single-tap equalizers are used but can be compensated through multitap equalizers. Perfect compensation may require a long equalizer. However, a 3- or 5-tap equalizer is sufficient to avoid a significant drop of SIR. We consider an SIR of above 30 dB as acceptable, given that, in typical practice, communication systems are often designed for output SINR values 20 dB or less. A margin of 10 dB should be sufficient to consider ISI/ICI due to timing drift as negligible. For the numerical example presented here, while a single-tap equalizer can preserve the SIR of above 30 dB for the first 400 CMT symbols, the use of a 3-tap and

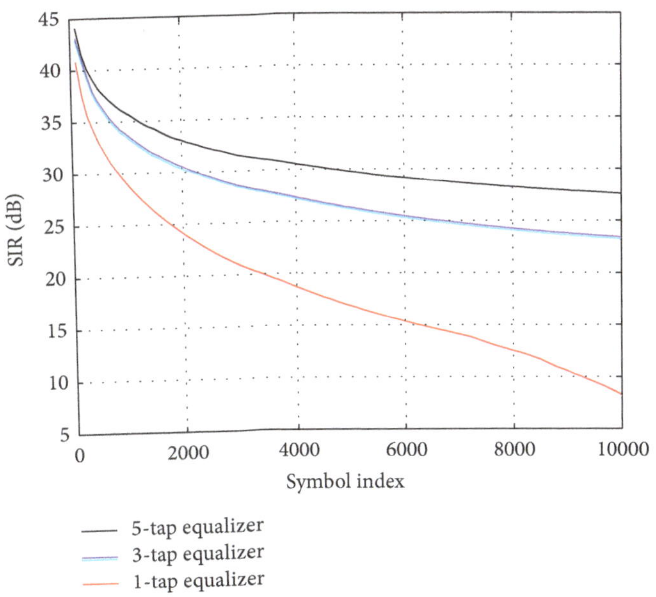

FIGURE 26: SIR variation as a function of symbol index (equivalently, timing drift) in a CMT transceiver.

5-tap equalizer increases this range to 1200 and 2500 CMT symbols, respectively.

Another possible solution that may be adopted to resolve the difference between the DAC clock at the transmitter and ADC clock at the receiver may be proposed by taking note of the following observation. An FBMC waveform (being CMT, SMT, or FMT) is a cyclostationary signal whose statistics are repeated every T_c second, where T_c is an integer multiple of the symbol period T. In FMT and SMT, $T_c = T$, and in CMT, $T_c = 2T$. We conjecture that taking note of this property, one can design a digital resampler circuit that may be added to the output of ADC, at the receiver, for tuning the number of samples per each period of T_c equal to a predefined value. Such solution is yet to be found through a clever study of the underlying signals.

8. Applications

In [19], OFDM and FBMC are contrasted with respect to their suitability for different applications. It is noted that while for many applications OFDM is a perfect choice, FBMC may be a superior choice for other applications. The applications in which FBMC may be preferred are (i) the uplink of multiuser multicarrier systems, (ii) cognitive radios, (iii) doubly dispersive channels, and (iv) digital subscriber lines and power line communications. Detailed discussion on why FBMC is a better choice in these applications is provided in [19]. The rest of this section emphasizes on an emerging class of communication systems, massive (multiuser) MIMO, whose significance as an enabling technology for 5G has recently been discussed in the literature and other forums.

Massive MIMO, in essence, is a multiuser technique, somewhat similar to code division multiple access (CDMA). In its simplest form, each mobile terminal (MT) has a single antenna, but the base station (BS) has a large number of antennas. The spreading code for each user is then the vector of channel gains between the respective MT antenna and the multiple antennas at the BS. Accordingly, by increasing the number of antennas at the BS, the processing gain of each user can be increased to an arbitrarily large value. In the pioneering work of Marzetta [95] it is noted that, in the limit, as the number of BS antennas tends to infinity, the processing gain of the system tends to infinity and, as a result, the effects of both noise and multiuser interference (MUI) are completely removed. Therefore, the network capacity (in theory) can be increased unboundedly by increasing the number of antennas at the BS [95].

Motivated by Marzetta's observations [95] multiple research groups in recent past have studied a variety of implementation issues related to massive MIMO systems, for example, [96–100]. Also, different groups have started the development of testbeds to confirm the theoretical observation of [95], in practice, for example. An assumption made by Marzetta [95] and followed by other researchers is that orthogonal frequency division multiplexing (OFDM) may be used to convert the frequency selective channels between each MT and the multiple antennas at the BS to a set of flat fading channels. Accordingly, the flat gains associated with the set of channels within each subcarrier constitute the spreading gain vector that is used for dispreading of the respective data in upstream as well as for precoding of the respective downstream data.

In a recent work [101], embarking on the above concepts, we have introduced the application of filter bank multicarrier (FBMC) to the area of massive MIMO communications. The interesting finding in this work is that unlike in conventional MIMO where applicability of FBMC (CMT and SMT, in particular) is found limited, FBMC offers a number of appealing properties in the application of massive MIMO. In particular, the linear combining of the signal components from different antennas at BS (equivalent to dispreading in CDMA) smooths channel distortion across each subcarrier band. This property, which we call *self-equalization*, allows one to relax on the requirement of having approximately flat gain for the subcarriers. This, in turn, allows one to reduce the number of subcarriers in an assigned bandwidth. Reducing the number of subcarriers has the following impacts, which all position FBMC as a strong candidate in the application of massive MIMO.

(1) The complexity of both transmitter and receiver will be reduced, since the underlying polyphase structures will be based on smaller size IFFT/FFT blocks.

(2) The system latency will be reduced. This is because the length and, thus, the corresponding group delay of the underlying prototype filter are reduced.

(3) The system sensitivity to CFO will be reduced.

(4) The PAPR of the transmit signal will be reduced.

The preliminary research results that are presented in [101], clearly, open up a new research direction for exploration of FBMC techniques in the area of massive MIMO.

9. Conclusions

This paper presented a tutorial review of the class of filter bank multicarrier (FBMC) communication systems, with emphasis on the recent results of the authors research laboratory. It was noted that by building FBMC systems based on well-designed prototype filters, the spectrum of each subcarrier can be contained within a limited bandwidth. This, in turn, allows transmission over noncontiguous bands, a property that makes FBMC an ideal choice for many applications, including the uplink of multiuser multicarrier networks and cognitive radios. It was also pointed out that FBMC systems can be adopted to doubly dispersive channels to achieve a performance far superior to that of OFDM. The polyphase structures that have been proposed for efficient implementation of FBMC systems were reviewed and their complexities were compared with that of a standard OFDM system. It was noted that the complexity of FBMC transmitters is close to (but less than) twice their OFDM counterpart, and the complexity of FBMC receivers is close to (but less than) three times that of their OFDM counterpart. However, it was also noted that, in some applications, for example, in the uplink of multiuser systems where synchronization cannot be made perfectly, hence, multiuser cancellation techniques should be applied, OFDM can be far more complex than FBMC. Moreover, it was noted that more research is needed to better understand the problems related to synchronization, equalization, and tracking of channel variations in FBMC systems. Possible application of FBMC in the emerging area of massive MIMO was also highlighted, and a number of advantages that FBMC offers in this application were identified.

Conflict of Interests

The author declares that there is no conflict of interests regarding the publication of this paper.

References

[1] W. Y. Chen, *DSL Simulation Techniques and Standards Development for Digital Subscriber Line Systems*, Macmillan Publishing, Indianapolis, Ind, USA, 1998.

[2] J. C. T. Starr and P. Silverman, *Understanding Digital Subscriber Line Technology*, Prentice Hall, Upper Saddle River, NJ, USA, 1999.

[3] R. V. Nee and R. Prasad, *OFDM for Wireless Multimedia Communications*, Arthec House, Boston, Mass, USA, 2000.

[4] Y. Li and G. Stüber, Eds., *Orthogonal Frequency Division Multiplexing for Wireless Communications*, Springer, New York, NY, USA, 2006.

[5] M. Morelli, C.-C. J. Kuo, and M.-O. Pun, "Synchronization techniques for orthogonal frequency division multiple access (OFDMA): a tutorial review," *Proceedings of the IEEE*, vol. 95, no. 7, pp. 1394–1427, 2007.

[6] D. Huang and K. B. Letaief, "An interference-cancellation scheme for carrier frequency offsets correction in OFDMA systems," *IEEE Transactions on Communications*, vol. 53, no. 7, pp. 1155–1165, 2005.

[7] K. Lee and I. Lee, "CFO compensation for uplink OFDMA systems with conjugated gradient," in *Proceedings of the IEEE International Conference on Communications (ICC '11)*, pp. 1–5, Kyoto, Japan, June 2011.

[8] K. Lee, S.-R. Lee, S.-H. Moon, and I. Lee, "MMSE-based CFO compensation for uplink OFDMA systems with conjugate gradient," *IEEE Transactions on Wireless Communications*, vol. 11, no. 8, pp. 2767–2775, 2012.

[9] A. Farhang, N. Marchetti, and L. E. Doyle, "Low complexity LS and MMSE based CFO compensation techniques for the uplink of OFDMA systems," in *Proceedings of the IEEE International Conference on Communications (ICC '13)*, pp. 5748–5753, IEEE, Budapest, Hungary, June 2013.

[10] H. Saeedi-Sourck, Y. Wu, J. W. M. Bergmans, S. Sadri, and B. Farhang-Boroujeny, "Complexity and performance comparison of filter bank multicarrier and OFDM in uplink of multicarrier multiple access networks," *IEEE Transactions on Signal Processing*, vol. 59, no. 4, pp. 1907–1912, 2011.

[11] S. Brandes, I. Cosovic, and M. Schnell, "Sidelobe suppression in OFDM systems by insertion of cancellation carriers," in *Proceedings of the 62nd IEEE Vehicular Technology Conference (VTC '05)*, vol. 1, pp. 152–162, September 2005.

[12] Z. Yuan, S. Pagadarai, and A. M. Wyglinski, "Cancellation carrier technique using genetic algorithm for OFDM sidelobe suppression," in *Proceedings of the Military Communications Conference (MILCOM '08)*, pp. 1–5, IEEE, 2008.

[13] A. Selim, I. Macaluso, and L. Doyle, "Efficient sidelobe suppression for OFDM systems using advanced cancellation carriers," in *Proceedings of the IEEE International Conference on Communications (ICC '13)*, pp. 4687–4692, Budapest, Hungary, June 2013.

[14] S. N. Premnath, D. Wasden, S. K. Kasera, B. Farhang-Boroujeny, and N. Patwari, "Beyond ofdm: best-effort dynamic spectrum access using filterbank multicarrier," in *Proceedings of the 4th International Conference on Communication Systems and Networks (COMSNETS '12)*, January 2012.

[15] S. N. Premnath, D. Wasden, S. K. Kasera, N. Patwari, and B. Farhang-Boroujeny, "Beyond OFDM: best-effort dynamic spectrum access using filterbank multicarrier," *IEEE/ACM Transactions on Networking*, vol. 21, no. 3, pp. 869–882, 2013.

[16] B. Farhang-Boroujeny and R. Kempter, "Multicarrier communication techniques for spectrum sensing and communication in cognitive radios," *IEEE Communications Magazine*, vol. 46, no. 4, pp. 80–85, 2008.

[17] B. Farhang-Boroujeny, "Prolate filters for nonadaptive multitaper spectral estimators with high spectral dynamic range," *IEEE Signal Processing Letters*, vol. 15, pp. 457–460, 2008.

[18] B. Farhang-Boroujeny, "Filter bank spectrum sensing for cognitive radios," *IEEE Transactions on Signal Processing*, vol. 56, no. 5, pp. 1801–1811, 2008.

[19] B. Farhang-Boroujeny, "OFDM versus filter bank multicarrier," *IEEE Signal Processing Magazine*, vol. 28, no. 3, pp. 92–112, 2011.

[20] A. Sahin, I. Guvenc, and H. Arslan, "A survey on multicarrier communications: prototype filters, lattice structures, and implementation aspects," *IEEE Communications Surveys and Tutorials*, vol. 16, no. 3, pp. 1312–1338, 2013.

[21] M. Bellanger, "FBMC physical layer: a primer," Tech. Rep. 06/2010, 2010.

[22] M. Payaró, A. Pascual-Iserte, and M. Nájar, "Performance comparison between FBMC and OFDM in MIMO systems under channel uncertainty," in *Proceedings of the European Wireless Conference (EW '10)*, pp. 1023–1030, April 2010.

[23] T. Ihalainen, A. Ikhlef, J. Louveaux, and M. Renfors, "Channel equalization for multi-antenna FBMC/OQAM receivers," *IEEE Transactions on Vehicular Technology*, vol. 60, no. 5, pp. 2070–2085, 2011.

[24] M. Tzannes, M. Tzannes, and H. Resnikoff, "The DWMT: a multicarrier transceiver for ADSL using M-band wavelet transforms," ANSI Contribution T1E1.4/93-067, 1993.

[25] M. Tzannes, M. Tzannes, J. Proakis, and P. Heller, "DMT systems, DWMT systems and digital filter banks," in *Proceedings of the IEEE International Conference on Communications (ICC '94)*, vol. 1, pp. 311–315, 1994.

[26] S. D. Sandberg and M. A. Tzannes, "Overlapped discrete multitone modulation for high speed copper wire communications," *IEEE Journal on Selected Areas in Communications*, vol. 13, no. 9, pp. 1571–1585, 1995.

[27] B. Farhang-Boroujeny, "Multicarrier modulation with blind detection capability using cosine modulated filter banks," *IEEE Transactions on Communications*, vol. 51, no. 12, pp. 2057–2070, 2003.

[28] S. Galli and O. Logvinov, "Recent developments in the standardization of power line communications within the IEEE," *IEEE Communications Magazine*, vol. 46, no. 7, pp. 64–71, 2008.

[29] F. Sjoberg, R. Nilsson, M. Isaksson, P. Odling, and P. Borjesson, "Asynchronous Zipper [subscriber line duplex method]," in *Proceedings of the IEEE International Conference on Communications (ICC '99)*, vol. 1, pp. 231–235, Vancouver, Canada, June 1999.

[30] "Very-high bit-rate digital subscriber lines (vdsl) metallic interface, part 3: technical specification of a multi-carrier modulation transceiver," Working Group RIE1.4, Document T1E1.4/2000-013R3, 2000.

[31] D. J. G. Mestdagh, M. R. Isaksson, and P. Ödling, "Zipper VDSL: A solution for robust duplex communication over telephone lines," *IEEE Communications Magazine*, vol. 38, no. 5, pp. 90–96, 2000.

[32] M. Nava and G. Ökvist, "The zipper prototype: a complete and flexible VDSL multicarrier solution," *IEEE Communications Magazine*, vol. 40, no. 12, pp. 92–105, 2002.

[33] R. Nilsson, F. Sjöberg, M. Isaksson, J. M. Cioffi, and S. K. Wilson, "Autonomous synchronization of a DMT-VDSL system in unbundled networks," *IEEE Journal on Selected Areas in Communications*, vol. 20, no. 5, pp. 1055–1063, 2002.

[34] G. Cherubini, E. Eleftheriou, and S. Ölçer, "Filtered multitone modulation for very high-speed digital subscriber lines," *IEEE Journal on Selected Areas in Communications*, vol. 20, no. 5, pp. 1016–1028, 2002.

[35] "Very high speed digital subscriber line transceivers 2 (VDSL2)," ITU-T Recommendation G.993.2, 2006.

[36] M. Bellec and P. Pirat, "OQAM performances and complexity," IEEE P802.22 Wireless Regional Area Network, 2006, http://www.ieee802.org/22/Meeting_documents/2006_Jan/index.html.

[37] "Wideband air interface isotropic orthogonal transform algorithm," Public Safety Wideband Data Standards Project Digital Radio Technical Standards, TIA Committee TR-8.5, 2003.

[38] B. Farhang-Boroujeny and C. Yuen, "Cosine modulated and offset QAM filter bank multicarrier techniques: a continuous-time prospect," *EURASIP Journal on Advances in Signal Processing*, vol. 2010, Article ID 165654, 16 pages, 2010.

[39] R. Chang, "High-speed multichannel data transmission with bandlimited orthogonal signals," *Bell System Technical Journal*, vol. 45, pp. 1775–1796, 1966.

[40] B. Saltzberg, "Performance of an efficient parallel data transmission system," *IEEE Transactions on Communication Technology*, vol. 15, no. 6, pp. 805–811, 1967.

[41] M. G. Bellanger and J. L. Daguet, "TDM-FDM transmultiplexer: digital polyphase and FFT," *IEEE Transactions on Communications*, vol. 22, no. 9, pp. 1199–1205, 1974.

[42] B. Hirosaki, "An analysis of automatic equalizers for orthogonally multiplexed QAM systems," *IEEE Transactions on Communications*, vol. 28, no. 1, pp. 73–83, 1980.

[43] B. Hirosaki, "An orthogonally multiplexed qam system using the discrete fourier transform," *IEEE Transactions on Communications*, vol. 29, no. 7, pp. 982–989, 1981.

[44] B. Le Floch, M. Alard, and C. Berrou, "Coded orthogonal frequency division multiplex," *Proceedings of the IEEE*, vol. 83, no. 6, pp. 982–996, 1995.

[45] G. Cariolaro and F. C. Vagliani, "OFDM scheme with a half complexity," *IEEE Journal on Selected Areas in Communications*, vol. 13, no. 9, pp. 1586–1599, 1995.

[46] P. Siohan and C. Roche, "Cosine-modulated filterbanks based on extended Gaussian functions," *IEEE Transactions on Signal Processing*, vol. 48, no. 11, pp. 3052–3061, 2000.

[47] B. Hirosaki, S. Hasegawa, and A. Sabato, "Advanced groupband data modem using orthogonally multiplexed QAM technique," *IEEE Transactions on Communications*, vol. 34, no. 6, pp. 587–592, 1986.

[48] S. Govardhanagiri, T. Karp, P. Heller, and T. Nguyen, "Performance analysis of multicarrier modulation systems using cosine modulated filter banks," in *Proceedings of the IEEE International Conference on Acoustics, Speech, and Signal Processing (ICASSP '99)*, vol. 3, pp. 1405–1408, March 1999.

[49] P. Vaidyanathan, *Multirate Systems and Filter Banks*, Prentice Hall, Englewood Cliffs, NJ, USA, 1993.

[50] N. J. Fliege, *Multirate Digital Signal Processing*, John Wiley & Sons, New York, NY, USA, 1994.

[51] N. J. Fliege, "Modified DFT polyphase SBC filter banks with almost perfect reconstruction," in *Proceedings of the IEEE International Conference on Acoustics, Speech, and Signal Processing (ICASSP '94)*, vol. 3, pp. III/149–III/152, Adelaide, Australia, April 1994.

[52] P. N. Heller, T. Karp, and T. Q. Nguyen, "A general formulation of modulated filter banks," *IEEE Transactions on Signal Processing*, vol. 47, no. 4, pp. 986–1002, 1999.

[53] T. Karp and N. J. Fliege, "Modified DFT filter banks with perfect reconstruction," *IEEE Transactions on Circuits and Systems II: Analog and Digital Signal Processing*, vol. 46, no. 11, pp. 1404–1414, 1999.

[54] R. Bregović and T. Saramaki, "A systematic technique for designing linear-phase FIR prototype filters for perfect-reconstruction cosine-modulated and modified DFT filterbanks," *IEEE Transactions on Signal Processing*, vol. 53, no. 8, part 2, pp. 3193–3201, 2005.

[55] S. Salcedo-Sanz, F. Cruz-Roldan, C. Heneghan, and X. Yao, "Evolutionary design of digital filters with application to subband coding and data transmission," *IEEE Transactions on Signal Processing*, vol. 55, no. 4, pp. 1193–1203, 2007.

[56] P. Siohan, C. Siclet, and N. Lacaille, "Analysis and design of OFDM/OQAM systems based on filterbank theory," *IEEE Transactions on Signal Processing*, vol. 50, no. 5, pp. 1170–1183, 2002.

[57] P. Amini, R.-R. Chen, and B. Farhang-Boroujeny, "Filterbank multicarrier communications for underwater acoustic channels," *IEEE Journal of Oceanic Engineering*, 2014.

[58] B. Farhang-Boroujeny, "A square-root Nyquist (M) filter design for digital communication systems," *IEEE Transactions on Signal Processing*, vol. 56, no. 5, pp. 2127–2132, 2008.

[59] K. W. Martin, "Small side-lobe filter design for multitone data-communication applications," *IEEE Transactions on Circuits and Systems II: Analog and Digital Signal Processing*, vol. 45, no. 8, pp. 1155–1161, 1998.

[60] S. Mirabbasi and K. Martin, "Overlapped complex-modulated transmultiplexer filters with simplified design and superior stopbands," *IEEE Transactions on Circuits and Systems II: Analog and Digital Signal Processing*, vol. 50, no. 8, pp. 456–469, 2003.

[61] M. Rossi, J.-Y. Zhang, and W. Steenaart, "Iterative least squares design of perfect reconstruction QMF banks," in *Proceedings of the Canadian Conference on Electrical and Computer Engineering (CCECE '96)*, pp. 762–765, May 1996.

[62] W. Kozek and A. F. Molisch, "Nonorthogonal pulseshapes for multicarrier communications in doubly dispersive channels," *IEEE Journal on Selected Areas in Communications*, vol. 16, no. 8, pp. 1579–1589, 1998.

[63] G. Matz, D. Schafhuber, K. Gröchenig, M. Hartmann, and F. Hlawatsch, "Analysis, optimization, and implementation of low-interference wireless multicarrier systems," *IEEE Transactions on Wireless Communications*, vol. 6, no. 5, pp. 1921–1930, 2007.

[64] G. Matz, H. Bolcskei, and F. Hlawatsch, "Time-frequency foundations of communications: concepts and tools," *IEEE Signal Processing Magazine*, vol. 30, no. 6, pp. 87–96, 2013.

[65] B. Boashash, Ed., *Time Frequency Signal Analysis and Processing: A Comprehensive Reference*, Elsevier, Oxford, UK, 2003.

[66] R. Haas and J.-C. Belfiore, "A time-frequency well-localized pulse for multiple carrier transmission," *Wireless Personal Communications*, vol. 5, no. 1, pp. 1–18, 1997.

[67] M. Alard, "Construction of a multicarrier signal," Patent WO 96/35278, 1996.

[68] P. Amini, R.-R. Chen, and B. Farhang-Boroujeny, "Filterbank multicarrier for underwater communications," in *Proceedings of the 49th Annual Allerton Conference on Communication, Control, and Computing (Allerton '11)*, pp. 639–646, September 2011.

[69] P. Amini, C. H. Yuen, R.-R. Chen, and B. Farhang-Boroujeny, "Isotropic filter design for MIMO filter bank multicarrier communications," in *Proceedings of the IEEE Sensor Array and Multichannel Signal Processing Workshop (SAM '10)*, pp. 89–92, October 2010.

[70] J. Du and S. Signell, "Classic OFDM systems and pulse shaping OFDM/OQAM systems," Tech. Rep., KTH—Royal Institute of Technology, Stockholm, Sweden, 2007.

[71] T. Strohmer and S. Beaver, "Optimal OFDM design for time-frequency dispersive channels," *IEEE Transactions on Communications*, vol. 51, no. 7, pp. 1111–1122, 2003.

[72] B. Farhang-Boroujeny, *Signal Processing Techniques for Software Radios*, Lulu, Raleigh, NC, USA, 2nd edition, 2009.

[73] Y. Dandach and P. Siohan, "FBMC/OQAM modulators with half complexity," in *Proceedings of the IEEE Global Telecommunications Conference (GLOBECOM '11)*, pp. 1–5, Houston, Tex, USA, December 2011.

[74] G. Cariolaro and F. C. Vagliani, "An OFDM scheme with a half complexity," *IEEE Journal on Selected Areas in Communications*, vol. 13, no. 9, pp. 1586–1599, 1995.

[75] L. Vangelista and N. Laurenti, "Efficient implementations and alternative architectures for OFDM-OQAM systems," *IEEE Transactions on Communications*, vol. 49, no. 4, pp. 664–675, 2001.

[76] J. S. R. W. Lucky and E. Weldon, *Principles of Data Communications*, McGraw-Hill, New York, NY, USA, 1965.

[77] E. H. Ong, J. Kneckt, O. Alanen, Z. Chang, T. Huovinen, and T. Nihtila, "IEEE 802.11ac: enhancements for very high throughput WLANs," in *Proceedings of the 22nd International Symposium on Personal, Indoor and Mobile Radio Communications (PIMRC '11)*, pp. 849–853, IEEE, September 2011.

[78] I. Barhumi, G. Leus, and M. Moonen, "Optimal training design for MIMO OFDM systems in mobile wireless channels," *IEEE Transactions on Signal Processing*, vol. 51, no. 6, pp. 1615–1624, 2003.

[79] M. Dong and L. Tong, "Optimal design and placement of pilot symbols for channel estimation," *IEEE Transactions on Signal Processing*, vol. 50, no. 12, pp. 3055–3069, 2002.

[80] D. Hu, L. Yang, Y. Shi, and L. He, "Optimal pilot sequence design for channel estimation in MIMO OFDM systems," *IEEE Communications Letters*, vol. 10, no. 1, pp. 1–3, 2006.

[81] C. K. Ho, B. Farhang-Boroujeny, and F. Chin, "Added pilot semi-blind channel estimation scheme for OFDM in fading channels," in *Proceedings of the IEEE Global Telecommunicatins Conference (GLOBECOM '01)*, vol. 5, pp. 3075–3079, November 2001.

[82] B. Farhang-Boroujeny and P. Amini, "Packet format design and decision directed tracking methods for filter bank multicarrier systems," *EURASIP Journal on Advances in Signal Processing*, vol. 2010, Article ID 307983, 2010.

[83] T. Fusco, A. Petrella, and M. Tanda, "Joint symbol timing and CFO estimation in multiuser OFDM/OQAM systems," in *Proceedings of the IEEE 10th Workshop on Signal Processing Advances in Wireless Communications (SPAWC '09)*, pp. 613–617, June 2009.

[84] T. Fusco, A. Petrella, and M. Tanda, "Data-aided symbol timing and CFO synchronization for filter bank multicarrier systems," *IEEE Transactions on Wireless Communications*, vol. 8, no. 5, pp. 2705–2715, 2009.

[85] M. Tanda, T. Fusco, and A. Petrella, "Joint symbol timing and CFO estimation for OFDM/OQAM systems in multipath channels," *Eurasip Journal on Advances in Signal Processing*, vol. 2010, Article ID 897607, 2010.

[86] H. Saeedi-Sourck, S. Sadri, Y. Wu, and B. Farhang-Boroujeny, "Near maximum likelihood synchronization for filter bank multicarrier systems," *IEEE Wireless Communications Letters*, vol. 2, no. 2, pp. 235–238, 2013.

[87] T. H. Stitz, A. Viholainen, T. Ihalainen, and M. Renfors, "CFO estimation and correction in a WiMAX-like FBMC system," in *Proceedings of the 10th IEEE Workshop on Signal Processing Advances in Wireless Communications (SPAWC '09)*, pp. 633–637, June 2009.

[88] T. H. Stitz, T. Ihalainen, A. Viholainen, and M. Renfors, "Pilot-Based synchronization and equalization in filter bank multicarrier communications," *EURASIP Journal on Advances in Signal Processing*, vol. 2010, article 9, Article ID 741429, 2010.

[89] M. Bellanger and T. Dao, "Multicarrier digital transmission system using an OQAM transmultiplexer," Granted patent, US7072412, 2006.

[90] B. Farhang-Boroujeny, "Discrete multitone modulation with blind detection capability," in *Proceedings of the Vehicular*

Technology Conference (VTC '02), vol. 1, pp. 376–380, IEEE, 2002.

[91] L. Lin and B. Farhang-Boroujeny, "Convergence analysis of blind equalizer in a filter-bank-based multicarrier communication system," *IEEE Transactions on Signal Processing*, vol. 54, no. 10, pp. 4061–4067, 2006.

[92] "Convergence an alysis of blind equalizer in a cosine modulated filter bank-based multicarrier communication system," in *Proceedings of the Workshop on Signal Processing Advances in Wireless Communications (SPAWC '03)*, pp. 368–372, IEEE, 2003.

[93] L. Lin and B. Farhang-Boroujeny, "Analytical study of the performance surface of blind equalizer in a cosine modulated multicarrier communication system," in *Proceedings of the IEEE International Conference on Acoustics, Speech, and Signal Processing (ICASSP '05)*, pp. IV337–IV340, March 2005.

[94] F. J. Harris and M. Rice, "Multirate digital filters for symbol timing synchronization in software defined radios," *IEEE Journal on Selected Areas in Communications*, vol. 19, no. 12, pp. 2346–2357, 2001.

[95] T. L. Marzetta, "Noncooperative cellular wireless with unlimited numbers of base station antennas," *IEEE Transactions on Wireless Communications*, vol. 9, no. 11, pp. 3590–3600, 2010.

[96] J. Hoydis, S. Ten Brink, and M. Debbah, "Massive MIMO in the UL/DL of cellular networks: how many antennas do we need?" *IEEE Journal on Selected Areas in Communications*, vol. 31, no. 2, pp. 160–171, 2013.

[97] J. Hoydis, C. Hoek, T. Wild, and S. Ten Brink, "Channel measurements for large antenna arrays," in *Proceedings of the International Symposium on Wireless Communication Systems (ISWCS '12)*, pp. 811–815, Paris, France, August 2012.

[98] J. Jose, A. Ashikhmin, T. L. Marzetta, and S. Vishwanath, "Pilot contamination and precoding in multi-cell TDD systems," *IEEE Transactions on Wireless Communications*, vol. 10, no. 8, pp. 2640–2651, 2011.

[99] X. Gao, O. Edfors, F. Rusek, and F. Tufvesson, "Linear precoding performance in measured very-large MIMO channels," in *Proceedings of the IEEE Vehicular Technology Conference (VTC Fall '11)*, pp. 1–5, IEEE, San Francisco, Calif, USA, September 2011.

[100] S. Payami and F. Tufvesson, "Channel measurements and analysis for very large array systems at 2.6 GHz," in *Proceedings of the 6th European Conference on Antennas and Propagation (EuCAP '12)*, pp. 433–437, March 2012.

[101] A. Farhang, N. Marchetti, L. Doyle, and B. Farhang-Boroujeny, "Filter bank multicarrier for massive MIMO," in *Proceedings of the IEEE International Conference on Vehicular Technology (VTC-Fall '14)*, Vancouver, Canada, September 2014.

10

A Hybrid Short-Term Power Load Forecasting Model Based on the Singular Spectrum Analysis and Autoregressive Model

Hongze Li, Liuyang Cui, and Sen Guo

School of Economics and Management, North China Electric Power University, Beijing 102206, China

Correspondence should be addressed to Sen Guo; guosen324@163.com

Academic Editor: Mamun B. I. Reaz

Short-term power load forecasting is one of the most important issues in the economic and reliable operation of electricity power system. Taking the characteristics of randomness, tendency, and periodicity of short-term power load into account, a new method (SSA-AR model) which combines the univariate singular spectrum analysis and autoregressive model is proposed. Firstly, the singular spectrum analysis (SSA) is employed to decompose and reconstruct the original power load series. Secondly, the autoregressive (AR) model is used to forecast based on the reconstructed power load series. The employed data is the hourly power load series of the Mid-Atlantic region in PJM electricity market. Empirical analysis result shows that, compared with the single autoregressive model (AR), SSA-based linear recurrent method (SSA-LRF), and BPNN (backpropagation neural network) model, the proposed SSA-AR method has a better performance in terms of short-term power load forecasting.

1. Introduction

Short-term power load forecasting is one of the most important issues in economic and reliable operation of power system. Many operating decisions related to electricity power system such as unit commitment, dispatch scheduling of the generating capacity, reliability analysis, security assessment, and maintenance scheduling of the generators are based on the short-term power load forecasting.

In recent years, domestic and foreign scholars have done many studies in the field of short-term power load forecasting. Currently, the short-term power load forecasting method can be divided into two categories, that is, load-series-based forecasting method and affecting-factors-based forecasting method. Although the power load shows the random and uncertain characteristic, it also has an apparent tendency. Therefore, the load-series-based forecasting method is based upon the internal structure of the short-term power load series, which includes ARIMA, ARMAX [1, 2], neural networks [3, 4], gray prediction model [5, 6], wavelet analysis [7, 8], and other forecasting methods. However, these methods have some shortcomings: the load-series-based forecasting method can only be used for data fitting and is not suitable for the treatment of regularity; the neural network method has the problem that the relation between the input variables cannot be expressed explicitly; the grey prediction model is used for the case of the little sample data; the wavelet analysis forecasting method transforms the original sequence by the orthogonal wavelets to get the subsequences of different frequency-domain and then uses the subsequences to predict and reconstruct, which has a higher accuracy than using the original series directly. The affecting-factors-based method is from the perspective of the factors affecting the power load, which uses the regression analysis to perform the prediction on the base of determining the relation between different variables [9]. However, it is very difficult to find all the factors accurately and comprehensively, which are influenced by the geographical factors, economic situation, and climate and other factors that are often quite different in different periods to some extent. So, it is too hard to find an equation which can be suited to all the forecasting cases. Although the affecting-factors-based method is better than the load-series-based in theory, it has much less practical operability. Moreover, the affecting-factors-based method can be put into use for forecasting the medium- and long-term power load. To our knowledge and related work [10, 11], the key of short-term power load forecasting is to grasp the primary ingredients reflecting the variation tendency and it is quite necessary to find out a method that can depict the fluctuation characteristics of power load series.

The singular spectrum analysis (SSA) technology is a typical time-series-based analysis method which has been used for industrial production forecasting [12], signals detection [13], electricity price forecasting [14], murmur detection from heart sounds [15], and so on. The aim of SSA is to make a decomposition of the original time series into the sum of a small number of independent and interpretable components such as a slowly varying trend, oscillatory components, and a structureless noise. Then, some of these components are used for time series forecasting. At present, the SSA method has been widely applied to cope with the problem in many domains such as geography and sociology [16, 17].

From the perspective of the fluctuation characteristics the power load shows randomness, trend, and periodicity, which can be extracted by using the SSA method. That is to say, the stochastic noise components which influence the forecasting accuracy can be eliminated. And Afshar and Briceño have already applied this method to load forecasting with linear recurrent formulae (LRF) [18, 19], but, as we all know, the relation of power load in different time is not simply linear; usually complicated nonlinear relationship is presented. So the SSA-LRF is not so perfect if we take this into consideration. Meanwhile, using the time-series-based model to forecast power load can not only overcome the problem of invalid linear fitting, but also make up for the shortage of failing to handle the regularity of time series. Therefore, in this paper, the SSA approach and autoregressive (AR) model are combined to forecast the short-term power load, that is, SSA-AR forecasting model.

The rest of this paper is organized as follows. In Section 2, the SSA methodology and the AR model are described briefly, and the hybrid SSA-AR power load forecasting model is introduced; the empirical analysis is performed in Section 3, and the forecasting results of several different forecasting models are presented and discussed; Section 4 concludes this paper.

2. The Basic Principle of SSA-AR Model

2.1. A Brief Introduction to SSA Method. Singular spectrum analysis (SSA) method contains two phases named decomposition and reconstruction. The former phase arranges the original sequence in a form of time-delay matrix before decomposing the original time series. Then, the time series are reconstructed via grouping and diagonal averaging, which is called reconstruction. The reconstructed series is then used for forecasting the new data points.

2.1.1. Decomposition. Given a time series $Y = (Y_1, Y_2, \ldots, Y_T)$, $T > 2$. In the first step, the embedding procedure is applied to construct a sequence of $X_i = (y_{i-1}, \ldots, y_{i+L-2})^T$ with L-dimension vectors from the time series Y_T:

$$X = \begin{bmatrix} y_1 & \cdots & y_K \\ \vdots & \ddots & \vdots \\ y_L & \cdots & y_T \end{bmatrix}, \quad (1)$$

where X is a Hankel matrix with equal elements on the diagonals, $i + j = \text{const}$, $K = T - L + 1$.

In the second step, the $L \times L$ matrix XX^T is calculated and its Eigen triples (s_i, U_i, and V_i) are determined by singular value decomposition (SVD). Denote the Eigen values of XX^T by λ_i $(i = 1, 2, \ldots, L)$ in descending order and let U_i and V_i be the ith left and right Eigen vectors of XX^T, respectively. Set $d = \text{rank}(X)$. Then, the trajectory matrix X can be rewritten as

$$X = X_1 + X_2 + \cdots + X_d, \qquad X_i = \sqrt{\lambda_i} U_i V_i, \quad (2)$$

where s_i is the ith singular value of X and X_i $(i = 1, \ldots, d)$ are the matrices of rank one.

2.1.2. Reconstruction. After decomposition, the time series is reconstructed via grouping and diagonal averaging. In grouping step, the indices $J = i = 1, \ldots, d$ are grouped into M disjoint subsets $I_1, \ldots, I_M$ corresponding to splitting the elementary matrices X_i $(i = 1, \ldots, d)$ into M groups. Each group contains a set of indices as $I = \{i_1, \ldots, i_p\}$. Then, the resultant matrix X_i is defined as $X_{I_i} = X_{i_1} + X_{i_2} + \cdots + X_{i_p}$, so

$$X = X_{I_1} + X_{I_2} + \cdots + X_{I_M}, \quad (3)$$

where the trajectory matrix X is represented as a sum of M resultant matrices. Next, the diagonal averaging transfers each matrix X_{I_n} $(n = 1, \ldots, M)$ into a time series.

Let X be a $(L \times K)$ matrix with elements x_{ij}. Set $L^* = \min(L, K)$, $K^* = \max(L, K)$, and

$$Y^* = \begin{cases} Y, & L \le K \\ Y^T, & L > K, \end{cases} \quad (4)$$

where $Y^* \in R^{L^* \times K^*}$.

Diagonal averaging transfers the matrix X to a series $G = (g_0, \ldots, g_{T-1})$ according to the following:

$$g_k = \begin{cases} \dfrac{1}{k+1} \displaystyle\sum_{m=1}^{k+1} x^*_{m,k-m+2}, & 0 \le k < L^* - 1 \\ \dfrac{1}{L^*} \displaystyle\sum_{m=1}^{L^*} x^*_{m,k-m+2}, & L^* - 1 \le k < K^* \\ \dfrac{1}{T-k} \displaystyle\sum_{m=k-K^*+2}^{T-k+1} x^*_{m,k-m+2}, & K^* - 1 \le k < T. \end{cases} \quad (5)$$

2.2. A Brief Introduction to AR Model. The principle of autoregressive (AR) model is to use the current interference and the limited past observations to predict the present value. Given a time series $Y = (y_1, y_2, \ldots, y_t)$, the mathematical representation of the p-order autoregressive model is as follows:

$$y_t = \alpha_1 y_{t-1} + \alpha_2 y_{t-2} + \cdots + \alpha_p y_{t-p} + \varepsilon_t, \quad (6)$$

where α_i $(i = 1, 2, \ldots, p)$ is the undetermined coefficients of the model (also called autoregressive coefficient) and ε_t is a

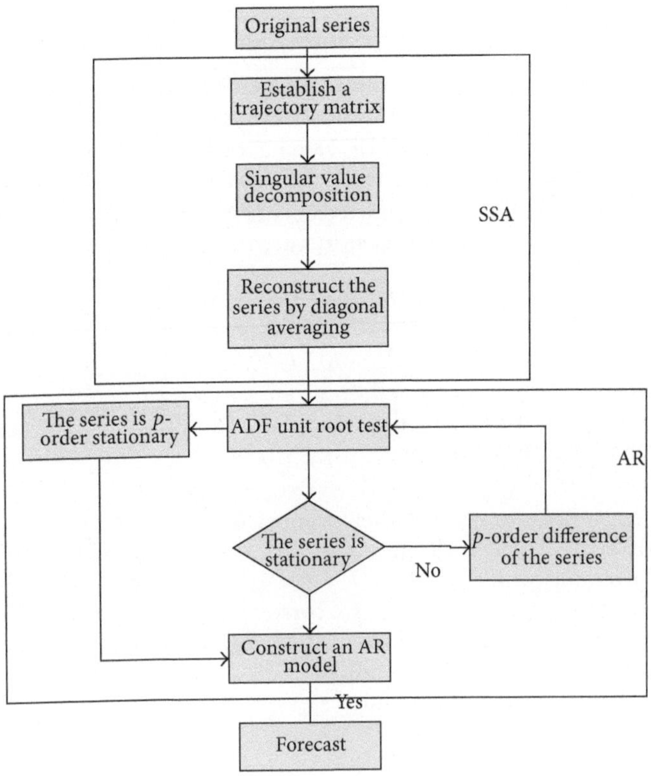

FIGURE 1: The specific flow chart of the SSA-AR model.

random disturbance whose mean value is zero but variance is not equal to zero.

If the lag operator is defined as φ, then $\varphi y_t = y_{t-1}, \varphi^k y_t = y_{t-k}$. The p-order autoregressive model can be presented as follows:

$$y_t = \left(\alpha_1 \varphi + \alpha_2 \varphi^2 + \cdots + \alpha_p \varphi^p\right) y_t + \varepsilon_t \triangleq \Phi(\varphi) y_t + \varepsilon_t. \quad (7)$$

Only the reciprocal of all roots of lag operator's polynomial is less than one (both fall within the unit circle); the AR (p) process is called covariance stationary process.

2.3. Introduction to SSA-AR Model. The application process of SSA-AR model for short-term power load forecasting is mainly divided into two steps: first, using the SSA method to decompose and reconstruct the original power load series and then using the AR model to forecast with the reconstructed sequences. The specific calculation procedure of SSA-AR model for short-term power load forecasting is shown in Figure 1.

3. Empirical Analysis

The hourly power load series of Mid-Atlantic region in the PJM electricity market from 18 June to 18 July 2013 containing 720 sample points is employed in the experimental data. The mentioned data versus time have been shown in Figure 2.

3.1. Decomposing and Reconstructing the Original Power Load Series Based on SSA. As mentioned earlier, the window

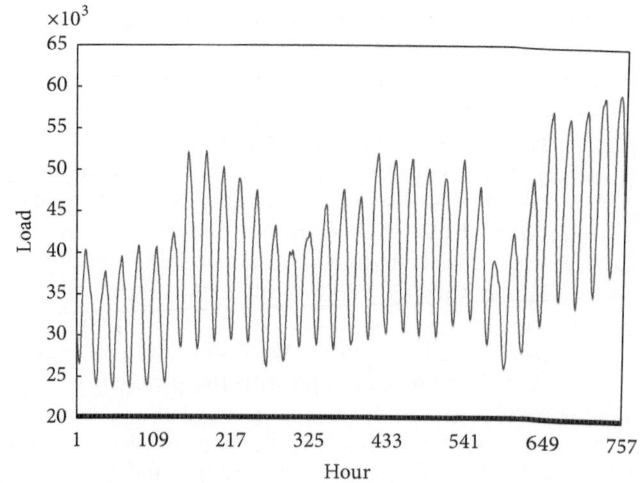

FIGURE 2: The hourly load curve of the Mid-Atlantic region from June 18 to July 18 in 2013.

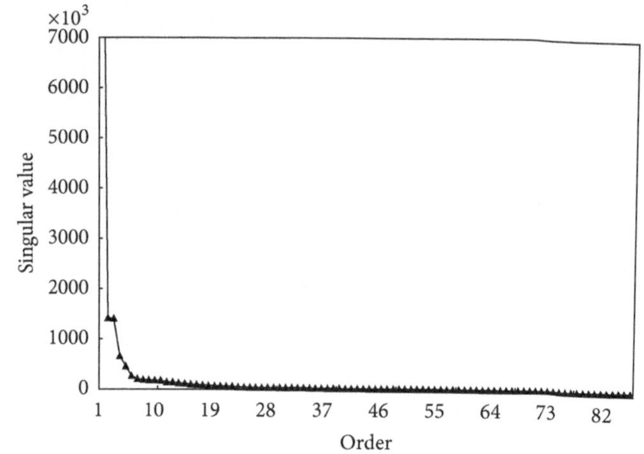

FIGURE 3: Singular values curve (in descending order).

length L is the only parameter in the decomposition stage. If the time series has a periodic component, the window length is taken proportional to that period to get better separability. Therefore, $L = 24 \times 7 = 168$ h is assumed here, which corresponds to weekly variations of power load time series. This window length results in 168 Eigen triples. Then, the singular value decomposition (SVD) is applied to the time-delay matrix by *MATLAB* programming. Figure 3 illustrates the trend of the 168 singular point values.

As shown in Figure 3, the convergence rate of the singular value is very fast. The first 30 singular values have a share of 99.6% of the power load series which can nearly be considered as the weekly trend. The singular values from the 30th one are basically close to zero, which comprise the random component of the original power load serials. Therefore, the original serials should be omitted when reconstructed. The first 30 singular values in descending order (taken to the base-10 logarithm) are listed in Table 1.

The decomposition helps us to make the proper groups to extract the trend, the harmonic component, and random

TABLE 1: The first 30 singular values in descending order (taken to the base-10 logarithm).

Order	log(x)	Order	log(x)	Order	log(x)
1	16.31	11	12.06	21	11.13
2	14.15	12	11.87	22	11.07
3	14.15	13	11.87	23	11.07
4	13.39	14	11.73	24	10.80
5	13.03	15	11.67	25	10.76
6	12.49	16	11.48	26	10.68
7	12.23	17	11.43	27	10.61
8	12.16	18	11.36	28	10.61
9	12.11	19	11.26	29	10.60
10	12.08	20	11.13	30	10.60

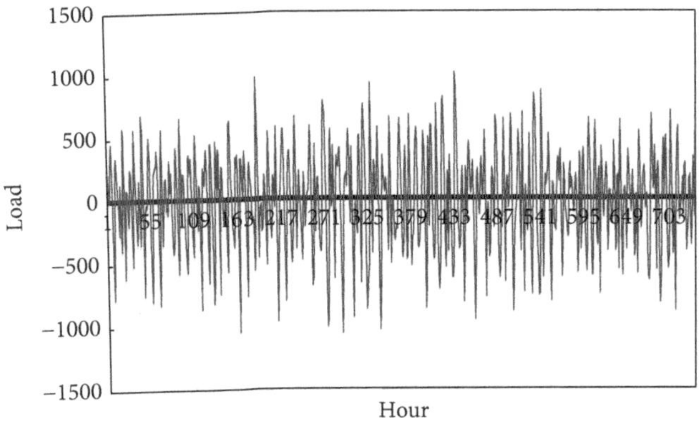

FIGURE 4: The gap between the reconstructed power load series (RS) and the original power load series.

noise component. Based on the relevant mathematical theory about the singular value decomposition, there are three basic conclusions: (1) the first Eigen triple almost represents the whole trend of the time series; (2) every harmonic component with a different frequency produces two Eigen triples with close singular values; (3) as a rule, a pure noise series produces a slowly decreasing sequence of singular values the explicit plateau in the Eigen value spectra is caused by two Eigen triples with close singular values. As shown in Table 1, it is obvious that the Eigen triples of $(2, 3)$, $(12, 13)$, $(22, 23)$, $(27, 28)$, and $(29, 30)$ produce the main harmonics of power load series.

Based on the above analysis, this paper tries to reconstruct the power load series using the first 30 Eigen values. Figure 4 shows the gap between the original power load series (OS). Just as shown in Figure 4, the reconstruction has been done with a satisfactory error.

Figure 4 reflects that the reconstructed power load series fit well with the original series, of which the correlation coefficient reaches 0.9988. Therefore, the decomposition and reconstruction of the original power load series not only greatly reduce the matrix dimension, but also extract the main ingredient of the original series. So, we can describe its dynamic change trend in a better way and also get a better predicting outcome.

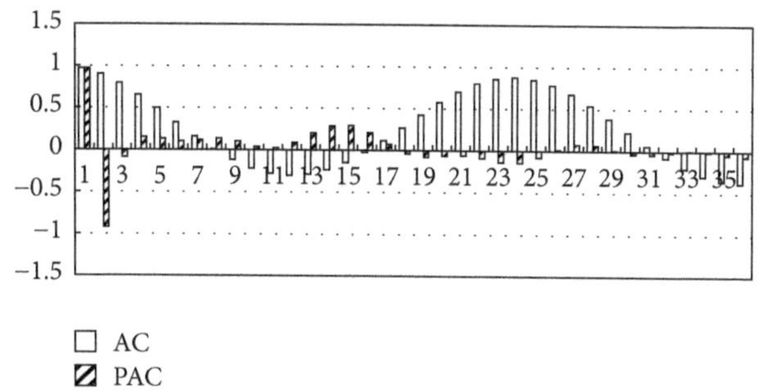

FIGURE 5: The autocorrelation and partial correlation of the reconstructed series.

TABLE 2: The result of ADF test on first order difference power load series.

Item	ADF statistics	Prob.
Original series	−2.77525	0.0623
First order difference	−3.72147	0.0040

TABLE 3: The regression result of AR (3) model of reconstructed power load series.

Variable	Coefficient	Std. error	t-statistic	Prob.
AR (1)	2.067776	0.035534	58.19094	0.0000
AR (2)	−1.396917	0.066505	−21.00462	0.0000
AR (3)	0.263791	0.035598	7.410290	0.0000
Parament			Value	
Adjusted R-squared			0.986002	
Schwarz criterion			13.56932	
Durbin-Watson stat			1.967281	

3.2. Forecasting the Reconstructed Power Load Series Based on AR Model. Because the AR model is only applicable to the stationary time series, the first thing that needs to be done is to examine the stationarity of the reconstructed power load series with the ADF unit root test. By employing the software *Eviews* 6.0, we can get the P value equal to $0.0623 > 0.05$, which means that the null hypothesis should not be rejected. That is to say, the reconstructed power load series is a nonstationary time series. Then, consider the stationarity of its first order difference series, and the calculation result is listed in Table 2. It can be seen that the first order difference series is stationary.

Observe the autocorrelation coefficient and partial correlation coefficient of the first order difference sequence which are shown in Figure 5; we can see the autocorrelation coefficient gradually decreases with fluctuation and the partial correlation coefficient tends to zero after the third order. Therefore, the AR (3) model can be established, which is shown as follows:

$$\Delta y_t = \alpha \Delta y_{t-1} + \beta \Delta y_{t-2} + \gamma \Delta y_{t-3} + \varepsilon_t. \quad (8)$$

Then, estimate (8) with the ordinary least squares (OLS) method, and the regression result is listed in Table 3.

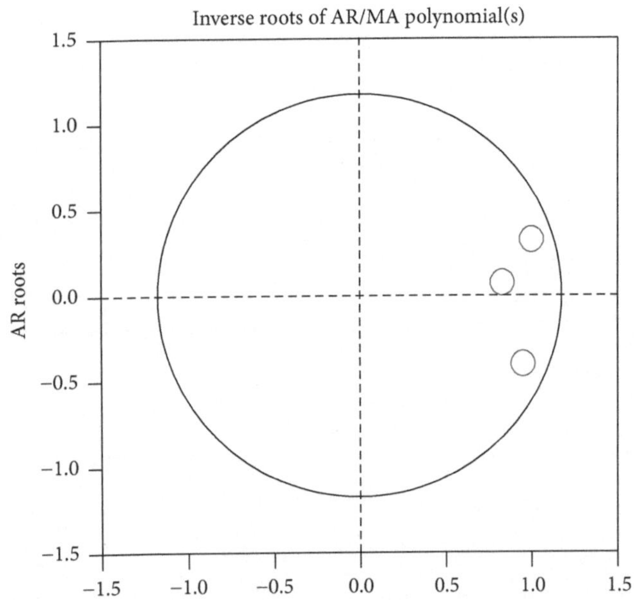

FIGURE 6: Unit circle test of the covariance stationarity.

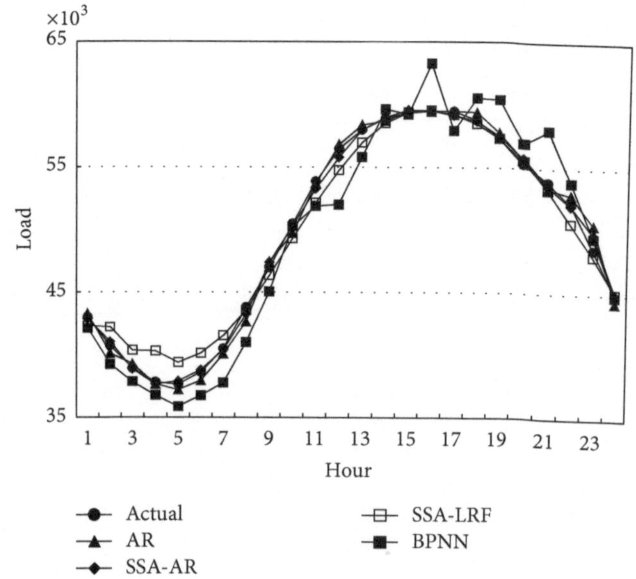

FIGURE 7: The comparison of prediction results of different models and the actual value.

From Figure 6, we can see that the reciprocals of two roots of lag operator's polynomial of AR (3) both fall within the unit circle, which indicates that this process is covariance stationary.

Then, expand the sample size to 745 from 744, and use the AR (3) model to forecast its value. By calculation, the 745th first-order differential value equals −3620.273. Therefore, the 745th point value can be forecasted, which equals 42867.027. In the same way, the following value at any point of a certain period can also be forecasted.

3.3. Comparing the Forecasting Results of Different Models. The AR model, SSA-LRF model, and BPNN model are selected as the comparative models. The AR model is applied to the original power load series without any treatment that can extract the main trend. Although AR model can well represent the whole tendency of the original series, the predicted value is not perfect in a way. The SSA-LRF model is the combination of SSA method and linear recurrent formula (LRF), in which LRF is a simple linear combination of the known data and its coefficients are determined by SSA method. BPNN (backpropagation neural network) is made of neuron. Not only are sufficient neurons connected to net properly, but the BPNN model should be trained appropriately before it can simulate all types of nonlinear characteristics. BPNN model is applied most widespread among all artificial neural network models, which has been widely applied to many fields related to forecasting. The predicted hourly power load results in 24 hours of Mid-Atlantic region in PJM market on July 19 by employing the above four forecasting methods are depicted in Figure 7.

In order to measure the performance of the three forecasting methods, two indices, that is, hourly mean error (HME) and the hourly peak error (HPE), have been employed in this paper. The HME and HPE are the well-known statistical indices for evaluating prediction methods defined as follows:

$$\text{HME} = \frac{1}{24}\sum_{i=1}^{24}\frac{\left|L_{i\text{ACT}} - L_{i\text{FOR}}\right|}{L_{i\text{ACT}}},$$
$$\text{HPE} = \operatorname*{Max}_{1\le i\le 24}\left(\frac{\left|L_{i\text{ACT}} - L_{i\text{FOR}}\right|}{L_{i\text{ACT}}}\right), \tag{9}$$

where $L_{i\text{ACT}}$ and $L_{i\text{FOR}}$ are the actual and predicted power load of hour i, respectively.

The calculation results of absolute error rate, HME, and HPE are shown in Figure 8. It can be seen that the value of SSA-AR model in terms of HME and HPE is 0.58 and 2.57, respectively, and the absolute error rate of SSA-AR model is the smallest compared with the other three models. These indicate that the SSA-AR model shows better performance than SSA-LRF, AR, and BPNN model in terms of short-term power load forecasting.

4. Conclusions

In this paper, a hybrid short-term power load forecasting model based on the singular spectrum analysis (SSA) and autoregressive (AR) model is proposed. As we all know, the short-term power load forecasting is vital in the fundamental operational functions of electricity market, such as unit commitment, economic dispatch, interchange evaluation, scheduled maintenance, and security assessment. In this paper, the SSA-AR model has been employed as a tool for short-term power load forecasting. Firstly, the power load series is analyzed with the SSA method to obtain the effective and predictable components of power load series. Then, the AR method is used to forecast the future values of the power load series. The hybrid SSA-AR power load forecasting

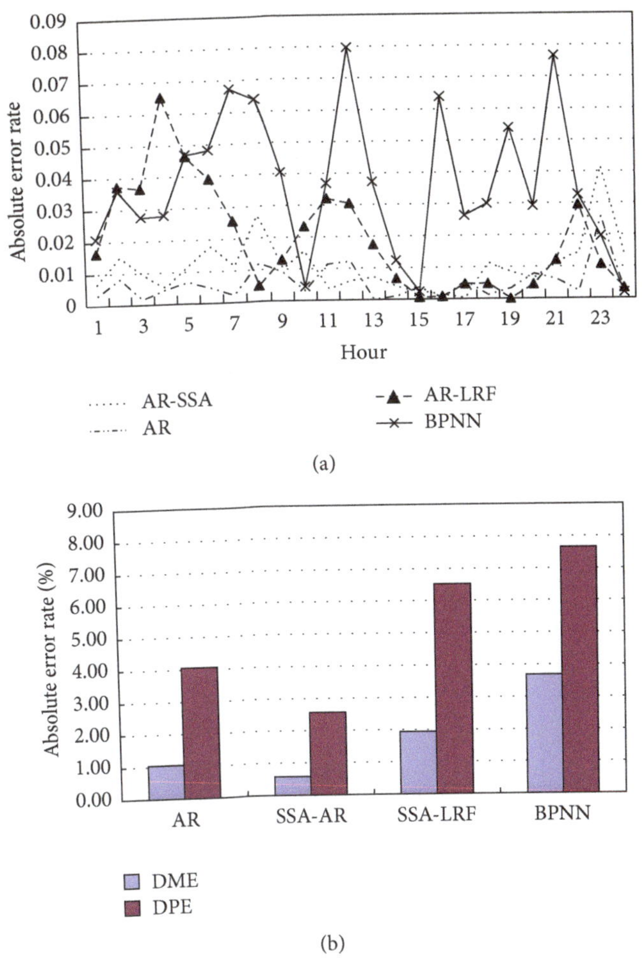

FIGURE 8: The forecasting error comparison of four methods: (a) absolute error rate; (b) HME and HPE.

method is examined by using the experimental data of Mid-Atlantic region in PJM electricity market. The obtained results show that the proposed method has a good ability in the prediction of the desired power load series. However, there is one point that needs emphasis: this method does not take the factors influencing the power load fluctuation into account. Once the outside situation, such as political, economic, and climate condition, has a sudden change, this method may not work. Therefore, this method is applicable to the short-term power load forecasting without the tremendous changes of outside situation.

Conflict of Interests

The authors declare that they have no conflict of interests.

Acknowledgment

This study is supported by the Beijing Philosophy and Social Science Planning Project (no. 11JGB070).

References

[1] H. A. Amarawickrama and L. C. Hunt, "Electricity demand for Sri Lanka: a time series analysis," *Energy*, vol. 33, no. 5, pp. 724–739, 2008.

[2] S. S. Pappas, L. Ekonomou, D. C. Karamousantas, G. E. Chatzarakis, S. K. Katsikas, and P. Liatsis, "Electricity demand loads modeling using AutoRegressive Moving Average (ARMA) models," *Energy*, vol. 33, no. 9, pp. 1353–1360, 2008.

[3] D.-X. Niu, H.-F. Shi, and D. D. Wu, "Short-term load forecasting using bayesian neural networks learned by Hybrid Monte Carlo algorithm," *Applied Soft Computing Journal*, vol. 12, no. 6, pp. 1822–1827, 2012.

[4] D.-M. Zhou, X.-H. Guan, J. Sun, and Y. Huang, "A short-term load forecasting system based on BP artificial neural network," *Power System Technology*, vol. 26, pp. 10–14, 2002.

[5] G.-D. Li, C.-H. Wang, S. Masuda, and M. Nagai, "A research on short term load forecasting problem applying improved grey dynamic model," *International Journal of Electrical Power & Energy Systems*, vol. 33, no. 4, pp. 809–816, 2011.

[6] D.-M. Shi, L.-C. Li, and J.-W. Song, "Power system load forecasting based upon combination of gray forecast and artificial neural network," *Power System Technology*, vol. 25, pp. 13–16, 2001.

[7] G. Jie, "Application of wavelet analysis to short-term load forecasting of power system," *Proceedings of the EPSA*, vol. 15, pp. 40–45, 2003.

[8] H. M. Al-Hamadi and S. A. Soliman, "Long-term/mid-term electric load forecasting based on short-term correlation and annual growth," *Electric Power Systems Research*, vol. 74, no. 3, pp. 353–361, 2005.

[9] P. Peng and J.-H. Peng, "Research on the prediction of power load based on multiple linear regression model," *Journal of Safety Science and Technology*, vol. 7, pp. 158–161, 2011.

[10] H.-Z. Li, S. Guo, C.-J. Li, and J.-Q. Sun, "A hybrid annual power load forecasting model based on generalized regression neural network with fruit fly optimization algorithm," *Knowledge-Based Systems*, vol. 37, pp. 378–387, 2013.

[11] H. Li, S. Guo, H. Zhao, C. Su, and B. Wang, "Annual electric load forecasting by a least squares support vector machine with a fruit fly optimization algorithm," *Energies*, vol. 5, no. 11, pp. 4430–4445, 2012.

[12] H. Hassani, S. Heravi, and A. Zhigljavsky, "Forecasting UK industrial production with multivariate singular spectrum analysis," *Journal of Forecasting*, vol. 32, no. 5, pp. 395–408, 2013.

[13] M. Rui, C. Yushu, and S. Huagang, "Detection of weak signals based on empirical mode decomposition and singular spectrum analysis," *IET Signal Processing*, vol. 7, no. 4, pp. 269–276, 2013.

[14] A. Miranian, M. Abdollahzade, and H. Hassani, "Day-ahead electricity price analysis and forecasting by singular spectrum analysis," *IET Generation Transmission & Distribution*, vol. 7, no. 4, pp. 337–346, 2013.

[15] S. Sanei, M. Ghodsi, and H. Hassani, "An adaptive singular spectrum analysis approach to murmur detection from heart sounds," *Medical Engineering & Physics*, vol. 33, no. 3, pp. 362–367, 2011.

[16] L. Telesca, T. Matcharasvili, T. Chelidze, and N. Zhukova, "Relationship between seismicity and water level in the Enguri high dam area (Georgia) using the singular spectrum analysis," *Natural Hazards and Earth System Sciences*, vol. 12, pp. 2479–2485, 2012.

[17] H. Hassani, A. S. Soofi, and A. A. Zhigljavsky, "Predicting daily exchange rate with singular spectrum analysis," *Nonlinear Analysis: Real World Applications*, vol. 11, no. 3, pp. 2023–2034, 2010.

[18] K. Afshar and N. Bigdeli, "Data analysis and short term load forecasting in Iran electricity market using singular spectral analysis (SSA)," *Energy*, vol. 36, no. 5, pp. 2620–2627, 2011.

[19] H. Briceño, C. M. Rocco, and E. Zio, "Singular spectrum analysis for forecasting of electric load demand," *Chemical Engineer Transactions*, vol. 33, pp. 919–924, 2013.

11

Nonconvex Economic Dispatch Using Particle Swarm Optimization with Time Varying Operators

Vinay Kumar Jadoun, Nikhil Gupta, K. R. Niazi, and Anil Swarnkar

Malaviya National Institute of Technology, Jaipur 302017, India

Correspondence should be addressed to Vinay Kumar Jadoun; vjadounmnit@gmail.com

Academic Editor: Nikos D. Lagaros

This paper presents a particle swarm optimization (PSO) to solve hard combinatorial constrained optimization problems such as nonconvex and discontinuous economic dispatch (ED) problem of large thermal power plants. Several measures have been suggested in the control equation of the classical PSO by modifying its operators for better exploration and exploitation. The inertia operator of the PSO is modulated by introducing a new truncated sinusoidal function. The cognitive and social behaviors are dynamically controlled by suggesting new exponential constriction functions. The overall methodology effectively regulates the velocity of particles during their flight and results in substantial improvement in the classical PSO. The effectiveness of the proposed method has been tested for economic load dispatch of three standard test systems considering various operational constraints like valve-point loading effect, prohibited operating zones (POZs), network power loss, and so forth. The application results show that the proposed PSO method is very promising.

1. Introduction

The economic dispatch (ED) aims at determining the optimal scheduling of thermal generating units so as to minimize the fuel cost while satisfying several operational and power system network constraints. The generator fuel cost functions are invariably nonlinear and also exhibit discontinuities due to prohibited operating zones (POZs). In addition, the valve-point loading effect causes nonconvex characteristic with multiple minima in the generator fuel cost functions and thus imposes challenges of obtaining the global optima for high dimensional ED problems. Thus, ED is a highly nonlinear, complex combinatorial, nonconvex, and multiconstraint optimization problem with continuous decision variables.

The classical mathematical methods like gradient, Lagrange relaxation methods, and so forth, except dynamic programming, are not suitable for such complex optimization problems. The modern metaheuristic search techniques such as particle swarm optimization (PSO), genetic algorithms (GAs), biogeography-based optimization (BBO), differential evolution (DE), ant colony optimization (ACO), artificial bee colony (ABC), and hybrid swarm intelligent based harmony search algorithm (HHS) [1, 2] have shown potential to solve such complex ED problems due to their ability to obtain global or near global solution but are computationally demanding especially for modern power systems which are large and complex.

The PSO has several advantages over other metaheuristic techniques in terms of simplicity, convergence speed, and robustness [3]. It provides convergence to the global or near global optima, irrespective of the shape or discontinuities of the cost function [4]. The potential of PSO to handle nonsmooth and nonconvex ELD problem was demonstrated by [5, 6]. However, the performance of the PSO greatly depends on its parameters and it often suffers from the problems such as being trapped in local optima due to premature convergence [6], lack of efficient mechanism to treat the constraints [7], and loss of diversity and performance in optimization process [8]. PSO is a population based metaheuristic optimization technique in which the movement of the particles is governed by the two stochastic acceleration coefficients, that is, cognitive and social components and the inertia component [5]. In order to enhance the exploration and exploitation capabilities of PSO, the components affecting velocity of particles should be properly managed and controlled.

Several methods have been reported in the recent past to enhance the computational efficiency of the classical PSO. A constriction factor was suggested in the velocity updating equation to assure convergence of PSO [9–11]. However, the exact determination of this factor is computationally demanding. Selvakumar and Thanushkodi [12] modified cognitive behavior of the swarm by communicating with the worst particle. This method provides some additional diversity to the particle by the worst experience component, but showing poor local searching ability unless it is hybridized with certain other heuristic approaches. Roy and Ghoshal [13] proposed crazy PSO (CPSO), where the particle velocity is randomized within predefined limits. The idea was to randomize the velocity of some of the particles, referred to as "crazy particles," by applying a predefined probability of craziness to maintain the diversity for global search and better convergence. However, the value of predefined probability of craziness can only be achieved after several experimentations. Some attempts [14–18] have been made to vary the cognitive and social behavior of the swarm during the search process by dynamically controlling the acceleration coefficients within maximum and minimum bounds. Again the determination of limiting values of these acceleration coefficients is a difficult task, as it required many simulations. Coelho and Lee [19] randomized cognitive and social behavior of the swarm using chaotic sequences and Gaussian distribution, respectively. Selvakumar and Thanushkodi [20] proposed civilized swarm optimization (CSO), by combining society-civilization algorithm (SCA) with PSO to improve communication. The proposed algorithm provides clustered search that results in better exploration and exploitation of the search space but needs several experimentations to determine the optimum values of the control parameters of CSO. Efforts have also been made to suggest a new formulation of the control equation [6, 7]. Safari and Shayeghi [6] proposed iteration PSO (IPSO), where one additional velocity component pertaining to the best fitness of the current iteration is added in the control equation of the classical PSO to avoid local trap, but parameter setting is essential. Vlachogiannis and Lee [7] suggested new control equation in improved coordinated aggregation PSO (ICAPSO) for better communication among particles to enhance local search. They allowed particles to interact with its own best experience along with all other particles have better experience on aggregate basis, instead of the global best experience. However, the authors accepted that the performance of the proposed method is quite sensitive to various parameters setting and their tuning is essential. Chaotic PSO (CPSO) of [21] proposed adapted inertia weight which varies dynamically with fitness value for exploration and chaotic local search was used to determine the particle position for better exploitation. The improved PSO (IPSO) of [22] suggested chaotic inertia weight which decreases and oscillates simultaneously under the decreasing line in a chaotic manner. In this way, additional diversity is introduced, but it requires tuning of chaotic control parameters.

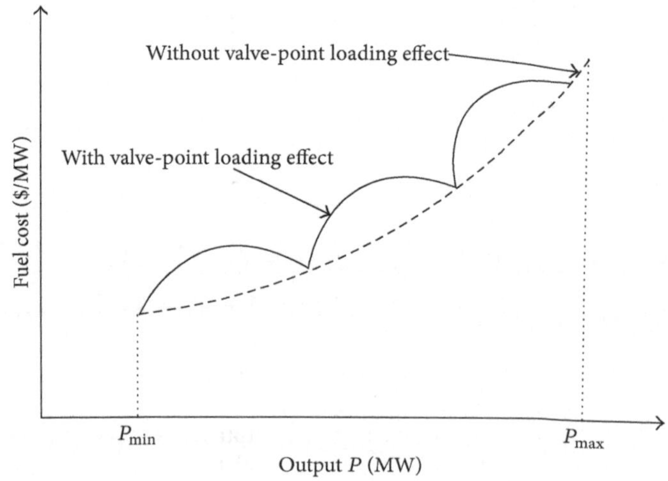

FIGURE 1: Fuel cost function with and without valve-point loading effect.

This paper attempts to overcome drawbacks of some existing PSO methods and presents a modified version of PSO for economic load dispatch of power systems. Several measures have been incorporated in the control equation by modifying operators of the classical PSO by introducing new constriction functions. The proposed method effectively controls and regulates the components affecting velocity of particles so as to ensure better exploration (searching new areas) and exploitation (fine tuning of the current solution). A correction algorithm is also suggested to repair infeasible solutions whenever appeared in the computational process. The proposed method is self-adjusting and does not require experimentations to obtain the optimal values of control parameters and thus overcome the drawbacks of existing PSO methods. The effectiveness of the proposed method has been investigated on three standard test systems considering various operational constraints like valve-point loading effect, prohibited operating zones (POZs), network power loss, and so forth. The application results show that the proposed PSO method is very promising.

2. Problem Formulation

The generator cost function is usually considered as quadratic, when valve-point loading effects are neglected. The large turbine generators usually have a number of fuel admission valves which are operated in sequence to meet out increased generation. The opening of a valve increases the throttling losses rapidly and thus the incremental heat rate rises suddenly. This valve-point loading effect introduces ripples in the heat-rate curves which introduces nonconvexity in the generator fuel cost function as shown in Figure 1. The effect of valve-point loading effects can be modeled as sinusoidal function in the cost function. Therefore, the objective function for the nonconvex ED problem may be stated as

$$\text{Minimize } F(P_{Gi}) = \sum_{i=1}^{N_G} \left(a_i + b_i P_{Gi} + c_i P_{Gi}^2\right) + \left|e_i \sin\left(f_i \left(P_{Gi\min} - P_{Gi}\right)\right)\right|, \quad (1)$$

where a_i, b_i, and c_i are the cost coefficients of the ith generator, e_i and f_i are the valve-point effect coefficients, P_{Gi} is the real power output of the ith generator, and N_G is the number of generating units in the system.

Subject to the following constraints:

(1) *Power Balance Constraint.* The total power generation of all generators must be equal to the sum of total power demand plus the network power loss. The network power loss can be evaluated using B-coefficient loss formula [21, 23]. Therefore, the generator power balance equation may be stated as follows:

$$\sum_{i=1}^{N_G} P_i = PD + \sum_{i=1}^{N_G} \sum_{j=1}^{N_G} P_{Gi} B_{ij} P_{Gj} + \sum_{i=1}^{N_G} P_{Gi} B_{i0} + B_{00}, \quad (2)$$

where B_{ij} is the transmission loss coefficient $i = 1, 2, \ldots, N_G$ and $j = 1, 2, \ldots, N_G$, B_{i0} is the ith element of the loss coefficient vector. B_{00} is the loss coefficient constant.

(2) *Generator Constraint.* For stable operation, power output of each generator is restricted within its minimum and maximum limits. The generator power limits are expressed as follows:

$$P_{Gi}^{\min} \le P_{Gi} \le P_{Gi}^{\max}. \quad (3)$$

(3) *Prohibited Operating Zones.* Prohibited operating zones lead to discontinuities in the input output relation of generators. Prohibited zones divide the operating region between minimum and maximum generation limits into disjoint convex subregions [14, 20]. The generation limits for the ith unit with j number of prohibited zones can be expressed as follows:

$$\begin{aligned} &P_{Gi}^{\min} \le P_{Gi} \le P_{Gi,1}^{L}, \\ &P_{Gi,j-1}^{U} \le P_{Gi} \le P_{Gi,j}^{L}, \\ &P_{Gi,N_{PZi}}^{U} \le P_{Gi} \le P_{Gi}^{\max}; \end{aligned} \quad (4)$$

$$i \in \{1, 2, \ldots, N_{GPZ}\}, \quad j \in \{2, 3, \ldots, N_{PZi}\},$$

where superscripts L and U stand for the lower and upper limit of prohibited operating zones of generators. N_{GPZ} and N_{PZi} denote the total number of generators with prohibited zones and the total number of prohibited zones for the ith generator, respectively.

3. Proposed PSO

The classical PSO is initialized with a population of random solutions and searches for optima by updating particle positions. The velocity of the particle is influenced by the three components: initial, cognitive, and the social component. Each particle updates its previous velocity and position vectors according to the following model [3, 24, 25]:

$$\begin{aligned} v_i^{k+1} &= W v_i^k + c_1 \times \text{rand}_1() \times \frac{pbest_i - s_i^k}{\Delta t} + c_2 \times \text{rand}_2() \times \frac{gbest_i - s_i^k}{\Delta t}, \\ s_j^{k+1} &= s_j^k + v_j^{k+1} \times \Delta t, \end{aligned} \quad (5)$$

where v_i^k is the velocity of ith particle at kth iteration, $\text{rand}_1()$ and $\text{rand}_2()$ are random numbers between 0 and 1, s_i^k is the position of ith particle at kth iteration, c_1, c_2 are the acceleration coefficients, $pbest_i$ is the best position of ith particle achieved based on its own experience, $gbest_i$ is the best particle position based on overall swarm experience, Δt is the time step, usually set to 1 second, and W is the inertia weight which is allowed to decrease linearly as follows:

$$W = W_{\min} + \frac{(W_{\max} - W_{\min}) \times (\text{itr}_{\max} - \text{itr})}{\text{itr}_{\max}}, \quad (6)$$

where $W_{\min}$ and $W_{\max}$ are the minimum and maximum value of inertia weight, respectively, $\text{itr}_{\max}$ is the maximum number of iterations and itr is the current number of iteration.

For better performance of PSO, the particles must fly with higher velocities during the early flights to enhance global search and should be relatively slow during later flights of the journey to improve local search. Therefore, with appropriate regulation of particle's velocity during the journey, the performance of PSO could be improved. Initially, the impact of cognitive component must be high and that of the social component be less to ensure global exploration of the search space by all particles without trapping into a local minima. During later search, the impact of social component must increase and that of the cognitive component must decrease to divert all particles towards global best to improve the convergence. This is essential for a good balance between exploration and exploitation as suggested by [15].

In classical PSO, only the initial velocity component using inertia weight is regulated dynamically. However, the cognitive and social behavior of the swarm, though randomized to ensure diversity, is statically controlled by assigning constant values to acceleration coefficients. These cognitive and social components of velocity are added in the regulated initial velocity component to decide the movement of particles. This probably results in uncontrolled particle velocities during the whole computation process and thus causes insufficient exploration and exploitation of the search space. This results in poor convergence due to local trapping. Therefore, a modified control equation (7) is suggested for dynamically regulating particle's velocity during their whole course of the flight. The modifications suggested in the control equation are explained as follows:

$$v_i^{k+1} = W \times v_i^k + \zeta_1 \times C_{1b} \times \text{rand}_1() \times \frac{pbest_i - s_i^k}{\Delta t} + (1-\zeta_1) \times C_{1p} \times \text{rand}_2() \times \frac{s_i^k - ppoor_i}{\Delta t} + \zeta_2 \times C_2 \times \text{rand}_3() \times \frac{gbest_i - s_i^k}{\Delta t}. \quad (7)$$

In (7), the inertia weight is modified to regulate the trade-off between the global exploration and the local exploitation of the swarm. The poor experience $ppoor_i$ has been added to improve the cognitive component. Further, dynamic acceleration coefficients have been introduced using constriction functions ζ_1 and ζ_2 to regulate the cognitive and social behaviors of the swarm. These modifications are discussed in the following sections.

3.1. Inertia Weight Update. In [25], Shi and Eberhart suggested linear modulation of the inertia weight. This trend is followed to solve ELD problems using PSO by many researchers till date and some of them can be mentioned as [4, 6, 8, 12, 13, 15, 19, 20, 25, 26], and so forth. In the proposed method, the inertia weight has been allowed to vary in accordance with a truncated sinusoidal function rather than to decrease linearly. The modulations suggested to update the inertia weight is governed by the following relation:

$$W = W_{min} + (W_{max} - W_{min}) \cos^2\left(\frac{\theta}{2}\right); \quad 0 \le \theta \le \pi, \quad (8)$$

where $\theta = X \times \text{itr} + Y$ and the coefficients X and Y are given by (9). itr is the iteration count which is in general varied from itr_{min} to itr_{max}

$$X = \frac{\pi}{(\text{itr}_{max} - \text{itr}_{min})}, \quad Y = \frac{-\pi \times \text{itr}_{min}}{(\text{itr}_{max} - \text{itr}_{min})}. \quad (9)$$

Figure 2 shows a comparison of the conventional linear modulation and sinusoidal modulation for the inertia weight to be employed in the proposed PSO. It can be depicted from the figure that using the sinusoidal variations in the inertia weight, the inertia component of the velocity of particles maintained always higher during the early half and lower during the later half of the search, when compared with its linear variations. Therefore, using sinusoidal modulations the coarse search is enhanced during the early half by exploring larger search space with higher values assigned to particle velocities. And, during the later half, the fine search is enhanced by assigning lower values to particle velocities. This facilitates particles to explore the regions in the close proximity of near global solution.

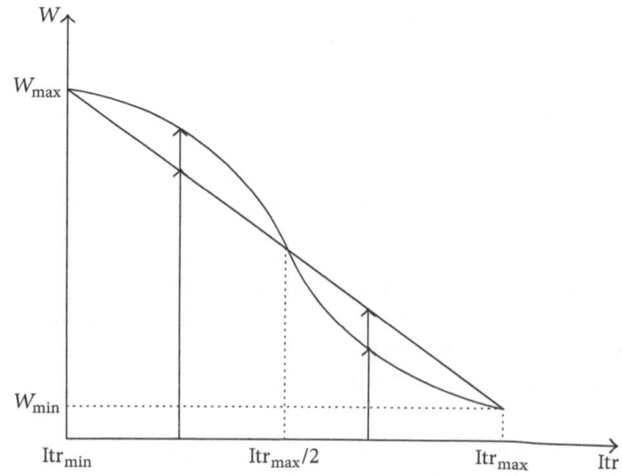

FIGURE 2: Comparison of linear and sinusoidal modulations of inertia weight.

3.2. Updating of Poor Experience. The cognitive behavior was split in [12] by considering the worst experience, in addition to the best experience, of the particle. Though this modification provides additional diversity, it still demands a local random search to enhance exploitation potential of the PSO. This occurs as the particle's velocity is not well regulated during later part of the search. Therefore, the concept of poor experience $ppoor_i$ is suggested, instead of the worst experience, to improve cognitive behavior of the swarm. Here the current fitness of each particle is compared with its fitness value in the preceding iteration, and if it is found less, it will be treated as the poor experience. This concept is different than that of [12], where the worst particle is determined by considering the whole past experience of the particle movement. The poor particle produces much less diversity than the worst particle and thus exploit the region near global optima, during later iterations, in much better way without the support of any local random search.

3.3. Dynamic Control of Acceleration Coefficients. In classical PSO, the cognitive and social behaviors are governed by assigning static values to acceleration coefficients. Many researchers, as discussed earlier, suggested that these acceleration coefficients must be dynamically controlled to regulate particle's velocity during the whole computation process. In the present work, the acceleration coefficients are dynamically controlled by suggesting new exponential constriction functions ζ_1 and ζ_2. These constriction functions dynamically regulate the cognitive and social behaviors of the swarm, thus limiting particles' velocities during their whole course of the flight and are given by

$$\zeta_1 = e^{-\mu_1 \eta}, \quad \zeta_2 = k e^{\mu_2 \eta}, \quad (10)$$

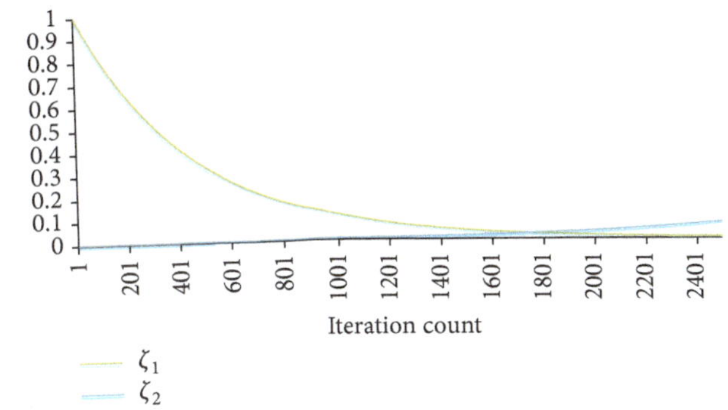

FIGURE 3: Proposed exponential constriction functions.

TABLE 1: Particle encoding for the proposed PSO.

P_{G1}	P_{G2}	$\cdots$	P_{Gi}	$\cdots$	P_{GN}

where

$$\eta = \frac{\text{itr}}{\text{itr}_{\max}}; \quad \text{itr}_{\min} \le \text{itr} \le \text{itr}_{\max}, \qquad k = \frac{\zeta_1 C_{1b}}{\zeta_2 C_2}. \tag{11}$$

The coefficient of exponent μ_1 has been considered −5.5, as the term $e^{-\mu_1\eta}$ is not perceptible at the end of search. With μ_2 as 4, the coefficient k is determined for exact match of ζ_1 and ζ_2 at two-third of the search. The variation in ζ_1 and ζ_2 with iterations is shown in Figure 3 for the above mentioned value of the exponent coefficients. It can be depicted from the figure that the dominance of cognitive behavior falls sharply and that of the social behavior rises gradually as the search progresses. Thus, during the early part of the computational search, the cognitive behavior is well dominated over the social behavior of the swarm to enhance the global search for most probable area having the global optima. However, during the later part, it is the social behavior of the swarm that dominates over its cognitive counterpart. This may enhance local exploitation by the swarm to search global or near global optima.

These alterations in the control equation of the classical PSO regulate particle velocity within predefined bounds without any additional formulation as reported in many improved versions of PSO [4, 6–8, 10, 13–16, 18, 21, 26], yet preserving diversity due to the stochastic nature of cognitive and social behaviors of the swarm.

3.4. Particle Encoding and Initialization. The solution of an ED problem is the set of most optimal generations for the desired objective(s) bounded by certain operational constraints. In the proposed PSO, the particles are encoded in real numbers as the set of current generations in MW, as shown in Table 1.

Where P_{Gi} denotes generation of the ith generator in MW, the initial population is randomly created with predefined number of particles to maintain diversity. Each of these particles satisfies problem constraints defined by (2)–(4). Infeasible particle, if appeared, is not rejected but corrected using a correction algorithm as described later in the section. This improves the computational efficiency of the PSO. The fitness of each particle is evaluated using (1) and then *pbest*, *ppoor*, and *gbest* are initialized. The initial velocity of particles is assumed to be zero.

3.5. Correction Algorithm. The velocity and position update may create infeasible solutions. Infeasible individuals are not rejected but are corrected to feasible individuals by using a correction algorithm. For the purpose, the generations of all generators are adjusted by their respective bounded generation limits and then the error is calculated from the power balance equation. The error in the power is equally distributed among all generators and the procedure is repeated till the error is reduced to a predefined mismatch value ϵ. In this work the mismatch is considered as 0.001. This reduces the computational burden of PSO.

3.6. Elitism and Termination Criterion. In stochastic based algorithms like PSO, the solution with the best fitness in the current iteration may be lost in the next iteration. Therefore, the particle with the best fitness is kept preserved for the next iteration. The algorithm is terminated when either all particles reach to the best position or the predefined maximum iteration number is reached. The flow chart of the proposed method is shown in Figure 4.

4. Simulation Results

The proposed algorithm is tested on 13-generator system [23] and 40-generator system [23]. The control parameters used for all these systems to solve the ED problem using classical and proposed PSO are considered as mentioned in Table 2. The proposed algorithm has been developed using MATLAB and simulations have been carried on a personal computer of Intel i5, 3.2 GHz, and 4 GB RAM.

4.1. Case Study 1: 13-Generator System. The proposed method is applied on 13 thermal generating units which consist of valve-point effect and network power losses. The thermal generating units' data and B-coefficient power loss data are referred from [23]. The ED problem is solved for a power demand of 2520 MW. The simulation results obtained for the best and average fuel cost, total power output, and power losses after 100 independent trails using proposed PSO are presented in Table 3. The table shows that the proposed PSO is capable of obtaining better best, average, and best fuel costs with smaller power loss than other available existing stochastic methods in reasonable CPU time. Thus, the proposed method provides good quality solution to solve complex nonconvex ED problems. The best generating schedule obtained using the proposed PSO is presented in Appendix.

4.2. Case Study 2: 40-Generator System with Valve-Point Effect. The effectiveness of the proposed method is now investigated

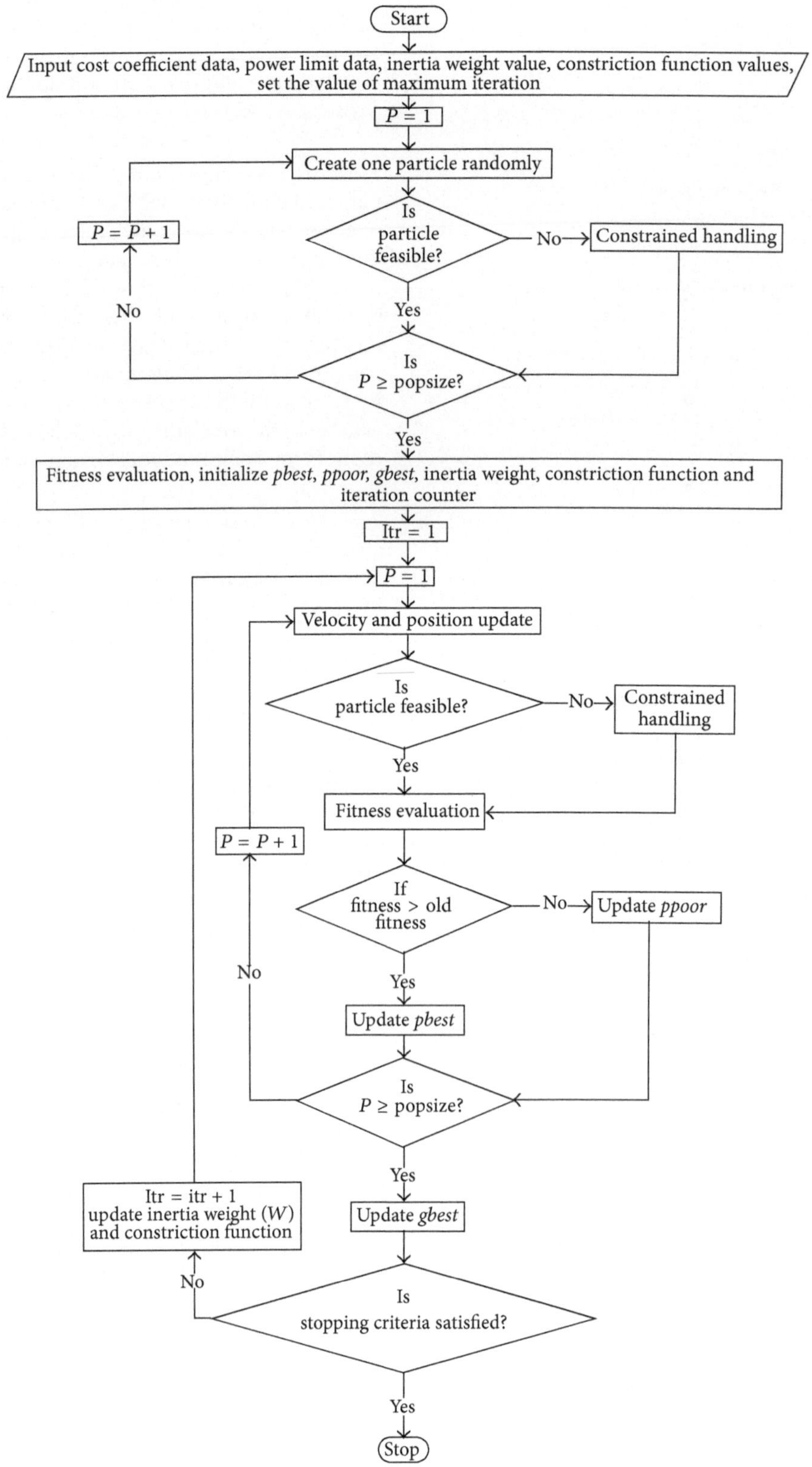

FIGURE 4: Flow chart of the proposed PSO.

TABLE 2: Various parameters for classical PSO and proposed PSO.

Method	W_{min}/W_{max}	C_{1b}	C_{1p}	C_2	μ_1	μ_2	itr_{max}	Population size
Classical PSO	0.1/0.9	2	—	2	—	—	2500	100
Proposed PSO	0.1/0.9	1.5	0.5	2	−5.5	4	2500	100

TABLE 3: Comparison results for case study 1.

Method	Best fuel cost ($/hr)	Average fuel cost ($/hr)	Worst fuel cost ($/hr)	Total power (MW)	Power loss (MW)	CPU time (s)
GA [27]	24632.42	24874.93	25188.59	2559.87	39.87	2.25
DE [27]	24819.32	25217.64	25656.40	2562.34	42.34	2.58
HDE [27]	24591.76	24739.53	25074.90	2559.16	39.16	3.57
STHDE [27]	24560.08	24706.63	24872.44	2564.33	44.33	2.98
ICA-PSO [7]	24540.06	24561.46	24589.45	2559.05	39.05	21.5
SDE [1]	24514.88	24516.31	—	2560.43	40.43	—
Proposed PSO	24514.46	24514.58	24515.26	2558.07	38.07	2.96

TABLE 4: Comparison results for case study 2.

Method	Best fuel cost ($/hr)	Average fuel cost ($/hr)	Worst fuel cost ($/hr)	CPU time (s)
SQP [28]	122904.4243	124883.7692	126585.2290	10.80
EP-SQP [29]	122323.9700	122379.6300	—	997.73
PSO-SQP [29]	122094.6700	122245.2500	—	733.97
PSO-LRS [12]	122035.7946	123382.0000	125740.6300	31.61
NPSO [12]	121704.7391	122221.3697	122995.0976	8.23
NPSO-LRS [12]	121664.4308	122981.5913	122209.3185	20.74
DEC-SQP [30]	121741.9800	123367.6500	125397.9600	925.63
DEC(2)-SQP(1) [28]	121741.9793	122295.1278	122839.2941	14.26
ACO [31]	121532.4100	121606.4500	121679.6400	52.45
FCASO [32]	121516.4700	122082.5900	—	145.2
SOH-PSO [14]	121501.1400	121853.5700	122446.3000	—
TSARGA [33]	121463.0700	122928.3100	124296.5400	696.0
CPSO-SQP [34]	121458.5400	122028.1600	—	
GA-PS-SQP [29]	121458.0000	122039.0000	—	46.98
ABC [35]	121441.0300	121995.8200	—	30.02
CCPSO [22]	121412.5362	121445.3269	121525.4934	19.3
ICA-PSO [7]	121422.1000	—	—	139.9
DE/BBO [36]	121420.8948	—	—	12
HHS [37]	121415.5920	121615.8544	—	16.39
IPSO [2]	121412.8660	121509.5223	121546.8420	42.89
NAPSO [8]	121412.5700	—	—	12.7
CSA [38]	121412.5355	121520.4106	121810.2538	3.03
Proposed PSO	121412.5355	121432.3215	121564.3454	9.99

TABLE 5: Comparison results for case study 3.

Method	Best fuel cost ($/hr)	Average fuel cost ($/hr)	Worst fuel Cost ($/hr)	CPU time (s)
PSO [8]	124875.8523	125162.7011	—	—
FAPSO [8]	122261.3706	122471.0751	122597.5196	19.6
NAPSO [8]	121491.0662	121491.2756	121491.5261	12.7
CSA [38]	121487.7727	121611.3170	122162.9295	14.7
Proposed PSO	121487.7718	121511.3114	121753.7157	8.4

on the most popular test generating system taken from [23]. This system consists of 40 thermal units with nonconvexity in cost function due to valve-point loading effects. The expected power demand for this test system is 10500 MW. The results obtained after 100 independent trials of the proposed PSO are presented and compared with a variety of other available existing deterministic and population based or their hybrid techniques in Table 4. The table validates the effectiveness of the proposed PSO as it generates either comparable or better best fuel cost than other several established techniques including hybrid techniques. The table also shows that the proposed PSO is less computationally demanding than many other references including some latest ones. Although, NPSO [12] and CSA [38] demand less CPU time than the proposed PSO, but the proposed method is capable of generating better quality solution. Thus the proposed PSO is promising to solve nonconvex ED problems. The optimal dispatch of thermal generators obtained by the proposed PSO can be referred to in Appendix.

4.3. Case Study 3: 40-Generator System with Valve-Point and POZs. Finally, the effectiveness of the proposed method is investigated on the 40 generators test generating system with discontinuities in the cost function due to prohibited operating zones. The units 10–14 have POZs as given in [8] (POZ 2). The expected power demand for this test system is 10500 MW. The results obtained after 100 trials of the proposed PSO are presented and compared with other available existing population based techniques in Table 5. The table shows that the proposed PSO is capable of generating comparable or better result in less computational time than other established available methods. The better value of average fuel cost is obtained by proposed method than other methods. This shows robustness of the proposed PSO. Thus the high dimensional nonconvex discrete ED problems can be effectively and efficiently solved using the proposed PSO. The optimal dispatch of thermal generators obtained by the proposed PSO can be referred to in Appendix.

5. Discussion

In order to appreciate and understand the performance of the proposed method a comparison of cognitive and social behavior of particle in PSO and the proposed PSO is shown in Figures 5 and 6, respectively. Figure 5 shows that, in the classical PSO, the cognitive and social behaviors of particle velocity vary randomly throughout the computational process within limits of 0 to 2. The proposed constriction functions used to guide the cognitive and social behaviors of the swarm are allowed to vary exponentially as shown in Figure 6. The lower and upper limits of these behaviors are governed by (10). However, the sum of the best and poor cognitive behavior of the swarm remains constant during the computation process. This plays an important role in providing sufficient diversity by the poor experience during the whole flight of the swarm. It can be seen from Figure 6 that, using proposed PSO, the modulations of cognitive (best), cognitive (poor), and social behaviors though randomly distributed are dynamically controlled within exponential bounds of 1.5, 0.5, and 0.15, respectively. This constitutes a marked difference with other versions of existing PSO. Thus the particles experience entirely different cognitive and social behaviors during their flights and need no additional mechanism to bind their velocities.

Any stochastic based search technique must be designed to accomplish global exploration and tends to facilitate local exploitation. In order to investigate the effectiveness of each of these modifications, a set of convergence characteristics for the best and average fuel cost obtained during a sample trial for 40 generators system is shown in Figures 7 and 8, respectively. In Figure 7, the characteristic "a" is for the conventional PSO, "b" refers to "a" with sinusoidal modulation in inertia weight, "c" refers to "b" with improved cognitive behavior due to poor experience, and "d" refers to the proposed PSO. It can be observed from the figure that the performance of the PSO is somewhat improved when inertia weight is sinusoidally modulated and is further improved with a good margin when poor experience of particles is also considered. However, these two modifications do not seem to be sufficient to exploit the promising region effectively and efficiently. This leads to premature convergence due to local trappings which can be depicted from "d". In d, the proposed constriction functions regulate particles' velocities so that they can fly more comprehensively in the search space. In fact, due to higher initial cognitive component than the social component, the proposed PSO becomes more competent to explore wider search space during the initial phase and thus identify the promising region in about 1000 iterations. However, particles move with strong communication and thus intensively exploit the region near the global optima during later part of the search owing to high values of social component. Finally, all particles converge towards the global minima as can be observed from Figure 8. Thus, the proposed PSO provides better exploration and exploitation of the search space and produces better quality solutions. These results also highlight that the modifications suggested in the control equation of the classical PSO are very effective as it makes the proposed PSO perform much better.

The proposed method offers better exploration and exploitation of the search space because the velocity of particles is regulated throughout their flight. The movement of a sample particle in the classical PSO and the proposed PSO is illustrated in Figures 9 and 10, respectively. These figures show the traces of initial, cognitive, and social components of particle's velocity and also the overall velocity imparted to it during a sample trial.

The classical PSO searches for about 400 iterations, as shown in Figure 9. After this, all the three components of particle's velocity became insignificant and thus the particle gets trapped into local minima. Figure 10(b) shows the cognitive component for the best experience which is then superimposed by its poor experience as in Figure 10(c) to obtain the overall cognitive component as in Figure 10(d). It can be concluded from Figure 10(d) that the poor experience is contributing to tune the cognitive behavior of the swarm. The social component, as shown in Figure 10(e), is providing

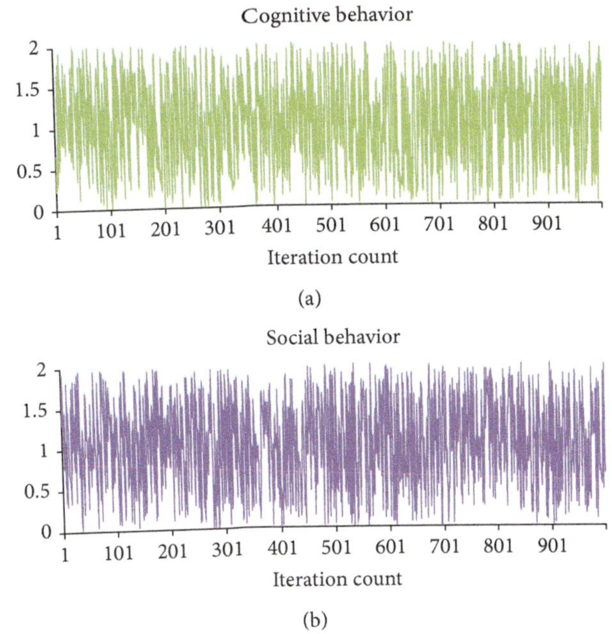

FIGURE 5: (a) Cognitive behavior and (b) social behavior in classical PSO.

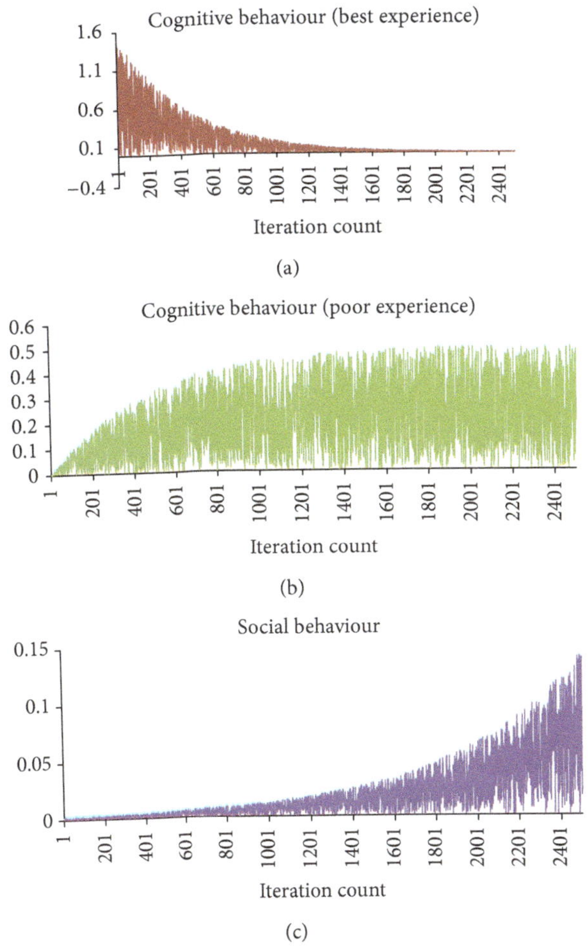

FIGURE 6: (a) Cognitive behavior (best experience), (b) cognitive behavior (poor experience), and (c) social behavior in proposed PSO.

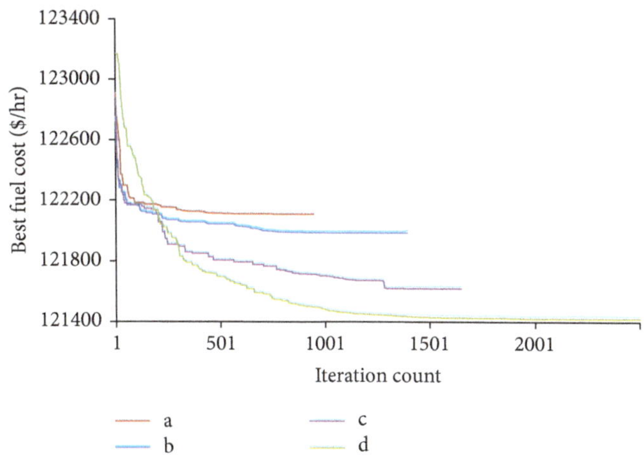

FIGURE 7: Effect on the convergence for best fuel cost by suggested modifications in the proposed PSO.

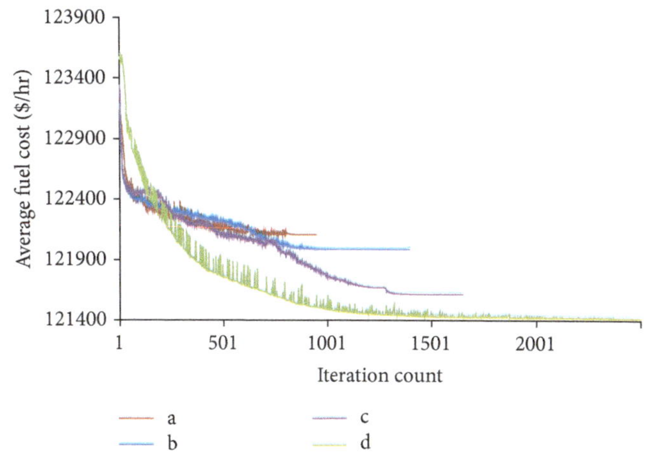

FIGURE 8: Effect on the convergence for average fuel cost by suggested modifications in the proposed PSO.

fine tuning as desired in high dimensional optimization problem. It should be noted that the social component has been kept quite weak in this work, as compared to other published literature till date and is one of the keys to obtain high quality solutions. In addition, the proposed modulation in inertia weight intends particles for better exploration and exploitation of the search space by imparting suitable velocity during the flight, as seen from Figure 10(a). The impact of improved initial, cognitive, and social components of particle's velocity is shown in Figure 10(f). The figure shows a marked improvement in particle movement during the whole computation while compared with Figure 9(d). In the proposed PSO, during early part of the search, the particles widely travelled in the search space yet their velocity is regulated by the poor experience as the social component is almost negligible. This facilitates the swarm to explore the region of global optima. However, in later part of the search, both poor and the social components are driving the swarm toward the global optima, as the cognitive best experience has been made quite weak during this part of the search.

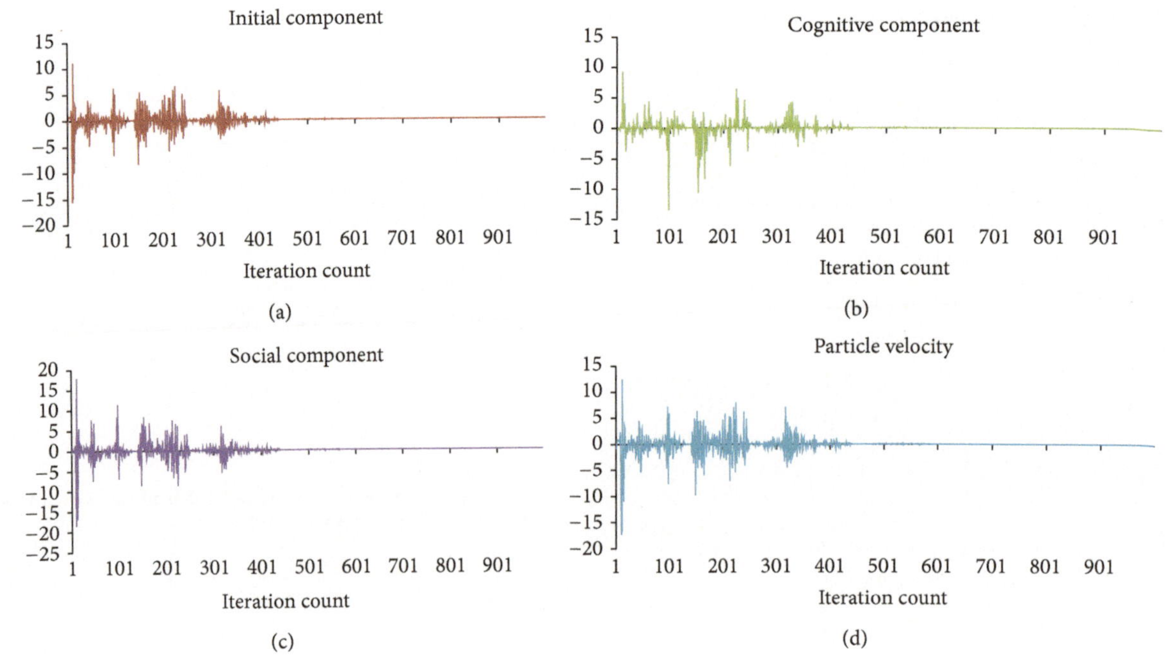

FIGURE 9: Particle velocity and its components in PSO.

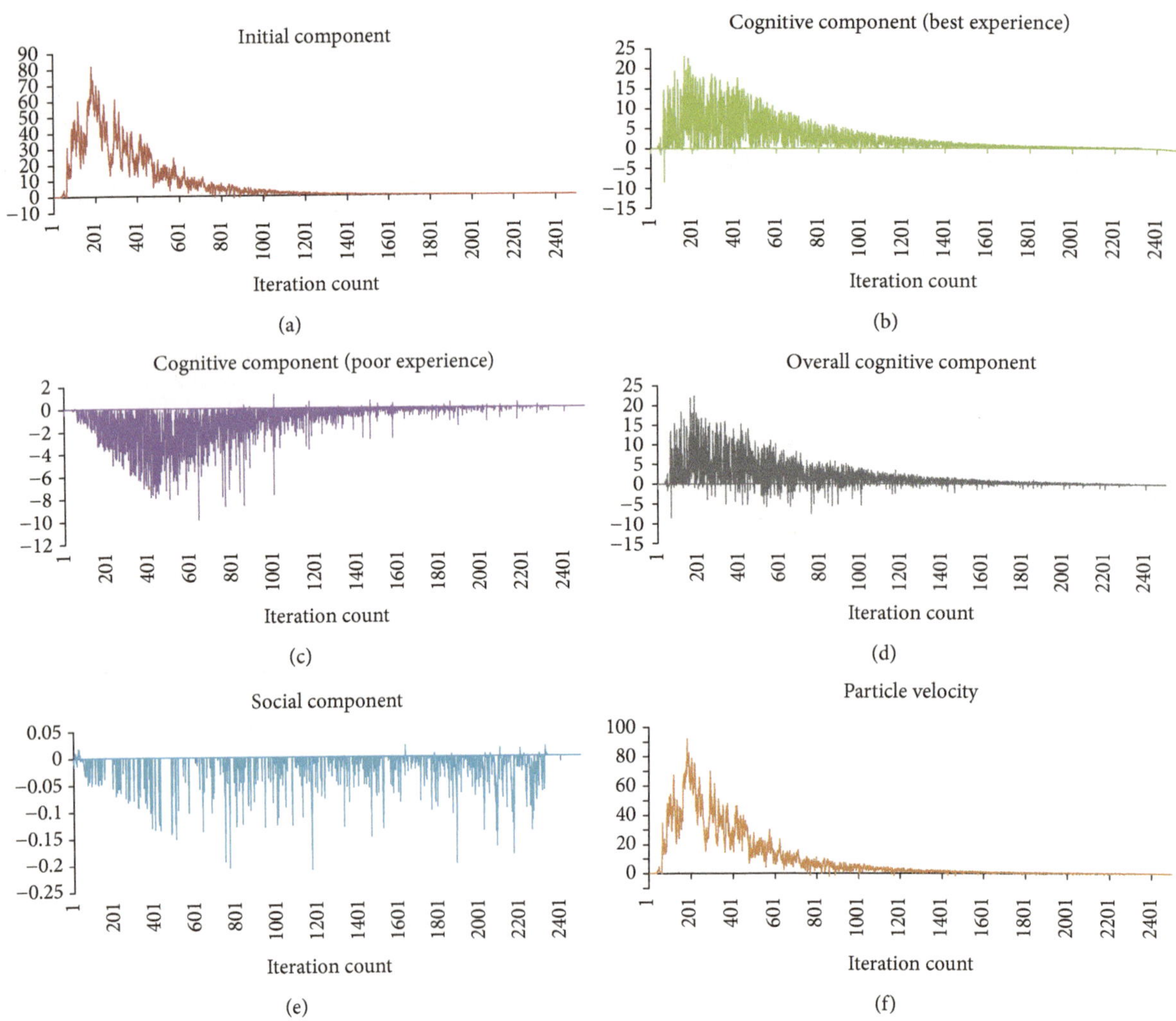

FIGURE 10: Particle velocity and its components in the proposed PSO.

This improves exploitation potential of the PSO for local search. Thus, the proposed PSO provides better exploration and exploitation of the search space and thus produces better quality solutions than the classical PSO or other existing stochastic based methods.

6. Conclusions

The economic dispatch is a highly complex combinatorial constrained optimization problem with continuous decision variables. The classical PSO has proven potential to solve such hard combinatorial constraints optimization problem, but it usually gets trapped into local minima while dealing with high dimensional ED problems. This paper presents a modified version of PSO to make it suitable for solving highly complex ED problems. The proposed method has been tested to solve ED problems of three different test systems of different dimensions with a variety of operational and network constraints. The application results are also compared with available existing PSO methods. The application results show that the proposed method is efficient and is usually not trapped in local minima. The comparison shows that proposed method is capable of giving better results than the existing PSO and other stochastic based methods. This may be due to the fact that proposed PSO essentially aims to regulate particle velocity during its whole course of flight in such a fashion so as to enhance exploration and exploitation potentials of the PSO. The operators in the proposed PSO are made to vary dynamically by introducing new truncated sinusoidal and exponential functions. The concept of poor particle is introduced to improve the cognitive behavior of the swarm and also maintain a good balance between cognitive and social behavior of the swarm during the whole course of the flight. These modifications guide the swarm to identify the area where the global optima may exist. Thereafter, particles have suitable velocities to wandering within in this area to explore global or near global solution. Further, it has been observed that in the proposed PSO, the particle is accelerated more comprehensively during whole of its flight than in the classical PSO. This causes better exploration of the search space during the early part and better exploitation during the later part of the search. It is noteworthy that the proposed PSO is free from any mechanism to avoid local trapping and does not require any empirical formula to bound particle's velocity. Moreover, the proposed algorithm is robust as it generates better quality solutions irrespective of the initial position of the particles. The proposed PSO can be extended to solve ED problems with the inclusion of more objectives and constraints like environmental issues, reserve capacity, network security, network congestion management, and so forth.

Appendix

See Table 6.

TABLE 6: Optimal generating schedule for case studies 1, 2, and 3.

Unit	Case study 1 Power (MW)	Case study 2 Power (MW)	Case study 3 Power (MW)
1	628.3185	110.799825	110.799789
2	298.8000	110.799825	110.799807
3	298.8000	97.3999130	97.3998080
4	159.7400	179.733100	179.733093
5	159.7400	87.7999050	87.7998250
6	159.7400	140.000000	140.000000
7	159.7400	259.599650	259.599600
8	159.7300	284.599650	284.599496
9	159.7400	284.599650	284.599700
10	76.20000	130.000000	130.000000
11	113.3200	94.0000000	168.798140
12	92.10000	94.0000000	168.041419
13	92.10000	214.759790	125.000000
14	—	394.279370	400.000000
15	—	394.279370	394.279018
16	—	394.279370	394.279205
17	—	489.279370	489.279397
18	—	489.279370	489.279380
19	—	511.279370	511.279377
20	—	511.279370	511.279299
21	—	523.279370	523.279354
22	—	523.279370	523.279373
23	—	523.279370	523.279372
24	—	523.279370	523.279365
25	—	523.279369	523.279377
26	—	523.279370	523.279400
27	—	10.0000000	10.0000000
28	—	10.0000000	10.0000000
29	—	10.0000000	10.0000000
30	—	87.799902	87.7998910
31	—	190.000000	190.000000
32	—	190.000000	190.000000
33	—	190.000000	190.000000
34	—	164.799825	164.799766
35	—	194.397782	164.799800
36	—	200.000000	164.799803
37	—	110.000000	110.000000
38	—	110.000000	110.000000
39	—	110.000000	109.998798
40	—	511.279370	511.279348

Conflict of Interests

The authors declare that there is no conflict of interests regarding the publication of this paper.

Acknowledgment

The authors would like to thank the editor and reviewers for their valuable comments and recommendations.

References

[1] A. Srinivasa Reddy and K. Vaisakh, "Shuffled differential evolution for large scale economic dispatch," *Electric Power Systems Research*, vol. 96, pp. 237–245, 2013.

[2] A. K. Barisal, "Dynamic search space squeezing strategy based intelligent algorithm solutions to economic dispatch with multiple fuels," *International Journal of Electrical Power & Energy Systems*, vol. 45, no. 1, pp. 50–59, 2013.

[3] J. Kennedy and R. Eberhart, *Swarm Intelligence*, Morgan Kaufmann, 2001.

[4] D. N. Jeyakumar, T. Jayabarathi, and T. Raghunathan, "Particle swarm optimization for various types of economic dispatch problems," *International Journal of Electrical Power and Energy Systems*, vol. 28, no. 1, pp. 36–42, 2006.

[5] A. Mahor, V. Prasad, and S. Rangnekar, "Economic dispatch using particle swarm optimization: a review," *Renewable and Sustainable Energy Reviews*, vol. 13, no. 8, pp. 2134–2141, 2009.

[6] A. Safari and H. Shayeghi, "Iteration particle swarm optimization procedure for economic load dispatch with generator constraints," *Expert Systems with Applications*, vol. 38, no. 5, pp. 6043–6048, 2011.

[7] J. G. Vlachogiannis and K. Y. Lee, "Economic load dispatch—a comparative study on heuristic optimization techniques with an improved coordinated aggregation-based PSO," *IEEE Transactions on Power Systems*, vol. 24, no. 2, pp. 991–1001, 2009.

[8] T. Niknam, H. D. Mojarrad, and H. Z. Meymand, "Non-smooth economic dispatch computation by fuzzy and self adaptive particle swarm optimization," *Applied Soft Computing Journal*, vol. 11, no. 2, pp. 2805–2817, 2011.

[9] B. Yu, X. Yuan, and J. Wang, "Short-term hydro-thermal scheduling using particle swarm optimization method," *Energy Conversion and Management*, vol. 48, no. 7, pp. 1902–1908, 2007.

[10] G. Baskar and M. R. Mohan, "Security constrained economic load dispatch using improved particle swarm optimization suitable for utility system," *International Journal of Electrical Power and Energy Systems*, vol. 30, no. 10, pp. 609–613, 2008.

[11] L. Wang and C. Singh, "Stochastic economic emission load dispatch through a modified particle swarm optimization algorithm," *Electric Power Systems Research*, vol. 78, no. 8, pp. 1466–1476, 2008.

[12] A. I. Selvakumar and K. Thanushkodi, "A new particle swarm optimization solution to nonconvex economic dispatch problems," *IEEE Transactions on Power Systems*, vol. 22, no. 1, pp. 42–51, 2007.

[13] R. Roy and S. P. Ghoshal, "A novel crazy swarm optimized economic load dispatch for various types of cost functions," *International Journal of Electrical Power & Energy Systems*, vol. 30, no. 4, pp. 242–253, 2008.

[14] K. T. Chaturvedi, M. Pandit, and L. Srivastava, "Self-organizing hierarchical particle swarm optimization for nonconvex economic dispatch," *IEEE Transactions on Power Systems*, vol. 23, no. 3, pp. 1079–1087, 2008.

[15] K. T. Chaturvedi, M. Pandit, and L. Srivastava, "Particle swarm optimization with time varying acceleration coefficients for non-convex economic power dispatch," *International Journal of Electrical Power and Energy Systems*, vol. 31, no. 6, pp. 249–257, 2009.

[16] K. K. Mandal and N. Chakraborty, "Daily combined economic emission scheduling of hydrothermal systems with cascaded reservoirs using self organizing hierarchical particle swarm optimization technique," *Expert Systems with Applications*, vol. 39, no. 3, pp. 3438–3445, 2012.

[17] Y. Wang, J. Zhou, C. Zhou, Y. Wang, H. Qin, and Y. Lu, "An improved self-adaptive PSO technique for short-term hydrothermal scheduling," *Expert Systems with Applications*, vol. 39, no. 3, pp. 2288–2295, 2012.

[18] B. Mohammadi-Ivatloo, "Combined heat and power economic dispatch problem solution using particle swarm optimization with time varying acceleration coefficients," *Electric Power Systems Research*, vol. 95, pp. 9–18, 2013.

[19] L. D. S. Coelho and C.-S. Lee, "Solving economic load dispatch problems in power systems using chaotic and Gaussian particle swarm optimization approaches," *International Journal of Electrical Power and Energy Systems*, vol. 30, no. 5, pp. 297–307, 2008.

[20] A. I. Selvakumar and K. Thanushkodi, "Optimization using civilized swarm: solution to economic dispatch with multiple minima," *Electric Power Systems Research*, vol. 79, no. 1, pp. 8–16, 2009.

[21] J. Cai, X. Ma, L. Li, and P. Haipeng, "Chaotic particle swarm optimization for economic dispatch considering the generator constraints," *Energy Conversion and Management*, vol. 48, no. 2, pp. 645–653, 2007.

[22] J.-B. Park, Y.-W. Jeong, J.-R. Shin, and K. Y. Lee, "An improved particle swarm optimization for nonconvex economic dispatch problems," *IEEE Transactions on Power Systems*, vol. 25, no. 1, pp. 156–166, 2010.

[23] N. Sinha, R. Chakrabarti, and P. K. Chattopadhyay, "Evolutionary programming techniques for economic load dispatch," *IEEE Transactions on Evolutionary Computation*, vol. 7, no. 1, pp. 83–94, 2003.

[24] J. Kennedy and R. Eberhart, "Particle swarm optimization," in *Proceedings of the IEEE International Conference on Neural Networks (ICNN '95)*, pp. 1942–1948, December 1995.

[25] Y. Shi and R. C. Eberhart, "Empirical study of particle swarm optimization," in *Proceedings of the Congress on Evolutionary Computation (CEC '99)*, pp. 1945–1950, Piscataway, NJ, USA, July 1999.

[26] Z.-L. Gaing, "Particle swarm optimization to solving the economic dispatch considering the generator constraints," *IEEE Transactions on Power Systems*, vol. 18, no. 3, pp. 1187–1195, 2003.

[27] S. K. Wang, J. P. Chiou, and C. W. Liu, "Non-smooth/non-convex economic dispatch by a novel hybrid differential evolution algorithm," *IET Generation, Transmission and Distribution*, vol. 1, no. 5, pp. 793–803, 2007.

[28] L. dos Santos Coelho and V. C. Mariani, "Combining of chaotic differential evolution and quadratic programming for economic dispatch optimization with valve-point effect," *IEEE Transactions on Power Systems*, vol. 21, no. 2, pp. 989–996, 2006.

[29] J. S. Alsumait, J. K. Sykulski, and A. K. Al-Othman, "A hybrid GA-PS-SQP method to solve power system valve-point economic dispatch problems," *Applied Energy*, vol. 87, no. 5, pp. 1773–1781, 2010.

[30] P. Subbaraj, R. Rengaraj, and S. Salivahanan, "Enhancement of Self-adaptive real-coded genetic algorithm using Taguchi method for Economic dispatch problem," *Applied Soft Computing Journal*, vol. 11, no. 1, pp. 83–92, 2011.

[31] S. Pothiya, I. Ngamroo, and W. Kongprawechnon, "Ant colony optimisation for economic dispatch problem with non-smooth cost functions," *International Journal of Electrical Power and Energy Systems*, vol. 32, no. 5, pp. 478–487, 2010.

[32] J. Cai, Q. Li, L. Li, H. Peng, and Y. Yang, "A fuzzy adaptive chaotic ant swarm optimization for economic dispatch," *International Journal of Electrical Power & Energy Systems*, vol. 34, no. 1, pp. 154–160, 2012.

[33] P. Subbaraj, R. Rengaraj, and S. Salivahanan, "Enhancement of self-adaptive real-coded genetic algorithm using Taguchi method for economic dispatch problem," *Applied Soft Computing Journal*, vol. 11, no. 1, pp. 83–92, 2011.

[34] J. Cai, Q. Li, L. Li, H. Peng, and Y. Yang, "A hybrid CPSO-SQP method for economic dispatch considering the valve-point effects," *Energy Conversion and Management*, vol. 53, no. 1, pp. 175–181, 2012.

[35] S. Hemamalini and S. P. Simon, "Artificial bee colony algorithm for economic load dispatch problem with non-smooth cost functions," *Electric Power Components and Systems*, vol. 38, no. 7, pp. 786–803, 2010.

[36] A. Bhattacharya and P. K. Chattopadhyay, "Hybrid differential evolution with biogeography-based optimization for solution of economic load dispatch," *IEEE Transactions on Power Systems*, vol. 25, no. 4, pp. 1955–1964, 2010.

[37] V. R. Pandi, B. K. Panigrahi, R. C. Bansal, S. Das, and A. Mohapatra, "Economic load dispatch using hybrid swarm intelligence based harmony search algorithm," *Electric Power Components and Systems*, vol. 39, no. 8, pp. 751–767, 2011.

[38] D. N. Vo, P. Schegner, and W. Ongsakul, "Cuckoo search algorithm for non-convex economic dispatch," *IET Generation, Transmission and Distribution*, vol. 7, no. 6, pp. 645–654, 2013.

12

Characterization and Modeling of Received Signal Strength and Charging Time for Wireless Energy Transfer

Uthman Baroudi,[1] Amin-ud-din Qureshi,[2] and Samir Mekid[3]

[1] *Computer Engineering Department, King Fahd University of Petroleum and Minerals, Dhahran 31261, Saudi Arabia*
[2] *Systems Engineering, King Fahd University of Petroleum and Minerals, Dhahran 31261, Saudi Arabia*
[3] *Mechanical Engineering Department, King Fahd University of Petroleum and Minerals, Dhahran 31261, Saudi Arabia*

Correspondence should be addressed to Uthman Baroudi; ubaroudi@kfupm.edu.sa

Academic Editor: Abdulkareem Adinoyi

Wireless sensor networks can provide effective means for monitoring and controlling a wide range of applications. Recently, tremendous effort was directed towards devising sensors powered from ambient sources such as heat, wind, and vibration. Wireless energy transfer is another source that has attractive features that make it a promising candidate for supplying power to wireless sensor nodes. This paper is concerned with characterizing and modeling the charging time and received signal strength indicator for wireless energy transfer system. These parameters play a vital role in deciding the geometry of sensor network and the routing protocols to be deployed. The development of communication protocols for wireless-powered wireless sensor networks is also improved with the knowledge of such models. These two quantities were computed from data acquired at various coordinates of the harvester relative to a fixed position of RF energy source. Data was acquired for indoor and outdoor scenarios using the commercially available PowerCast energy harvester and evaluation board. Mathematical models for both indoor and outdoor environments were developed and analyzed. A few guidelines on how to use these models were suggested. Finally, the possibility of harvesting the energy from the ambient RF power to energize wireless sensor nodes was also investigated.

1. Introduction

The previous two decades have witnessed an unprecedented advancement in radio frequency (RF) based equipment, ranging from personal and medical devices to complex civil structures' monitoring and military systems, all with reliably precise specifications. Several wireless systems have replaced their wired counterparts, for example, personal communication using cellular phones (significantly reducing the use of landline phones) and wireless data and computer communication networks (WLANS). The advent of extremely low power processors and energy efficient RF devices has further assisted drastic development and deployment of wireless sensor networks (WSN). Examples of such applications include structural health monitoring (SHM), healthcare systems, habitat monitoring, and precision agriculture [1]. A periodical battery charging/replacement of individual sensor nodes is, however, always needed to ensure the continuous operation of WSN. In practice, there are several WSN applications where the human approach to deployed sensor node is either too difficult or expensive. By this very nature of WSN applications, it is very cumbersome and sometimes very expensive to periodically attain this action of battery charging and/or replacement. Examples of such scenarios are SHM and border surveillance systems using unattended acoustic and seismic sensors. This problem has made the researchers look for alternatives for powering wireless sensor nodes [2]. The literature survey suggests that there are two possible alternative solutions to cope with the above-mentioned problem. The first alternative is to use some ambient sources of energy. Different energy harvesting schemes have been under consideration; they mainly include solar [3], piezoelectric [4], and ambient RF energy harvesting systems [5–9]. In cases where such energy harvesting is not possible, the second alternative is to charge a battery using directed transmission of RF energy as a wireless battery charging scheme [10].

This paper is concerned with the investigation of some important aspects of RF energy harvesting using both the ambient RF energy and directed RF energy transmission.

In both of the aforementioned alternatives, since the charging process is not instant, there are time intervals when a sensor node has not enough amount of energy to send data packets, and hence the communication is intermittent. Such a scenario creates the need for development of new protocols for communication among different nodes. The charging time T_c and received signal strength RSSI are the major variables to influence the new communication protocols. Recently several papers have discussed different aspects of wireless powering of the WSNs and some new routing protocols [9, 11, 12]. However, to the best of our knowledge, there is no research work available in the literature considering the mathematical relationship of these two parameters with the spatial coordinates.

Our main contribution is the development of a 3D mathematical model that presents the real behavior of RSSI and T_c at indoor and outdoor environments. Our mathematical models relate these variables or parameters (RSSI and T_c) to spatial coordinates. In order to study any communication protocol, we need a model to represent the signal behavior between the transmitter and receiver. Generally, a random channel model (Gaussian, Rician, etc.) is used. However, this approach does not always catch the real behavior of the suggested protocols that make them irrelevant. Consider designing a routing protocol for RF energy harvesting based WSN. In this case, our model can easily be used to quantify/estimate the charging time needed for each node and we can select the next hop accordingly. On the other hand, RSSI models can be exploited to test certain protocol under different channel coding, for example. Once, the expected RSSI for a given scheme is determined, we can use our model to find out the expected charging time at the intended receiver and neighboring nodes as well.

For data acquisition, a series of experiments were performed on P2110-EVAL-01 PowerCast Energy Harvesting Development Kit for Wireless Sensors. RF survey was also conducted to assess the capability of ambient RF energy for the purpose of energy harvesting by PowerCast energy harvester.

The rest of the paper is organized as follows. Section 2 describes the ambient RF survey. Section 3 describes the experimental setup. Section 4 analyzes the data of outdoor experiments. Section 5 describes the indoor experiments and its outcomes. Mathematical modeling is considered in Section 6. The paper is concluded in Section 7.

2. RF Survey

The RF survey is performed to investigate the possibility of harvesting the energy from the ambient RF energy due to several sources like cellular mobile transmitters, radio stations, Wi-Fi networks, and so forth. This survey was performed by scanning the available RF power spectrum at six different locations inside the King Fahd University campus using GW Instek spectrum analyzer. Instead of showing the spectral peaks recorded at all six locations, we have shown them only for one location. In the following data, Table 1 lists the observed spectral peaks while powers along several bands are mentioned separately. It should be noticed here that the minimum power requirement for a PowerCast power harvester is −10 dBm and that it is optimally designed to receive power in 902–928 MHz band.

Table 1: Spectral power peaks.

Peak number	Peak frequency (MHz)	Peak power (dBm)
	952	−37
2	939.5998	−39
3	922.24	−43.9
4	177	−42.7
5	178	−42.7
6	179	−42.7
7	181	−42.9
8	183	−43.2
9	184	−43.4

All of the following readings are taken using PowerCast omnidirectional (dipole) antenna.

Power through the system:

(1) whole spectrum (1 MHz–2.7 GHz) = −14.4 dBm;

(2) band (900 MHz–950 MHz) = −31.0 dBm;

(3) band (902 MHz–928 MHz) = −33.8 dBm;

(4) band (500 MHz–1500 MHz) = −20.0 dBm.

From the spectral peaks and powers across several bands, it was observed that none of them is even close to −10 dBm from which we may conclude that PowerCast harvester is not capable of harvesting from ambient RF energy.

It is important to note that these measurements were collected via an antenna which is optimally designed for the band 902 MHz–928 MHz. Promising measurements can be observed if an array of antennas is used, expecting the power of magnitude to be −10 dBm or even more, making the energy harvesting possible from ambient RF energy. The interested readers may read the recent paper [8] and references therein for a review of attempts made in this regard.

3. Experimental Setup

The following subsections describe the hardware, experiment scenarios, data acquisition, and analysis tools followed by outdoor and indoor experiments. The parameters of interest are shown graphically as well as in tabular form.

3.1. Hardware. The hardware components used in the experiments are listed as follows:

(1) P2110-EVAL-01 PowerCast Energy Harvesting Development Kit for Wireless Sensors;

(2) LabJack U6 data acquisition card;

(3) notebook computer.

The constituent components of P2110-EVAL-01 Power-Cast Energy Harvesting Development Kit are listed as follows:

(1) 915 MHz, 3 W transmitter;

(2) 915 MHz directional (patch) antenna;

(3) 915 MHz omnidirectional (dipole) antenna;

(4) wireless sensor board (WSN-EVAL-01);

(5) access point: Microchip 16-bit XLP Development Board (DM240311);

(6) 2.4 GHz, IEEE 802.15.4 radio: Microchip MRF24J40 PICtail/PICtail Plus Daughter Board (AC164134-1).

Figure 1 shows the hardware listed above. For further details and operation of the P2110-EVAL-01 Development Kit, the interested reader should refer to the corresponding user manual at http://www.powercastco.com/.

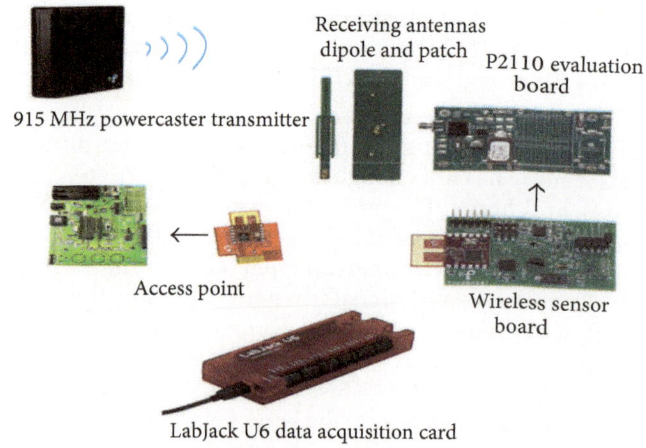

Figure 1: Hardware components used in data acquisition.

3.2. Scenarios. The data acquisition was performed for two different scenarios: outdoor free-field and indoor reverberant environment.

The outdoor scenario is straightforward to imagine where there is no source of reflection for the transmitted energy reaching the harvester, except the negligible ground reflections. Such situation can be termed as an ideal one for the RF energy harvesting with no obstacle amid transmitter-receiver and zero reflections.

The indoor data acquisition is performed in two rooms of different dimensions. The smaller room has dimensions 20.5 × 9 × 8.5 ft while the larger room has dimensions 40 × 25 × 8.5 ft. Both rooms are carpeted. Two faces of each room are concrete walls with the other two faces made of steel partitions. All of these four faces are enameled. Ceilings of both rooms are made of stainless steel. Both rooms with such five faces out of total six have certain reflection coefficients with significant magnitudes, thereby making them reverberant for RF signals.

Before the description of experimental details for both scenarios, in the following, the spatial coordinates are explained where the data acquisition was performed.

In order to characterize the power harvesting, a spherical coordinate system was chosen, since the radiation patterns are well described in this coordinate system. The characteristics are more meaningful without any confusion.

The power harvester system was tested along certain radial lines, that is, for fixed azimuth (φ) and elevation (θ) with different radial distances from origin.

As compared to [13], new radial lines are also included for enhanced data acquisition. Details of these radial lines are given in Table 2 and visualized in Figure 2. It should be noted that the radiation pattern of the transmitter is symmetric about x-y plane and hence it can be argued that the results obtained for radial lines corresponding to azimuth 16.29° and 30° can be equally associated with the radial lines corresponding to azimuth −16.29° and −30°. Hence there are effective twelve radial lines for data acquisition.

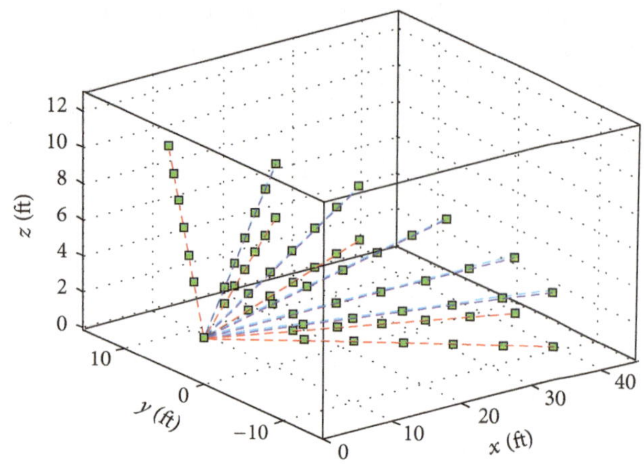

Figure 2: Spherical coordinates of test points (red lines for $\theta = 0°$ and blue lines for $\theta = 5°$).

3.3. Data Acquisition and Analysis Tools. In order to characterize the power harvester the time taken by the capacitors to get charged, denoted by T_c, and the signal strength received by the harvester, denoted by RSSI, are needed, both for changing harvester coordinates. Experiments were performed for two types of antennas used by the harvester: the low-gain omnidirectional antenna and high-gain directional antenna. These two parameters are obtained from the data packets sent by the sensor node attached to the harvester to the access point through HyperTerminal as text file. A portion of such a file is shown in Figure 3. A packet is sent whenever the capacitors are charged with an amount corresponding to enough energy to send a packet of data. This packet includes, among others, the T_c and RSSI, which are parameters of interest to us. For further analysis, the mean value of RSSI obtained from several packets was taken. However for T_c it was preferred to exploit the charging waveform of the capacitors acquired through LabJack data acquisition card and its software utility. Power spectrum (which is the FFT of the autocorrelation) of this waveform gave a direct measure of the T_c by finding the frequency of dominant peak along the spectrum and then inverting it. Power spectrum is a universally accepted tool for finding the fundamental or dominant time period (T_c in this case) of a signal (the charging waveform in our case). The

TABLE 2: Spherical coordinates for testing harvester (r, Θ, φ).

Point number	Line number					
	1	2	3	4	5	6
1	(10, 0, 0)	(10, 5, 0)	(10, 0, 16.29)	(10, 0, 30)	(10, 5, 16.29)	(10, 5, 30)
2	(10, 0, 0)	(15, 5, 0)	(15, 0, 16.29)	(15, 0, 30)	(10, 5, 16.29)	(15, 5, 30)
3	(10, 0, 0)	(20, 5, 0)	(20, 0, 16.29)	(20, 0, 30)	(10, 5, 16.29)	(20, 5, 30)
4	(10, 0, 0)	(25, 5, 0)	(25, 0, 16.29)	(25, 0, 30)	(10, 5, 16.29)	(25, 5, 30)
5	(10, 0, 0)	(30, 5, 0)	(30, 0, 16.29)	(30, 0, 30)	(10, 5, 16.29)	(30, 5, 30)
6	(10, 0, 0)	(35, 5, 0)	(35, 0, 16.29)	(35, 0, 30)	(10, 5, 16.29)	(35, 5, 30)

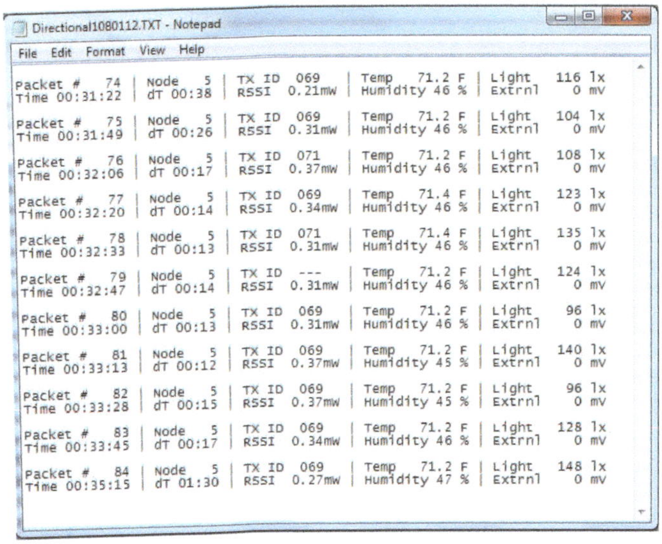

```
Directional1080112.TXT - Notepad
File  Edit  Format  View  Help
Packet #   74 | Node    5 | TX ID  069    | Temp    71.2 F | Light   116 lx
Time 00:31:22 | dT 00:38  | RSSI  0.21mW | Humidity 46 %  | Extrnl    0 mV
Packet #   75 | Node    5 | TX ID  069    | Temp    71.2 F | Light   104 lx
Time 00:31:49 | dT 00:26  | RSSI  0.31mW | Humidity 46 %  | Extrnl    0 mV
Packet #   76 | Node    5 | TX ID  071    | Temp    71.2 F | Light   108 lx
Time 00:32:06 | dT 00:17  | RSSI  0.37mW | Humidity 46 %  | Extrnl    0 mV
Packet #   77 | Node    5 | TX ID  069    | Temp    71.4 F | Light   123 lx
Time 00:32:20 | dT 00:14  | RSSI  0.34mW | Humidity 46 %  | Extrnl    0 mV
Packet #   78 | Node    5 | TX ID  071    | Temp    71.4 F | Light   135 lx
Time 00:32:33 | dT 00:13  | RSSI  0.31mW | Humidity 46 %  | Extrnl    0 mV
Packet #   79 | Node    5 | TX ID  ---    | Temp    71.2 F | Light   124 lx
Time 00:32:47 | dT 00:14  | RSSI  0.31mW | Humidity 46 %  | Extrnl    0 mV
Packet #   80 | Node    5 | TX ID  069    | Temp    71.2 F | Light    96 lx
Time 00:33:00 | dT 00:13  | RSSI  0.31mW | Humidity 46 %  | Extrnl    0 mV
Packet #   81 | Node    5 | TX ID  069    | Temp    71.2 F | Light   140 lx
Time 00:33:13 | dT 00:12  | RSSI  0.37mW | Humidity 45 %  | Extrnl    0 mV
Packet #   82 | Node    5 | TX ID  069    | Temp    71.2 F | Light    96 lx
Time 00:33:28 | dT 00:15  | RSSI  0.37mW | Humidity 45 %  | Extrnl    0 mV
Packet #   83 | Node    5 | TX ID  069    | Temp    71.2 F | Light   128 lx
Time 00:33:45 | dT 00:17  | RSSI  0.34mW | Humidity 46 %  | Extrnl    0 mV
Packet #   84 | Node    5 | TX ID  069    | Temp    71.2 F | Light   148 lx
Time 00:35:15 | dT 01:30  | RSSI  0.27mW | Humidity 47 %  | Extrnl    0 mV
```

FIGURE 3: A snapshot of data received at access point.

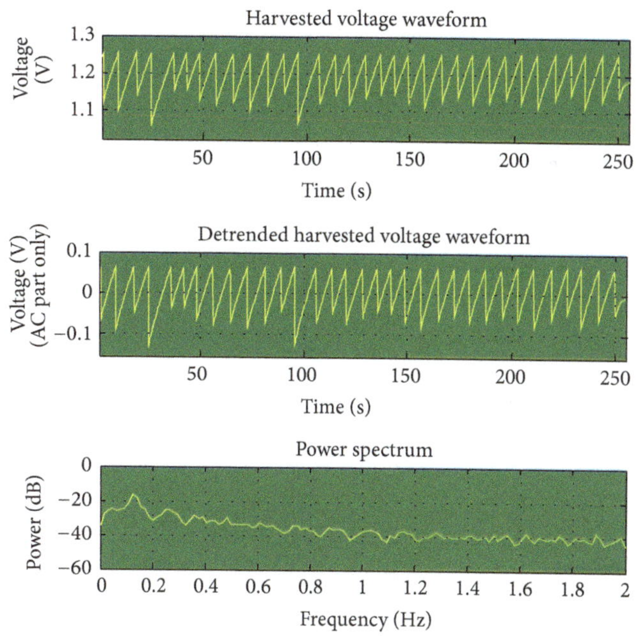

FIGURE 4: Charging waveform and its power spectrum.

waveform is acquired at a sampling interval of 10 ms, much smaller than the fastest charging time, as will be seen in the next sections. A charging waveform and the corresponding spectrum are shown in Figure 4 when the omnidirectional antenna was used. It should be emphasized here that since the waveform is always positive, the signal is preprocessed by detrending it, thereby showing the true peak other than the DC component by power spectrum. The signal processing was performed offline in MATLAB.

4. Outdoor Experiments

The outdoor experiments were performed in one of the football grounds inside King Fahd University. The data was acquired for both types of antennas. In the following, we analyze the acquired data for each radial line one by one. Numerical values of the evaluated parameters are shown in tables and their trends are shown graphically. The standard deviations for mean RSSI values are also mentioned in brackets just below them in the tables. The reader should follow the convention in Table 2 for naming different coordinates, that is, radial line number and point number. The acronym "NC" stands for "no charging" which means that "at this point the harvester could not charge." Although the data was analyzed for both of the antennas, results in the graphical form will be shown only for the directional (patch) antenna for the sake of brevity. A plot corresponding to a particular radial line is sketched to the point beyond which it could not charge.

FIGURE 5: The outdoor experiment setup. The RF harvester with omnidirectional antenna and 14.1 mF charging capacitor.

It will be found in Tables 3, 4, 5, 6, 7, 8, 9, 10, 11, 12, 13, and 14 that sometimes numerical values of T_c and RSSI are shown for only one point, which means that the antenna could charge only at that very point. Figure 5 shows the outdoor experiment setup.

TABLE 3: Parameters for radial line 1 in the outdoor experiment, azimuth = elevation = 0°.

Serial number	Radial distance r [ft]	RSSI [mW] (directional antenna)	RSSI [mW] (omnidirectional antenna)	T_c [s] (directional antenna)	T_c [s] (omnidirectional antenna)
1	10	1.0003 (0.00025)	0.4000 (0.00033)	3.7	9.1
2	15	0.4191 (0.00047)	0.2012 (0.00013)	8	21.3
3	20	0.2770 (0.0002)	NC*	16	NC
4	25	0.1857 (0.00006)	NC	32	NC
5	30	0.13 (0)	NC	197.8	NC

*No charging.

TABLE 4: Parameters for radial line 2 in the outdoor experiment, azimuth = 0°, elevation = 5°.

Serial number	Radial distance r [ft]	RSSI [mW] (directional antenna)	RSSI [mW] (omnidirectional antenna)	T_c [s] (directional antenna)	T_c [s] (omnidirectional antenna)
1	10	0.8311 (0.047)	0.3317 (0.00029)	3.7	12.8
2	15	0.4063 (0.00093)	NC	8	NC
3	20	0.2216 (0.00011)	NC	25.6	NC

TABLE 5: Parameters for radial line 3 in the outdoor experiment, azimuth = 16.29°, elevation = 5°.

Serial number	Radial distance r [ft]	RSSI [mW] (directional antenna)	RSSI [mW] (omnidirectional antenna)	T_c [s] (directional antenna)	T_c [s] (omnidirectional antenna)
1	10	0.9309 (0.0027)	0.2522 (0.00039)	3.2	16
2	15	0.3778 (0.0012)	0.155 (0.00068)	10.6	64
3	20	0.2050 (0.00011)	NC	21.3	NC

TABLE 6: Parameters for radial line 4 in the outdoor experiment, azimuth = 30°, elevation = 5°.

Serial number	Radial distance r [ft]	RSSI [mW] (directional antenna)	RSSI [mW] (omnidirectional antenna)	T_c [s] (directional antenna)	T_c [s] (omnidirectional antenna)
1	10	0.6528 (0.001)	0.2036 (0.0001)	5.3	32
2	15	0.2379 (0.000012)	NC	21.3	NC
3	20	NC	NC	NC	NC

TABLE 7: Parameters for radial line 5 in the outdoor experiment, azimuth = 16.29°, elevation = 5°.

Serial number	Radial distance r [ft]	RSSI [mW] (directional antenna)	RSSI [mW] (omnidirectional antenna)	T_c [s] (directional antenna)	T_c [s] (omnidirectional antenna)
1	10	0.8885 (0.0607)	0.320 (0.0155)	3.5	17.3
2	15	0.4557 (0.0231)	0.1667 (0.0115)	17.2	150.3
3	20	0.2000 (0.01)	NC	95.5	NC
4	21	0.19 (0.01)	NC	187	NC

TABLE 8: Parameters for radial line 6 in the outdoor experiment, azimuth = 30°; elevation = 5°.

Serial number	Radial distance r [ft]	RSSI [mW] (directional antenna)	RSSI [mW] (omnidirectional antenna)	T_c [s] (directional antenna)	T_c [s] (omnidirectional antenna)
1	10	0.75 (0.0346)	0.2175 (0.0126)	4.3	31.5
2	12	0.5672 (0.0222)	0.1875 (0.0096)	7.6	27.3
3	14	0.3843 (0.0222)	0.1100 (0)	10.9	114
4	15	0.2929 (0.0111)	NC	12.5	NC
5	20	0.222 (0.011)	NC	20.6	NC
6	22	0.15 (0.01)	NC	94.5	NC

TABLE 9: Parameters for radial line 1 in the indoor experiment, azimuth = elevation = 0°.

Serial number	Radial distance r [ft]	RSSI [mW] (directional antenna)	RSSI [mW] (omnidirectional antenna)	T_c [s] (directional antenna)	T_c [s] (omnidirectional antenna)
1	10	5.0410 (0.26)	1.5561 (0.0087)	1.6	0.5
2	15	0.9243 (0.0078)	0.3683 (0.00092)	10.6	2.7
3	20	0.2506 (0.000711)	0.17 (0.00013)	64	16
4	25	0.2818 (0.0014)	0.3807 (0.0015)	10.6	12.8
5	30	0.2116 (0.00083)	0.7040 (0.0020)	5.3	21.3
6	35	0.7767 (0.0031)	0.9802 (0.0040)	3.7	4

TABLE 10: Parameters for radial line 2 in the indoor experiment, azimuth = 0°, elevation = 5°.

Serial number	Radial distance r [ft]	RSSI [mW] (directional antenna)	RSSI [mW] (omnidirectional antenna)	T_c [s] (directional antenna)	T_c [s] (omnidirectional antenna)
1	10	0.5809 (0.0027)	0.4023 (0.0024)	5.8	9.1
2	15	0.2452 (0.00027)	0.19 (0.2111)	18.2	32
3	20	0.2182 (0.00025)	NC	21.3	NC
4	25	0.2107 (0.00025)	NC	32	NC
5	30	0.2146 (000.11)	0.244 (0.00035)	32	16
6	35	NC	NC	NC	NC

TABLE 11: Parameters for radial line 3 in the indoor experiment, azimuth = 16.29°, elevation = 5°.

Serial number	Radial distance r [ft]	RSSI [mW] (directional antenna)	RSSI [mW] (omnidirectional antenna)	T_c [s] (directional antenna)	T_c [s] (omnidirectional antenna)
1	10	0.2614 (0.0006)	0.3713 (0.0007)	9.1	21.3
2	15	0.1975 (0.0024)	0.2127 (0.0002)	21.3	128
3	20	0.5585 (0.0016)	0.2071 (0.0001)	21.3	5.8
4	25	0.2783 (0.0015)	0.2327 (0.0006)	21.3	16
5	30	0.6685 (0.0043)	0.4124 (0.0008)	9.1	4.5

TABLE 12: Parameters for radial line 4 in the indoor experiment, azimuth = 30°; elevation = 5°.

Serial number	Radial distance r [ft]	RSSI [mW] (directional antenna)	RSSI [mW] (omnidirectional antenna)	T_c [s] (directional antenna)	T_c [s] (omnidirectional antenna)
1	10	2.4960 (0.0164)	0.2256 (0.00031)	1	21.3
2	15	0.9080 (0.0011)	0.1487 (0.0008)	3.3	64
3	20	NC	0.4183 (0.00047)	NC	8

TABLE 13: Parameters for radial line 5 in the indoor experiment, azimuth = 16.29°, elevation = 0°.

Serial number	Radial distance r [ft]	RSSI [mW] (directional antenna)	RSSI [mW] (omnidirectional antenna)	T_c [s] (directional antenna)	T_c [s] (omnidirectional antenna)
1	10	1.0604 (0.0504)	NC	2.5	NC
2	15	0.64 (0.0390)	0.4000 (0.0232)	5	8.1
3	20	0.2318 (0.0272)	0.1367 (0.0115)	26.7	92.3
4	25	NC	NC	NC	NC
5	30	NC	NC	NC	NC
6	39	0.1786 (0.0146)	0.2980 (0.0179)	46.5	14.5

TABLE 14: Parameters for radial line 6 in the indoor experiment, azimuth = 30°; elevation = 0°.

Radial distance r [ft]	RSS [mW] (directional antenna)	T_c [s] (directional antenna)	Radial distance r, [ft]	RSSI [mW] (omnidirectional antenna)	T_c [s] (omnidirectional antenna)
10	0.4900 (0.0230)	6.61	10	0.3 (0.0132)	13.4
15	0.2880 (0.0132)	14	11	0.48 (0.0150)	7.1
16	0.6900 (0.0339)	6.3	12	0.23 (0.0141)	17.5
17	0.8462 (0.0375)	3.7	16	0.1940 (0.0114)	25.25
18	0.2325 (0.0175)	23.25	18	0.222 (0.011)	19.6
19	0.2590 (0.0173)	18.4	20	0.1310 (0.0208)	99.6

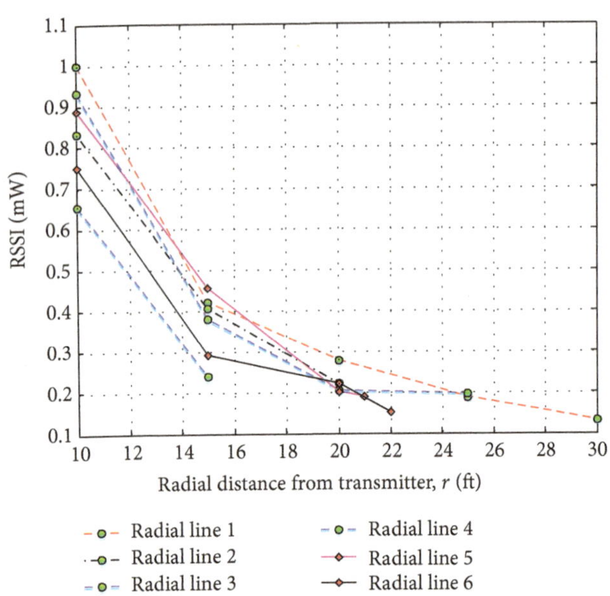

FIGURE 6: RSSI versus radial distance using directional antenna in the outdoor scenario.

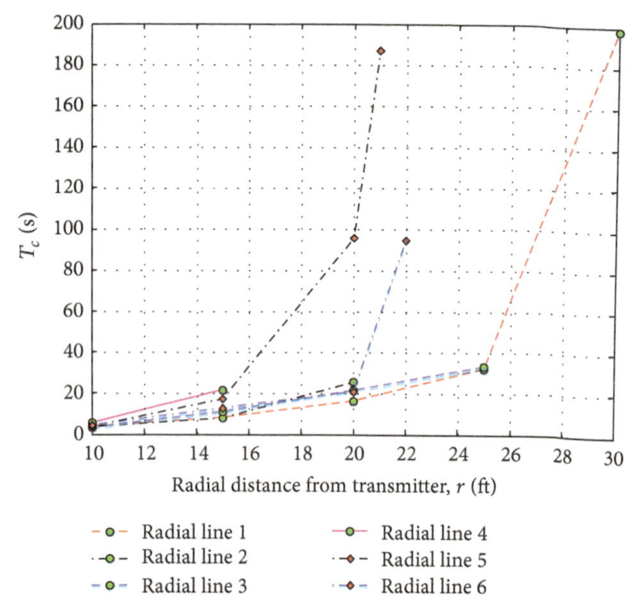

FIGURE 7: Charging time T_c versus radial distance using directional antenna in the outdoor scenario.

Tables 3 through 8 show the numerical values of the parameters of interest while Figures 6 and 7 show the variations of RSSI and T_c versus radial distance. We have observed that charging time and RSSI are affected by radial distance as well as azimuth and elevations. It is observed that charging time is significantly increased when patch type directional antenna is used. The maximum harvesting range is found to be 30 ft when directional antenna is used along the zero azimuth and elevation, while it is only 15 ft in the case of dipole antenna. At azimuths other than zeros, it is only 15 ft. In addition, as it can be expected, RSSI is inversely proportional to the distance between transmitter and harvester. From Figure 6, we can infer that the parameters such as the charging time and RSSI follow trends similar to the inverse square law.

5. Indoor Experiments

In a room with the dimensions 40 × 25 × 8.5 ft, the experiments were performed along the same radial lines as in the case of outdoor experiments and the results are documented in the same way in the following. Tables 9 through 14 show the numerical values of the parameters of interest obtained from indoor experiments, while Figures 8 and 9 show the variations of RSSI and T_c versus radial distance.

The indoor experiments exhibit very interesting observations. As compared to outdoor experiments, the maximum range is 35 ft for both antennas. The trends of these parameters with respect to radial distance are not as regular as those for the outdoor experiments. The parameters have very poor values at some points nearer to the transmitter and very good values at farther points. This can be attributed to the fact that five faces of the room out of six are highly reflective and caused multiple reflections of the transmitted waves. These reflections might cause complete cancellation at a nearer point while allowing enhancement of the signal at some farther points.

6. Mathematical Modeling of T_c and RSSI

As compared to battery powered WSN, the placement of individual sensor nodes in wireless-powered WSN requires special attention. Firstly, the charging range of wireless power transmitter is limited by its power as well as radiation pattern. RSSI at a certain node is pragmatically supposed to be significantly different from those of others. This difference is further propagated into several communication parameters like data rate, range of transmission, and so forth. Secondly, wireless charging of the sensor nodes contributes to the intersample delay of a sensor node because a node cannot transmit a packet of data unless it is charged with enough energy, which takes a certain amount of time. Keeping these facts in view, the mathematical models of T_c and RSSI are of significant importance in WSN applications because of the following reasons.

(a) *Geometry of the Sensor Network*. With the knowledge of a closed form mathematical model, describing RSSI as a function of Cartesian coordinates, locations of sensor nodes can be optimally decided.

(b) *Sampling Interval of Data*. The mathematical model of T_c as a function of geometrical coordinates will give a measure of the charging time of a certain node and thus interval between two data points can be anticipated prior to the actual operation.

It should be emphasized here that our work is much different from channel modeling. Channel modeling deals with the variation of different parameters of a signal in due course of propagation from transmitter to the receiver. Channel models are simply unable to give the information required in the scenario of wireless-powered WSNs. For example, it is not possible to compute charging time from channel model.

The models were obtained through data from directional antenna as described previously. A separate model is developed for each set of data with constant elevation, both for T_c and RSSI. For the sake of clarity and easiness in modeling, spherical coordinates of data points are converted to Cartesian coordinates. The surface fitting tool of MATLAB is utilized to fit our data into mathematical models. Different

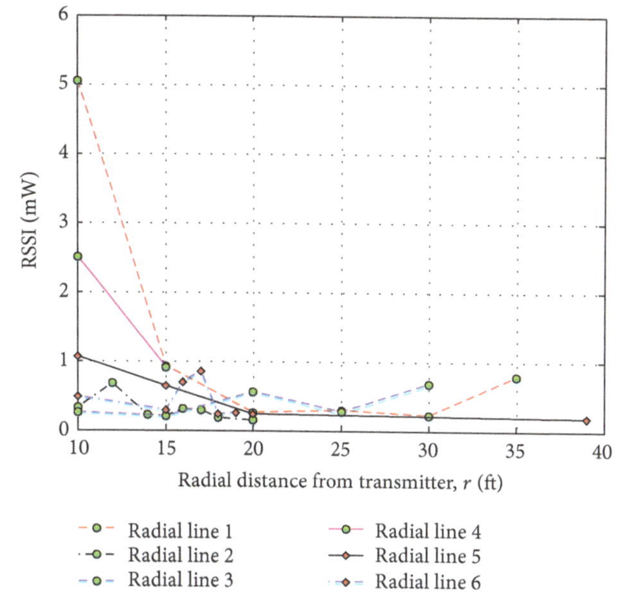

FIGURE 8: RSSI versus radial distance using directional antenna in the indoor scenario.

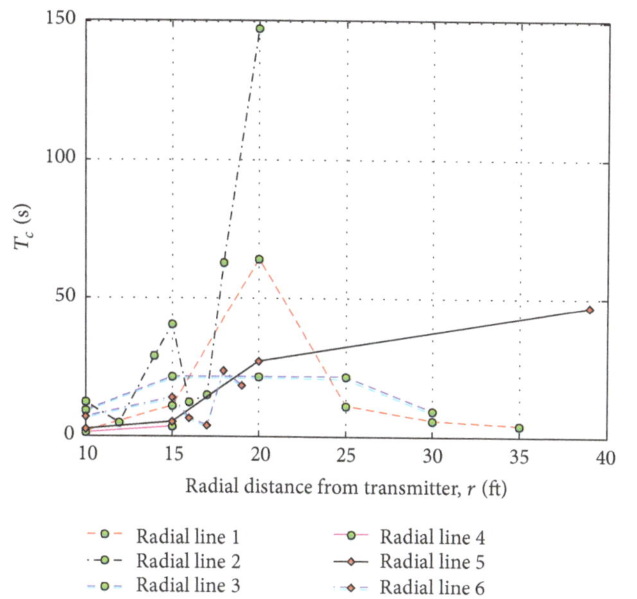

FIGURE 9: T_c versus radial distance using directional antenna in the indoor scenario.

TABLE 15: Mathematical model of $RSSI_0$ (outdoor, elevation = 0°).

Model	
$RSSI_0 = p_{00} + p_{20}x^2 + p_{02}y^2 + p_{30}x^3 + p_{12}xy^2 + p_{40}x^4 + p_{22}x^2y^2$	
Coefficients	
$p_{00} = 1.929$	$p_{22} = -0.0001521$
$p_{02} = -0.05405$	$p_{30} = 0.0009966$
$p_{12} = 0.005695$	$p_{40} = -1.542e-5$
$p_{20} = -0.01803$	
Goodness of fit	
SSE 0.009726	R-square 0.9939
Adjusted R-square 0.9913	RMSE 0.02636

TABLE 16: Mathematical model of T_{c0} (outdoor, elevation = 0°).

Model	
$T_{c0} = p_{00} + p_{10}x + p_{20}x^2 + p_{02}y^2 + p_{30}x^3 + p_{12}xy^2 + p_{40}x^4 p_{22}x^2y^2 + p_{04}y^4$	
Coefficients	
$p_{00} = -3381$	$p_{12} = -2.927$
$p_{10} = 866.2$	$p_{40} = -0.04471$
$p_{20} = -79.7$	$p_{22} = 0.105$
$p_{02} = 20.75$	$p_{04} = -0.0129$
$p_{30} = 3.138$	
Goodness of fit	
SSE 1.991	R-square 0.998
Adjusted R-square 0.997	RMSE 0.5334

TABLE 17: Mathematical model of $RSSI_5$ (outdoor, elevation = 5°).

Model	
$RSSI_5 = p_{00} + p_{10}x + p_{01}y + p_{20}x^2 + p_{02}y^2 + p_{30}x^3 + p_{12}xy^2 + p_{40}x^4 + p_{04}y^4$	
Coefficients	
$p_{00} = -11.38$	$p_{30} = 0.01511$
$p_{10} = 3.535$	$p_{12} = -0.00748$
$p_{20} = -0.3588$	$p_{40} = -0.0002276$
$p_{02} = 0.01644$	$p_{04} = -8.596e-5$
Goodness of fit	
SSE 0.009726	R-square 0.9939
Adjusted R-square 0.9913	RMSE 0.02636

TABLE 18: Mathematical model of T_{c5} (outdoor, elevation = 5°).

Model	
$T_{c5} = p_{00} + p_{10}x + p_{20}x^2 + p_{02}y^2 + p_{30}x^3 + p_{12}xy^2 + p_{40}x^4 + p_{22}x^2y^2 + p_{04}y^4$	
Coefficients	
$p_{00} = 114.4$	$p_{12} = 0.2142$
$p_{10} = -30.14$	$p_{40} = -0.002207$
$p_{20} = 2.957$	$p_{22} = -0.007837$
$p_{02} = -1.307$	$p_{04} = 0.002548$
$p_{30} = -0.1278$	
Goodness of fit	
SSE 1.163	R-square 0.9993
Adjusted R-square 0.9987	RMSE 0.3813

TABLE 19: Mathematical model of RSSI (indoor, elevation = 0°).

Model	
$RSSI_0 = p_{00} + p_{10}x + p_{20}x^2 + p_{02}y^2 + p_{30}x^3 + p_{12}xy^2 + p_{40}x^4 + p_{22}x^2y^2 + p_{04}y^4$	
Coefficients	
$p_{00} = -10.3$	$p_{12} = 0.01243$
$p_{10} = 3.283$	$p_{40} = -0.0001045$
$p_{20} = -0.2823$	$p_{22} = -0.0002286$
$p_{02} = -0.1685$	$p_{04} = 0.000234$
$p_{30} = 0.0093$	
Goodness of fit	
SSE 10.53	R-square 0.52219993
Adjusted R-square 0.2972	RMSE 0.7873813

TABLE 20: Mathematical model of T_c (indoor, elevation = 0°).

Model	
$RSSI_0 = p_{00} + p_{10}x + p_{20}x^2 + p_{02}y^2 + p_{30}x^3 + p_{12}xy^2 + p_{40}x^4 + p_{22}x^2y^2 + p_{04}y^4$	
Coefficients	
$p_{00} = 380.6$	$p_{12} = -0.1329$
$p_{10} = -97.3$	$p_{40} = 0.003456$
$p_{20} = 8.502$	$p_{22} = 0.002338$
$p_{02} = 1.042$	$p_{04} = 0.004663$
$p_{30} = -0.2932$	
Goodness of fit	
SSE 1513	R-square 0.7605
Adjusted R-square 0.6477	RMSE 9.433

model forms can be selected to fit the data. The polynomial form was chosen as it is the most general form and encompasses a large number of functions because most of the functions can be decomposed into polynomials using Taylor's series, including exponentials and sinusoids. The models of RSSI and T_c are generally described as

$$RSSI_\theta = f^{\theta}_{RSSI}(x, y), \quad T_{c\theta} = f^{\theta}_{T_c}(x, y), \qquad (1)$$

where the subscript θ indicates that the model corresponds to that elevation only.

In the modeling process, an attempt is made to reduce the number of coefficients to a possible minimum with acceptable goodness of fit. The following tables detail the obtained models. Tables 15, 16, 17, 18, 19, 20, 21, and 22 give the numerical details of the models while Figures 10–13 give the pictorial description of the fitted models. Note that SSE and RMSE stand for sum of squared errors and root mean square error, respectively.

TABLE 21: Mathematical model of RSSI (indoor, elevation = 5°).

Model	
$\mathrm{RSSI}_0 = p_{00} + p_{10}x + p_{20}x^2 + p_{02}y^2 + p_{30}x^3 + p_{12}xy^2 + p_{40}x^4 + p_{22}x^2y^2 + p_{04}y^4$	
Coefficients	
$p_{00} = 24.41$	$p_{12} = -0.01498$
$p_{10} = -5.708$	$p_{40} = 0.000199$
$p_{20} = 0.4696$	$p_{22} = 0.0003446$
$p_{02} = 0.1692$	$p_{04} = -0.0001921$
$p_{30} = -0.01618$	
Goodness of fit	
SSE 0.886	R-square 0.8981
Adjusted R-square 0.8165	RMSE 0.2977

TABLE 22: Mathematical model of T_c (indoor, elevation = 5°).

Model	
$\mathrm{RSSI}_0 = p_{00} + p_{10}x + p_{20}x^2 + p_{02}y^2 + p_{30}x^3 + p_{12}xy^2 + p_{40}x^4 + p_{22}x^2y^2 + p_{04}y^4$	
Coefficients	
$p_{00} = -271.1$	$p_{12} = -0.01982$
$p_{10} = 63.45$	$p_{40} = 0.002266$
$p_{20} = -5.129$	$p_{22} = 0.0005403$
$p_{02} = 0.5736$	$p_{04} = 0.01159$
$p_{30} = -0.01159$	
Goodness of fit	
SSE 7.647	R-square 0.9922
Adjusted R-square 0.9922	RMSE 0.8745

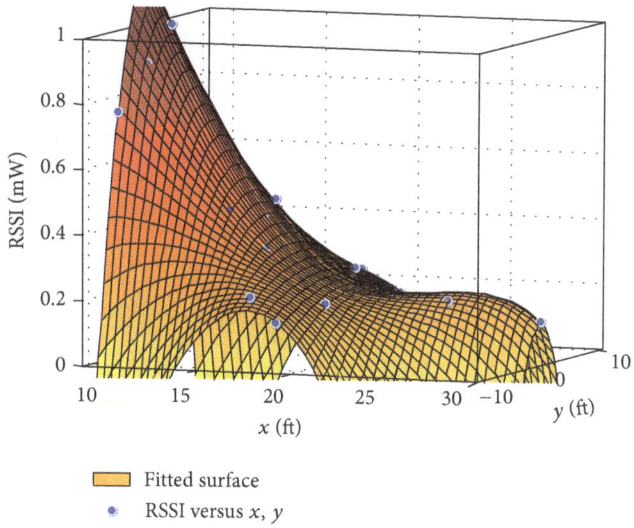

FIGURE 10: The fitted surface for RSSI_0 (outdoor, elevation = 0°).

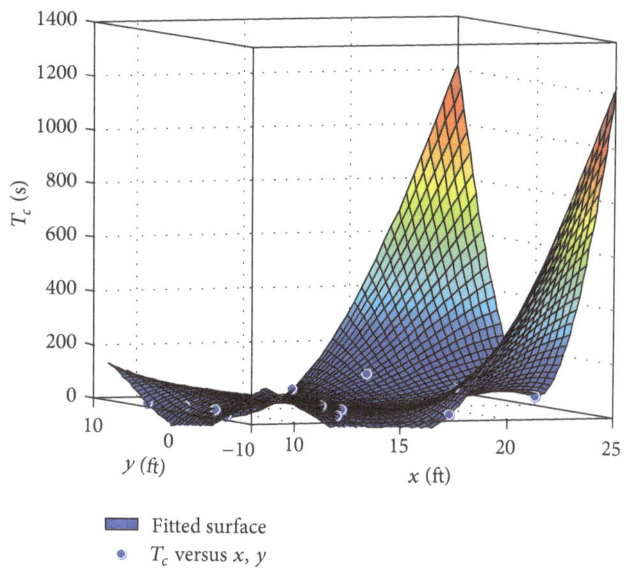

FIGURE 11: The fitted surface for T_{c0} (outdoor, elevation = 0°).

6.1. Outdoor Model. The RSSI model for outdoor environment is illustrated in Figure 10. The reciprocal relationship between the signal strength and the distance from the emitter is obvious. In addition, for the same elevation, we can observe huge variations in the signal strength due to the unequal gain of directional antenna used in the experiment. Hence, in order to obtain the same RSSI, we need to place the sensor closer to the antenna compared to the direct situation. Figure 11 describes the relationship between the charging time and the sensor position from the emitter and its complements in Figure 10. Again, we can easily observe the reciprocal relationship between the charging time and the distance. The farther the distance from the emitter, the longer the time needed to obtain enough energy to activate/transmit the sensor board.

Figures 12 and 13 illustrate the obtained models for RSSI and charging time for 5° elevation scenarios, respectively. Again, the reciprocal relation is very clear. However, we can observe local minima (e.g., $x = 16$) and even maxima (e.g., $x = 22$). Knowing this behavior helps the sensor network architect in placing the sensors in the best positions to obtain the best performance. Moreover, this knowledge helps also in designing/selecting the routes for delivering data over such network. We can also notice that due to lower RSSI compared to 0° elevation, the feasible charging region is smaller but smoother. As expected, the signal strength at 0° elevation shows higher level (>1 mw) than 5° elevation (<0.9 mw).

Figures 14 and 19 provide a comparison of experimental and fitted parameters for a couple of radial lines. The fitted curves show perfect matching with experimental data.

6.2. Indoor Model. Indoor modeling is very challenging and complex due to the surrounding environment. Our model, as described above, is simple and it can lend itself easily to similar environment like wholesale yard and so on. Figure 15 shows an interesting behavior where closer positions to the emitter suffer lower RSSI than some farther positions.

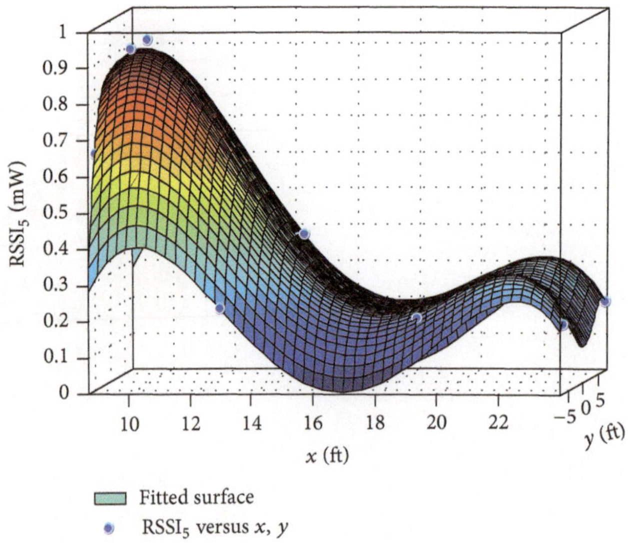

FIGURE 12: The fitted surface for RSSI$_5$ (outdoor, elevation = 5°).

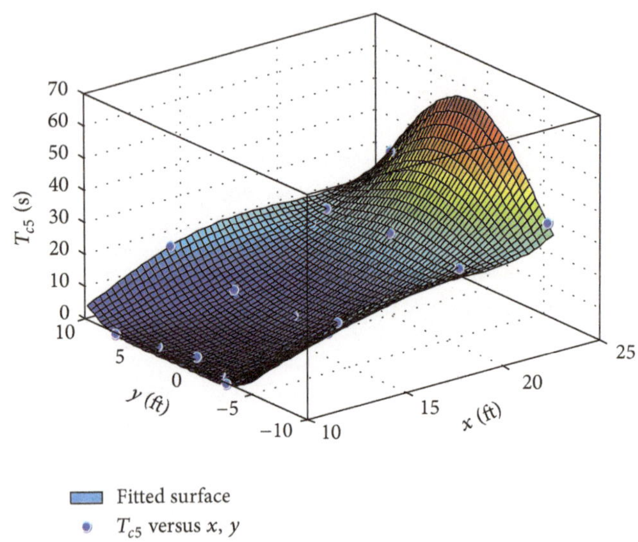

FIGURE 13: The fitted surface for T_{c5} (outdoor, elevation = 5°).

Furthermore, the farther positions near the walls exhibit higher RSSI than the middle of the room.

On the other hand, though we have carried out several trials to enhance the correlation coefficient, the best obtained correlation coefficient is the one presented in Table 19 that is low indicating a poor data fitting. The same thing applies on Figure 16, but with a better coefficient. These results manifest the difficulty in modeling the indoor RSSI/T_c behavior.

Considering the RSSI/T_c behavior for indoor with elevation 5°, Figures 17 and 18 depict these two relationships. It is interesting to notice that we were able to reach good fitness models that have almost unity correlation coefficients. Moreover, we have a better RSSI and lower charging time compared to the 0° elevation. This result can be attributed to the elevation where we minimize the ground floor reflections and the received signal has lower attenuation and consequently higher RSSI. In addition, we can observe almost a perfect symmetry. Figure 15 shows a good matching between the fitting model and the experimental data points.

6.3. Comparison between Indoor and Outdoor Behaviours. Having studied the indoor and outdoor models, we will summarize the differences between the two models in the following points.

(i) The P2110-EVAL-01 Development Kit works well for both environments.

(ii) However, the kit is optimized to work for indoor as it is observed in the obtained readings for RSSI and charging time.

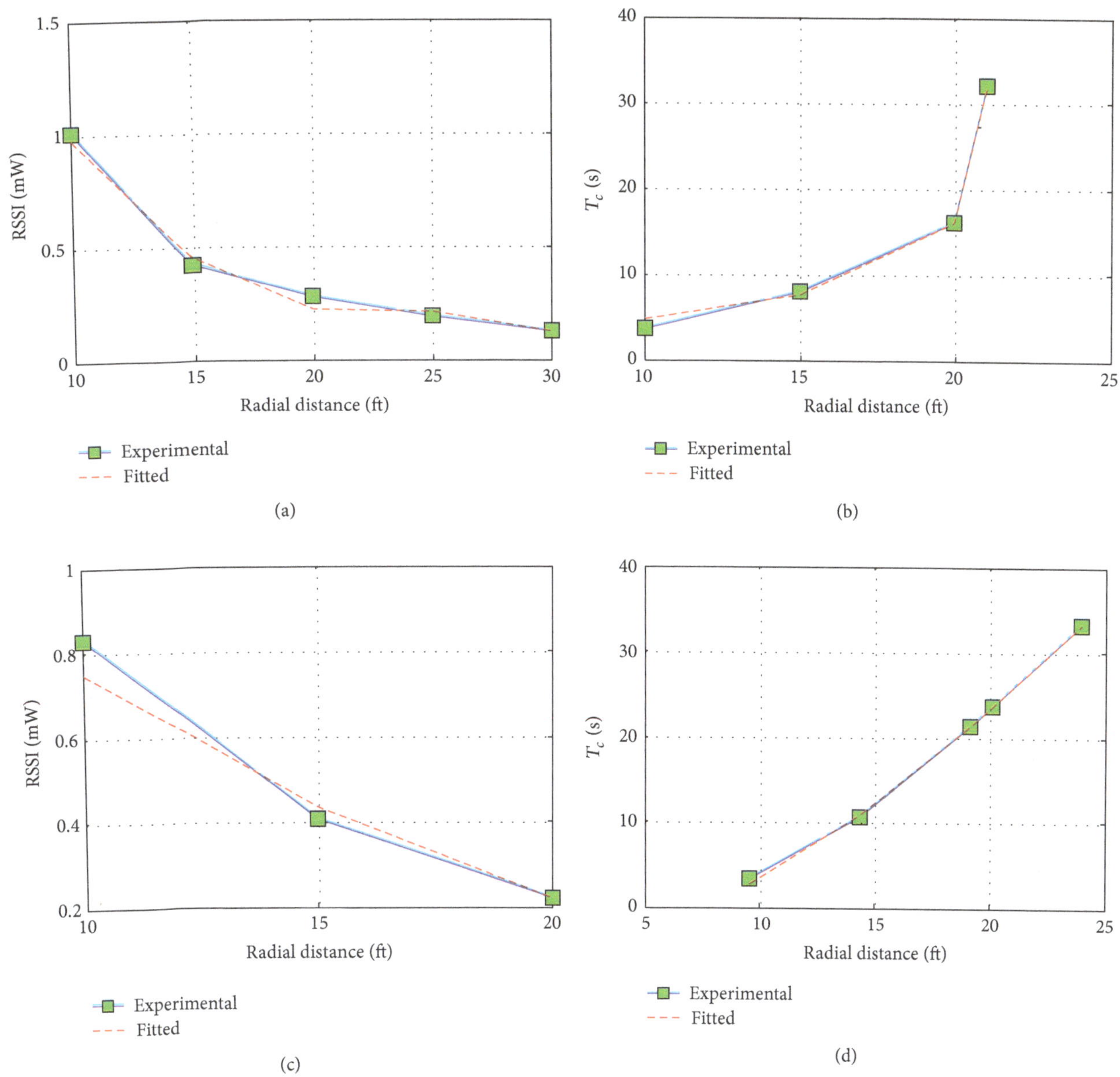

FIGURE 14: Comparison of (outdoor) experimental and fitted values of the parameters for radial line 1 (a, b) and radial line 5 (c, d).

(iii) The harvesting can cover larger outdoor area compared to the indoor area. Nevertheless, the quality of the signal for indoor is better over the short range.

6.4. Guidelines to Use These Models. In the section, we present a few guidelines to use the above models.

(i) The RSSI models complement the T_c models.

(ii) For given position coordinates (x, y, θ), we can determine the corresponding RSSI value and T_c value using both models.

(iii) If we want to use the same kit, we can use these models to find the optimal positions for placing the sensors by simply differentiating RSSI with respect to x, y or both and then we find the corresponding T_c. Alternatively, we can do the opposite, which is more practical.

(iv) For routing design, we can integrate our models with the route selection function.

(v) For a given RSSI, which can be extracted from any channel model, we can easily find the corresponding T_c. In this case, the position coordinates (x, y, θ) are not relevant. We just used them to establish the relationship between RSSI and T_c. This makes our model applicable to a wide range of channel models.

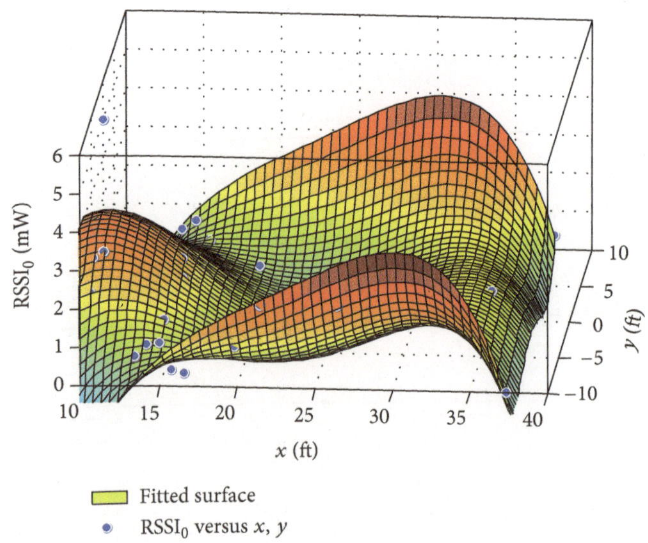

FIGURE 15: The fitted surface for RSSI$_0$ (indoor, elevation = 0°).

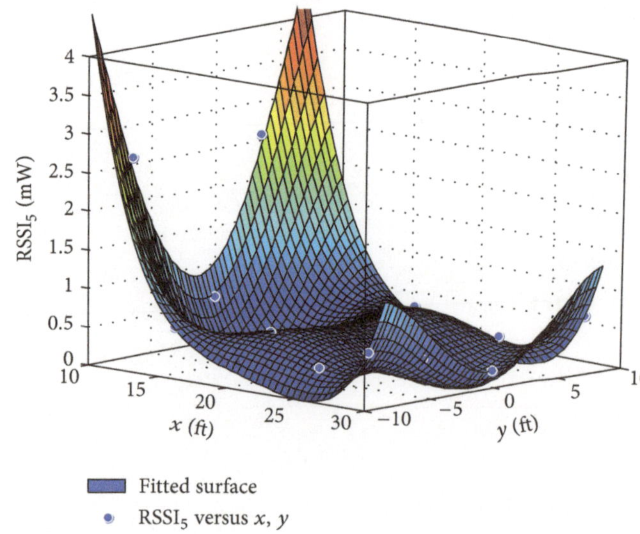

FIGURE 17: The fitted surface for RSSI$_5$ (indoor, elevation = 5°).

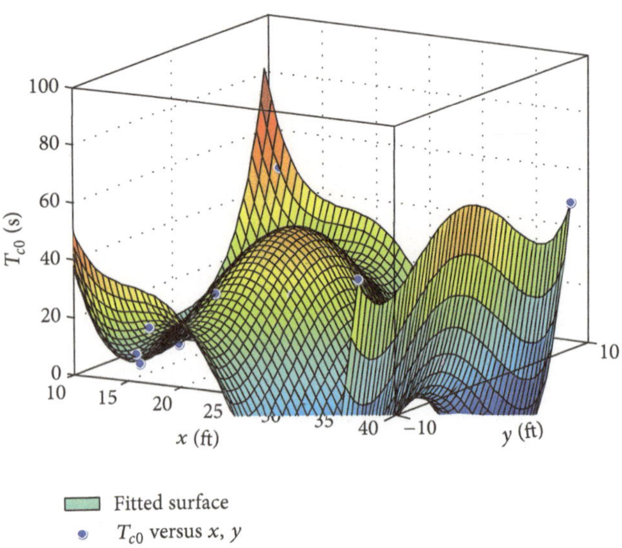

FIGURE 16: The fitted surface for T_{c0} (indoor, elevation = 0°).

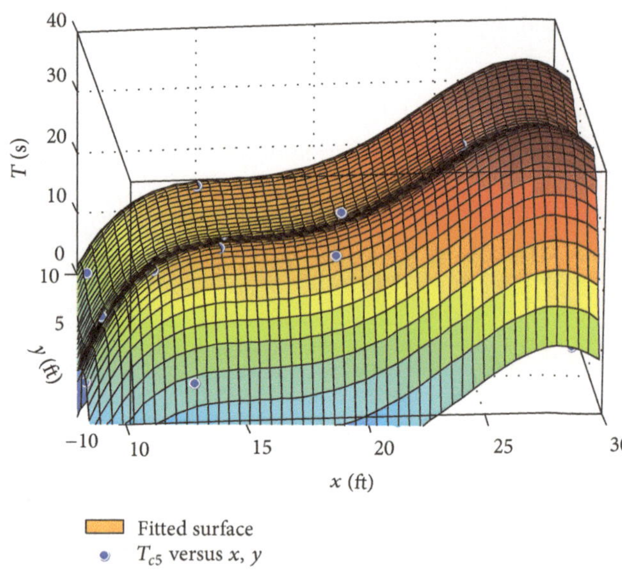

FIGURE 18: The fitted surface for T_{c5} (indoor, elevation = 5°).

7. Conclusion

This paper describes a series of data acquisition experiments and modeling of two important parameters for RF powered WSNs. These parameters are the charging time of a capacitor/battery powering the sensor node and the received signal strength indicator at a node. The mathematical models of these parameters are important in designing the routing protocols and WSN geometry. Extensive data are acquired in both indoor and outdoor environments. A detailed experimental procedure is explained in the paper. Modeling results are presented and discussed in detail. Then, a few guidelines for the usage of these models are suggested.

Additionally, the RF survey and the experimental results using PowerCast power harvester suggested that it is practically impossible to harvest sufficient energy for running the associated application with PowerCast. However, an optimized multiband antenna with further improvements in the electronics of the harvesting circuitry can be realized in the future to improve the level of harvested energy.

Our plan is to incorporate electromagnetic properties of reflecting walls to come up with similar models for the indoor scenario, which is much complicated as compared to the outdoor scenario.

Conflict of Interests

The authors declare that there is no conflict of interests regarding the publication of this paper.

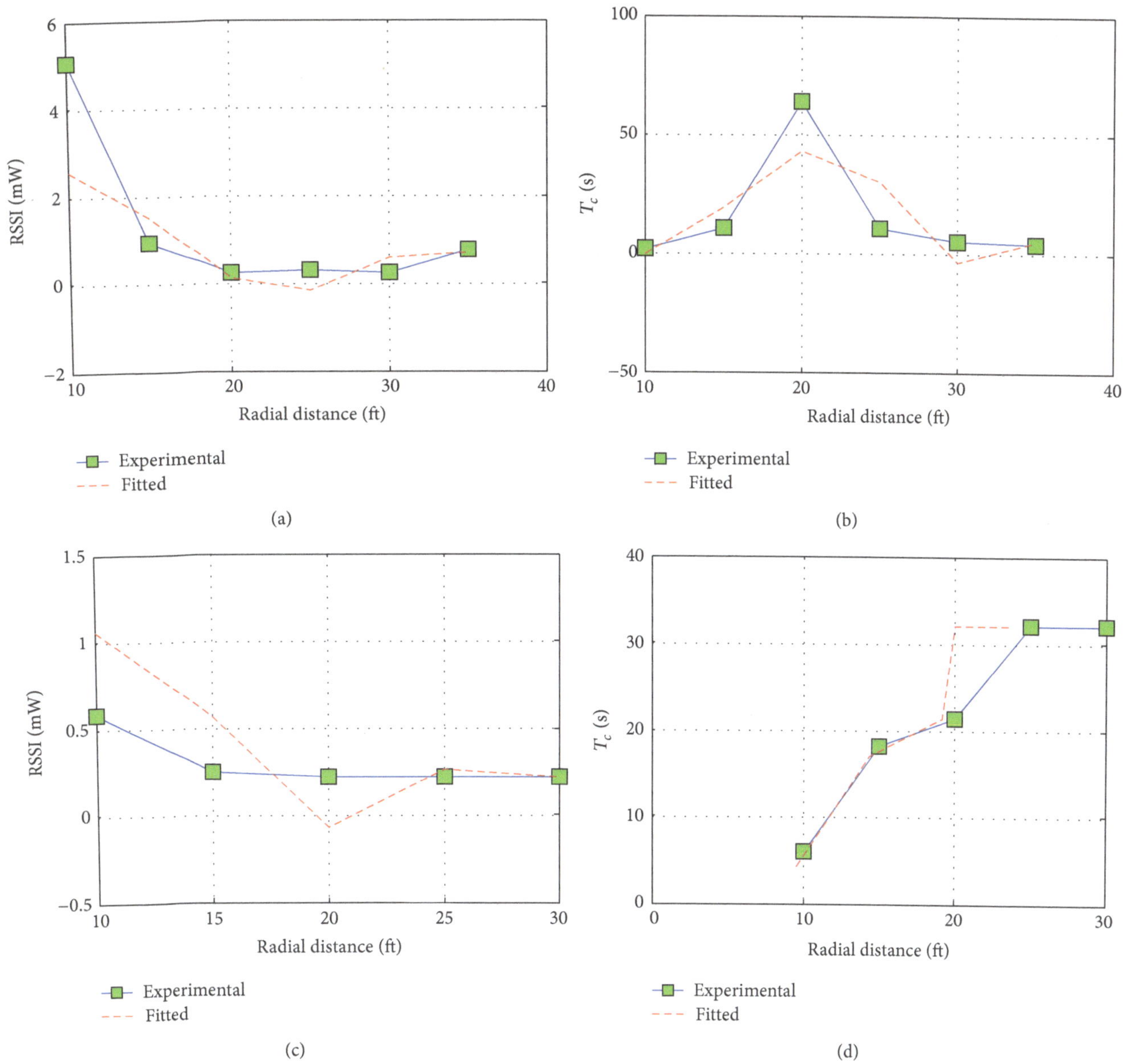

FIGURE 19: Comparison of (indoor) experimental and fitted values of the parameters for radial line 1 (a, b) and radial line 5 (c, d).

Acknowledgment

The authors would like to acknowledge the support provided by King Abdulaziz City for Science and Technology (KACST) through King Fahd University of Petroleum and Minerals (KFUPM) for funding this Project, no. 09ELE758-04, as part of the National Science Technology and Innovation Plan.

References

[1] Z. W. Sim, R. Shuttleworth, M. J. Alexander, and B. D. Grieve, "Compact patch antenna design for outdoor rf energy harvesting in wireless sensor networks," *Progress in Electromagnetics Research*, vol. 105, pp. 273–294, 2010.

[2] D. Niyato, E. Hossain, M. M. Rashid, and V. K. Bhargava, "Wireless sensor networks with energy harvesting technologies: a game-theoretic approach to optimal energy management," *IEEE Wireless Communications*, vol. 14, no. 4, pp. 90–96, 2007.

[3] P. T. V. Bhuvaneswari, R. Balakumar, V. Vaidehi, and P. Balamuralidhar, "Solar energy harvesting for wireless sensor networks," in *Proceedings of the 1st International Conference on Computational Intelligence, Communication Systems and Networks (CICSYN '09)*, pp. 57–61, July 2009.

[4] G. K. Ottman, H. F. Hofmann, A. C. Bhatt, and G. A. Lesieutre, "Adaptive piezoelectric energy harvesting circuit for wireless remote power supply," *IEEE Transactions on Power Electronics*, vol. 17, no. 5, pp. 669–676, 2002.

[5] D. Bouchouich, F. Dupont, M. Latrach, and L. Ventura, "Ambient RF energy harvesting," in *Proceedings of the International Conference on Renewable Energies and Power Quality (ICREPQ '10)*, Granada, Spain, March 2010.

[6] S. Kim, "RF energy harvesting techniques for wirelessly powered devices," in *Proceedings of the IEEE MTT-S International Microwave Workshop Series on Intelligent Radio for Future Personal Terminals (IMWS-IRFPT '11)*, 2011.

[7] M. Mi, M. H. Mickle, C. Capelli, and H. Swift, "RF energy harvesting with multiple antennas in the same space," *IEEE Antennas and Propagation Magazine*, vol. 47, no. 5, pp. 100–106, 2005.

[8] S. Mandal and S. K. Giri, "Comparison of antennas for radio frequency energy harvesting in 0.2–2.4 GHz range," in *Proceedings of the 3rd International Conference on Electronics Computer Technology (ICECT '11)*, pp. 93–97, April 2011.

[9] M. Minami, T. Morito, H. Morikawa, and T. Aoyama, "Solar Biscuit. A battery-less wireless sensor network system for environmental monitoring applications," in *Proceedings of the 2nd International Workshop on Networked Sensing Systems*, June 2005.

[10] L. Tang and C. Guy, "Radio frequency energy harvesting in wireless sensor network," in *Proceedings of the International Conference on Wireless Communications and Mobile Computing (IWCMC '09)*, 2009.

[11] L. Xie, Y. Shi, Y. T. Hou, W. Lou, H. D. Sherali, and S. F. Midkiff, "On renewable sensor networks with wireless energy transfer: the multi-node case," in *Proceedings of the 9th Annual IEEE Communications Society Conference on Sensor, Mesh and Ad Hoc Communications and Networks (SECON '12)*, pp. 10–18, Seoul, Republic of Korea, June 2012.

[12] M. Erol-Kantarci and H. T. Mouftah, "Suresense: sustainable wireless rechargeable sensor networks for the smart grid," *IEEE Wireless Communications*, vol. 19, no. 3, pp. 30–36, 2012.

[13] U. Baroudi, A.-U. Qureshi, S. Mekid, and A. Bouhraoua, "Radio frequency energy harvesting characterization: an experimental study," in *Proceedings of the 11th IEEE International Conference on Trust, Security and Privacy in Computing and Communications (TrustCom '12)*, pp. 1976–1981, Liverpool, UK, June 2012.

13

Single Active Element Based Voltage-Mode Multifunction Filter

Jitendra Mohan,[1] Bhartendu Chaturvedi,[1] and Sudhanshu Maheshwari[2]

[1] *Department of Electronics and Communication Engineering, Jaypee Institute of Information Technology, Noida 201304, India*
[2] *Department of Electronics Engineering, Z. H. College of Engineering and Technology, Aligarh Muslim University, Aligarh 202002, India*

Correspondence should be addressed to Jitendra Mohan; jitendramv2000@rediffmail.com

Academic Editor: Richard T. Lauer

The paper presents a new voltage-mode multifunction filter. The proposed filter employs single modified fully differential second generation current conveyor (FDCCII), two grounded capacitors, and three resistors. The proposed circuit enjoys the employment of two grounded capacitors (attractive for absorbing shunt parasitic capacitance and ideal for IC implementation). The proposed circuit provides all five generic filter responses (low pass (LP), high pass (HP), band pass (BP), notch (NH), and all pass (AP) filter responses) simultaneously with single input. The novel proposed filter has low active and passive sensitivities. A number of time domain and frequency domain simulation results depicted through PSPICE using 0.18 μm TSMC process parameters are included to validate the theory. The proposed circuit is expected to enhance the existing knowledge on the subject.

1. Introduction

Analog filters are the basic building blocks and widely used for continuous-time signal processing. Applications of analog filters employing current-mode active elements extend over a large number of areas [1–5]. In the literature, several voltage-mode (VM) universal biquadratic filters with single input and multioutputs (SIMO) have been reported [6–10]. A SIMO circuit configuration reported in [6] employs five current feedback amplifiers (CFAs), two grounded capacitors, and six resistors. It realizes all the standard filter functions, namely, low pass (LP), band pass (BP), high pass (HP), notch (NH), and all pass (AP), simultaneously, and also enjoys the features of using only grounded capacitors and orthogonal control between the resonance angular frequency and quality factor. In [7], two voltage-mode universal filter circuits are presented. Each circuit employs two FDCCII, two resistors, and two capacitors. The filter circuits reported in [8, 9] use single FDCCII, three resistors, and two grounded capacitors and realize all generic filter responses in VM. Two VM universal filters are reported in [10]. The first universal filter employs three DVCCs together with two grounded capacitors and three grounded resistors and provides five outputs with single input but it requires matching condition to realize all the generic filter responses. The second reported VM universal filter with three inputs and one output employs three DVCCs, two grounded capacitors, and three grounded resistors. Over the last decades, numerous voltage-mode biquadratic filters using current conveyors are also presented [11–18]. However, all of these reported filters cannot realize five filtering responses simultaneously.

In this paper, a new voltage-mode multifunction filter has been presented. The proposed filter employs single modified FDCCII as an active element, two grounded capacitors, and three resistors. The proposed circuit realizes all generic filter functions (low pass (LP), band pass (BP), high pass (HP), Notch (NH), and all pass (AP)) simultaneously. The proposed circuit of multifunction filter enjoys the features of minimum number of active and passive components and no requirement of component choice except for to realize all pass filter and low active and passive sensitivities.

2. Circuit Description

Fully differential second-generation current conveyor (FDCCII) as a current-mode active device was proposed to improve the dynamic range especially in mixed-mode applications where fully differential signal processing is

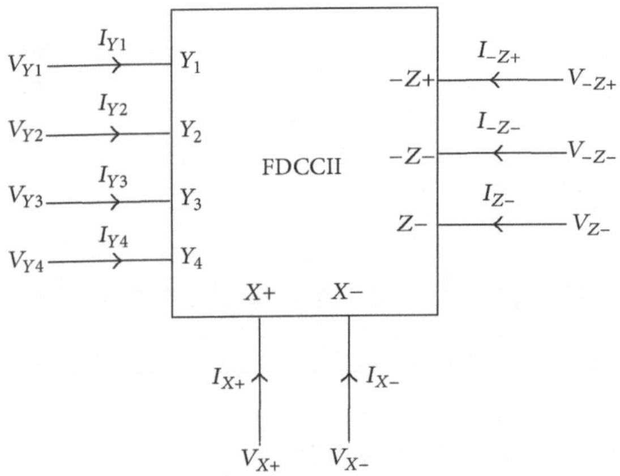

FIGURE 1: Symbol of modified FDCCII.

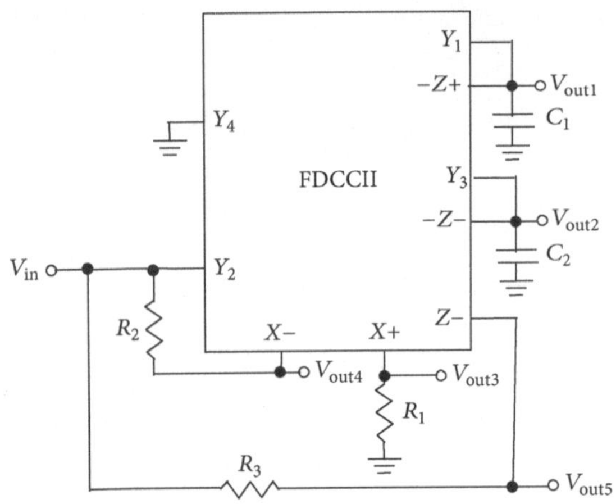

FIGURE 2: Proposed voltage mode multifunction filter.

required [19]. This active element is a versatile building block whose applications exist in the literature [7–9, 11, 13–17]. One of the most important applications is the design of analog filters.

The input-output relationship of the modified FDCCII whose symbol is shown in Figure 1 is characterized by the following matrix equation:

$$\begin{bmatrix} V_{X+} \\ V_{X-} \\ I_{-z+} \\ I_{-z-} \\ I_{z-} \end{bmatrix} = \begin{bmatrix} 1 & -1 & 1 & 0 & 0 & 0 \\ -1 & 1 & 0 & 1 & 0 & 0 \\ 0 & 0 & 0 & 0 & -1 & 0 \\ 0 & 0 & 0 & 0 & 0 & -1 \\ 0 & 0 & 0 & 0 & 0 & 1 \end{bmatrix} \begin{bmatrix} V_{Y1} \\ V_{Y2} \\ V_{Y3} \\ V_{Y4} \\ I_{X+} \\ I_{X-} \end{bmatrix}. \quad (1)$$

The proposed circuit employing single modified FDCCII, two grounded capacitors, and three resistors is shown in Figure 2. The use of grounded capacitors is particularly attractive for integrated circuit implementation because grounded capacitors compensate stray capacitances at their nodes [20].

Routine analysis of the proposed circuit using (1) yields the following transfer functions:

$$\begin{aligned} \frac{V_{out1}}{V_{in}} &= \frac{sC_2R_2}{\Delta_1}, \\ \frac{V_{out2}}{V_{in}} &= \frac{1}{\Delta_1}, \\ \frac{V_{out3}}{V_{in}} &= -\frac{s^2C_1C_2R_1R_2}{\Delta_1}, \\ \frac{V_{out4}}{V_{in}} &= \frac{s^2C_1C_2R_1R_2+1}{\Delta_1}, \\ \frac{V_{out5}}{V_{in}} &= \frac{s^2C_1C_2R_1R_2+sC_2(R_2-R_3)+1}{\Delta_1}, \end{aligned} \quad (2)$$

where $\Delta_1 = s^2C_1C_2R_1R_2 + sC_2R_2 + 1$.

It can be seen from (2) that a noninverting BP response is obtained from V_{out1}, a noninverting LP response is obtained from V_{out2}, an inverting HP response is obtained from V_{out3}, a noninverting NH response is obtained from V_{out4}, and if $R_3 = 2R_2$, a noninverting AP response is obtained from V_{out5}. It is to be noted that there is no need to impose component choice except in the realization of AP filter response.

The resonance angular frequency (ω_o) and the quality factor (Q) are given by

$$\omega_o = \sqrt{\frac{1}{C_1C_2R_1R_2}}, \qquad Q = \sqrt{\frac{C_1R_1}{C_2R_2}}. \quad (3)$$

The sensitivities of ω_o and Q with respect to passive components are given as

$$S^{\omega_o}_{R_1,R_2,C_1,C_2} = -\frac{1}{2}, \qquad S^{Q}_{R_1,C_1} = -S^{Q}_{R_2,C_2} = \frac{1}{2}. \quad (4)$$

It is to be noted from (4) that all the sensitivities are less than unity in magnitude.

3. Nonideal Analysis

To account for non ideal sources, two sets of parameters α_i and β_j are introduced where α_i ($i = 1,2,3$) and β_j ($j = 1,2,3,4,5,6$) represent the current and voltage transfer gains of the internal current and voltage followers of the FDCCII, respectively. They can be approximated by the first order of low pass functions, which can be considered to have unity values for frequencies much less than their corner frequencies [10]. By assuming, the circuit is working at the frequencies much less than the corner frequencies of α_i and β_j. More specifically, $\alpha_i = 1-\delta_i$, ($|\delta_i| \ll 1$) where δ_1 (from $X+$ to $-Z+$), δ_2 (from $X-$ to $-Z-$) and δ_3 (from $X-$ to $Z-$) are current tracking errors. Similarly, $\beta_j = 1 - \varepsilon_j$, ($|\varepsilon_j| \ll 1$) where ε_1 (from Y_1 to $X+$), ε_2 (from Y_2 to $X+$), ε_3 (from Y_3 to $X+$), ε_4

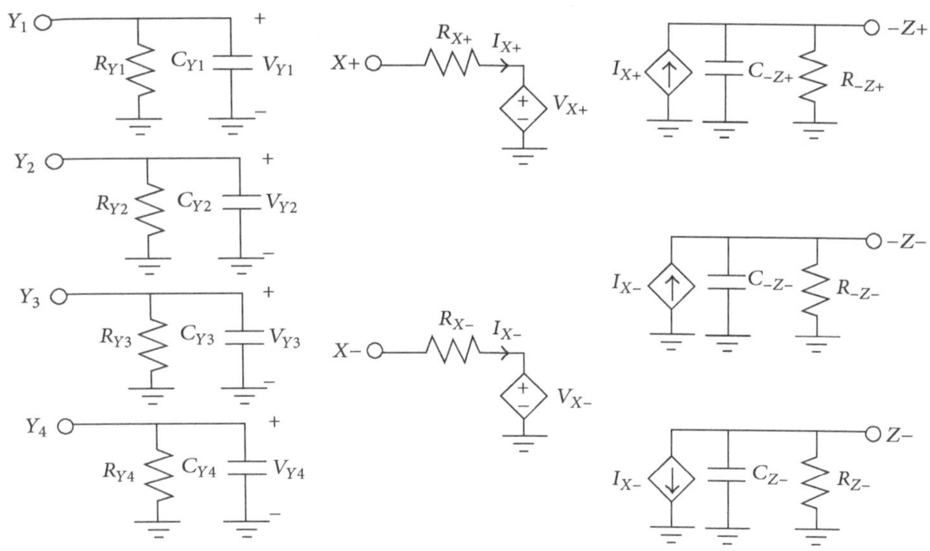

FIGURE 3: Parasitic model of modified FDCCII.

(from Y_1 to $X-$), ε_5 (from Y_2 to $X-$), and ε_6 (from Y_4 to $X-$), are voltage tracking errors. By incorporating these two non ideal sources onto the ideal input-output matrix relationship, the modified FDCCII now leads to:

$$\begin{bmatrix} V_{X+} \\ V_{X-} \\ I_{-Z+} \\ I_{-Z-} \\ I_{Z-} \end{bmatrix} = \begin{bmatrix} \beta_1 & -\beta_2 & \beta_3 & 0 & 0 & 0 \\ -\beta_4 & \beta_5 & 0 & \beta_6 & 0 & 0 \\ 0 & 0 & 0 & 0 & -\alpha_1 & 0 \\ 0 & 0 & 0 & 0 & 0 & -\alpha_2 \\ 0 & 0 & 0 & 0 & 0 & \alpha_3 \end{bmatrix} \begin{bmatrix} V_{Y1} \\ V_{Y2} \\ V_{Y3} \\ V_{Y4} \\ I_{X+} \\ I_{X-} \end{bmatrix}. \quad (5)$$

The proposed circuit of Figure 2 is reanalyzed using (5) so as to obtain nonideal transfer functions as

$$\frac{V_{\text{out1}}}{V_{\text{in}}} = \frac{sC_2R_2\alpha_1\beta_2 + \alpha_1\alpha_2\beta_3(1-\beta_5)}{\Delta_2},$$

$$\frac{V_{\text{out2}}}{V_{\text{in}}} = \frac{\alpha_1\alpha_2\beta_2\beta_4 - (sC_1R_1 + \alpha_1\beta_1)\alpha_2(1-\beta_5)}{\Delta_2},$$

$$\frac{V_{\text{out3}}}{V_{\text{in}}} = -\left(\frac{s^2C_1C_2R_1R_2\beta_2 + sC_1R_1\alpha_2\beta_3(1-\beta_5)}{\Delta_2}\right),$$

$$\frac{V_{\text{out4}}}{V_{\text{in}}} = \frac{s^2C_1C_2R_1R_2\beta_5 - sC_2R_2\alpha_1(\beta_2\beta_4 - \beta_1\beta_5) + \alpha_1\alpha_2\beta_3\beta_4(2\beta_5 - 1)}{\Delta_2},$$

$$\begin{aligned}\frac{V_{\text{out5}}}{V_{\text{in}}} = \Bigg(& s^2C_1C_2R_1R_2\left[\alpha_3(1-\beta_5) + 1\right] \\ &+ sC_2R_2\alpha_1\left[1 + \alpha_3\left(1 - \beta_5 - \beta_2\beta_4\frac{R_3}{R_2}\right)\right] \\ &+ \alpha_1\alpha_2\beta_3\beta_4\left[\alpha_3(1-\beta_5)\left(\frac{1}{R_2} - 1\right) + 1\right]\Bigg) \\ &\times (\Delta_2)^{-1},\end{aligned} \quad (6)$$

where $\Delta_2 = s^2C_1C_2R_1R_2 + sC_2R_2\alpha_1\beta_1 + \alpha_1\alpha_2\beta_3\beta_4$.

The resonance angular frequency (ω_o) and quality factor (Q) are obtained by

$$\omega_o = \sqrt{\frac{\alpha_1\alpha_2\beta_3\beta_4}{C_1C_2R_1R_2}}, \qquad Q = \frac{1}{\beta_1}\sqrt{\frac{R_1C_1\alpha_2\beta_3\beta_4}{R_2C_2\alpha_1}}. \quad (7)$$

The active and passive sensitivities of ω_o and Q are given as

$$S^{\omega_o}_{R_1,R_2,C_1,C_2} = -\frac{1}{2}, \qquad S^{\omega_o}_{\alpha_1,\alpha_2,\beta_3,\beta_4} = \frac{1}{2},$$
$$S^{Q}_{R_1,C_1} = -S^{Q}_{R_2,C_2} = \frac{1}{2}, \qquad S^{Q}_{\alpha_2,\beta_3,\beta_4} = -S^{Q}_{\alpha_1} = \frac{1}{2}, \quad (8)$$
$$S^{Q}_{\beta_1} = -1.$$

It is to be noted from (8) that all the active and passive sensitivities are less than or equal to unity in magnitude.

4. Effects of FDCCII Parasitic

A parasitic model of FDCCII is shown in Figure 3.

When the parasitic equivalent circuit model of modified FDCCII is used instead of the ideal one, the parasitics of modified FDCCII appear as the undesirable factors in (2), which lead to modified transfer functions as follows:

$$\frac{V_{\text{out1}}}{V_{\text{in}}} = \frac{sa_1 + a_0}{s^2b_2 + sb_1 + b_0},$$

$$\frac{V_{\text{out2}}}{V_{\text{in}}} = \frac{R_{Y1}R_{Y3}R_{-Z+}R_{-Z-}}{s^2b_2 + sb_1 + b_0},$$

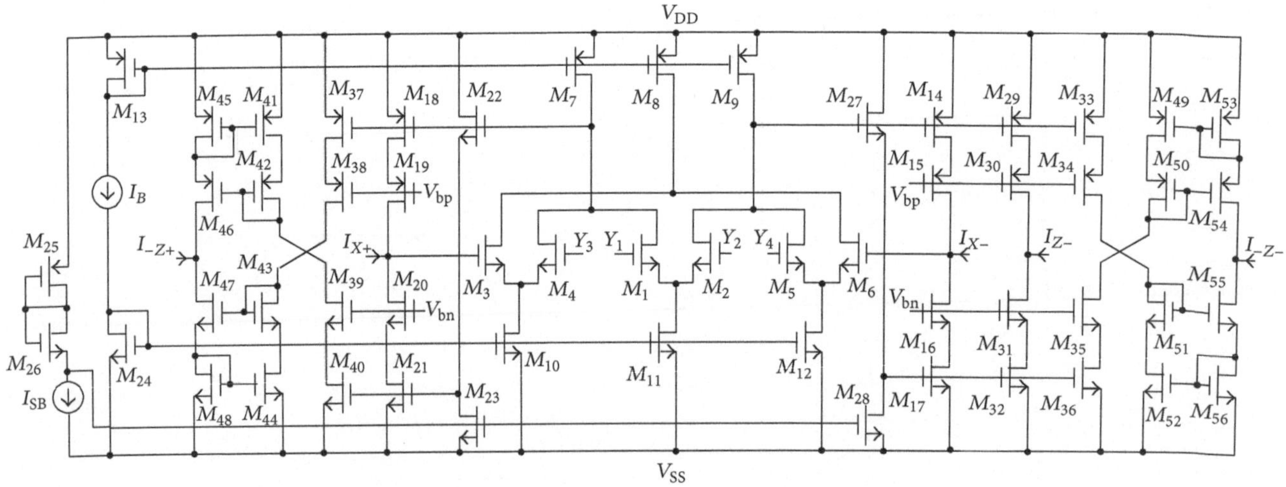

FIGURE 4: CMOS implementation of the modified FDCCII.

$$\frac{V_{out3}}{V_{in}} = -\left(\frac{s^2 c_2 + s c_1 + c_0}{s^2 b_2 + s b_1 + b_0}\right),$$

$$\frac{V_{out4}}{V_{in}} = \frac{s^2 b_2 + s d_1 + d_0}{s^2 b_2 + s b_1 + b_0},$$

$$\frac{V_{out5}}{V_{in}} = \left(s^2 b_2 + s\left(b_1 - R'_3 C'_2 R_{Y1} R_{Y3} R_{-Z+} R_{-Z-}\right) + \left(b_0 - R'_3 R_{Y1} R_{-Z+}\left(R_{Y3} + R_{-Z-}\right)\right)\right) \times \left(s^2 b_2 + s b_1 + b_0\right)^{-1}, \tag{9}$$

where

$$a_1 = R'_2 C'_2 R_{Y1} R_{Y3} R_{-Z+} R_{-Z-},$$

$$a_0 = R'_2 R_{Y1} R_{-Z+}\left(R_{Y3} + R_{-Z-}\right),$$

$$b_2 = R'_1 R'_2 C'_1 C'_2 R_{Y1} R_{Y3} R_{-Z+} R_{-Z-},$$

$$b_1 = C'_2 R'_2 R_{Y3} R_{-Z-}\left(R_{Y1} R_{-Z+} + R'_1 R_{Y1} + R'_1 R_{-Z+}\right) + C'_1 R'_1 R'_2 R_{Y1} R_{-Z+}\left(R_{Y3} + R_{-Z-}\right),$$

$$b_0 = R'_1 R'_2\left(R_{-Z-} R_{-Z+} + R_{Y3} R_{-Z+} + R_{Y1} R_{-Z-} + R_{Y1} R_{Y3}\right) + R_{Y1} R_{-Z+}\left(R_{Y3} R_{-Z-} + R'_2 R_{Y3} + R'_2 R_{-Z-}\right),$$

$$c_2 = R'_1 R'_2 C'_1 C'_2 R_{Y1} R_{Y3} R_{-Z+} R_{-Z-},$$

$$c_1 = C'_1 R'_1 R'_2 R_{Y1} R_{-Z+}\left(R_{Y3} + R_{-Z-}\right),$$

$$c_0 = R'_1 R'_2\left(R_{-Z-} R_{-Z+} + R_{Y3} R_{-Z+} + R_{Y1} R_{-Z-} + R_{Y1} R_{Y3}\right),$$

$$d_1 = C'_2 R'_2 R'_1 R_{Y3} R_{-Z-}\left(R_{Y1} + R_{-Z+}\right) + C'_1 R'_1 R'_2 R_{Y1} R_{-Z+}\left(R_{Y3} + R_{-Z-}\right),$$

$$d_0 = R'_1 R'_2\left(R_{-Z-} R_{-Z+} + R_{Y3} R_{-Z+} + R_{Y1} R_{-Z-} + R_{Y1} R_{Y3}\right) + R_{Y1} R_{-Z+} R_{Y3} R_{-Z-},$$

$$R'_1 = R_{X+} + R_1,$$

$$R'_2 = R_{X-} + R_2 + \left(R_{Y2} // C_{Y2}\right),$$

$$R'_3 = R_3 + \left(R_{Y2} // C_{Y2}\right) + \left(R_{Z-} // C_{Z-}\right),$$

$$C'_1 = C_1 // C_{Y1} // C_{-Z+},$$

$$C'_2 = C_2 // C_{Y3} // C_{-Z-}. \tag{10}$$

In practical FDCCII, the external resistors R_k (where, $k = 1, 2, 3$) can be chosen very much smaller than the parasitic resistors at the Y and Z terminals of FDCCII and very much greater than the parasitic resistors at the X terminals of FDCCII, that is, $(R_Y, R_Z) \gg R_k \gg R_X$. Moreover, the external capacitances C_1 and C_2 can be chosen very much greater than the parasitic capacitances at the Y and Z terminals of FDCCII, that is, $(C_1, C_2) \gg (C_Y, C_Z)$.

5. Simulation Results

To verify the given theoretical analysis, the proposed voltage-mode multifunction filter has been simulated using PSPICE. The CMOS implementation of modified FDCCII is shown in Figure 4 and is realized using TSMC 0.18 μm CMOS process model parameters. The supply voltages and currents have been selected as $V_{DD} = -V_{SS} = 1.1$ V, $V_{bp} = V_{bn} = 0$ V, $I_B = 1.4\,\mu$A, and $I_{SB} = 1.3\,\mu$A.

The proposed circuit shown in Figure 2 is designed with the passive component values as $C_1 = C_2 = 50$ pF, $R_1 = R_2 = 1\,\text{k}\Omega$ and $R_3 = 2\,\text{k}\Omega$ to obtain angular frequency of 3.18 MHz and a quality factor of $Q = 1$. All the magnitude responses

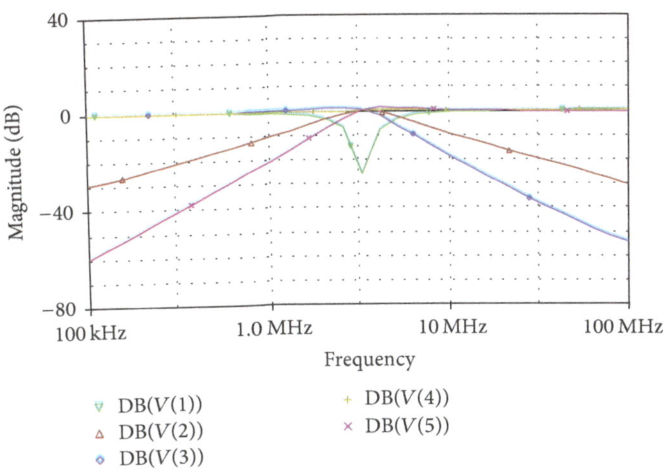

FIGURE 5: Magnitude responses for all generic filters (LP, HP, BP, NH, and AP) for $Q = 1$ at 3.18 MHz.

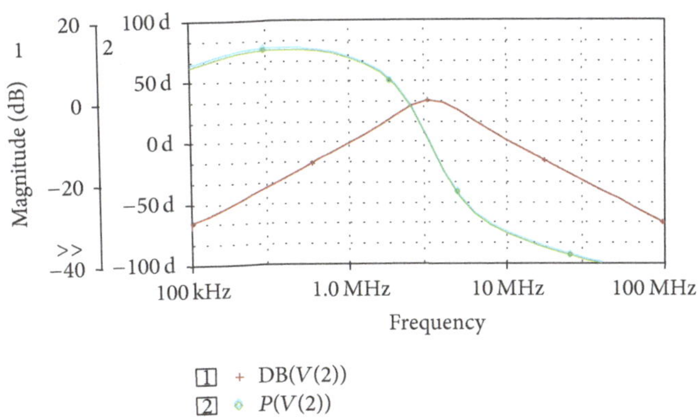

FIGURE 6: Simulated magnitude and phase response of the proposed BP filter.

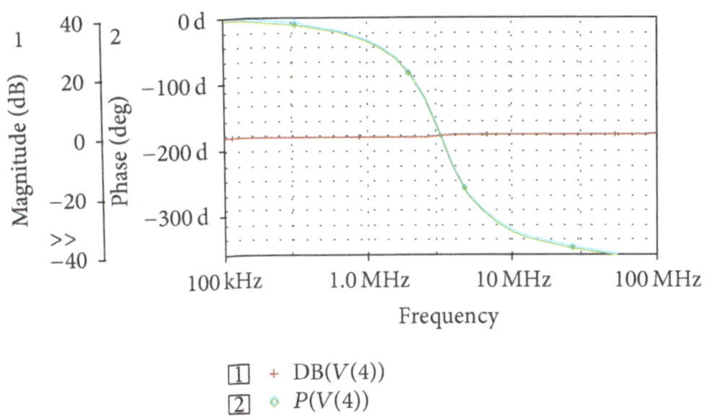

FIGURE 7: Simulated magnitude and phase response of the proposed AP filter.

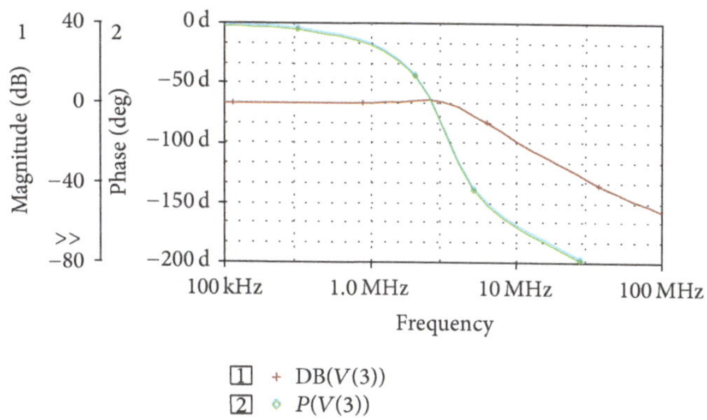

FIGURE 8: Simulated magnitude and phase response of the proposed LP filter.

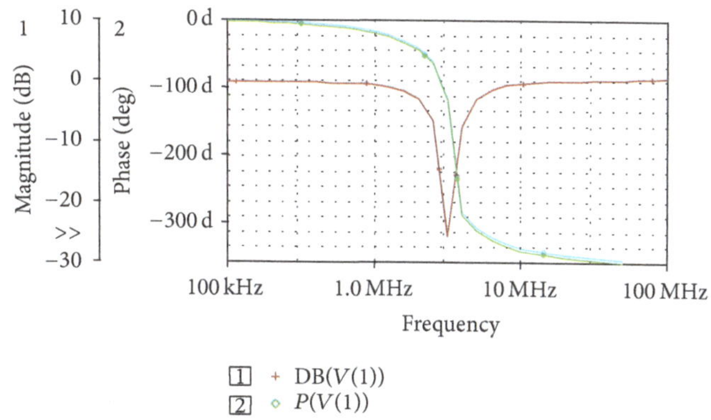

FIGURE 9: Simulated magnitude and phase response of the proposed NH filter.

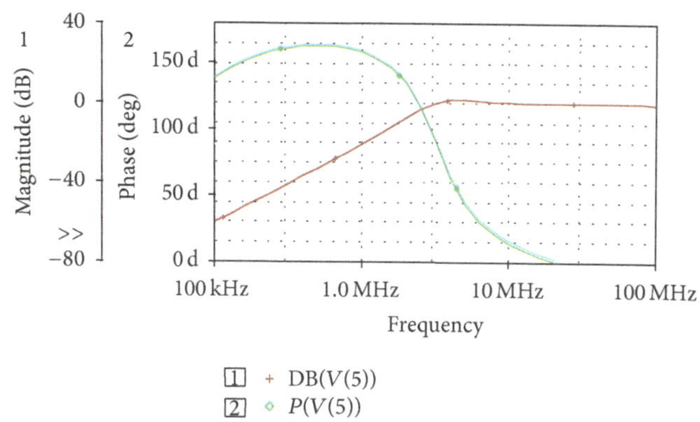

FIGURE 10: Simulated magnitude and phase response of the proposed HP filter.

of the LP, HP, BP, NH, and AP in dB are simultaneously shown in Figure 5. In addition, the magnitude and phase responses of all the generic filter responses are shown in Figures 6, 7, 8, 9, and 10, separately. To test the input dynamic range of the proposed filter, the sinusoidal input signal of an amplitude 300 mV (p-p) at 3.18 MHz is applied. Figure 11 shows the input and output sinusoidal signals of the band pass response. Figure 12 shows the noise behavior of the HP filter using the INOISE and ONOISE statements. Another study is carried out on the temperature performance of the proposed multifunction filter. Figure 13 shows the simulated frequency responses of the AP filter at different operating temperature. (−30°C to 90°C).

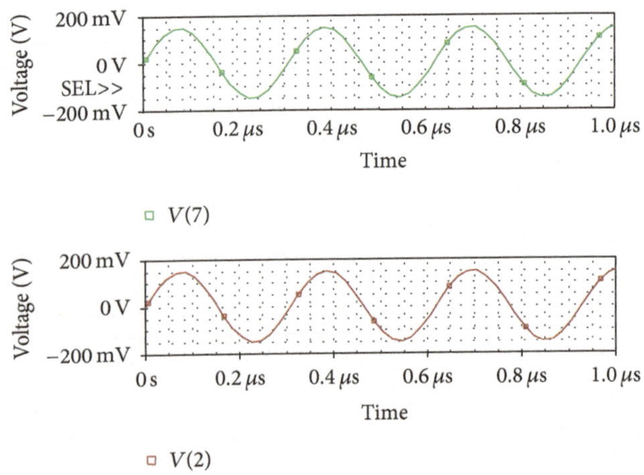

FIGURE 11: Time domain input and output waveforms for BP filter.

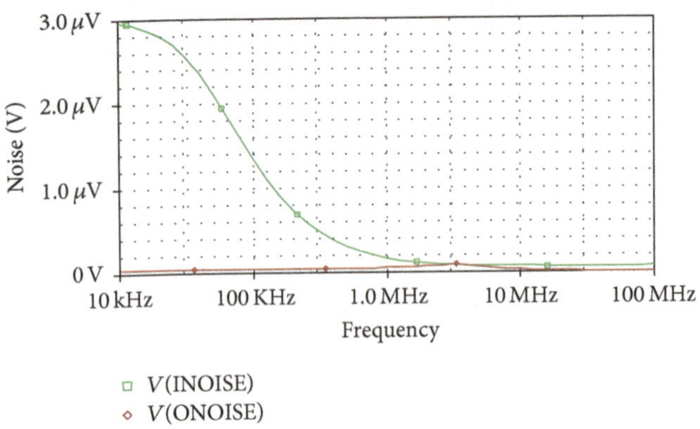

FIGURE 12: The input and output noise spectral densities for HP filter.

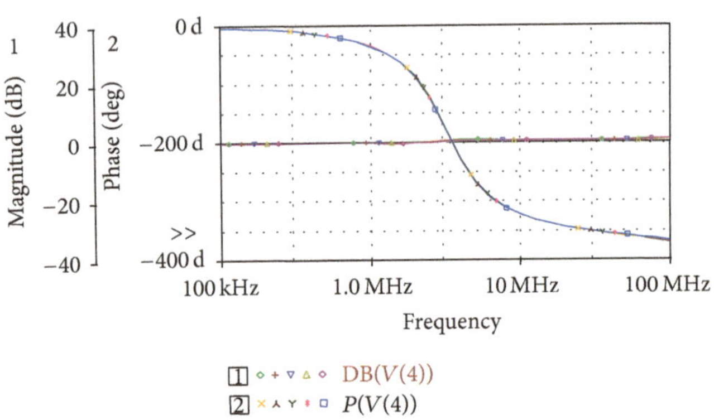

FIGURE 13: Frequency responses of the AP filter at different operating temperatures.

6. Conclusion

A new voltage-mode multifunction filter is presented. The proposed circuit realizes all generic filter responses simultaneously. Nevertheless, the proposed circuit needs a single matching constraint only to realize an AP response. The proposed circuit is simple and contains a minimum number of active components required to achieve a second-order transfer function. However, with the advantage of circuit simplicity, pole-ω_0 and Q are not independently adjustable. PSPICE simulations using 0.18 μm TSMC process parameters are used to support the validity and practical utility of the proposed circuit.

Conflict of Interests

The authors declare that there is no conflict of interests regarding the publication of this paper.

Acknowledgment

The authors thank Academic Editor for recommending this paper. At the time of paper submission paper processing charges for the journal were waived off.

References

[1] B. Wilson, "Recent developments in current conveyors and current-mode circuits," *IEE Proceedings Circuits, Devices and Systems*, vol. 137, no. 2, pp. 63–77, 1990.

[2] A. Fabre, O. Saaid, F. Wiest, and C. Boucheron, "Low power current mode second-order bandpass IF filter," *IEEE Transactions on Circuits and Systems II: Analog and Digital Signal Processing*, vol. 44, no. 6, pp. 436–446, 1997.

[3] M. A. Ibrahim, S. Minaei, and H. Kuntman, "A 22.5 MHz current-mode KHN-biquad using differential voltage current conveyor and grounded passive elements," *International Journal of Electronics and Communications*, vol. 59, no. 5, pp. 311–318, 2005.

[4] D. Biolek, R. Senani, V. Biolkova, and Z. Kolka, "Active elements for analog signal processing: classification, review, and new proposals," *Radioengineering*, vol. 17, no. 4, pp. 15–32, 2008.

[5] S. Maheshwari, "Analogue signal processing applications using a new circuit topology," *IET Circuits, Devices and Systems*, vol. 3, no. 3, pp. 106–115, 2009.

[6] M. T. Abuelma'atti and H. A. Al-Zaher, "New universal filter with one input and five outputs using current-feedback amplifiers," *Analog Integrated Circuits and Signal Processing*, vol. 16, no. 3, pp. 239–244, 1998.

[7] H. P. Chen, "Voltage-mode FDCCII-based universal filters," *AEU: International Journal of Electronics and Communications*, vol. 62, no. 4, pp. 320–323, 2008.

[8] H. P. Chen, "Single FDCCII-based universal voltage-mode filter," *International Journal of Electronics and Communications (AEU)*, vol. 63, no. 9, pp. 713–719, 2009.

[9] C. N. Lee and C. M. Chang, "Single FDCCII-based mixed-mode biquad filter with eight outputs," *AEU—International Journal of Electronics and Communications*, vol. 63, no. 9, pp. 736–742, 2009.

[10] S. Minaei and E. Yuce, "All-grounded passive elements voltage-mode DVCC-based universal filters," *Circuits, Systems, and Signal Processing*, vol. 29, no. 2, pp. 295–309, 2010.

[11] C. M. Chang, B. M. Al-Hashimi, C. L. Wang, and C. W. Hung, "Single fully differential current conveyor biquad filters," *IEE Proceedings: Circuits, Devices and Systems*, vol. 150, no. 5, pp. 394–398, 2003.

[12] M. A. Ibrahim, H. Kuntman, and O. Cicekoglu, "Single DDCC biquads with high input impedance and minimum number

of passive elements," *Analog Integrated Circuits and Signal Processing*, vol. 43, no. 1, pp. 71–79, 2005.

[13] C. M. Chang and H. P. Chen, "Single FDCCII-based tunable universal voltage-mode filter," *Circuits, Systems, and Signal Processing*, vol. 24, no. 2, pp. 221–227, 2005.

[14] H. P. Chen and Y. Z. Liao, "High-input and low-output impedance voltage-mode universal biquadratic filter using FDCCIIs," in *Proceedings of the 9th International Conference on Solid-State and Integrated-Circuit Technology (ICSICT '08)*, pp. 1794–1798, Beijing, China, October 2008.

[15] F. Kaçar and A. Yeşil, "Voltage mode universal filters employing single FDCCII," *Analog Integrated Circuits and Signal Processing*, vol. 63, no. 1, pp. 137–142, 2010.

[16] S. Maheshwari, J. Mohan, and D. S. Chauhan, "Novel Cascadable All-Pass/Notch Filters Using a Single FDCCII and Grounded Capacitors," *Circuits, Systems, and Signal Processing*, vol. 30, no. 3, pp. 643–654, 2011.

[17] F. Kaçar and A. Yeşil, "FDCCII-based electronically tunable voltage-mode biquad filter," *International Journal of Circuit Theory and Applications*, vol. 40, no. 4, pp. 377–383, 2012.

[18] S. Maheshwari and B. Chaturvedi, "Additional high input low output impedance analog networks," *Active and Passive Electronic Components*, vol. 2013, Article ID 574925, 9 pages, 2013.

[19] A. A. El-Adawy, A. M. Soliman, and H. O. Elwan, "A novel fully differential current conveyor and applications for analog VLSI," *IEEE Transactions on Circuits and Systems II: Analog and Digital Signal Processing*, vol. 47, no. 4, pp. 306–313, 2000.

[20] M. Bhushan and R. W. Newcomb, "Grounding of capacitors in integrated circuits," *Electronics Letters*, vol. 3, pp. 148–149, 1967.

14

Multimodel Robust Control for Hydraulic Turbine

Jakub Osuský and Stanislav Števo

Institute of Control and Industrial Informatics, Faculty of Electrical Engineering and Information Technology, Slovak University of Technology, Ilkovičova 3, 812 19 Bratislava, Slovakia

Correspondence should be addressed to Jakub Osuský; jakub.osusky@stuba.sk

Academic Editor: Mamun B. Ibne Reaz

The paper deals with the multimodel and robust control system design and their combination based on $M-\Delta$ structure. Controller design will be done in the frequency domain with nominal performance specified by phase margin. Hydraulic turbine model is analyzed as system with unstructured uncertainty, and robust stability condition is included in controller design. Multimodel and robust control approaches are presented in detail on hydraulic turbine model. Control design approaches are compared and used for derivation of new approaches which combine advantages of both.

1. Introduction

When hydraulic turbine power control is considered, the water inertia effect is a factor that makes it difficult to maintain stability under isolated operation or to have fast response in case of load change in the whole operational range. Nonlinear models of hydraulic turbine control systems are needed in those cases where the turbine power varies in the whole range. The turbine model parameters vary significantly with the unpredictable load variations. Such nonlinearities make the controller design a nontrivial task due to the fact that a controller designed for specific operating conditions may not perform well under different loads [1]. This fact leads to the adaptive [2] or hybrid control algorithms.

There are several approaches to model hybrid systems [3]. In a model of a large class of hybrid systems, [4] is considered as discrete event system and continuous dynamic modeled by differential or difference equation. Such models are used to formulate a general stability analysis and controller synthesis framework for hybrid systems. Results for modeling and stability analysis of hybrid systems have been presented in [3, 5, 6]. In this paper, we follow the class of hybrid system known as switched systems [7].

In this paper, hybrid control, often denoted by multimodel control, will be used. Several switching algorithms will be compared to show how the switching speed affects performance and stability. Well-known robust control method will be presented as a control solution which ensures stability on whole operating range, but with lower performance. Combination of multimodel control and robust control which ensure stability and good performance will be presented on numerical example of nonlinear hydraulic turbine model.

2. Preliminaries and Problem Formulation

Consider nonlinear system as uncertain system described by set of transfer functions $G_q(s)$ and controller $R_q(s)$ in standard configuration:

$$G_q(s) \in \Pi, \quad q = 1, 2, \ldots, N, \tag{1}$$

where $y(s)$, $u(s)$, and $w(s)$ are output, control, and set-point variables of system, respectively; $q = 1, 2, \ldots, N$ is the number of linear transfer functions describing uncertain system and also the number of controllers. Hence, for each identified transfer function, one controller is designed.

To deal with uncertainties instead of a single model, the behavior of a class of models Π is considered.

Multimodel control is based on several linear models identified in operating points. For each linear model, local

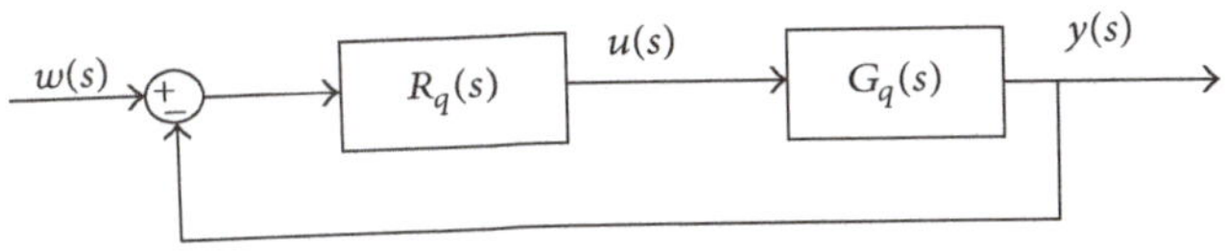

Figure 1: Standard feedback control configuration.

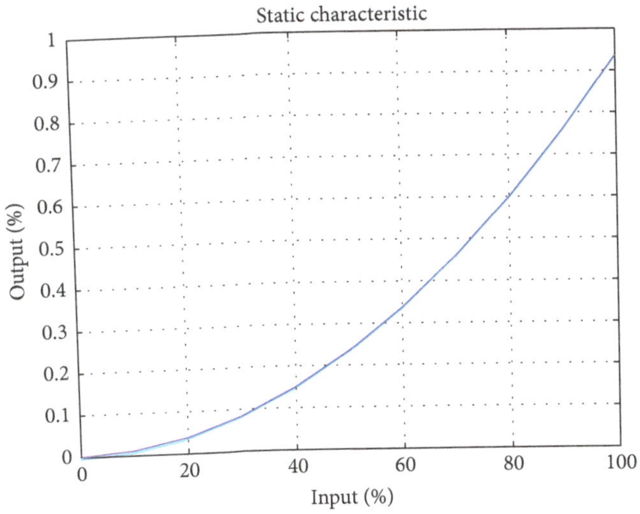

Figure 2: Static characteristic of nonlinear system.

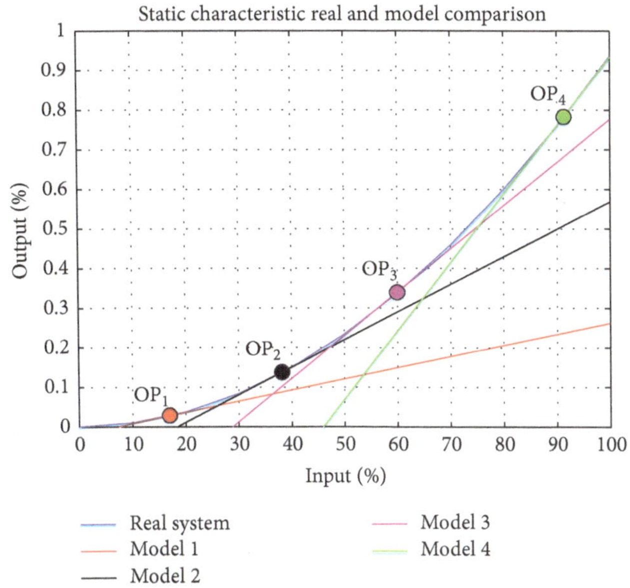

Figure 3: Static characteristic with linear models.

controller is designed which ensures stability and perhaps even performance only for this model. As nonlinear system moves from one operating point to another, controller switches from one local controller to another one. Performance of this control approach can be much better as using only one controller on whole range, but this method does not yield stability when nonlinear system is between operating points. Fast switching between local controllers can decrease performance and also lead to instability.

In robust control, several linear models identified in operating points are used for uncertainty model calculation. Then, for system control, one controller is used, which ensures stability on whole operating range, between all operating points. Performance (of robust control) is not so good as performance of multimodel control; however, there is no problem with stability by controllers switching. When nonlinearity is increased, uncertainty is increased too and performance of robust control decreases. If the system is strongly nonlinear, uncertainty could be too big to be able to control the system using robust control.

In this paper, designing a set of local controllers similar to those in multimodel control design will be done, but each local controller will be designed as robust to ensure stability between operating points. By robust design of local controller, instead of all operating points, only given and neighbourhood operating points will be considered. This type of control will be called robust multimodel control and will ensure stability and higher performance than robust control.

3. Theoretical Results

3.1. Multimodel Control Design. Consider a system with nonlinear static characteristic (Figure 2).

Static characteristic is nonlinear and could be approximate using m linear models (in our case four models, $m = 4$, Figure 3).

These models are simple to obtain using input-output identification. For each model (operating point (OP)), local PID controller can be tuned using any well-known and tested method. Depending on system output, multimodel controller switches between local PID controllers. Local controllers ensure stability only in operating point.

Characteristic equation of closed-loop feedback system (Figure 1) can be written as follows:

$$1 + R_q(s) G_q(s) = 0. \tag{2}$$

Equation (2) can be rewritten into polynomial form:

$$p_q(s) = p_0 + p_1 s + p_2 s^2 + \cdots + p_n s^n = 0, \tag{3}$$

where q is index of operating point.

Local controllers ensure only stability (negative roots) of (3) for $q = 1, 2, \ldots, m$. For stability of the whole system, segments

$$ps_i(s, \chi) = (1 - \chi) p_i(s) + \chi p_{i+1}(s), \quad \chi \in \langle 0, 1 \rangle, \quad i = 1, 2, m - 1 \tag{4}$$

between operation points must be stable too, independent of switching speed. One way how to reach segment stability is using of robust control.

3.2. Robust Control Design. A simple uncertainty model is obtained using unstructured model approach. In the sequel

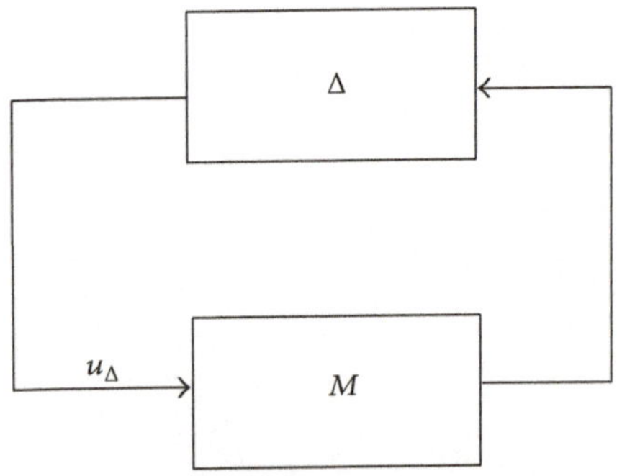

FIGURE 4: $M - \Delta$ structure.

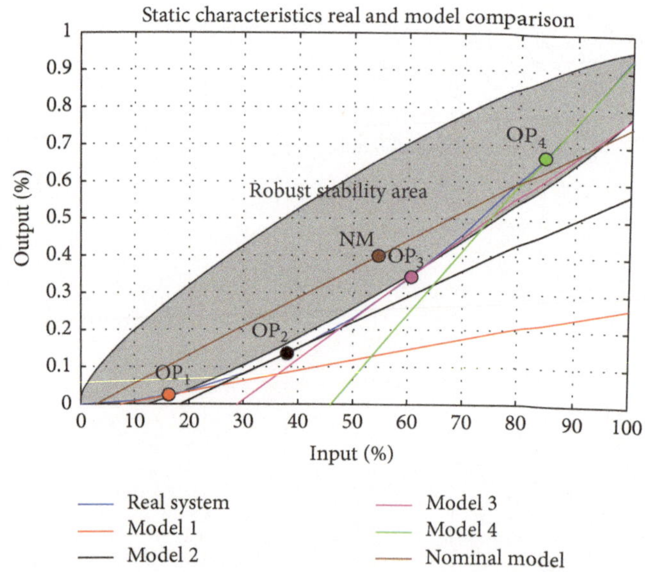

FIGURE 5: Illustration of how the full operating area of nonlinear system is covered by stability area of single robust controller.

(see Figure 5), two most common single uncertainty models are considered, additive and input multiplicative model type [8]. Corresponding classes of uncertain models are given as follows.

(1) Additive uncertainty model for operating points $q = 1, 2, \ldots, m$:

$$G_q(s) = G_0(s) + \Delta_a(s) W_a(s), \tag{5}$$

where

$$l_a(\omega) = \max_{G_q \in \Pi} \sigma_M \left[G_q(s) - G_0(s)\right], \quad q = 1, 2, \ldots, m, \quad |W_a(s)| \ge l_a(s). \tag{6}$$

(2) Input multiplicative uncertainty model for operating points $q = 1, 2, \ldots, m$:

$$G_q(s) = G_0(s)\left(I + \Delta_i(s) W_i(s)\right), \tag{7}$$

where

$$l_i(\omega) = \max_{G_q \in \Pi} \sigma_M \left[G_0(s)^{-1}\left(G_0(s) - G_q(s)\right)\right], \quad q = 1, 2, \ldots, m, \tag{8}$$

$$|W_i(s)| \ge l_i(s), \quad \sigma_M(\Delta_a(s)) \le 1, \quad \sigma_M(\Delta_i(s)) \le 1, \quad \forall s \in D, \tag{9}$$

where D is Nyquist contour [8].

Standard feedback configuration with unstructured uncertainty of any type can be rearranged to obtain the general $M - \Delta$ structure in Figure 4 where $M(s)$ represents the transfer function matrix (containing controller $R(s)$ nominal model $G_0(s)$) and $\Delta(s)$; $\sigma_{\max}[\Delta(j\omega)] \le 1$ is the normalized perturbation.

Robust stability condition for unstructured perturbations is formulated in terms of stability of the $M - \Delta$ system: if both the nominal system $M(s)$ (nominal stability) and the normalized perturbation $\Delta(s)$ are stable, closed-loop stability is guaranteed for

$$\sigma_{\max}[M(j\omega)] < 1, \ \forall \omega \tag{10}$$

(i) for additive uncertainty:

$$M_a(s) = -[I + R(s) G_0(s)]^{-1} R(s), \tag{11}$$

(ii) for input multiplicative uncertainty:

$$M_i(s) = -[I + R(s) G_0(s)]^{-1} R(s) G_0(s). \tag{12}$$

In view of (11) and (12), condition (10) reads as follows:

$$\sigma_{\max}[M_k(s)] < \frac{1}{|l_k(s)|}, \quad k = a, i. \tag{13}$$

Robust control approach based on unstructured uncertainty and with robust stability ensured using $M - \Delta$ structure can be summarized as follows.

(a) Nominal model calculation: nominal model $G_0(s)$ is calculated as average transfer function from all m operating points $G_q(s)$, $q = 1, 2, \ldots, m$.

(b) For nominal model standard feedback controller is designed: good choice could be controller design method with desired phase margin.

(c) Robust stability verification according to (13): if robust stability is not passed, part (b) must be done again. Increasing phase margin by controller design is possible to improve robust properties of controller.

Note: segments (4) are inclusive of robust stability area.

3.3. Robust Multimodel Control Design. In this case, for each operating point of switched system, with arbitrary switch algorithm, robust controller $R_q(s)$, $q = 1, 2, \ldots, m$ is designed.

Controller guarantees the robust stability and performance for all operating points and over the entire operating range specified by uncertain models (5) or (7).

Let derive robust stability condition for operating point q. The nominal model G_0 and uncertainty model will be calculated from transfer functions $G_{qn}(s)$, $qn = q-1, q, q+1$; $q = 1, 2, \dots, m$. Because of switching the process from mode q, the plant may be switched to neighbourhood operating point $q+1 = 1, 2, \dots, m$ or $q-1 = 1, 2, \dots, m$; to guarantee the robust stability for such switching algorithm, (5) and (7) need to be modified as follows:

$$G_{qn}(s) = G_{0n}(s) + W_{qn\_a}(s)\,\Delta_{qn\_a}(s), \tag{14}$$

where

$$\begin{gathered} l_{qn\_a}(\omega) = \max_{G_{qn}(s)\in\Pi} \sigma_M\left(G_{0n}(s) - G_{qn}(s)\right), \\ qn = q-1, q, q+1; \quad q = 1, 2, \dots, m, \\ \left|W_{qna}(s)\right| \geq l_{qna}(\omega), \end{gathered} \tag{15}$$

and for input multiplicative case,

$$G_{qn}(s) = G_{0n}\left(I + \Delta_{qn\_i}(s)\,W_{qn\_i}(s)\right), \tag{16}$$

where

$$\begin{gathered} l_{qn\_i}(\omega) = \max_{G_{qn}(s)\in\Pi} \sigma_M\, G_{qn}(s)^{-1}\left(G_{0n} - G_{qn}(s)\right) \\ qn = q-1, q, q+1; \quad q = 1, 2, \dots, m. \end{gathered} \tag{17}$$

Robust controller design procedure for each operating point $q = 1, 2, \dots, m$ of switched system goes the same way as for robust control design but instead of all operating points only given and neighbourhood operating points are considered.

Robust multimodel control approach based on unstructured uncertainty and with robust stability ensured using $M-\Delta$ structure can be summarized as follows.

(a) For each operation point, nominal model is calculated using not all but only neighborhood operating points. Nominal model $G_{0n}(s)$ is for qth operating point calculated as average one from $G_q(s), G_{q-1}(s), G_{q+1}(s)$ (if $q = 1$ only $G_q(s)$ and $G_{q+1}(s)$ are considered. If $q = m$ only $G_q(s)$ and $G_{q-1}(s)$ are considered).

(b) For nominal model $G_{0n}(s)$ standard feedback controller is designed. Good choice could be controller design method with desired phase margin.

(c) Robust stability verification according to (13): if robust stability is not passed, part (b) must be done again. Increasing phase margin by controller design is possible to improve robust properties of controller.

These three steps have to be done m times once for each operating point.

Stability areas of local controllers (Figure 6) are overlapping so nonlinear system is stable also between operating points. Robust stability area is smaller like using only one robust controller which will yield less conservative solutions.

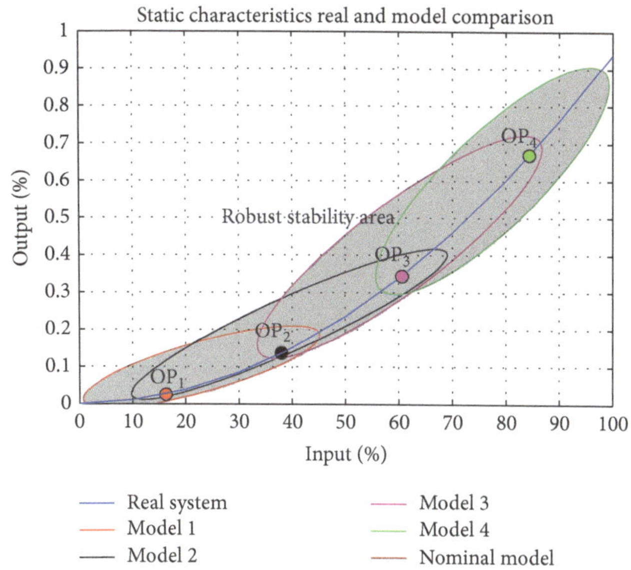

FIGURE 6: Illustration of how the full operating area of nonlinear system is covered by stability area of multimodel robust controller.

4. Case Study

Consider hydraulic turbine with power operating range (0–100)%. In our case, nonlinear model described in [1] will be identified using input-output identification in five operating points (output power of 20%, 40%, 60%, 80%, and 90%). As identification results linear models in transfer functions form will be obtained.

Verification of linear models identified in operating points can be found in appendix (Figures 16, 17, 18, 19, and 20). Transfer functions of these models are as follows:

$$\begin{aligned} G_1(s) &= \frac{-1.516s + 2.46}{s^2 + 2.597s + 2.474}, \\ G_2(s) &= \frac{-1.911s + 1.536}{s^2 + 1.994s + 1.554}, \\ G_3(s) &= \frac{-1.931s + 1.049}{s^2 + 1.653s + 1.069}, \\ G_4(s) &= \frac{-1.979s + 0.806}{s^2 + 1.493s + 0.8255}, \\ G_5(s) &= \frac{-2.002s + 0.72}{s^2 + 1.43s + 0.74}. \end{aligned} \tag{18}$$

Step 1 (multimodel control). In first step, multimodel control approach will be used for hydraulic turbine control. For each transfer function, PI controller will be designed using

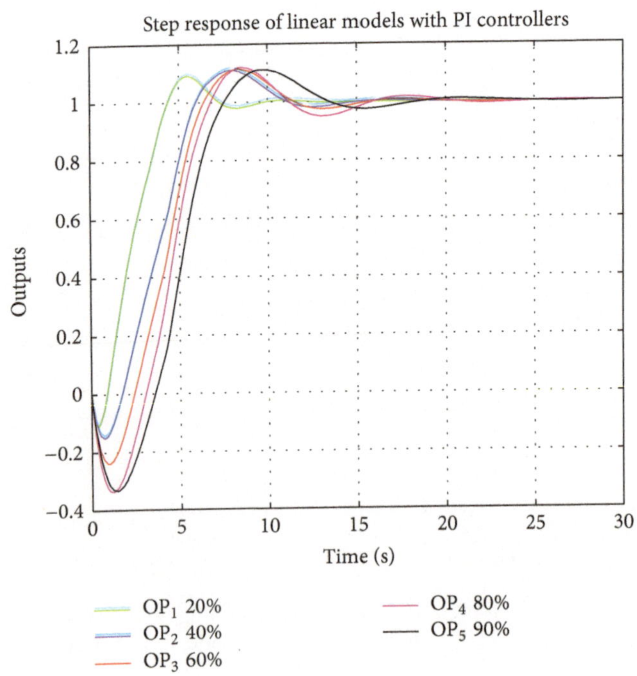

FIGURE 7: Step responses of linear models with PI controllers.

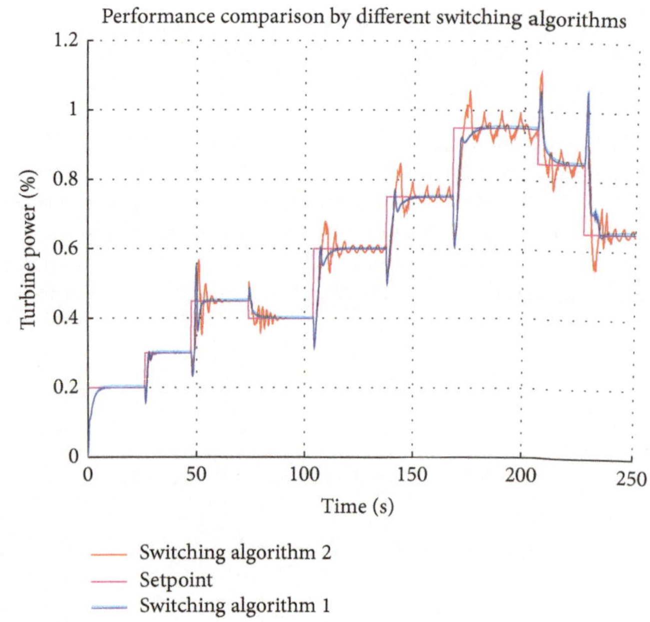

FIGURE 8: Control performance comparison by different switching algorithms.

TABLE 1: Switching algorithm based on system output.

Output [%]	<0, 30)	<30, 50)	<50, 70)	<70, 85)	<85, 100)
Active contr.	$R_1(s)$	$R_2(s)$	$R_3(s)$	$R_4(s)$	$R_5(s)$

D-partition method [9]. Aim of controller tuning will be keeping overshoot up to 10%:

$$R_1(s) = \frac{0.444s + 0.635}{s}, \qquad R_2(s) = \frac{0.263s + 0.36}{s},$$
$$R_3(s) = \frac{0.306s + 0.275}{s}, \qquad R_4(s) = \frac{0.354s + 0.225}{s}, \quad (19)$$
$$R_5(s) = \frac{0.29s + 0.196}{s}.$$

Figure 7 shows that overshoot condition is satisfied for all linear models (operating points). On the nonlinear system, two different switching algorithms will be tested.

First switching algorithm will be based on system output. Since system output will be lower than 30%, first controller will be active. If output will be more than 30% but less than 50%, second controller will be active and so on (Table 1).

Second algorithm will be based on comparison of system output y and model outputs y_q:

$$J_q = y - y_q, \quad q = 1, 2, \ldots, m. \quad (20)$$

In each sample time comparison will be done and $R_i(s)$ controller will be set as active, where $i \in q$ is index of model with smallest J (20).

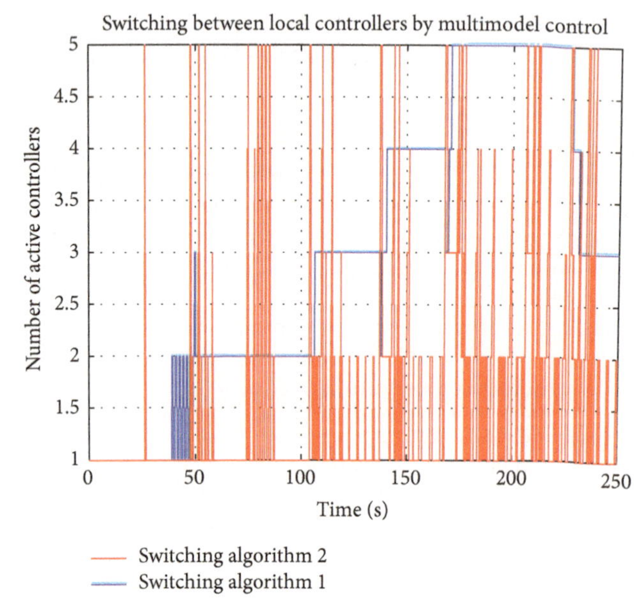

FIGURE 9: Switching between controllers $R_1(s) - R_5(s)$, comparison of two algorithms.

Nonlinear turbine model has strong nonminimal phase behavior. Performance comparison (Figure 8) by different switching algorithms shows that switching speed has significant influence on performance. First switching algorithm is slow and by second algorithm switching is very fast (Figure 9).

When high control performance is needed, it is recommended to avoid fast switching. Possible solutions could be increasing switching period (not at each sample time),

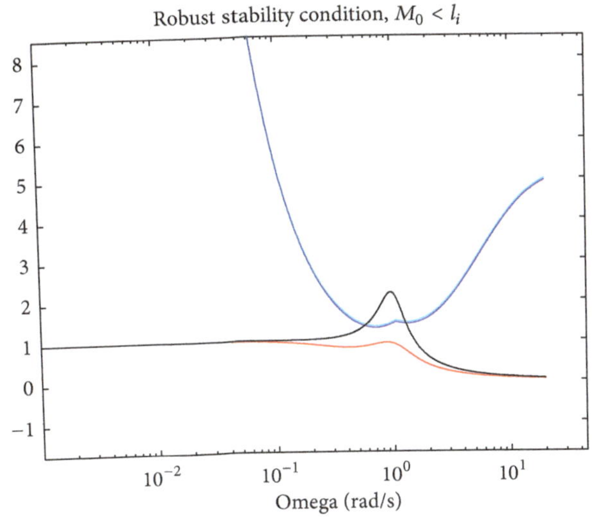

FIGURE 10: Robust stability condition for controllers ensuring phase margins PM = 50° and PM = 70°.

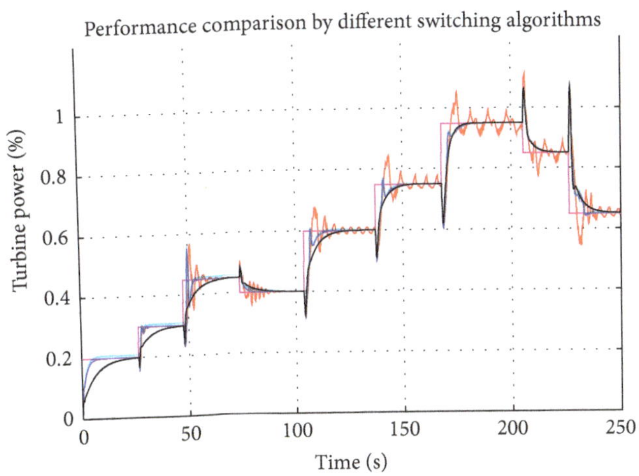

FIGURE 11: Control performance comparison multimodel and robust control.

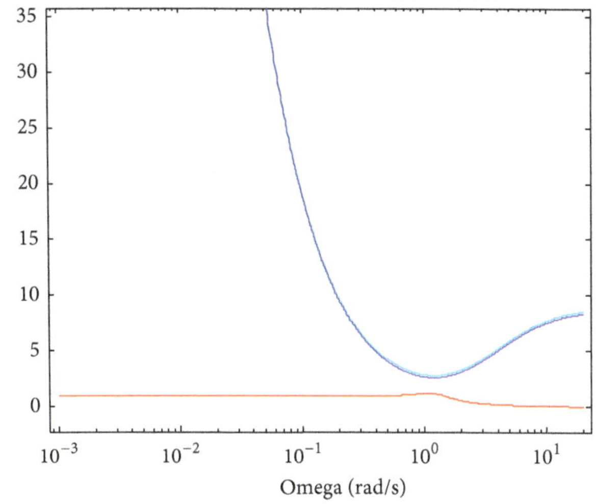

FIGURE 12: Robust stability condition, first operating point.

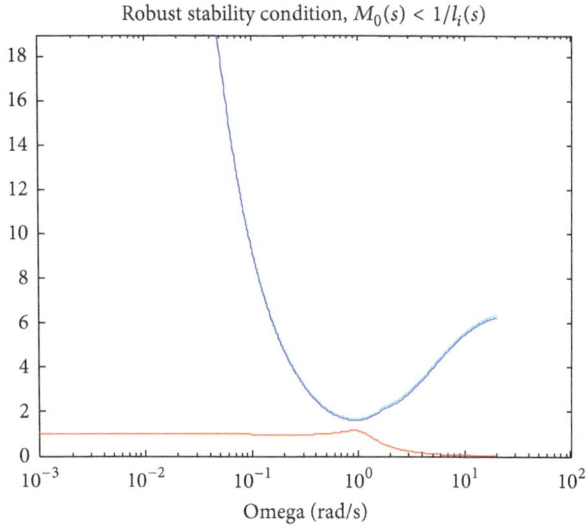

FIGURE 13: Robust stability condition, second operating point.

weighting of controller outputs, or, in this case, using first switching algorithm.

Step 2 (robust control). In second step by robust control design instructions from previous chapter will be used.

From transfer functions (18), nominal model $G_0(s)$ will be calculated as transfer function with the same structure and average values:

$$G_0(s) = \frac{-1.868s + 1.3145}{s^2 + 1.833s + 1.3324}. \quad (21)$$

For nominal model, PI controller was calculated using design method ensuring desired phase margin. Controller ensuring phase margin PM = 50° does not guarantee robust stability condition, because desired phase margin was increased to PM = 70° (Figure 10).

Transfer function of controller ensuring phase margin PM = 70° is

$$R(s) = \frac{0.3233s + 0.2287}{s}. \quad (22)$$

Comparison of multimodel and robust control (Figure 11) shows slower response of system with robust control in lower power output. By higher power output, system response with

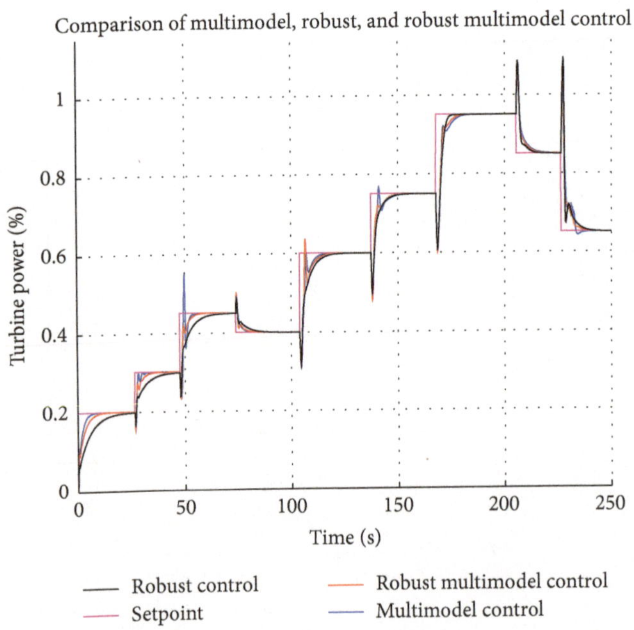

FIGURE 14: Comparison of multimodel, robust, and robust multimodel control.

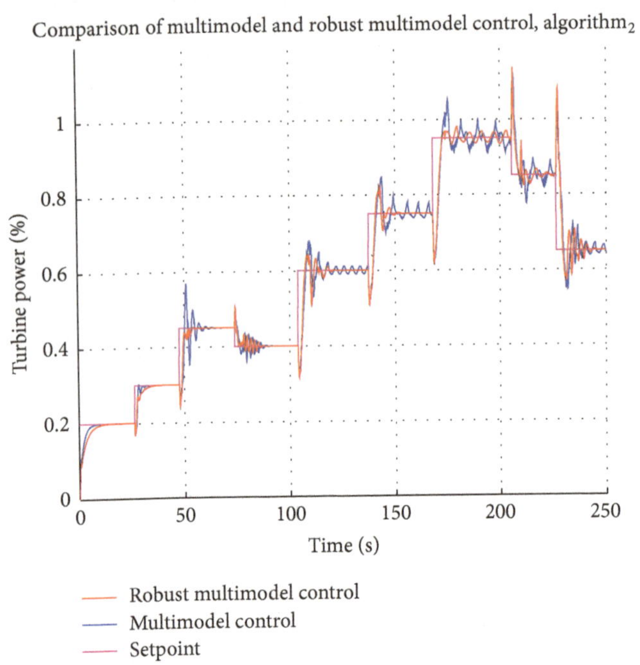

FIGURE 15: Comparison of multimodel and robust multimodel control using second switching algorithm.

TABLE 2: IAE comparison of multimodel control (two switching algorithms) and robust control.

Control	Multimodel 1	Multimodel 2	Robust
IAE	4,82	7,06	5,94

robust control is similar to system response with multimodel control. Performance of control approaches will be compared also numerically in Table 2, using IAE criterion.

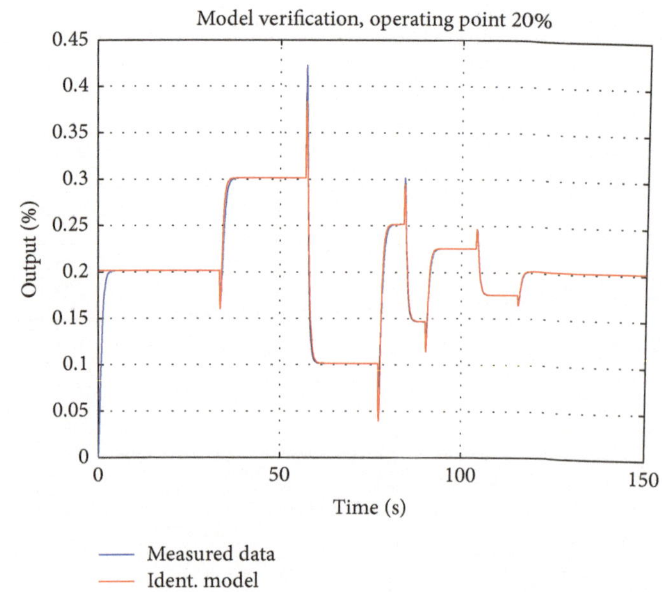

FIGURE 16: Comparison of data measured on nonlinear system and on identified linear model, OP 20%.

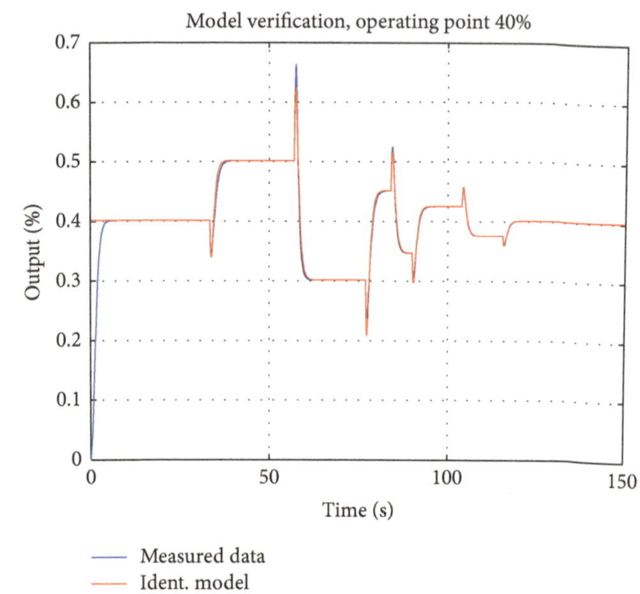

FIGURE 17: Comparison of data measured on nonlinear system and on identified linear model, OP 40%.

TABLE 3: IAE comparison of multimodel control, robust control, and robust multimodel control.

Control	Multimodel 1	Robust	Multimodel robust
IAE	4,82	5,94	4,92

Step 3 (robust multimodel control). In third step, robust multimodel control will be designed, which is a combination of robust and multimodel control (see Table 4).

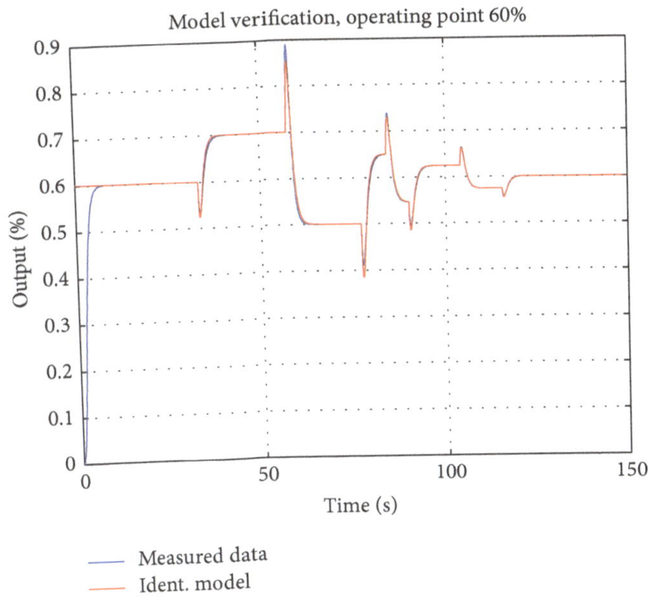

FIGURE 18: Comparison of data measured on nonlinear system and on identified linear model, OP 60%.

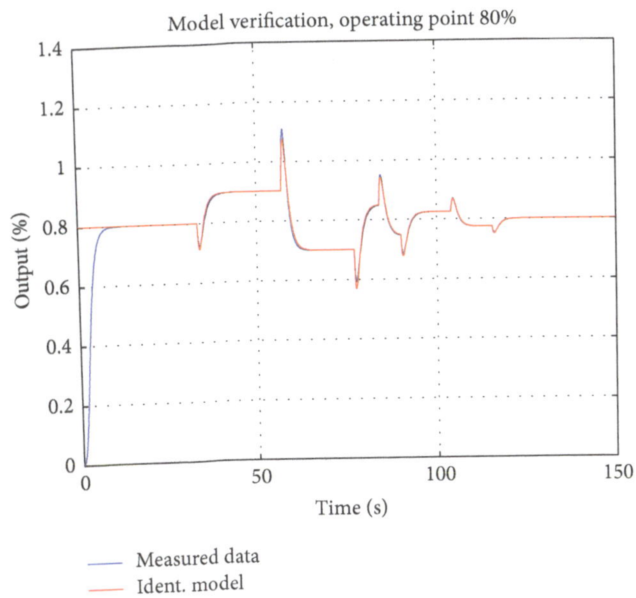

FIGURE 19: Comparison of data measured on nonlinear system and on identified linear model, OP 80%.

TABLE 4: IAE comparison of multimodel and robust multimodel control algorithm 2.

Control algorithm 2	Multimodel	Robust multimodel
IAE	7,06	5,7

By controller design for first operating point, nominal model $G_{0n}(s)$ is calculated only from $G_1(s)$ and $G_2(s)$ according to (18):

$$G_{0n1}(s) = \frac{-1.714s + 1.99}{s^2 + 2.29s + 2.014}. \tag{23}$$

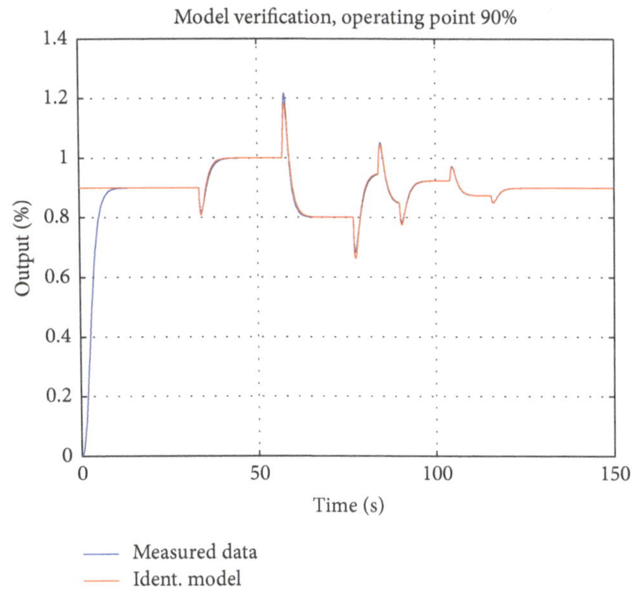

FIGURE 20: Comparison of data measured on nonlinear system and on identified linear model, OP 90%.

For this nominal model controller ensuring phase margin PM = 60° will be designed:

$$R_1(s) = \frac{0.4275s + 0.434}{s}. \tag{24}$$

Robust stability condition (Figure 12) proves robust properties of designed controller. From behaviour of $M_0(s)$ (in Figures 10 and 12), it is obvious that that robust multimodel control is less conservative as pure robust control.

By controller design for second operating point, nominal model $G_{0n}(s)$ is calculated from $G_1(s)$, $G_2(s)$, and $G_3(s)$ according to (18):

$$G_{0n2}(s) = \frac{-1.786s + 1.682}{s^2 + 2.08s + 1.7}. \tag{25}$$

For this nominal model controller ensuring phase margin PM = 60° will be designed:

$$R_2(s) = \frac{0.375s + 0.429}{s}. \tag{26}$$

Robust stability condition (Figure 13) proves robust properties of designed controller.

Controller design for the rest of operating points goes the same way. Controller parameters for operating points 3–5 are as follows:

$$R_3(s) = \frac{0.34s + 0.251}{s}, \qquad R_4(s) = \frac{0.297s + 0.23}{s}, \qquad R_5(s) = \frac{0.305s + 0.208}{s}. \tag{27}$$

Comparison of three control approaches (Figure 14) shows that robust multimodel control is almost so fast as multimodel control. Performance of multimodel, robust, and robust multimodel control is compared in Table 3 by first switching algorithm (described in Table 1).

Performance of multimodel and robust multimodel approaches will be compared also using second switching algorithm (Figure 15). Robust properties improved performance by this fast switching algorithm.

5. Conclusion

In this paper, hybrid control, often denoted by multimodel control, was used. Two switching algorithms were compared to show how the switching speed affects performance and stability. Robust control method based on $M - \Delta$ was presented as a control solution which ensures stability on whole operating range, but with lower performance. Combination of multimodel control and robust control which ensure stability and good performance was illustrated on nonlinear hydraulic turbine model. Designed control solution is less conservative as robust control but stable on whole operating range and almost with the same performance as using multimodel control.

Appendix

For identification results, in form of model verification, in all operating points , see Figures 16–20.

Conflict of Interests

The authors declare that there is no conflict of interests regarding the publication of this paper.

Acknowledgment

This work was supported by the Scientific Grant APVV 0772-12.

References

[1] T. Puleva, E. Garipov, and G. Ruzhekov, "Adaptive power control modeling and simulation of a hydraulic turbine," in *Proceedings of the16th International Conference on Intelligent System Applications to Power Systems (ISAP '11)*, September 2011.

[2] O. P. Malik and Y. Zeng, "Design of a robust adaptive controller for a water turbine governing system," *IEEE Transactions on Energy Conversion*, vol. 10, no. 2, pp. 354–359, 1995.

[3] J. Lunze and F. L. Lagarrigue, *Hanbook of Hybrid Systems Control, Theory, Tools, Applications*, Cambridge University Press, New York, NY, USA, 2009.

[4] M. S. Branicky, V. S. Borkar, and S. K. Mitter, "A unified framework for hybrid control: model and optimal control theory," *IEEE Transactions on Automatic Control*, vol. 43, no. 1, pp. 31–45, 1998.

[5] J. Lygeros, "An overview of research areas in hybrid control," in *Proceedings of the 44th IEEE Conference on Decision and Control, and European Control Conference (CDC-ECC '05)*, pp. 5600–5605, December 2005.

[6] R. Goebel and A. R. Teel, "Solutions to hybrid inclusions via set and graphical convergence with stability theory applications," *Automatica*, vol. 42, no. 4, pp. 573–587, 2006.

[7] D. Liberzon, *Switching in Systems and Control*, Birkhauser, Boston, Mass, USA, 2003.

[8] S. Skogestad and I. Postlethwaite, *Multivariable Feedback Control: Analysis and Design*, John Wiley & Sons, New York, NY, USA, 3rd edition, 1997.

[9] J. Osuský and V. Veselý, "Modification of Neimark D-partition method for desired phase margin," *Journal of Cybernetics and Informatics*, vol. 11, 2010.

15

Multiobjective Economic Load Dispatch Problem Solved by New PSO

Nagendra Singh[1] and Yogendra Kumar[2]

[1] *Department of Electrical Engineering, Mewar University, Chittorgarh, Rajasthan, India*
[2] *Department of Electrical Engineering, Maulana Azad National Institute of Technology, Bhopal, India*

Correspondence should be addressed to Nagendra Singh; nsingh007@rediffmail.com

Academic Editor: Nikos D. Lagaros

Proposed in this paper is a new particle swarm optimization technique for the solution of economic load dispatch as well as environmental emission of the thermal power plant with power balance and generation limit constraints. Economic load dispatch is an online problem to minimize the total generating cost of the thermal power plant and satisfy the equality and inequality constraints. Thermal power plants use fossil fuels for the generation of power; fossil fuel emits many toxic gases which pollute the environment. This paper not only considers the economic load dispatch problem to reduce the total generation cost of the thermal power plant but also deals with environmental emission minimization. In this paper, fuel cost and the environmental emission functions are considered and formulated as a multiobjective economic load dispatch problem. For obtaining the solution of multiobjective economic load dispatch problem a new PSO called moderate random search PSO was used. MRPSO enhances the ability of particles to explore in the search spaces more effectively and increases their convergence rates. The proposed algorithm is tested for the IEEE 30 bus test systems. The results obtained by MRPSO algorithm show that it is effective and efficient.

1. Introduction

Electrical power system is a very large interconnected system. It plays very important role in the economy of the country. For the efficient and reliable operation of such large interconnected power system, it required proper analysis and the way to operate such system economically. Economic load dispatch problem is an important optimization task in the electrical power system and study of economic load dispatch helps to operate power systems economically with an efficient way and provide power without any interruption. The economic load dispatch is an online process of allocating generation among the available generating units to minimize the total generation cost and satisfy the equality and inequality constraints. Since the civilization increases day by day the demand of electricity increases in the same ratio. For the satisfaction of the load demand large numbers of thermal power plants are installed and the capacity of coal burnt also increases. Due to burning large amounts of coal emitted many toxic gases like carbon dioxide (CO_x), sulphur dioxide (SO_x), and nitrogen oxides (NO_x) at thermal power plants and pollute the environment. Pollution is very harmful for the environment as well as living creatures. Environmental pollution increases the global warming and damage of the Ozone layer. So in recent trends it is required to generate the power with minimum cost and minimize the pollutant environment emission. The study of economic load dispatch helps to generate power on minimum cost and also reduce the environmental emission effects.

Many classical as well as modern techniques were used to solve economic load dispatch problem with environmental emission listed in the literature. Different methods have been reported in the literature for solving ELD problem as multiobjective problem. Talaq et al. [1] give a deep summary of economic load dispatch with environmental constraints. Linear programming techniques were proposed by Farag et al. [2] for the solution of ELD problem incorporated with environmental emission. They solved ELD problem as multiobjective problem with constraints. Authors of [3] proposed direct Newton-Raphson method for the solution of the multiobjective ELD problem in 2003. Goal programming approach was proposed by Nanda et al. [4] for the solution of ELD problem with emission constraints. Yokoyama et al. [5] presented multiobjective economic power generation dispatch based on probability security criteria. Multiobjective ELD problem with security constraint proposed by Chang et al. [6] was solved by using bicriterion global optimization techniques. Granelli et al. [7] presented emission constrained dynamic ELD. New stochastic search technique was proposed by Das and Patvardhan [8] for the multiobjective economic load dispatch problem. Some authors have proposed modern heuristic techniques such as fuzzy logic optimization technique [9] proposed for solution of multiobjective generation schedule. Genetic algorithm techniques were proposed by Xu et al. [10] to solve constrained multiobjective ELD problem. Particle swarm optimization techniques were proposed [11, 12] for obtaining multiple objectives.

Evolutionary programming technique was suggested by Suganya et al. [13] for multiobjective economic/emission load dispatch problem. Advanced MOEPSO-based multiobjective environmental, economic load dispatching was given by Mori and Okawa [14]. Abido [15] proposed multiobjective evolutionary algorithms for the electric power dispatch problem. Dutta and Sinha [12] suggested PSO technique for the solution of environmental economic dispatch problem with voltage stability constraint. Kennedy and Eberhart [16] for the first time in 1995 introduced particle swarm optimization (PSO) technique. It is a population-based evolutionary technique, inspired by the social behaviour of a flock of birds searching for food. The PSO algorithm simulates social behaviour among the particles flying in a multidimensional search space. In comparison with other evolutionary optimization techniques the PSO has a superior search performance with faster and more stable convergence rates. PSO is a very popular optimization technique between the researchers and many of the researchers used it for the solution of multiobjective economic load dispatch problem, but PSO has a drawback that it lacks global search ability in the last stage of iterations. So PSO is unable to give the global optimal solution for the multiobjective economic load dispatch problem. This problem of PSO may be overcome by using proposed MRPSO, because the MRPSO enhances the ability of particles to explore the solution spaces more effectively and increases their convergence rates. The proposed algorithm is tested on the IEEE 30 bus test systems. The results obtained by MRPSO algorithm show that it is practically efficient.

The multiobjective problem considered in this paper is solved by PSO and MRPSO with generation limit and power balance constraints. This study involves the solution of two objectives; the first of these is to minimize the total generation cost of generating units and second aspect is to minimize the environmental emission of thermal power plant. Effectiveness and efficiency of the proposed PSO technique were tested for the data of IEEE 30 bus network. Results obtained by PSO and MRPSO techniques were compared with other optimization techniques listed in the literature and it is found that MRPSO gives superior results compared to other techniques.

2. Mathematical Model of Objective Function and Constraints

In this paper two objective functions were considered. First objective is to minimize the total generation cost of generating power plant and the second objective is to minimize the environmental emission of the generating plants.

2.1. Objective I

Economic Generation Cost Function. Generation quadratic fuel cost characteristic of generating power plant is formulated as follows:

$$F_T = \text{Min}\, f\,(\text{FC}),$$
$$f\,(\text{FC}) = \sum_{i=1}^{N} a_i P_i^2 + b_i P_i + c_i, \tag{1}$$

where F_T is the total fuel cost, $f(\text{FC})$ is the fuel cost function, a_i, b_i, and c_i are the cost coefficients of the ith generator, P_i is the generated power of ith power plants, and N is the number of generators.

2.2. Objective II

Emission Objective Function. In this paper environmental emission was evaluated with consideration of NO_x gas. A typical NO_x emission at thermal power plants [12] can be formulated as shown in (2). Consider the following:

$$E_T = \text{Min} \sum_{i=1}^{N} f\,(E_i\,(P_i)), \tag{2}$$

$$E_i\,(P_i) = \left(\alpha_i + \beta_i P_i + \gamma_i P_i^2\right) + \xi_i \sin\left(\lambda_i P_i\right). \tag{3}$$

Now both objectives may be combined in a single objective as given in (4), (5), and (6). The generation cost of each generator was evaluated at its maximum output:

$$F_i\,(P_{i\max}) = \left(a_i P_{i\max}^2 + b_i P_{i\max} + c_i\right). \tag{4}$$

NO_x emission of each generator at its maximum output was evaluated:

$$E_i\,(P_{i\max}) = \left(\alpha_i + \beta_i P_{i\max} + \gamma_i P_{i\max}^2\right) + \xi_i \sin\left(\lambda_i P_{i\max}\right). \tag{5}$$

By (4) and (5) get

$$\frac{F_i\,(P_{i\max})}{E_i\,(P_{i\max})} = k_i. \tag{6}$$

So the final objective incorporated total generation cost and environmental emission generation which is given as

$$F_{\text{final\_object}} = F_T + k_i\,(E_T), \tag{7}$$

where E_T is the total emission, F_T is the total generation cost, $f(E_i(P_i))$ is the emission function, and α_i, β_i, γ_i, ξ_i, and λ_i are the emission coefficients of the ith generators.

2.3. Constraints

2.3.1. Power Balance Constraints. The total generated power should be equal to the sum of total load demand and line loss. It can be formulated as (8). Consider the following:

$$\sum_{i=1}^{n} P_i = P_D + P_L, \tag{8}$$

$$P_L = \sum_{i=1}^{n}\sum_{j=1}^{n} P_i B_{ij} P_j, \tag{9}$$

where P_D and P_L are the total system demand and line loss, respectively, and B_{ij} is the line loss elements.

2.3.2. Generator Limits. Generating output of each generating unit should lie between the maximum and minimum limits as given in

$$P_i^{\min} \le P_i \le P_i^{\max}, \tag{10}$$

where P_i is the output power of ith generator and $P_i^{\min}$ and $P_i^{\max}$ are the minimum and maximum generated power of ith generator, respectively.

3. Overview of PSO Strategies

This section represents a review of particle swarm optimization techniques which will serve as a performance measured for the PSO with moderate random search technique (MRPSO) [17] applied in this paper for solving of multiobjective ELD problem.

3.1. Particle Swarm Optimization. PSO is a very popular optimization technique and is used to solve optimization problems. It is a population-based optimization technique and it is motivated by the behaviour of social systems such as fish schooling and birds flocking. Particle swarm optimization was first introduced by Kennedy and Eberhart in the year 1995 [16]. It is a simple and powerful optimization tool which scatters random particles into the search space. Randomly initialized particles are called swarms; collect the information from each array of the problem constructed by their respective positions. The position of the particles is updated by using the velocityof particles. Both position and velocity are updated in a heuristic manner by using guidance from particles by their own experience and the experience of its neighbor's particles.

PSO randomly starts within the limits of maximum and minimum value of the power of the ith generator. The position and velocity of the ith particle of a d-dimensional search space can be represented in the following:

$$P_i = (P_{i1}, P_{i2}, \ldots, P_{id}),$$
$$V_i = (V_{i1}, V_{i2}, \ldots, V_{id}). \tag{11}$$

The best previous position of a particle is recorded and is represented as given in the following:

$$P_{\text{best}} = (P_{i1}, P_{i2}, \ldots, P_{id}). \tag{12}$$

If the gth particle is the best among all particles in the group so far, it is represented as

$$P_{g_{\text{best}}} = g_{\text{best}} = \left(P_{g1}, P_{g2}, \ldots, P_{gd}\right). \tag{13}$$

Velocity and position of the particle are updated by using

$$V_i^{(K+1)} = W * V_i^K + c_1 * \text{rand}_1 \times \left(P_{\text{best}i} - S_i^K\right)$$
$$+ c_2 * \text{rand}_2 \times \left(g_{\text{best}} - S_i^K\right), \tag{14}$$
$$S_i^{(K+1)} = S_i^K + V_i^{K+1},$$

where V_i^K is velocity of the particles at iteration K, W is the inertia weigh, c_1 and c_1 are the acceleration coefficients, rand_1 and rand_2 are the random numbers between 0 and 1, S_i^K is the current position of particle at iteration K, P_{best} is the best position of individual ith particle, and g_{best} is the global best position of the group.

The acceleration coefficients $c_1$and c_2 pull each particle towards P_{best} and g_{best} positions and W is the inertia weight parameter which provides a balance between global and local explorations. Since W decreases linearly from about 0.9 to 0.4 quite often during a run, the weighing function can be formulated as given in the following:

$$W = W_{\max} - \frac{W_{\max} - W_{\min}}{\text{iter}_{\max}} * \text{iter}, \tag{15}$$

where $W_{\max}$ and $W_{\min}$ are the initial and final inertia weight parameter, respectively, itermax is the maximum number of iterations, and iter is the current iteration position.

3.2. Moderate Random Search Particle Swarm Optimization (MRPSO). PSO is a very simple and popular optimization tool used for solving the ELD problem but it has some disadvantages also, such that PSO lacks global search ability at the last stage of iterations. So it is unable to give the global optimal solution of the ELD problem. In the year of 2011 [17] Gao and Xu first introduced the PSO with moderate random search technique called MRPSO. Moderate random search strategy enhances the global search ability of the particles. It can overcome the problem of PSO and gives the optimal solution for the proposed multiobjective ELD problem. In case of PSO, it is required to update the position and velocity of the particle, but after some iteration the velocity of particles should be zero so that in case of the MRPSO position of the particle can be updated as given in (17). If the particle's position is updated as given in (17) particles velocity does not change and it gives the global solution of the problem.

In MRPSO technique the particles are randomly generated for a population size within the range of 0-1 same as in case of basic PSO and it is located within the maximum and minimum operating range of the generators as given in the following:

$$P_{\text{Initial}} = P_{i\min} + \text{rand}\left(P_{i\max} - P_{i\min}\right), \tag{16}$$

where P_{Initial} is the initially generated particles, rand is random value between 0 and 1, and $P_{i\max}$ and $P_{i\min}$ are the maximum and minimum value of generated power of generator, respectively.

The position of the particles is updated for the ith particle at the $(K+1)$th iteration using (17). Consider the following:

$$S_i^{K+1} = P_d + \alpha\lambda\left(m_{\text{best}_i} - S_i^K\right), \quad (17)$$

$$m_{\text{best}_i} = \sum_{i=1}^{N} \frac{P_{\text{best}}}{N}, \quad (18)$$

where N is the population size in the MRPSO.

The parameter α used in (17) may be considered as a constant value between 0.45 and 0.35 oris obtained by changing α from 0.45 to 0.35 with the linear-decreasing method during iteration. P_d is theattractor moving direction of particles; it is given as

$$P_d = \text{rand}_0 P_{\text{best}} + (1 - \text{rand}_0)\, g_{\text{best}}, \quad (19)$$

where rand_0 is a uniformly distributed random variable within 0-1, P_{best} is the best value of particle, and g_{best} is the best value of P_{best} values:

$$\lambda = \frac{\text{rand}_1 - \text{rand}_2}{\text{rand}_3}, \quad (20)$$

where rand_1 and rand_2 are two random variables within $[0, 1]$ and rand_3 is a random variable within $[-1, 1]$.

4. Algorithm of MRPSO for Multiobjective Economic Load Dispatch Problem

In this paper PSO with the moderate random search technique (MRPSO) is used to solve multiobjective ELD problem. The study carries two objectives; the first objective is to minimize the total generation cost and the second objective is to reduce the environmental emission of the thermal power plant. The following steps are being used to solve the proposed multiobjective problem by using MRPSO.

Step 1. Select the constants.

Step 2. Initialize the swarm. First of all particles are randomly generated for a population size in the range 0-1 and located between the maximum and the minimum operating limits of the generators as given in (16).

Step 3. Initialize velocity and position for all particles randomly set to within their minimum and maximum limits.

Step 4. Set generation counter: counter = counter + 1.

Step 5. Evaluate the fitness for each particle according to the proposed objective function.

Step 6. Compare particles fitness evaluation with its personal best (P_{best}) and global best (g_{best}).

Step 7. Update position of particles by using (17).

Step 8. Apply stopping criteria. Number of iterations is the stopping criteria taken in this study. Means when number of iterations will completed the conversion of algorithm stopped.

Table 1: Cost coefficient with generation limits of IEEE 30 bus, 6 generating unit systems, and load demand = 283.4 MW.

Bus number	a_i	b_i	c_i	$P_{\min}$	$P_{\max}$
1	100	200	10	50	200
2	120	150	10	20	80
5	40	180	20	15	50
8	60	100	10	10	35
11	40	180	20	10	30
13	100	150	10	12	40

Table 2: Environmental emission coefficients of IEEE 30 bus systems.

Bus number	α_i	β_i	γ_i	ξ_i	λ_i
1	4.091	−5.554	6.490	$2 * 10^{-4}$	2.857
2	2.543	−6.047	5.638	$5 * 10^{-4}$	3.333
5	4.258	−5.094	4.586	$1 * 10^{-6}$	8.00
8	5.426	−3.556	3.380	$2 * 10^{-3}$	2.00
11	4.258	−5.094	4.586	$1 * 10^{-6}$	8.000
13	6.131	−5.555	5.151	$1 * 10^{-5}$	6.667

Table 3: Line loss coefficients of line loss of IEEE 30 bus systems.

0.0218	0.0107	−0.00036	−0011	0.00055	0.0033
0.0107	0.0107	−0.0001	−0079	0.00026	0.0028
−0.0004	−.0002	0.02459	−0133	−0.0118	−0.0079
−0.0011	−0.0018	−0.01328	0.0265	0.0098	0.0045
0.00055	0.00026	0.0118	0.0098	0.0216	−0.0001
0.0033	0.0028	−00792	0.0045	−.00012	0.02978

5. Problem Formulation and Results

The proposed algorithms are tested for the data of standard IEEE 30 bus, 6 generator systems [12]. The MRPSO has been applied for solving IEEE 30 bus system for the demand of 283.4 MW. Data of IEEE 30 bus cost coefficients are given in Table 1, and the data of emission coefficients are given in Table 2, respectively. Table 3 shows the value of line loss coefficients (B-coefficient).

Table 4 shows the results obtained by PSO [16], WIPSO [18], and MRPSO [17] without considering line loss.

Table 5 shows the result obtained by different PSO techniques considered by line loss. Constant used in this study for solving multiobjective ELD problem is shown in Table 6.

To assess the effectiveness and the efficiency of the proposed MRPSO algorithm for solving multiobjective ELD problem in this paper, a case study of IEEE 30 bus system with 6 generating units and their loss coefficients was taken. All data were considered from [12]. Different PSO techniques such as PSO, CPSO, WISPO, and MRPSO were used to solve the multiobjective ELD problem for the given data in Tables 1, 2, and 3. All PSO techniques were run on a 1.4 GHz, core-2 solo processor with 2 GB DDR RAM.

Each PSO technique was tested for 100 numbers of iteration and for 20 population sizes. The robustness of the proposed algorithms is judged in each case of 100 trials.

TABLE 4: Conversion results of IEEE 30 bus systems for the load of 283.4 MW without line loss.

Unit power output	PSO	CPSO	WIPSO	MRPSO
P1 (MW)	81.047	78.043	83.0324	73.0231
P2 (MW)	63.1092	63.0197	60.0947	62.1528
P5 (MW)	45.6863	48.632	44.023	46.8732
P8 (MW)	32.6824	34.0721	33.7961	34.0872
P11 (MW)	32.1054	27.0921	32.0823	28.4035
P13 (MW)	28.731	32.5401	30.432	38.8732
Total power output	283.361	283.399	283.460	283.413
Fuel cost ($/h)	1434747	1411729.	1429850	1364072
Environmental emission (Ton/h)	85137.024	83500.458	85265.256	79847.867
Total cost ($/h)	$16.69 * 10^5$	$16.44 * 10^5$	$16.66 * 10^5$	$15.87 * 10^5$
Time (sec)	0.4302	0.42804	0.42401	0.40702

TABLE 5: Conversion results of IEEE 30 bus systems for the load of 283.4 MW with line loss.

Unit power output	PSO	CPSO	WIPSO	MRPSO
P1 (MW)	147.03	146.034	147.581	145.7801
P2 (MW)	43.114	46.0732	46.889	43.0912
P5 (MW)	36.661	34.0742	47.0705	43.07654
P8 (MW)	23.019	26.0198	16.7863	24.0763
P11 (MW)	25.377	24.108	24.7219	23.1732
P13 (MW)	27.632	26.0911	19.8925	23.0453
Loss (MW)	19.435	19.0003	19.5407	18.8468
Total power output	302.835	302.400	302.9407	302.2468
Fuel cost ($/h)	2617397.1	2607622.2	2661326.7	2575426.4
Environmental emission (Ton/h)	163675.87	162228.52	167729.08	162153.591
Total cost ($/h)	$30.41 * 10^5$	$30.31 * 10^5$	$30.91 * 10^5$	$29.96 * 10^5$
Time (sec)	0.5165	0.5032	0.49541	**0.48203**

TABLE 6: Constant used to solve the proposed problem using MRPSO.

S. number	Constants	Value of constant
1	$C_1 = C_2$	2
2	A	0.35
3	W_{max}	0.9
4	W_{min}	0.4

Optimum results obtained by different PSO techniques are mentioned in Tables 4 and 5.

It is observed that for more than 100 trials the values obtained by different techniques were repeated and hence consider best results obtained in 100 trials. The optimal result obtained by MRPSO for the load of 283.4 MW without line loss is shown in Table 4: fuel cost of $1364072/h, environmental emission cost of 79847.867 Kg/h, and total cost including environmental emission of $\$15.87 * 10^5$/h. The best result obtained by using MRPSO with considering the line loss is shown in Table 5, getting a fuel cost of $2575430/h, environmental emission cost of 162153.96 Kg/h, total cost including environmental emission of $\$29.967 * 10^5$/h, and line loss of 18.8468 MW. Results obtained by MRPSO are less than other optimization techniques in both of the cases, that is, without line loss and with line loss. The convergence time taken by MRPSO is also less as compared to other PSO techniques.

6. Conclusion

Analysis of the economic load dispatch has a wide range of scope nowadays. Satisfying the load demand in minimum fuel cost is a great challenge for the power system. The optimal results obtained by the different PSO techniques are shown in Tables 4 and 5. Results obtained by different PSO techniques are shown in Tables 4 and 5. The fuel cost of the IEEE 30 bus system obtained by MRPSO is $1364072/h and environmental emission is 79847.867 Ton/h, and total cost (fuel cost + environmental emission) is $\$15.87 * 10^5$/h without considering line loss. Fuel cost is $1364072/h, environmental emission is 162153.591 Ton/h, and total cost is $\$29.96 * 10^5$/h with consideration of line loss. It shows that the results of MRPSO algorithm are better than other PSO techniques and it takes minimum convergence time. So MRPSO is the better approach when compared to other PSO techniques mentioned in this paper. Results obtained by MRPSO show that it is an effective and efficient method for obtaining the result of optimization problem.

Conflict of Interests

The authors declare that there is no conflict of interests regarding the publication of this paper.

References

[1] J. H. Talaq, F. El-Hawary, and M. E. El-Hawary, "A summary of environmental/economic dispatch algorithms," *IEEE Transactions on Power Systems*, vol. 9, no. 3, pp. 1508–1516, 1994.

[2] A. Farag, S. Al-Baiyat, and T. C. Cheng, "Economic load dispatch multiobjective optimization procedures using linear programming techniques," *IEEE Transactions on Power Systems*, vol. 10, no. 2, pp. 731–738, 1995.

[3] S.-D. Chen and J.-F. Chen, "A direct Newton-Raphson economic emission dispatch," *International Journal of Electrical Power & Energy System*, vol. 25, no. 5, pp. 411–417, 2003.

[4] J. Nanda, D. P. Kothari, and K. S. Lingamurthy, "Economic-emission load dispatch through goal programming technique," *IEEE Transactions on Energy Conversion*, vol. 3, no. 1, pp. 26–32, 1988.

[5] R. Yokoyama, S. H. Bae, T. Morita, and H. Sasaki, "Multiobjective generation dispatch based on probability security criteria," *IEEE Transactions on Power Systems*, vol. 3, no. 1, pp. 317–324, 1988.

[6] C. S. Chang, K. P. Wong, and B. Fan, "Security-constrained multiobjective generation dispatch using bicriterion global optimisation," *IEE Proceedings: Generation, Transmission and Distribution*, vol. 142, no. 4, pp. 406–414, 1995.

[7] G. P. Granelli, M. Montagna, G. L. Pasini, and P. Marannino, "Emission constrained dynamic dispatch," *Electric Power Systems Research*, vol. 24, no. 1, pp. 55–64, 1992.

[8] C. B. Das and C. Patvardhan, "New multi-objective stochastic search technique for economic load dispatch," *IEE Proceedings-Generation, Transmission and Distribution*, vol. 145, no. 6, pp. 747–752, 1998.

[9] D. Srinivasan, C. S. Chang, and A. C. Liew, "Multiobjective generation scheduling using fuzzy optimal search technique," *IEE Proceedings Generation, Transmission and Distribution*, vol. 141, no. 3, pp. 233–242, 1994.

[10] J. X. Xu, C. S. Chang, and X. W. Wang, "Constrained multiobjective global optimisation of longitudinal interconnected power system by genetic algorithm," *IEE Proceedings—Generation, Transmission and Distribution*, vol. 143, no. 5, pp. 435–446, 1996.

[11] C. A. Coello Coello, G. T. Pulido, and M. S. Lechuga, "Handling multiple objectives with particle swarm optimization," *IEEE Transactions on Evolutionary Computation*, vol. 8, no. 3, pp. 256–279, 2004.

[12] P. Dutta and A. K. Sinha, "Environmental economic dispatch constrained by voltage stability using PSO," in *Proceedings of the IEEE International Conference on Industrial Technology (ICIT '06)*, pp. 1879–1884, December 2006.

[13] G. Suganya, K. Balamurugan, and V. Dharmalingam, "Multi-objective evolutionary programming technique for economic/emission load dispatch," in *Proceedings of the 1st International Conference on Advances in Engineering, Science and Management (ICAESM '12)*, pp. 269–275, Tamil Nadu, India, March 2012.

[14] H. Mori and K. Okawa, "Advanced MOEPSO-based multi-objective environmental economic load dispatching," in *Proceedings of the IEEE Power and Energy Society General Meeting*, pp. 1–7, IEEE, Minneapolis, Minn, USA, July 2010.

[15] M. A. Abido, "Multiobjective evolutionary algorithms for electric power dispatch problem," *IEEE Transactions on Evolutionary Computation*, vol. 10, no. 3, pp. 315–329, 2006.

[16] J. Kennedy and R. C. Eberhart, "Particle swarm optimization," in *Proceedings of the IEEE International Conference on Neural Networks*, vol. 4, pp. 1942–1948, December 1995.

[17] H. Gao and W. Xu, "A new particle swarm algorithm and its globally convergent modifications," *IEEE Transactions on Systems, Man, and Cybernetics, Part B: Cybernetics*, vol. 41, no. 5, pp. 1334–1351, 2011.

[18] P. T. Vu, D. L. Le, N. D. Vo, and J. Tlusty, "A novel weight-improved particle swarm optimization algorithm for optimal power flow and economic load dispatch problems," in *Proceedings of the IEEE PES Transmission and Distribution Conference and Exposition*, pp. 1–7, April 2010.

16

Caching Eliminates the Wireless Bottleneck in Video Aware Wireless Networks

Andreas F. Molisch,[1] Giuseppe Caire,[1] David Ott,[2] Jeffrey R. Foerster,[2,3] Dilip Bethanabhotla,[1] and Mingyue Ji[1]

[1] *Department of Electrical Engineering, University of Southern California, Los Angeles, CA 90089, USA*
[2] *Intel Corporate Research, 5000 W Chandler Boulevard, Chandler, AZ 85226, USA*
[3] *Intel Corporate Research, 2111 NE 25th Avenue, Hillsboro, OR 97124, USA*

Correspondence should be addressed to Andreas F. Molisch; molisch@usc.edu

Academic Editor: Dimitrios Axiotis

Wireless video is the main driver for rapid growth in cellular data traffic. Traditional methods for network capacity increase are very costly and do not exploit the unique features of video, especially asynchronous content reuse. In this paper we give an overview of our work that proposed and detailed a new transmission paradigm exploiting content reuse and the widespread availability of low-cost storage. Our network structure uses caching in helper stations (femtocaching) and/or devices, combined with highly spectrally efficient short-range communications to deliver video files. For femtocaching, we develop optimum storage schemes and dynamic streaming policies that optimize video quality. For caching on devices, combined with device-to-device (D2D) communications, we show that communications within clusters of mobile stations should be used; the cluster size can be adjusted to optimize the tradeoff between frequency reuse and the probability that a device finds a desired file cached by another device in the same cluster. In many situations the network throughput increases linearly with the number of users, and the tradeoff between throughput and outage is better than in traditional base-station centric systems. Simulation results with realistic numbers of users and channel conditions show that network throughput can be increased by two orders of magnitude compared to conventional schemes.

1. Introduction

Demand for video content over wireless networks has grown significantly in recent years and shows no sign of letting up. According to the *Cisco Visual Networking Index* mobile forecast for 2012–2017, mobile video data is expected to grow at a compound annual growth rate of 75 percent to 7.4 exabyes (one million gigabytes) by 2017 [1]. By this time, it is expected to be 66.5 percent of global mobile traffic data (11.2 exabytes), up from 51 percent in 2012. We expect both broadcast and on-demand services will continue to expand, including traditional services like streaming TV content (e.g., sporting events) and newer services like video Twitter, video blogging, cloud-based live video broadcasting, and mobile-to-mobile video conferencing and sharing. Meanwhile, hardware platforms (smart phones, tablets, notebooks, television/set-top boxes, and in-vehicle infotainment systems) continue to push the envelope in performance and graphical quality. More capable processors, better performing graphics, increased storage capacities, and larger displays make devices more powerful and intelligent than ever before. With this increase in device capability comes a corresponding increase in demand for high-quality video data, for example, increasing demand for high-definition (HD) and 3D data types.

As demand for video traffic continues to grow, the quality of experience (QoE) delivered for this traffic becomes increasingly important. From a 2013 report by Conviva [2], 39.3% of video views experienced buffering, 4% of views failed to start, and 63% of the views experienced low resolution. In addition, other reports [3, 4] have shown that these poor QoE events directly impact a user's engagement in viewing the video and hence potential revenue from videos.

The implications of these trends for future wireless networks are significant. While continued evolution in spectral efficiency is to be expected, the maturity of MIMO, air interfaces using OFDM/OFDMA, and Shannon capacity-approaching codes mean that such spectral-efficiency improvements will not deliver the increased capacity needed to support future demand for video data. Additional measures like the brute force expansion of wireless infrastructure (number of cells) and the licensing of more spectrum, while clearly addressing the problem of network capacity, may be prohibitively expensive, require significant time to implement, or be infeasible due to prior spectrum allocations which are not easily modified.

Recognizing these challenges, Intel and several industry partners jointly developed a program to explore nonincremental, systems-level solutions through university research. Known as *video aware wireless networks* or simply VAWN, the program considers various approaches to enabling a higher capacity in future wireless networks and to enabling a higher quality of user experience for video and video-based services delivered over wireless networks to intelligent mobile devices. Broad strategies explored in the program include unconventional optimizations in video transport within the network, optimizations in video processing to reduce network transmission requirements and improve user experience, and novel network architectures better suited to address future capacity and quality of service challenges specific to video.

The approach taken by the group at the University of Southern California (including several of the authors) exploits a unique feature of wireless video, namely, the high degree of (asynchronous) content reuse. Based on the fact that storage is cheap and ubiquitous in today's wireless devices, this group developed a new network structure that is based on *replacing backhaul by caching*. This approach, first proposed by the USC group in 2010 [5] and expounded and refined in a series of papers [6–15], is at the center of the present overview.

A first approach for exploiting asynchronous content reuse, termed *Femtocaching*, uses dedicated "helper nodes" that can cache popular files and serve requests from wireless users by enabling localized wireless communication. Such helper nodes are similar to femto-BSs, but they have two key differences: they have *added* a large storage. while they *do not have or need* a high-speed backhaul (Note that storage space has become exceedingly cheap: 2 TByte of data storage capacity, enough to store 1000 movies, cost only about $100.). An even higher density of caching can be achieved by using devices themselves as video caches, in other words, using devices such as tablets and laptops (which nowadays have ample storage) as mobile helper stations [7]. The simplest way of using this storage would have each user cache the most popular files. However, this approach is not efficient because many users are interested in similar files, and thus the same videos will be duplicated on a large number of devices. On the other hand, the cache on each device is too small to cache a reasonably large number of files. Thus, it is preferable that the devices "pool" their caching resources, so that different devices cache different files and then exchange them, when the occasion arises, through short-range, highly spectrally efficient, device-to-device (D2D) communications. If a requesting device does not find the file in its neighborhood (or in its own cache), it obtains the file in the traditional manner from the base station (the base station can also control any occuring D2D communications).

The remainder of the paper is organized as follows: in Section 2, we describe video coding and video streaming techniques, as well as content reuse and viewing habits. The principle of the new network structure is described in Section 3. The placement of files in helper nodes and devices is discussed in Section 4. Fundamental results about throughput and outage in networks with helper stations and D2D communications are described in Sections 5 and 6, respectively. Conclusions in Section 7 round off the paper.

2. Dynamically Managing Video Quality of Experience

2.1. Video Streaming and Quality Management. Wireless channels are inherently dynamic and time-varying depending on a number of factors: (i) movement of device (walking, driving), (ii) changes in the reflectors in the environment (people moving, objects moving), (iii) changes in location (indoors, outdoors), (iv) changes in selected wireless network (WiFi, cellular), and (v) changes in the amount of traffic using the network (i.e., congestion). For data and web-based applications, some latency due to changes in available network capacity, while annoying, can be tolerated. However, for video-based applications (especially interactive video conferencing, but also—depending on buffering capability—for video playback), simply treating data communications as latency tolerant is not sufficient. In order to improve the end user quality of experience (QoE), it is often desirable to adapt the rate of the streamed video using techniques that take into account such factors as the type of video being streamed (fast motion, complex scenes, and interactive), the available capacity of the network, time variations in network and channel state, client device information (screen size, etc.), and playback buffer state. This section will describe some mechanisms for achieving this dynamic adaptation and the role of emerging standards.

Figure 1 illustrates the potential opportunities for managing streaming video traffic in intelligent ways. The figure shows a simplified view of an end-to-end system, including a video server (left), an end rendering device (right), and the network lying in between. Note that opportunities exist in all three domains of the end-to-end system. For example, the video server may accommodate different devices by supporting multiple streaming rates, or multiple copies (formats, bitrates) of the video content. Alternatively, a server may choose to transcode streamed video on the fly. The decision of whether to transcode or store multiple copies may depend on cost, complexity, and performance considerations. Video management opportunities in the network include, for example, support for content caching. How much and how often may depend on the popularity of the content, and whether the details of a particular network architecture make storing content feasible and inexpensive. At the rendering device, opportunities exist for choosing among multiple

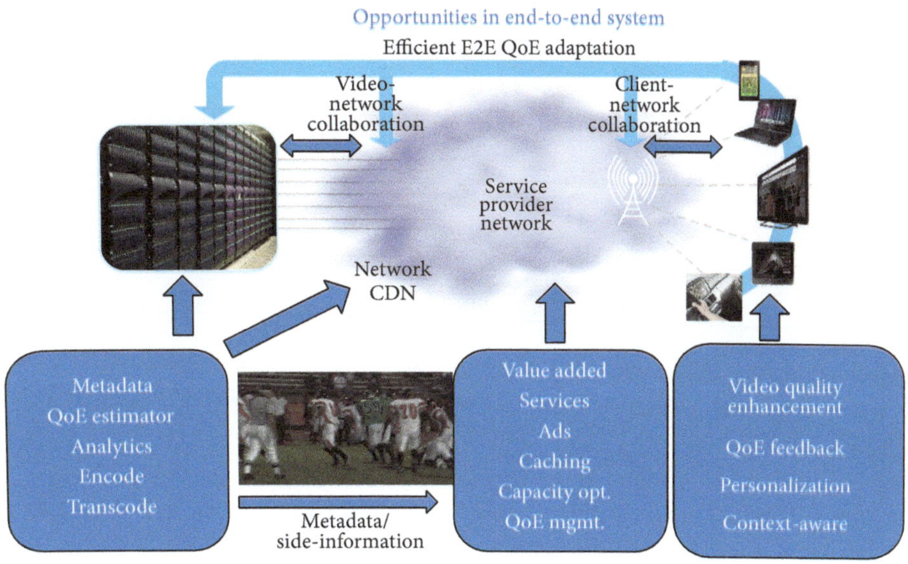

FIGURE 1: Simplified end-to-end system for video streaming.

video streaming rates given user preferences and making dynamic adjustments during a playback session in response to changes in wireless channel state. Such adjustments may select among the bitrates available from a video source and/or make changes to the display application's buffering strategy.

To better improve user playback experience and to improve the efficiency of data storage and transport, we believe quality of experience (QoE) will be a key metric in future video streaming management. Measures of QoE may take into account the quality of the displayed video (resolution, compression artifacts), rebuffering events, and lost packets. QoE metrics provide an alternative to throughput-based approaches which rely on the often mistaken assumption that higher bitrates invariably mean higher playback quality. A key challenge here, however, is effectively estimating video quality independent of bitrate. Fortunately, a great deal of progress has been made recently by researchers estimating video quality based on both device and content characteristics (see [16–18]). For instance, no-reference approaches to video quality assessment (VQA) can exploit natural video statistics (e.g., DCT) and movement coherency to predict perceived distortions [19]. This information, along with information on channel state, can be used to make automated adjustments in video bitrate at the server or buffering at the display device [17]. In general, QoE metrics enable new opportunities for tighter collaboration between each part of the end-to-end system shown in Figure 1 and for more intelligent control algorithms.

Enhancements to emerging standards are helping to promote QoE-based optimization within end-to-end systems. In particular, standards supporting Dynamic Adaptive Streaming over HTTP (DASH) are being developed by the MPEG and 3GPP standards bodies (see [20–31]). Two recent additions to these standards are (1) the inclusion of QoE feedback metrics from the device to the network and (2) support for providing QoE metrics along with video content that is sent to a device. (In some cases, video QoE metrics can also be computed directly by the end device.) These additions are important because they enable better system-wide optimization of video transport based on the end user QoE. For example, the device can decide which future segments to request based on the current status of its playback buffer and known quality levels of upcoming segments. This supports a more intelligent balancing of playback quality and rebuffering risk. The network can also make more informed decisions on how to allocate available bandwidth across multiple competing video flows by optimizing the quality jointly across all of them. Using rate-distortion information (a measure of video quality) and playback buffer state for each flow, for instance, a network scheduler can implement QoE-based resource allocation as an alternative to standard proportionally fair throughput schemes.

2.2. Content Reuse. Wireless video distinguishes itself from other wireless content through its strong content reuse; that is, the same content is seen by a large number of people. However, in contrast to TV, the bulk of wireless video traffic is due to asynchronous *video on demand*, where users request video files from some cloud-based server at arbitrary times. As indicated in Section 1, caching can exploit *content overlap*, even in the presence of *asynchronous requests*. In other words, a few popular videos (YouTube clips, sports highlights, and movies) account for a considerable percentage of video traffic on the Internet, even though they are viewed at different times by different people. Numerous experimental studies have indicated that Zipf distributions are good models for the measured popularity of video files [32–34]. Under this model, the frequency of the ith popular file, denoted by $P_r(f)$, is inversely proportional to its rank:

$$P_r(f) = \frac{1/f^{\gamma_r}}{\sum_{j=1}^{m} (1/j^{\gamma_r})}, \quad 1 \le f \le m. \tag{1}$$

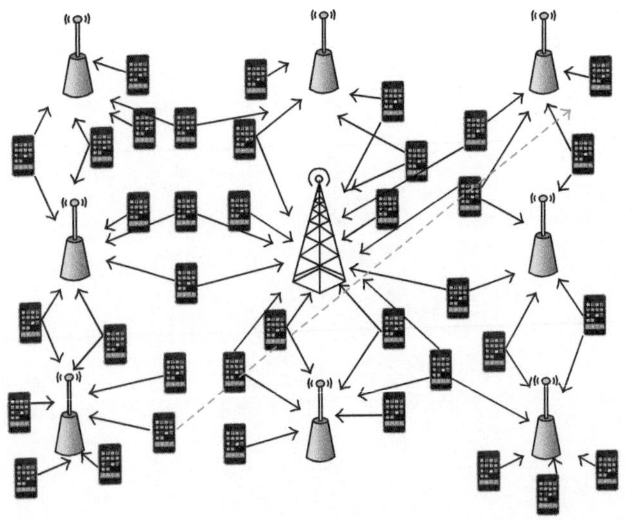

FIGURE 2: A sample network scenario with multiple helpers and users and a mobile user (green path).

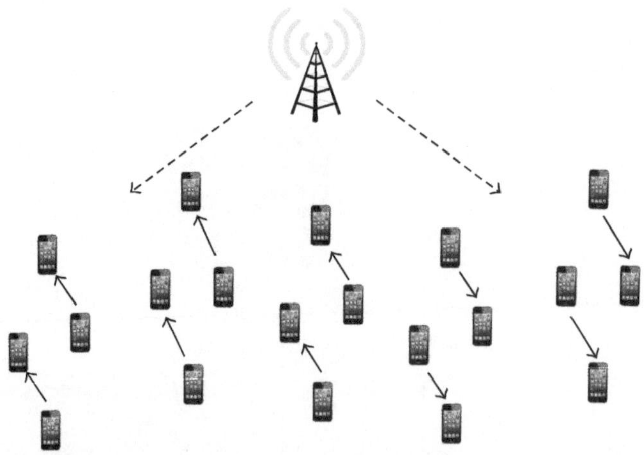

FIGURE 3: An illustration of D2D caching networks, where each user (device) can cache M files and we let users be served through high spectral-efficiency D2D links denoted by solid lines in the figure. The dotted lines denote the controlling signal from the base station to the devices.

The Zipf exponent γ_r characterizes the distribution by controlling the relative popularity of files. Larger γ_r exponents correspond to higher content reuse; that is, the first few popular files account for the majority of requests. Here, m is the size of the library of files that are of interest to the set of considered users (note that the library size can be a function of the number of considered users n; we assume in the following that m increases like n^α, where $\alpha \geq 0$).

A further important property of the library is that it changes only on a fairly slow timescale (several days or weeks); it can furthermore be shaped by content providers, for example, through pricing policies, or other means.

Note, however, some caveats concerning the general applicability of the work in the remainder of the paper. It applies principally to a setting where a content library of relatively large files (e.g., movies and TV shows) is refreshed relatively slowly (e.g., on a daily basis) and where the number of users consuming such a library is significantly larger than the number of items in the library. This may apply to a possible future implementation of movie services, while collections of short videos (like YouTube) show wider ranges of interests. In short, this paper reflects a set of results and approaches that are relevant in the case where the caching phase (placement of content in the caches) occurs with a clear time-scale separation with respect to the delivery phase (the process of delivering video packets for streaming to the users) and where the size of the content library is moderate with respect to the users' population.

3. Network Structure

3.1. Helper Stations and File Requests. We first consider the network structure with helper stations. The wireless network consists of multiple helper stations $\mathscr{H}$, talking to multiple users $\mathscr{U}$; a central base station may be present to serve users that cannot find the files they want in the helper stations. An example network is shown in Figure 2, which describes a sample network scenario with multiple helpers and users. Each user requests a video file from a library $\mathscr{F}$ of possible files. We denote the set of helpers in the vicinity of user u as $\mathscr{N}(u)$. Similarly, $\mathscr{N}(h)$ denotes the set of users in the vicinity of helper h. The helpers may not have access to the whole video library, because of backhaul constraints and/or caching constraints. In general, we denote by $\mathscr{H}(f)$ the set of helpers that contain file $f \in \mathscr{F}$. Hence, user u requesting file f_u can only download video chunks from helpers in the set $\mathscr{N}(u) \cap \mathscr{H}(f_u)$. In Section 5, we consider the problem of devising a dynamic scheduling scheme such that helpers feed the video files sequentially (chunk by chunk) to the requesting users. Given the high density of helpers, any user is typically in the range of multiple helpers. Hence, in order to cope with user-helper association, load balancing, and intercell interference, an efficient video streaming policy is described in Section 5 which allows the users to dynamically select the helper node to download from and determine adaptively the video quality level of the download.

3.2. Device-to-Device (D2D) Caching Networks. When users also have the ability of prefetching (video) files, instead of requesting the files from the base station or the helpers, we allow users to make requests from other users and get served via high-spectral-efficiency D2D links (see Figure 3). If the D2D links are not available for some users (see Section 6.2), then these unserved users are treated as in outage and in practice; they can be simply served by the base station or the helpers. To make the network model tractable, we consider the download of the video files instead of streaming and neglect the issue of rate adaptation. In addition, we consider a simple grid structure, which is formed by n user nodes $\mathscr{U} = \{1, \ldots, n\}$ placed on a regular grid on the unit square, with minimum distance $1/\sqrt{n}$. (See Figure 4(a); we will replace this grid structure by the random uniform distribution of the nodes when mentioned specifically.) Let each user $u \in \mathscr{U}$

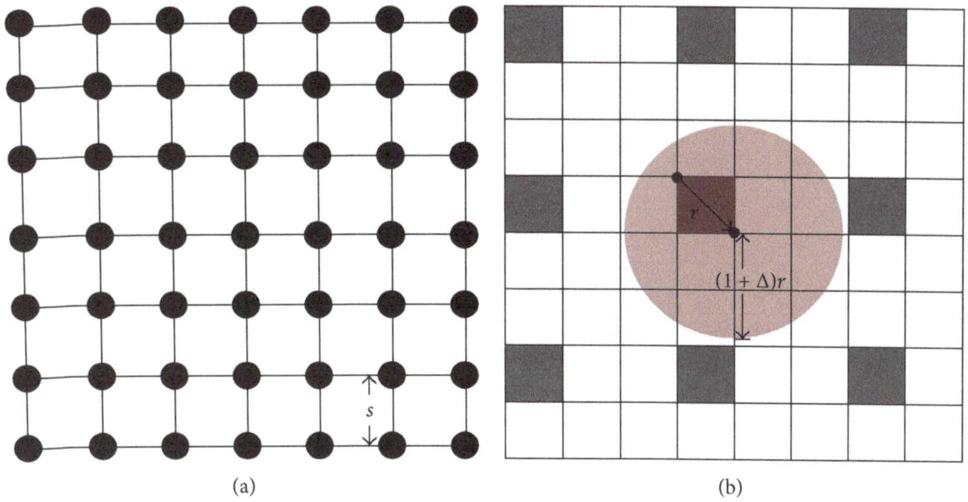

FIGURE 4: (a) Grid network with $n = 49$ nodes (black circles) with minimum separation $s = 1/\sqrt{n}$. (b) An example of single-cell layout and the interference avoidance TDMA scheme. In this figure, each square represents a cluster. The gray squares represent the concurrent transmitting clusters. The red area is the disk where the protocol model imposes no other concurrent transmissions. r is the worst case transmission range and Δ is the interference parameter. We assume a common r for all the transmitter-receiver pairs. In this particular example, the TDMA parameter is $K = 9$.

request a file $f \in \mathcal{F} = \{1,\ldots,m\}$ in an i.i.d. manner, according to a given request probability mass function $P_r(f)$, which is assumed to be a Zipf distribution given by (1) with parameter $0 < \gamma_r < 1$ [35]. Moreover, we let each user cache M files. The BS keeps track of which devices can communicate with each other and which files are cached on each device. Such BS-controlled D2D communication is more efficient (and more acceptable to spectrum owners if the communications occur in a licensed band) than traditional uncoordinated peer-to-peer communications.

Communications between nodes follow the protocol model [36] (In the simulations of Section 6.4, we relax the protocol model constraint and take interference into consideration by treating it like noise). Namely, transmission between user nodes u and v is possible if their distance $d(u, v)$ is less than or equal to some fixed transmission range r and if there is no other active transmitter within distance $(1 + \Delta)r$ from destination v, where $\Delta > 0$ is the interference control parameter. Successful transmissions can take place at rate C_r bit/s/Hz, which is a nonincreasing function of the transmission range r [9]. In this model, we do not consider power control (which would allow different transmit powers and thus transmission ranges), for each user. Moreover, we treat r as a design parameter that can be set as a function of m and n (Since the number of possibly requested files m typically varies with the number of users in the system n, and r can vary with n, r can also be a function of m). All communications are assumed to be single-hop (see also Section 6). These model assumptions allow for a sharp analytical characterization of the throughput scaling law including the leading constants. In Section 6, we will see that the schemes designed by this simple model yield promising performance also in realistic channel propagation and interference conditions.

For many of our derivations, we furthermore subdivide the cell into equal-sized, disjoint groups of users that we call "clusters" of size (radius) r, with g_c nodes in it. To further simplify the mathematical model, we assume that only nodes that are part of the same cluster can communicate with each other. If a user can find the requested file inside the cluster, we say there is one *potential link* in this cluster; when at least one link is scheduled, we say that the cluster is "active." We use an *interference avoidance* scheme, such that at most one link can be active in each cluster on one time-frequency resource.

4. File Placement

The proposed system operates in two steps: (i) file placement (caching) and (ii) delivery. These two processes happen on different timescales: the cache content needs to change only on a timescale of days, weeks, or months, that is, much slower than the actual delivery to the users. Thus, caches could be filled either through a very slow backhaul or through cellular connection at night time, when the spectral resources are not required for other purposes.

4.1. File Placement in Helper Stations. We start out with the case where complete files are stored in the helper stations. If the distance between helpers is large and each MS can connect only to a single helper, each helper should cache the most popular files, in sequence of popularity, until its cache is full. However, when each MS can communicate with multiple helpers, the question on how to best assign files to different helpers becomes more complicated. Consider the case in Figure 5, users U_1 and U_2 would prefer helper H_1 to cache the M most popular files since this minimizes their expected downloading time. Similarly, user U_4 would prefer that helper H_2 also caches the M most popular files. However U_3 would prefer H_1 to cache the M most popular files and H_2 the second M most popular (or the opposite), thus creating a distributed cache of size $2M$ for user U_3. Thus we can

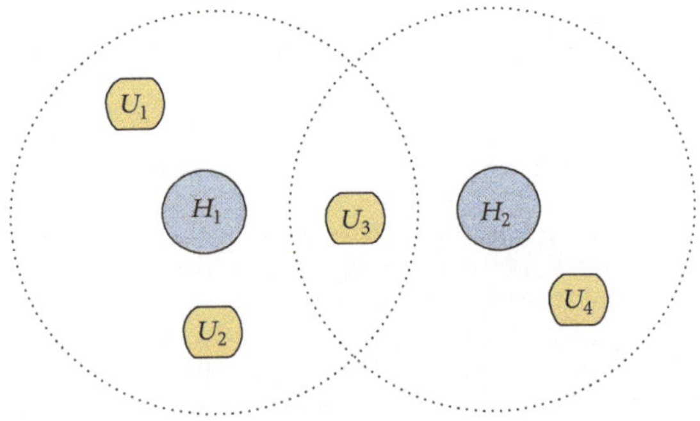

FIGURE 5: Example for caching when users have access to multiple helpers.

see that in the distributed caching problem, the individual objectives of different users may be in conflict, and we need sophisticated algorithms to find an optimum assignment.

Let us assume for the moment that (1) the network topology is known; (2) the long-term average link rates are known; (3) the user demand distribution (file popularity) is known. However, the actual demands are not known beforehand, so that caching placement must be done only based on the *statistics* of the user requests. Our goal is to minimize the average download time. We distinguish further between uncoded and coded caching. In the uncoded case, video-encoded files are cached directly (with the possibility of storing the same file in multiple locations). In the coded case, we consider placing coded chunks of the files on different helper stations, such that obtaining any sufficiently large number of these chunks allows reconstruction of the original video file (e.g., using the scheme in [37]).

In [9] we showed that the uncoded-placement problem is NP-complete. However, it can be formulated as the maximization of a monotone submodular function over matroid constraints, for which a simple greedy strategy achieves at least 1/2 of the optimum value. For the coded case, the optimum cache placement can be formulated as a convex optimization problem, for which optimum solutions can be found through efficient algorithms. In general, the optimum value of delay obtained with the coded optimization is better than the uncoded optimization because any placement matrix with integer entries is a feasible solution to the coded problem. In this sense, the coded optimization is a convex relaxation of the uncoded problem.

We conclude this section by mentioning that the conditions under which we derived the optimum caching are rarely fulfilled in practice. While the user demand distribution $P_r(f)$ may be well estimated and predicted, the network topology is typically time-varying with dynamics comparable or faster than the file transmission; therefore reconfiguring the caches at this time scale is definitely not practical. However, further computer experiments have also shown that the cache distribution obtained when the mobile stations are in "typical" distances from the helpers also provides good performance for various other realizations of random placement of nodes. Furthermore, distributed random caching turns out to be "good enough" as we shall see in Section 6. Hence, comparing optimal placement with random caching yields useful insight on the potential performance gap lost by a decentralized approach. Interestingly, in any reasonable network configuration it turns out that such a gap is very small.

4.2. File Placement for D2D Communications. Also for D2D communications, the question of which files should be cached by which user are essential. Building on the protocol model explained in Section 3.2, a critical question for each user is whether the file it is interested in can be found within the communication radius r from its current location. In other words, in order to enable D2D communication it is not sufficient that the distance between two users be less than r; users should also find their desired files in the cache of another device with which they can communicate. The decision of what to store can be taken in a centralized or distributed way, called deterministic and random

In deterministic caching a central control (typically the BS) orders the devices to cache specific files. Similar to the situation in femtocaching, we assume that the location of the caching nodes, and the demand distribution, is known. Finding the optimal deterministic file assignment for the general case follows the same principles as for femtocaching outlined above. A simplification occurs when the devices are grouped into *clusters* such that only communication within the cluster is possible (for more details see Section 6). In this case the deterministic caching algorithm is greatly simplified: the devices in the cluster should simply cache the most popular files in a disjoint manner; that is, no file should be cached twice in the cluster. Deterministic caching is only feasible, if the location of the nodes and the channel state information (CSI) are known a priori, and remains constant between the filling of the cache and the actual file transmission; thus it applies only if the caching nodes are fixed wireless devices. It is also useful for providing upper performance bounds for other caching strategies. In random caching, each device randomly and independently caches a set of files according to a common probability mass function. In our earlier papers, we assumed that the caching distribution is also a Zipf distribution, though with a parameter γ_c that is different from γ_r and which has to be optimized for a particular γ_r and r. Since the Zipf distribution is characterized by a single parameter, this description gives important intuitive insights about how concentrated the caching distribution should be.

In [14], we found that the optimal caching distribution P_c^* that maximizes the probability that any user finds its requested file inside its own cluster is given (for a node arrangement on a rectangular grid as described above) by

$$P_c^*(f) = \left[1 - \frac{\nu}{z_f}\right]^+, \quad f = 1, \ldots, m, \tag{2}$$

where $\nu = (m^* - 1)/\sum_{f=1}^{m^*}(1/z_f)$, $z_f = P_r(f)^{1/(M(g_c-1)-1)}$, $m^* = \Theta(\min\{(M/\gamma_r)g_c, m\})$, and $[\Lambda]^+ = \max[\Lambda, 0]$.

5. Adaptive Streaming from Helper Stations

We now turn to the delivery phase, in particular for the femtocaching (helper station). We furthermore concentrate on the case that the video files are *streamed*, that is, replay at the receiver starts before the complete file has been transmitted. Such streaming is widely used for standard video-on-demand systems, using protocols such as Microsoft Smooth Streaming (Silverlight), Apple HTTP Live Streaming, and 3GPP Dynamic Adaptive Streaming over HTTP (DASH). We have adapted such on-demand streaming to our caching architectures, in particular the network setup with helper stations. Dividing each video stream into chunks, we solve the problem of "which user should get a video 'chunk', at what quality, from which helper station."

5.1. Problem Formulation. We represent a video file as a sequence of chunks of equal duration. Each chunk may contain a different number of source-encoded bits, due to variable bit-rate (VBR) coding (see Section 2), and the same video file is encoded at different quality levels, such that lower quality levels correspond to fewer encoded bits. These quantities can vary across video files, and even for the same video they can vary across both chunks and quality levels. For example, the same compression level may produce a different user quality index as well as a different bit requirement from one chunk to the next, depending on if the video chunk is showing a constant blue sky or a busy city street.

In our system, the requested chunks are queued at the helpers, and each helper h maintains a queue Q_{hu} pointing at each of the users u in its vicinity. We pose the *network utility maximization* (NUM) problem of maximizing a concave and component-wise nondecreasing network utility function $\phi_u(\cdot)$ of the users' long-term average quality indices $\overline{D}_u$, subject to stability of the queues Q_{hu} at all the helpers. The concavity of the network utility function imposes some desired notion of fairness between the users. The problem formulation is given as follows:

$$\begin{aligned} &\text{maximize} \quad \sum_u \phi_u\left(\overline{D}_u\right) \\ &\text{subject to} \quad \overline{Q}_{hu} < \infty \quad \forall (h,u) . \end{aligned} \tag{3}$$

We solve this problem in [11] using the Lyapunov Drift Plus Penalty approach and obtain a policy that decomposes naturally into two distinct operations that can be implemented in a decentralized fashion: (1) congestion control; (2) transmission scheduling.

5.2. Congestion Control. Congestion control decisions are made at each streaming user, which decides from which helper to request the next chunk and at which quality index this shall be downloaded. For every time slot t, each $u \in \mathscr{U}$ chooses the helper in its neighborhood having the shortest queue; that is,

$$h_u^*(t) = \operatorname{argmin}\left\{Q_{hu}(t) : h \in \mathscr{N}(u) \cap \mathscr{H}(f_u)\right\}. \tag{4}$$

Then, it determines the quality level $m_u(t)$ of the requested chunk at time t as follows:

$$m_u(t) = \operatorname{argmin}\left\{Q_{h_u^*(t)u}(t)\, B_{f_u}(m,t) - \Theta_u(t)\, D_{f_u}(m,t)\right\}, \tag{5}$$

where $B_{f_u}(m,t)$ and $D_{f_u}(m,t)$ are the size in bits and the quality index (could be some subjective measure of video quality, for example, SSIM (structural similarity index)), respectively, of chunk t at quality level m. $\Theta_u(t)$ is a virtual queue introduced to solve the NUM problem. Notice that the streaming of the video file f_u may be handled by different helpers across the streaming session, but each individual chunk is entirely downloaded from a single helper. Notice also that in order to compute the above quantities, each user needs to know only *local information* formed by the queue backlog $Q_{hu}(t)$ and the locally computed virtual queue value $\Theta_u(t)$. This scheme is reminiscent of the current adaptive streaming technology for video on demand systems, referred to as DASH (Dynamic Adaptive Streaming over HTTP) [26, 38], where the client (user) progressively fetches a video file by downloading successive chunks and makes adaptive decisions on the quality level based on its current knowledge of the congestion of the underlying server-client connection. Our policy generalizes DASH by allowing the client to dynamically select the least backlogged server, for each chunk.

5.3. Transmission Scheduling. At time slot t, the general transmission scheduling consists of maximizing the weighted sum rate of the transmission rates achievable at scheduling slot t. Namely, the network of helpers must solve the max-weighted sum rate (MWSR) problem:

$$\begin{aligned} &\text{maximize} \quad \sum_{h\in\mathscr{H}} \sum_{u\in\mathscr{N}(h)} Q_{hu}(t)\, \mu_{hu}(t) \\ &\text{subject to} \quad \boldsymbol{\mu}(t) \in \mathscr{R}(t), \end{aligned} \tag{6}$$

where $\mathscr{R}(t)$ is the region of achievable rates supported by the network at time t and $\mu_{hu}(t)$ is the scheduled rate from helper h to user u in time slot t. We particularize the above general MWSR problem to a simple physical layer system.

Macrodiversity. In this physical layer system, referred to as "macrodiversity," the users can decode multiple data streams from multiple helpers if they are scheduled with nonzero rate on the same slot. In this case, the rate region $\mathscr{R}(t)$ is given by the Cartesian product of the following orthogonal access regions:

$$\sum_{u\in\mathscr{N}(h)} \frac{\mu_{hu}(t)}{C_{hu}(t)} \le 1, \quad \forall h \in \mathscr{H}, \tag{7}$$

where $C_{hu}(t)$ is the peak rate from helper h to user u in time slot t. In the macrodiversity system, the general MWSR problem (6) decomposes into individual problems, to be solved in a decentralized way at each helper node. The solution is given by each helper h independently choosing the user $u_h^*(t)$ given by

$$u_h^*(t) = \operatorname{argmax}\left\{Q_{hu}(t)\, C_{hu}(t) : u \in \mathscr{N}(h)\right\}, \tag{8}$$

with rate vector given by $\mu_{hu_h^*(t)}(t) = C_{hu_h^*(t)}(t)$ and $\mu_{hu}(t) = 0$ for all $u \neq u_h^*(t)$. Notice that here, unlike conventional cellular systems, we do not assign a fixed set of users to each helper. In contrast, the helper-user association is dynamic and results from the transmission scheduling decision. Notice also that despite the fact that each helper h is allowed to serve its queues with rates $\mu_{hu}(t)$ satisfying (7), the proposed policy allocates the whole tth downlink slot to a single user $u^* \in \mathcal{N}(h)$, served at its own peak-rate $C_{hu^*}(t)$.

5.4. Algorithm Performance. It can be shown that the time average utility achieved by the proposed policy comes within $O(1/V)$ of the utility of a genie-aided T-slot look ahead policy for any arbitrary sample path with a $O(V)$ tradeoff in time averaged backlog. Thus, the scheme provably achieves optimality of the network utility function under dynamic and arbitrarily changing network conditions; details of the proof can be found in [11].

5.5. Prebuffering and Rebuffering Chunks. The NUM problem formulation (3) does not take into account the possibility of stall events, that is, chunks that are not delivered within their playback deadline. This simplification has the advantage of yielding the simple and decentralized scheduling policy described in the previous sections. However, in order to make such a policy useful in practice we have to force the system to work in the *smooth streaming regime*, that is, in the regime where the stall events have small probability. This can be done by adaptively determining the prebuffering time T_u for each user u on the basis of an estimate of the largest delay of queues $\{Q_{hu} : h \in \mathcal{N}(u)\}$.

We define the size of the playback buffer Ψ_t as the number of playable chunks in the buffer not yet played. Without loss of generality, assume that the streaming session starts at $t = 1$. Then, Ψ_t is recursively given by the updating equation: ($1\{\mathcal{K}\}$ denotes the indicator function of a condition or event $\mathcal{K}$.)

$$\Psi_t = \max\{\Psi_{t-1} - 1\{t > T_u\}, 0\} + |a_t|, \quad (9)$$

where $|a_t|$ is the number of chunks that are completely downloaded in slot t. Let A_k denote the time slot in which chunk k arrives at the user and let W_k denote the delay with which chunk k is delivered. Note that the longest period during which Ψ_t is not incremented is given by the maximum delay to deliver chunks. Thus, each user u needs to adaptively estimate W_k in order to choose T_u. In the proposed method, at each time $t = 1, 2, \ldots$, user u calculates the maximum observed delay E_t in a sliding window of size Δ, by letting

$$E_t = \max\{W_k : t - \Delta + 1 \leq A_k \leq t\}. \quad (10)$$

Finally, user u starts its playback when Ψ_t crosses the level ξE_t, that is, $T_u = \min\{t : \Psi_t \geq \xi E_t\}$, where ξ is a tuning parameter. If a stall event occurs at time t, that is, $\Psi_t = 0$ for $t > T_u$, the algorithm enters a rebuffering phase in which the same algorithm presented above is employed again to determine the new instant $t + T_u + 1$ at which playback is restarted.

5.6. Extensions. In [12], we consider extensions and improvements of our work. In Sections 5.3 and 5.2, we treated the case of single-antenna base stations and, starting from a network utility maximization (NUM) formulation, we devised a "push" scheduling policy, where users place requests to sequential video chunks to possibly different base stations with adaptive video quality, and base stations schedule their downlink transmissions in order to stabilize their transmission queues. In [12], we consider a "pull" strategy, where every user maintains a request queue, such that users keep track of the video chunks that are effectively delivered. The pull scheme allows to download the chunks in the playback order without skipping or missing them. In addition, motivated by the recent/forthcoming progress in small cell networks (e.g., in wave-2 of the recent IEEE 802.11ac standard), we extend our dynamic streaming approach to the case of base stations capable of multiuser MIMO downlink, that is, serving multiple users on the same time-frequency slot by spatial multiplexing. By exploiting the "channel hardening" effect of high dimensional MIMO channels, we devise a low complexity user selection scheme to solve the underlying max-weighted rate scheduling (6), which can be easily implemented and runs independently at each base station.

5.7. Preliminary Implementation. As observed in Sections 5.3 and 5.2, users send their chunk requests to the helpers having the shortest queue pointing at them. Then, transmission scheduling decisions are made by each helper, which maximizes at each scheduling decision time its downlink weighted sum rate where the weights are provided by the queue lengths. The scheme can be implemented in a decentralized manner, as long as each user knows the lengths of the queues of its serving helpers, and each helper knows the individual downlink rate supported to each served user. Queue lengths and link rates represent rather standard protocol overhead information in any suitable wireless scheduling scheme. We have also implemented a version of such scheme on a testbed formed by Android smartphones and tablets, using standard WiFi MAC/PHY [10].

6. Performance of D2D Caching Networks

We now turn to D2D networks, that is, architectures where the devices themselves act as caches. In contrast to our analysis of femtocaching, we consider here only the download of video *files* (i.e., no streaming) and also neglect the issue of video rate adaptation (these are topics of ongoing research). In this section, we first outline the principle and intuitive insights. We then discuss the fundamental scaling laws, both for the sum throughput in the cell (disregarding any fairness considerations) and for the tradeoff between throughput and outage. Combining D2D transmission with coding and multicasting is also discussed.

6.1. Principle and Mathematical Model. As outlined in Section 3.2, we consider a network where each device can cache a fixed number M video files and send them, upon request, to other devices nearby. If a device cannot obtain a file through D2D communications, it can obtain it from a macrocellular base station (BS) through conventional cellular transmission.

Consider a setup in which clustering is used (see Section 3.2) and assume furthermore deterministic caching. The main performance factor that can be influenced by the system designer is the cluster size; this is regulated through the transmit power (we assume that it is the same for all users in a cell but can be optimized as a function of user density, library size, and size of the caches). Increasing cluster size increases the probability for finding the desired file in the cluster, while it decreases the (spatial) reuse of time-frequency transmission resources.

There are a number of different criteria for optimizing the system parameters. One obvious candidate is the total network throughput. It is maximized by maximizing the number of active clusters. In [39], we showed that, for deterministic caching, the expected throughput can be computed as

$$E\{T\} = \frac{1}{r^2} \sum_{k=0}^{n} \left(1 - \prod_{i=1}^{k} \left(1 - \left(P_{\text{CVC}}(k) - P_{\text{r}}(f_i)\right)\right) \right) \times \Pr[K = k], \tag{11}$$

where $P_{\text{CVC}}(k)$ is the probability that the requested file is in the Common Virtual Cache (the union of all caches in the cluster), that is, among the k most popular files. $\Pr[K = k]$, the probability that there are k users in a cluster, is deterministic for the rectangular grid arrangement, and

$$\Pr[K = k] = \binom{n}{k} \left(r^2\right)^k \left(1 - r^2\right)^{n-k}, \tag{12}$$

for random node placement.

6.2. Theoretical Scaling Laws Analysis. We now turn to scaling laws, that is, determine how the capacity scales up as more and more users are introduced into the network. We are dealing with "dense" networks, such that the user density increases, while the area covered by a cell remains the same. As mentioned in Section 4.2, for the achievable caching scheme, we consider a simple "decentralized" random caching strategy, where each user caches M files chosen independently on the library $\mathcal{F}$ with probability $P_c^*(f)$ given by (2).

We furthermore deal again with the "clustered" case; that is, the network is divided into clusters of equal size $g_c(m)$. A system admission control scheme decides whether to serve potential links or ignore them. The served potential links in the same cluster are scheduled with equal probability (or, equivalently, in round robin), such that all admitted user requests have the same average throughput $\mathbb{E}[T_u] = \overline{T}_{\min}$ (see [14] for formal definitions), for all users u, where expectation is with respect to the random user requests, random caching, and the link scheduling policy (which may be randomized or deterministic, as a special case). To avoid interference between clusters, we use a time-frequency reuse scheme [40, Ch. 17] with parameter K as shown in Figure 4(b). In particular, we can pick $K = (\lceil \sqrt{2}(1+\Delta) \rceil + 1)^2$, where Δ is the interference parameter defined in the protocol model.

In [8] we established lower and upper bounds for the throughput of D2D communications (this was done under the assumption of random node distribution and caching according to a Zipf distribution). The main conclusion from the scaling law is that for highly concentrated demand distribution, $\gamma_r > 1$, the throughput scales linearly with the number of users, or equivalently the per-user throughput remains constant as the user density increases; the number of users in a cluster also stays constant. For heavy-tailed demand distributions, the throughput of the system increases only sublinearly, as the clusters have to become larger (in terms of number of nodes in the cluster), to be able to find requested files within the caches of the cluster members.

In [14] we tightened the bounds and extended them to the case of throughput-outage tradeoff. Qualitatively (for formal definition see [14]), we say that a user is in outage if the user cannot be served in the D2D network. This can be caused by the fact that (i) the file requested by the user cannot be found in the user's own cluster and (ii) that the system admission control decides to ignore the request. We define the outage probability p_o as the average fraction of users in outage. At this point, we can define the throughput-outage tradeoff as follows.

Definition 1 (throughput-outage tradeoff). For a given network and request probability mass function $\{P_r(f) : f \in \mathcal{F}\}$, an outage-throughput pair (p, t) is *achievable* if there exists a cache placement scheme and an admission control and transmission scheduling policy with outage probability $p_o \le p$ and minimum per-user average throughput $\overline{T}_{\min} \ge t$. The outage-throughput achievable region $\mathcal{T}(P_r, n, m)$ is the closure of all achievable outage-throughput pairs (p, t). In particular, we let $T^*(p) = \sup\{t : (p, t) \in \mathcal{T}(P_r, n, m)\}$.

Notice that $T^*(p)$ is the result of the optimization problem:

$$\begin{aligned} &\text{maximize} \quad \overline{T}_{\min} \\ &\text{subject to} \quad p_o \le p, \end{aligned} \tag{13}$$

where the maximization is with respect to the cache placement and transmission policies. Hence, it is immediate to see that $T^*(p)$ is nondecreasing in p.

The following results are proved in [14] and yield scaling law of the optimal throughput-outage tradeoff under the clustering transmission scheme defined above.

Although the results of [14] are more general, here we focus on the most relevant regime of the scaling of the file library size with the number of users, referred to as "small library size" in [14]. Namely, we assume that $\lim_{n\to\infty}(m^{\alpha}/n) = 0$, where $\alpha = (1-\gamma_r)/(2-\gamma_r)$. Since $\gamma_r \in (0, 1)$, we have $\alpha < 1/2$. This means that the library size m can grow even faster than quadratically with the number of users n. In practice, however, the most interesting case is where m is sublinear with respect to n (see [14] for justifications.).

Remarkably, any scaling of m versus n slower than $n^{1/\alpha}$ is captured by the following result.

Theorem 2. *Assume* $\lim_{n\to\infty}(m^{\alpha}/n) = 0$. *Then, the throughput-outage tradeoff achievable by one-hop D2D network with random caching and clustering behaves as follows:*

$$T^{*}(p) \geq \begin{cases} \dfrac{C_r}{K}\dfrac{M}{\rho_1 m} + \delta_1(m), \\ \quad p = (1-\gamma_r)\, e^{\gamma_r - \rho_1}, \\ \dfrac{C_r A}{K}\dfrac{M}{m(1-p)^{1/(1-\gamma_r)}} + \delta_2(m), \\ \quad p = 1 - \gamma_r^{\gamma_r}\left(\dfrac{Mg_c}{m}\right)^{1-\gamma_r}, \\ \dfrac{C_r B}{K} m^{-\alpha} + \delta_3(m), \\ \quad 1 - \gamma_r^{\gamma_r} M^{1-\gamma_r} \rho_2^{1-\gamma_r} m^{-\alpha} \\ \quad\quad \leq p \leq 1 - a(\gamma_r)\, m^{-\alpha}, \\ \dfrac{C_r D}{K} m^{-\alpha} + \delta_4(m), \\ \quad p \geq 1 - a(\gamma_r)\, m^{-\alpha}, \end{cases} \tag{14}$$

where $a(\gamma_r)$, A, B, *and* D *are constants depending on* γ_r *and* M, *which can be found in [14] and where* ρ_1 *and* ρ_2 *are positive parameters satisfying* $\rho_1 \geq \gamma_r$ *and* $\rho_2 \geq ((1-\gamma_r)/\gamma_r^{\gamma_r} M^{1-\gamma_r})^{1/(2-\gamma_r)}$. *The cluster size* g_c *is any function of* m *satisfying* $g_c = \omega(m^{\alpha})$ *and* $g_c \leq \gamma_r m/M$. *The functions* $\delta_i(m)$, $i = 1, 2, 3, 4$ *are vanishing for* $m \to \infty$ *with the following orders* $\delta_1(m) = o(M/m)$, $\delta_2(m) = o(M/m(1-p)^{1/(1-\gamma_r)})$, *and* $\delta_3(m)$, $\delta_4(m) = o(m^{-\alpha})$.

The dominant term in (14) can accurately capture the system performance even in the finite-dimensional case shown by simulations in Figure 6. Further, also in [14], we can show that the achievable throughput-outage tradeoff given by (14) is order optimal. When $Mn \geq m$ (the whole library can be cached in the network), for arbitrarily small outage probability, by using (14), the per-user throughput scales as $T^{*}(p) = \Theta(M/m)$. This means that the per-user throughput is independent of the number of users (or in other words, the network throughput increases linearly with the number of users, as already indicated above). Furthermore, the throughput grows linearly with M. This can be very attractive since, for example, in order to double the throughput, instead of increasing the bandwidth or power, we can just double the (cheap) storage capacity per user.

Interestingly, our result shown by (14) coincides with the achievable throughput by using the subpacketized caching and coded multicasting algorithms in [13, 41]. However, in realistic channel assumptions, the result is shown in Section 6.4.

6.3. Coded Caching and Multicasting. From the previous analysis of the D2D caching network, one important property

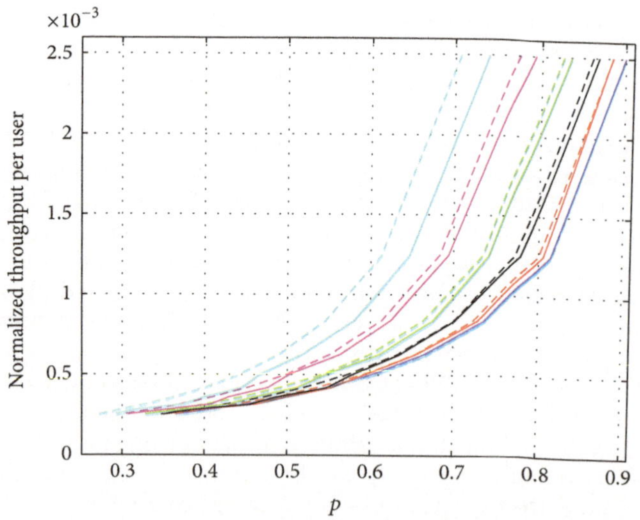

FIGURE 6: Comparison between the normalized theoretical result (solid lines) and normalized simulated result (dashed lines) in terms of the minimum throughput per user versus outage probability. The throughput is normalized by C_r, so that it is independent of the link rate. We assume $m = 1000$, $n = 10000$, $M = 1$, and reuse factor $K = 4$. The parameter γ_r for the Zipf distribution varies from 0.1 to 0.6, which are shown from the right (blue) to the left (cyan). The theoretical curves show the plots of the dominating term in (14) divided by C_r [14].

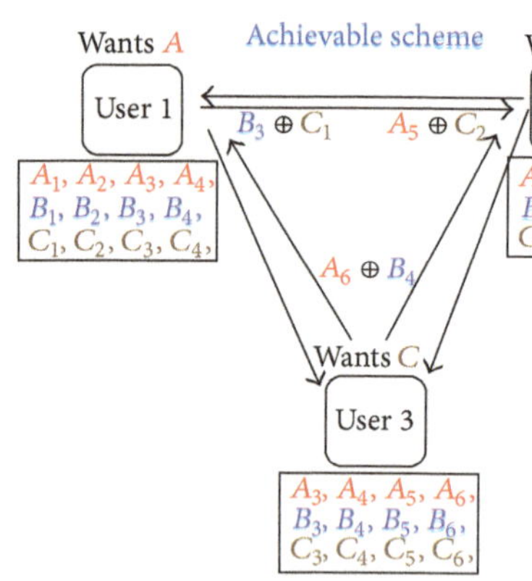

FIGURE 7: Illustration of the example of 3 users, 3 files, and $M = 2$, achieving $1/2$ transmissions in terms of file. We divide each file into 6 packets (e.g., A is divided into $A_1, \ldots, A_6$.) User 1 requests A; user 2 requests B; and user 3 requests C. The cached packets are shown in the rectangles under each user. For the delivery phase, user 1 transmits $B_3 \oplus C_1$; user 2 transmits $A_5 \oplus C_2$; and user 3 transmits $A_6 \oplus B_4$. The normalized number of transmissions is $3 \cdot (1/6) = 1/2$, which is also information theoretically optimal for this network [13].

of the proposed scheme is that in both the caching phase and the delivery phase, an uncoded approach is applied. The gain of the throughput is mainly obtained by spatial reuse (TDMA). At this point, a natural question to ask is whether coded multicasting for D2D transmissions can provide an additional gain or whether the coding gain and the spatial reuse gain can accumulate. In [13], we designed a subpacketized caching and a network-coded delivery scheme for the D2D caching networks. The schemes are best to be explained by the example shown in Figure 7, where we assume no spatial reuse can be used, or only one transmission per time-frequency slot is allowed but the transmission range can cover the whole network. This scheme can be generalized to any n, m, M. Without using spatial reuse, for zero outage, the achievable normalized number of transmissions such that every user can successfully decode is $(m/M)(1 - M/m)$ (We normalize the number of transmissions by the file size, which is assumed to be the same for all the files). Which is surprisingly almost the same as the result shown in [41], where instead of D2D communications, one central server (base station) which has access to all the files multicasts coded packets. In addition, it also has the same scaling law as the throughput by using our previously proposed decentralized caching and uncoded delivery scheme. (Notice that the reciprocal of the number of transmissions is proportional to the throughput under our protocol model assumption.) Moreover, it can be shown that there is no further gain when spatial reuse is also exploited. In other words, the gains of spatial reuse and coding cannot accumulate. Intuitively, if spatial reuse is not allowed, a complicated caching scheme can be designed such that one transmission can be useful for as many users as possible. While if we reduce transmission range and perform our scheme in one cluster as shown in Figure 4(b), then the number of users benefitted by one transmission is reduced but the D2D transmissions can operate simultaneously at a higher rate. Moreover, the complexity of caching subpacketization and coding can also be reduced. Hence, the benefit of coding depends on the actual physical layer throughput (bits/s/Hz) and the caching/coding complexity rather than throughput scaling laws.

6.4. Simulation Results. To see the difference between the performance of the proposed D2D caching network and the state-of-the-art schemes for video streaming, we need to consider the realistic propagation and interference channel model instead of the protocol model. One reason is that as mentioned in Section 6.2, for small outage probability, the throughput of the proposed D2D scheme has the same scaling laws as the coded multicasting scheme in [41]. The state-of-the-art schemes that will be compared with are conventional unicasting, harmonic broadcasting, and coded multicasting, whose details can be found in [15]. In the following, for practical considerations, the proposed uncoded D2D scheme discussed in Section 6.2 is used for simulations.

For simulations, we considered a network of size 600 m × 600 m, where we relax the grid structure of the users' distribution and let $n = 10000$ users distributed uniformly. The file library has size $m = 300$ (e.g., 300 popular movies and TV shows to be refreshed on a daily basis at off-peak times by the cellular network). The storage capacity in each user is $M = 20$ and the parameter for the Zipf distribution is $\gamma_r = 0.4$ [35]. We considered a regular pattern of buildings of size 50 m × 50 m, separated by streets of widths 10 m [15], with indoor, outdoor, indoor-to-outdoor, and outdoor-to-indoor pathloss and shadowing models taken from [42], assuming that D2D links operate at 2.4 GHz (WiFi Direct). We assumed a channel bandwidth of 20 MHz in

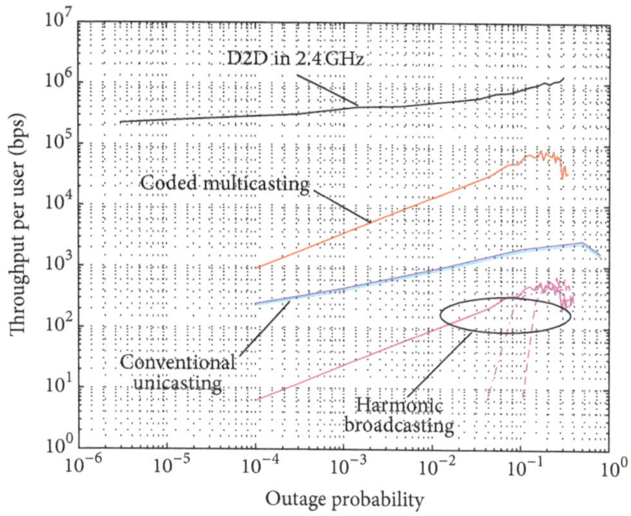

FIGURE 8: Simulation results for the throughput-outage tradeoff for different schemes under the realistic indoor/outdoor propagation environment (for details, see [15]). For harmonic broadcasting with only the m' most popular files, solid line: $m' = 300$; dash-dot line: $m' = 280$; dash line: $m' = 250$. We have $n = 10000$, $m = 300$, $M = 20$, and $\gamma_r = 0.4$. From [15].

order to provide throughput in bit/s. All the details of the simulation parameters, including the pathloss and shadowing models, can be found in [15]. The simulation results of the throughput-outage tradeoff for different schemes are given in Figure 8. We observe that in this realistic propagation scenario the D2D single-hop caching network can provide both large throughput, sufficient for streaming video at standard definition quality, and low outage probability. Also, the D2D caching scheme significantly outperforms the other schemes in the regime of low outage probability. This performance gain is particularly impressive with respect to conventional unicasting and harmonic broadcasting from the base station, which are representative of the current technology. We also note the distinct performance advantages compared to coded multicasting, despite the fact that the two schemes have the same scaling laws. The main reason for this development is that the capacity of multicasting is limited by the "weakest link" between BS and the various MSs, while, for the D2D transmission scheme, short distance transmission (which usually has high SNR, shallow fading, and thus high capacity) determines the overall performance.

It is also worthwhile to notice that the scheduling scheme used in the simulations is based on the clustering structure and the interference avoidance (TDMA) discussed in Section 6.2 without using any advanced interference management scheme such as FlashLinQ [43] and ITLinQ [44], which may provide an even higher gain in terms of throughput for the D2D caching networks.

7. Conclusions

As user demand for video data continues to increase sharply in cellular networks, new approaches are needed to dramatically expand network capacity. This paper has provided an overview of an approach explored by the University of Southern California as part of the industry-sponsored research program, video aware wireless networks (VAWN). The approach exploits a key feature of wireless video, namely, the high degree of (asynchronous) content reuse across users. To exploit this feature, we propose replacing expensive backhaul infrastructure with inexpensive caching capabilities. This can be realized in two ways: the use of femtocaching or dedicated helper nodes that cache popular files and serve nearby user requests, and the use of user devices themselves to cache and exchange files using device-to-device (D2D) communications. Simulations with realistic settings show that even for relatively low-density deployment of helper stations, throughput can be increased by a factor of five. D2D networks allow in many situations a throughput increase that is linear with the number of users (thus making the per-user throughput independent of the number of users). Simulations in realistic propagation channels, storage capacity settings, video popularity distributions, and user densities show that (for constant outage) the throughput can be two orders of magnitude or more higher than the state-of-the-art multicast systems.

A key issue in our caching approach is that of file placement. In the helper node approach, we show that the problem of minimizing average file downloading time in the uncoded-placement case (video-encoded files are cached directly on help nodes) is NP-complete but can be reformulated and is solvable as a monotone submodular function over matroid constraints. For the coded case (coded chunks of files are placed on different helper stations), optimum cache placement can be formulated and is solvable as a convex optimization problem. Also for the D2D approach, the question of which files to cache is key. Two approaches are deterministic caching in which a BS instructs devices which files to cache (i.e., the most popular and in a disjoint manner) and random caching in which each device randomly caches a set of files according to a probability mass function. It is remarkable that the simple random caching is not only optimum from a scaling law point of view, but also in numerical simulations provides throughputs that are close to the deterministic caching (which is ideal but difficult to realize for time-varying topologies).

An important area of future work is that of predicting user requests. The effectiveness of caching schemes depends not only on the degree of content reuse, but also on our ability to understand and predict request behavior across clusters of users. Furthermore, the approach is predicated on a "time-scale decomposition," namely, that request distributions change much more slowly (over days or weeks) than the time it takes to stream a video (minutes to a couple of hours). For femtocaching, it is noteworthy that the type of users (and thus the requests) within range of a helper station might change over the course of a day; more research on how such spatiotemporal aspects can be predicted and accommodated is required. Similarly, the impact of social networks on user preferences could be exploited. It is noteworthy that not exploiting the space-time correlation of the demands yields only loss of *potential further performance gains* over those already demonstrated here. In short, if we have a correlated demand process with Zipf first-order statistics, we could further gain by taking into account the correlation structure.

In the D2D sphere, research on new approaches for incentivizing users to participate in cooperative caching schemes is needed. Both femtocaching and D2D caching schemes would benefit from research into multihop cache retrieval schemes and PHY schemes that better exploit advances in wireless communication technology (e.g., multiuser MIMO). In the D2D area, we are/will be investigating how to optimize neighbor discovery, estimating channel conditions and then using the information to make scheduling optimizations, and transmission schemes closely tuned to existing communications standards like WiFi Direct.

Conflict of Interests

The authors declare that there is no conflict of interests regarding the publication of this paper.

References

[1] "Cisco visual networking index: global mobile data traffic forecast update, 2012–2017," http://www.cisco.com/c/en/us/solutions/collateral/service-provider/visual-networking-index-vni/VNI-Forecast_QA.html.

[2] "Conviva viewer experience report," http://www.conviva.com/vxr/.

[3] F. Dobrian, V. Sekar, A. Awan et al., "Understanding the impact of video quality on user engagement," in *Proceedings of the ACM SIGCOMM Conference (SIGCOMM '11)*, pp. 362–373, August 2011.

[4] M. Z. Shafiq, J. Erman, L. Ji, A. X. Liu, J. Pang, and J. Wang, "Understanding the impact of network dynamics on mobile video user engagement," *SIGMETRICS Performance Evaluation Review*, vol. 42, no. 1, pp. 367–379, 2014.

[5] https://software.intel.com/en-us/articles/video-aware-wireless-networks.

[6] N. Golrezaei, A. G. Dimakis, A. F. Molisch, and G. Caire, "Wireless video content delivery through distributed caching and peer-to-peer gossiping," in *Proceedings of the 45th Asilomar Conference on Signals, Systems and Computers (ASILOMAR '11)*, pp. 1177–1180, Pacific Grove, Calif, USA, November 2011.

[7] N. Golrezaei, A. Molisch, A. G. Dimakis, and G. Caire, "Femtocaching and device-to-device collaboration: a new architecture for wireless video distribution," *IEEE Communications Magazine*, vol. 51, no. 4, pp. 142–149, 2013.

[8] N. Golrezaei, A. G. Dimakis, and A. F. Molisch, "Scaling behavior for device-to-device communications with distributed caching," *IEEE Transactions on Information Theory*, vol. 60, no. 7, pp. 4286–4298, 2014.

[9] K. Shanmugam, N. Golrezaei, A. G. Dimakis, A. F. Molisch, and G. Caire, "FemtoCaching: wireless content delivery through distributed caching helpers," *IEEE Transactions on Information Theory*, vol. 59, no. 12, pp. 8402–8413, 2013.

[10] J. Kim, F. Meng, P. Chen et al., "Adaptive video streaming for device-to-device mobile platforms," in *Proceedings of the 19th Annual International Conference on Mobile Computing & Networking (MobiCom '13)*, pp. 127–129, October 2013.

[11] D. Bethanabhotla, G. Caire, and M. J. Neely, "Adaptive video streaming for wireless networks with multiple users and helpers," http://arxiv.org/abs/1304.8083.

[12] D. Bethanabhotla, G. Caire, and M. J. Neely, "Adaptive video streaming in MU-MIMO networks," Submitted to ISIT 2014, http://arxiv.org/abs/1401.6476.

[13] M. Ji, G. Caire, and A. F. Molisch, "Fundamental limits of distributed caching in D2D wireless networks," in *Proceedings of the IEEE Information Theory Workshop (ITW '13)*, September 2013.

[14] M. Ji, G. Caire, and A. F. Molisch, "The throughput-outage tradeoff of wireless one-hop caching networks," http://arxiv.org/abs/1312.2637.

[15] M. Ji, G. Caire, and A. F. Molisch, "Wireless device-to-device caching networks: basic principles and system performance," http://arxiv.org/abs/1305.5216.

[16] J. Foerster and M. Gong, "Industrial column: special Issue on Video Aware Wireless Networks," invited special issue for IEEE Multimedia Communications Technical Committee (MMTC) E-Letters, September 2013.

[17] L. K. Choi, Y. Liao, and A. C. Bovik, "Video QoE models for the compute continuum," Invited Paper for IEEE Multimedia Communications Technical Committee (MMTC) E-Letters, 2013.

[18] Y. Liao, A. Younkin, J. Foerster, and P. Corriveau, "Achieving high QoE across the compute continuum: how compression, content, and devices interact," in *Proceedings of the 7th International Workshop on Video Processing and Quality Metrics for Consumer Electronics*, Scottsdale, Ariz, USA, 2013.

[19] M. A. Saad, A. C. Bovik, and C. Charrier, "Blind prediction of natural video quality," *IEEE Transactions on Image Processing*, vol. 23, no. 3, pp. 1352–1365, 2014.

[20] ISO, "Information technology—dynamic adaptive streaming over http (dash)—part 1: media presentation description and segment formats," ISO/IEC 23009-1:2014, 2014.

[21] "Transparent end-to-end packet switched streaming service (pss); progressive download and dynamic adaptive streaming over http (3gp-dash), version 12.3.0," Tech. Rep. G. T. 26.247, 2014.

[22] ISO, "Information technology—dynamic adaptive streaming over http (dash)—part 2: conformance and reference software," ISO/IEC ISO/IEC 23009-2:2014, 2014.

[23] ISO/IEC DTR 23009-3, "Information technology—dynamic adaptive streaming over HTTP (DASH)—part 3: implementation guidelines".

[24] "Information technology—dynamic adaptive streaming over http (dash)—part 4: segment encryption and authentication," Tech. Rep. ISO/IEC 23009-4, 2013.

[25] I. Sodagar, "The MPEG-dash standard for multimedia streaming over the internet," *IEEE Multimedia*, vol. 18, no. 4, pp. 62–67, 2011.

[26] T. Stockhammer, "Dynamic adaptive streaming over HTTP: standards and design principles," in *Proceedings of the 2nd Annual ACM Conference on Multimedia Systems (MMSys '11)*, pp. 133–144, ACM, 2011.

[27] O. Oyman and S. Singh, "Quality of experience for HTTP adaptive streaming services," *IEEE Communications Magazine*, vol. 50, no. 4, pp. 20–27, 2012.

[28] V. Ramamurthi and O. Oyman, "Video-QoE aware radio resource allocation for HTTP adaptive streaming," in *Proceedings of the IEEE ICC 2014-Communication QoS, Reliability and Modeling Symposium*, 2014.

[29] O. Oyman, U. Kumar, V. Ramamurthi, M. Rehan, and M. Morsi, "Dynamic adaptive streaming over http: standards and technology," Invited Paper for IEEE Multimedia Communications Technical Committee (MMTC) E-Letters, 2013.

[30] O. Oyman, "Dynamic adaptive streaming over http (dash) standardization at mpeg and 3gpp," in *Proceedings of the IEEE Multimedia Communications Technical Committee (MMTC '13)*, E-Letters, 2013.

[31] "Optimizing dash delivery over wireless networks," in *Proceedings of the IEEE Multimedia Communications Technical Committee (MMTC '13)*, E-Letters, March 2013.

[32] M. Cha, H. Kwak, P. Rodriguez, Y.-Y. Ahnt, and S. Moon, "I tube, you tube, everybody tubes: analyzing the world's largest user generated content video system," in *Proceedings of the 7th ACM SIGCOMM Internet Measurement Conference (IMC '07)*, pp. 1–14, October 2007.

[33] http://traces.cs.umass.edu/index.php/network/network.

[34] M. Hefeeda and O. Saleh, "Traffic modeling and proportional partial caching for peer-to-peer systems," *IEEE/ACM Transactions on Networking*, vol. 16, no. 6, pp. 1447–1460, 2008.

[35] L. Breslau, P. Cao, L. Fan, G. Phillips, and S. Shenker, "Web caching and Zipf-like distributions: evidence and implications," in *Proceedings of the 18th Annual Joint Conference of the IEEE Computer and Communications Societie (INFOCOM '99)*, vol. 1, pp. 126–134, March 1999.

[36] P. Gupta and P. R. Kumar, "The capacity of wireless networks," *IEEE Transactions on Information Theory*, vol. 46, no. 2, pp. 388–404, 2000.

[37] S. Pawar, N. Noorshams, S. El Rouayheb, and K. Ramchandran, "DRESS codes for the storage cloud: Simple randomized constructions," in *Proceedings of the IEEE International Symposium on Information Theory Proceedings (ISIT '11)*, pp. 2338–2342, IEEE, August 2011.

[38] A. Zambelli, *IIS Smooth Streaming Technical Overview*, Microsoft Corporation, 2009.

[39] N. Golrezaei, A. F. Molisch, and A. G. Dimakis, "Base-station assisted device-to-device communications for high-throughput wireless video networks," in *Proceedings of the IEEE International Conference on Communications (ICC '12)*, pp. 7077–7081, June 2012.

[40] A. Molisch, *Wireless Communications*, John Wiley & Sons, 2011.

[41] M. Maddah-Ali and U. Niesen, "Fundamental limits of caching," http://arxiv.org/abs/1209.5807.

[42] WINNER-II, "D1. 1.2, WINNER II channel models," 2007.

[43] X. Wu, S. Tavildar, S. Shakkottai, and ETAL, "FlashLinQ: A synchronous distributed scheduler for peer-to-peer ad hoc networks," in *Proceedings of the 48th Annual Allerton Conference on Communication, Control, and Computing (Allerton '10)*, pp. 514–521, October 2010.

[44] N. Naderializadeh and A. S. Avestimehr, "ITLinQ: a new approach for spectrum sharing in device-to-device communication systems," *IEEE Journal on Selected Areas in Communications*, vol. 32, no. 6, pp. 1139–1151, 2014.

17

Development of Power Electronics Based Test Platform for Characterization and Testing of Magnetocaloric Materials

Deepak Elamalayil Soman, Jelena Loncarski, Lisa Gerdin, Petter Eklund, Sandra Eriksson, and Mats Leijon

Department of Engineering Sciences, Ångström Laboratory, Uppsala University, Lägerhyddsvägen 1, P.O. Box 534, 751 21 Uppsala, Sweden

Correspondence should be addressed to Deepak Elamalayil Soman; deepak.soman@angstrom.uu.se

Academic Editor: Gorazd Stumberger

Magnetocaloric effects of various materials are getting more and more interesting for the future, as they can significantly contribute towards improving the efficiency of many energy intensive applications such as refrigeration, heating, and air conditioning. Accurate characterization of magnetocaloric effects, exhibited by various materials, is an important process for further studies and development of the suitable magnetocaloric heating and cooling solutions. The conventional test facilities have plenty of limitations, as they focus only on the thermodynamic side and use magnetic machines with moving bed of magnetocaloric material or magnet. In this work an entirely new approach for characterization of the magnetocaloric materials is presented, with the main focus on a flexible and efficient power electronic based excitation and a completely static test platform. It can generate a periodically varying magnetic field using superposition of an ac and a dc magnetic field. The scale down prototype uses a customized single phase H-bridge inverter with essential protections and an electromagnet load as actuator. The preliminary simulation and experimental results show good agreement and support the usage of the power electronic test platform for characterizing magnetocaloric materials.

1. Introduction

In 1976 it was proved by Brown that the magnetocaloric effect could be used for substantial cooling effect near room temperature [1]. Magnetic refrigeration has two main advantages: it can reach higher efficiency compared to conventional techniques and there is no need for refrigerant gases [2]. The efficiency of magnetic refrigeration can reach 30%–60% of the Carnot cycle, whereas vapor compression refrigeration reaches only 5%–10% of the Carnot cycle [3].

When a magnetocaloric material is subjected to an external magnetic field the magnetic dipoles of the atoms align, causing a reduction in the spin-entropy. If this is done in an adiabatic manner, the entropy of the lattice must increase with an increase in the material's temperature. In a cooling cycle this heat is removed. When the magnetic field is removed, thermal entropy transforms into magnetic entropy, which results in cooling of the material. Hence heat is transferred from the surroundings to the material. This phenomenon can be used in the construction of refrigerators [4, 5] and the materials usually considered for such applications are referred to as magnetocaloric materials.

The most important efforts made in experimental work during the last decades are summarized in a few review papers [1, 3, 6, 7]. The magnetocaloric effect is determined by the entropy change, ΔS, if the magnetic field is changed isothermally, and by the adiabatic temperature change, ΔT_{ad}, if the magnetic field is changed adiabatically. These two parameters are not directly coupled, which is why they have to be optimized simultaneously. When investigating magnetic refrigeration a suitable value for the magnetic flux density is around 1 T [2].

There are two generations of magnetic refrigerators and heat pumps according to Yu et al. The first generation's machines use superconducting magnets and operate at a low frequency. The breaking point between the two generations came with a machine using permanent magnets and a rotating disc of magnetocaloric materials, which was constructed by Astronautics Corporation of America [3]. All magnetic machines in both first and second generation are devices containing either moving beds of magnetocaloric material, which enter and exit a static magnetic field, or magnets moving over a static bed of magnetocaloric material [3, 6]. According to Yu et al. periodic movements, linear or rotational, are required in order to obtain continuous powers of a magnetic refrigerator or heat pump. All researchers inventing these magnetic machines have faced more or less the same problems regarding design and operation of the machines [3].

This work shows the preliminary and entirely new approach of combining the rather conventional power electronics with essential control and the novel idea to employ a specific magnetic circuit for the testing of the magnetocaloric materials. In particular, the generation of the periodically varying magnetic flux density is obtained using superposition of an ac magnetic field and a dc magnetic field. The focus of most previous research projects within magnetic refrigeration has been on the thermodynamic part, whereas, in this paper, focus lies on the possibilities of using power electronics in order to obtain a varying magnetic flux density. To obtain a varying magnetic flux density a simple conventional power electronics setup with additional protection has been employed to control a very unusual and purely inductive load. The main benefit of using power electronics is that both magnet and magnetocaloric material can be stationary. Furthermore, density and frequency of the magnetic field can easily be changed by the control algorithm.

Preliminary measurements and simulations of a scaled down prototype of magnetocaloric setup, which can generate a biased alternating magnetic field for the purpose of testing of magnetocaloric materials in terms of testing the magnetocaloric properties on a small sample material, have been the main focus of this paper. The ac magnetic field is generated by an electromagnet, which is powered by an external dc-source connected to a single phase H-bridge inverter. The dc magnetic field is generated by a permanent magnet, connected to the electromagnet in a magnetic circuit according to Figure 1. Hence, the magnetic circuit consists of an E-formation iron core with a coil and a permanent magnet placed on the outer branches and an 8 mm air gap on the middle branch. The ac and dc magnetic field are superposed, creating a biased magnetic field in the air gap.

A review and comparison of the efficiency of different magnetic devices states that use of the permanent magnets is more efficient compared to the use of electromagnets or superconducting electromagnets [7]. Since no device combining a permanent magnet and an electromagnet has been constructed before [3, 7], it is not possible to say how efficient such a device would be in relation to existing devices. An electromagnet requires power to produce a magnetic field, which is why a permanent magnet is preferable [7]. However, the device investigated in this paper does not need a motor in order to magnetize and demagnetize the magnetocaloric

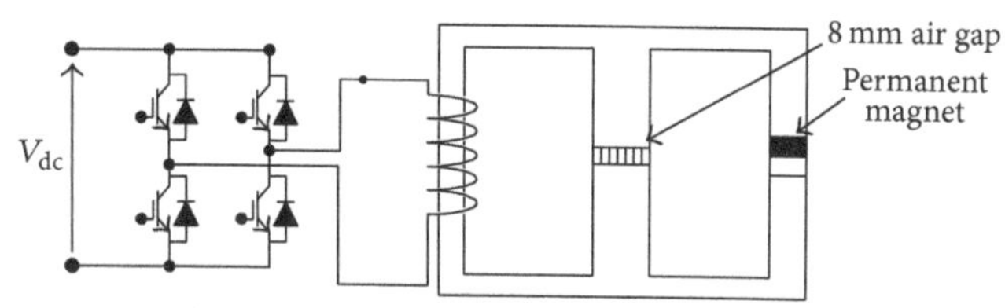

FIGURE 1: Simple representation of the test platform.

material, which might open up for new ways of designing efficient magnetic refrigerators or heat pumps.

2. Power Electronics and Protection

Experimental setups used for the study of magnetocaloric materials are mostly based on rotating disc type mechanisms, where the induced flux density frequency depends on the speed of the rotating disc. These arrangements need complex control mechanisms to obtain the desired speed simultaneously as the desired flux. A power converter based flux induction method is more compact and is able to provide easier control for magnetization and demagnetization of the material.

In this section, the power converter system used for the magnetocaloric experiment is described. A voltage source inverter (VSI) is used for powering up the electromagnetic coil. The VSI converts dc power to corresponding ac at the desired frequency.

Due to the requirements of the experimental setup the inverter has to be able to operate in a low frequency area (around 5 Hz) and manage a purely inductive load. The frequency of the magnetic field should be variable in order to enable investigation of how magnetocaloric materials react at different time scales; hence the operating frequency of the inverter has to be variable as well.

The power converter has to be designed to satisfy all the requirements of the electromagnet and at the same time it should be able to cope with the highly inductive load stresses. As the operating frequency can vary from very low to high during testing all the protections should work during the entire range of operation. The custom made inverter thus has four main stages: accurate gate drive circuit with optical isolation, insulated-gate bipolar transistor (IGBT) based H-bridge power stage, load current limit circuit, and Labview CompactRIO based field programmable gate array (FPGA) digital controller.

2.1. IGBT Gate Drive Circuit. The gate drive circuit of the inverter has been customized to satisfy all the operating requirements. Primary protection features are also included in this part. The complete circuit representation of the gate driver is shown in Figure 2. Optically isolated gate driver IC HCPL-316J from Avago is selected as the core component for the circuit [8]. Soft IGBT turn off, Desat protection, and under voltage lock-out (UVLO) protection features are ensured by this IC. Moreover the optical isolation of PWM pulses and fault feedback signal are incorporated in the chip.

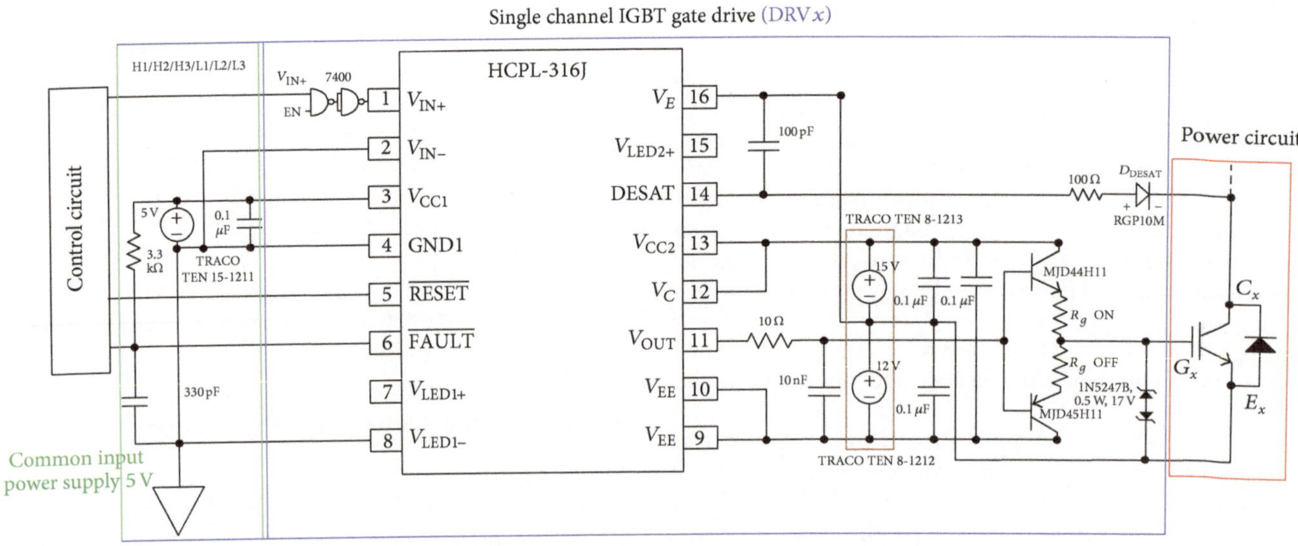

FIGURE 2: IGBT gate driver circuit using HCPL-316J.

The higher gate current and negative drive requirements of the IGBTs are addressed by designing a totem pole arrangement using matched NPN-PNP transistors as shown in Figure 2. The figure shows matched NPN and PNP transistors to provide push pull IGBT gate drive current amplified, to satisfy the gate drive requirements. This ensures fast turning ON and turning OFF of the IGBT. While boosting the output currents from IC Drivers, shoot through phenomenon can be avoided in this case. The pair of transistors protects each other's base-emitter junctions and handles current surges quite well. Complete isolation of the digital and power stages is ensured using this driver circuit to facilitate a fully safe operation of the digital controller even during the high voltage stresses in the output stage of the inverter.

2.2. Single Phase H-Bridge Power Stage. A single phase H-bridge inverter power stage is constructed to drive the electromagnet with sufficient amount of current as shown in Figure 3. The purely inductive nature of the load is not common for typical inverter applications and it can produce large voltage stress on the IGBT switches [9]. To avoid any type of overrated voltage spikes on the inverter switches, both snubber capacitors and varistors are connected in parallel to each switch.

2.3. Load Current Limit Circuit. The driver circuit ensures that the inverter bridge is protected against over loads and voltage stresses. The load should be also protected from over current to safe guard the electromagnet. This is achieved using a series current limiting circuit along with an accurate hall effect current sensor as shown in Figure 4. This allows the user to set the current limits for the load and control the inverter output accordingly. If the specified limit is crossed, the series current limiting circuit detects a fault and shuts down the inverter operation.

2.4. Labview-FPGA Based Digital Controller. The complete control logic is implemented in an FPGA which is programmed using Labview as shown in Figure 5. The system is sufficiently fast and all the feedback signals can be incorporated into a single FPGA. This helps to sense the fault status output from the inverter driver and execute any tertiary control action in case of any fault condition in the output load.

3. Simulation and Prototype Development

The aim of this project was to employ the power converter topology, described in Section 2, in the experimental setup which generates the alternating magnetic field in an air gap for testing the magnetocaloric properties of the materials. A solution with combined ac and dc field has been investigated and proposed. In this way, the magnetocaloric material does not need to be rotated or moved in and out of the air gap since the magnetic field is time-varying. It was of interest to obtain the field whose intensity and frequency can be easily changed, which is not the case with a pure PM machine, commonly used for the testing and characterizing of the magnetocaloric materials.

The solution with the combined ac and dc field can be realized by a three legged iron core, having the coil in one leg (creating the ac field), air gap in the middle leg, and permanent magnet in the third leg (creating the dc field). The alternating field in the air gap should be the superposition of the two fields from the other legs and positively biased, which is achieved by using permanent magnets creating a dc field. In this way, both the effects from both ac and dc field could be separately studied. The biased magnetic field should be achieved in the case when the two fields (ac and dc) are contributing in the same direction to the field in the air gap. The frequency of the magnetic field should be variable (0–50 Hz), to enable investigation of how magnetocaloric materials react in different time scales.

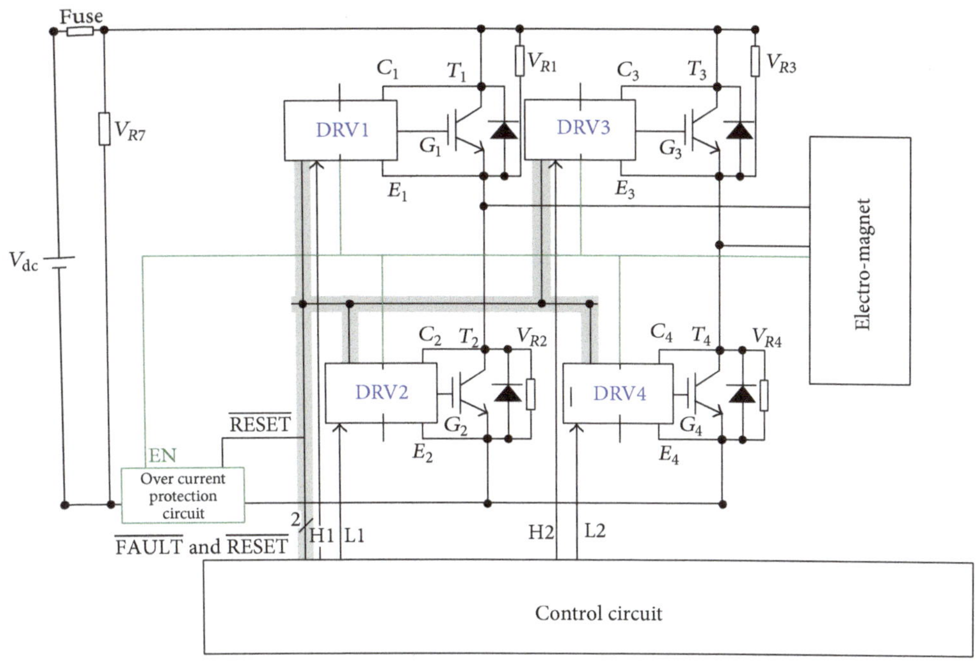

FIGURE 3: Single phase H-bridge inverter power stage.

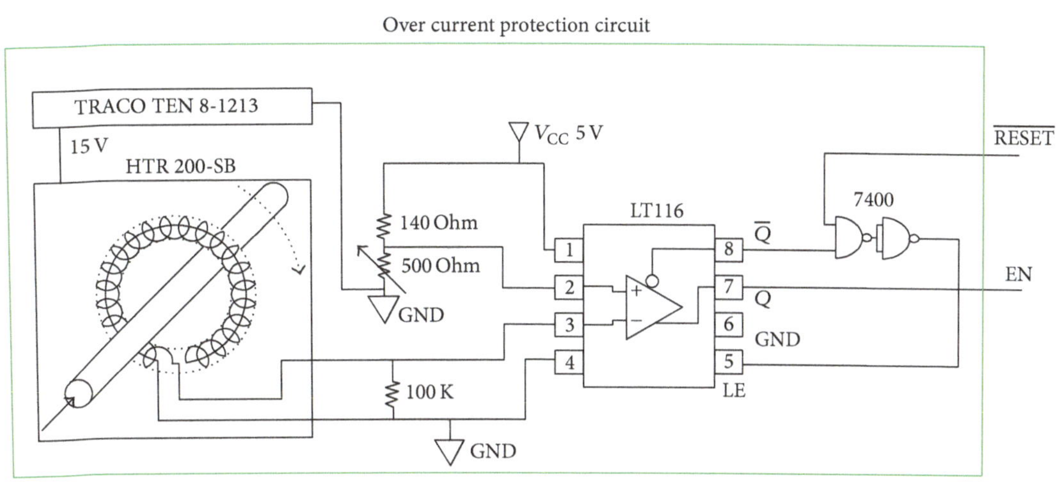

FIGURE 4: Load current limiting and protection circuit.

For the purpose of the initial experimental verifications, a prototype design of the three legged iron core has been proposed, having the dimensions and magnetic properties according to Figures 6 and 7. The coil leg is supposed to be slightly thinner than other legs, in order to reach saturation in that leg, forcing the dc field through the air gap.

To calculate the area A corresponding to a specific magnetic flux, Φ, and magnetic flux density, B, the following equation was used:

$$\Phi = \int_A B \cdot dA. \quad (1)$$

The number of ampere-turns needed in the coil to generate the desired magnetic flux density was found using the formula bellow, neglecting the leakage flux:

$$NI = H_g l_g + H_c l_c, \quad (2)$$

where H_g is the magnetic field intensity in the air gap (B_g/μ_0), l_g is the length of the air gap, H_c is the magnetic field intensity in the core ($H_c = H_g$), and l_c is the length of the core. NI was found to be approximately 1970 AT (leakage is neglected), which can be realized with the coil having 786 turns and the current of 2.5 A, being reasonable and under the limit of the protection circuit.

The data of the proposed iron core are introduced in Comsol simulation tool. The first leg of the iron core is

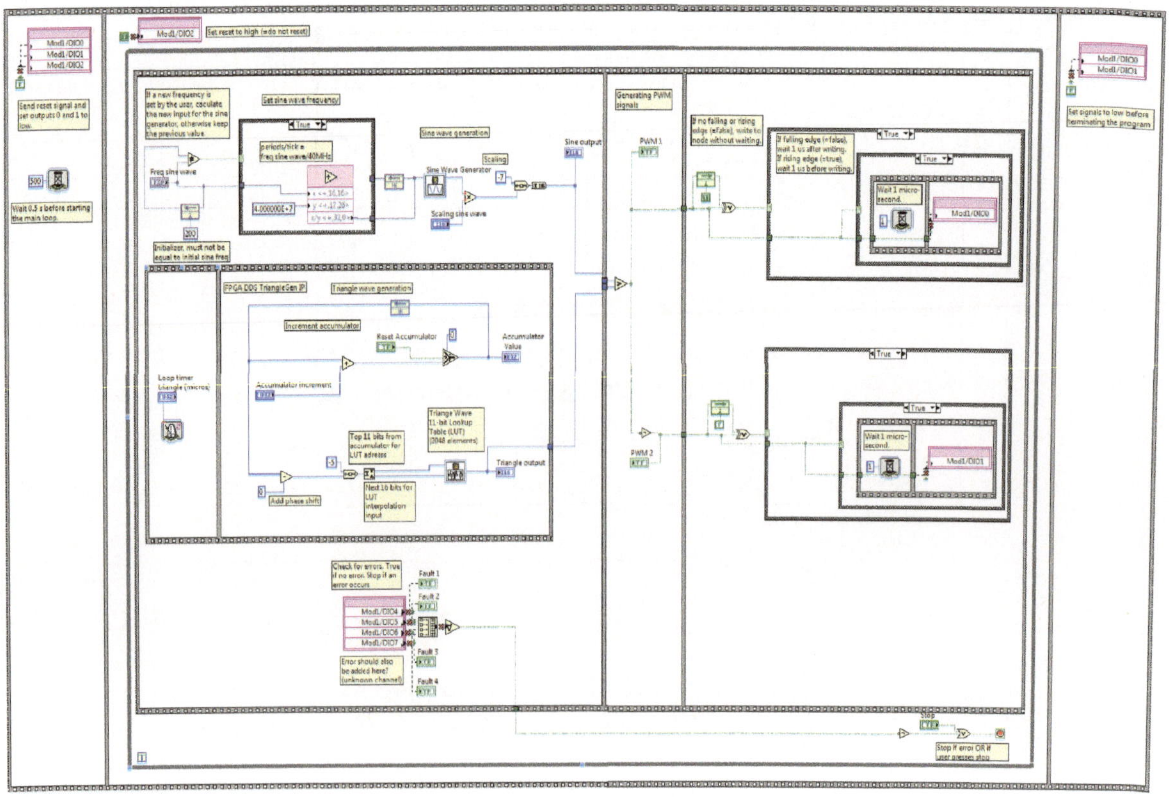

Figure 5: Control loop in Labview.

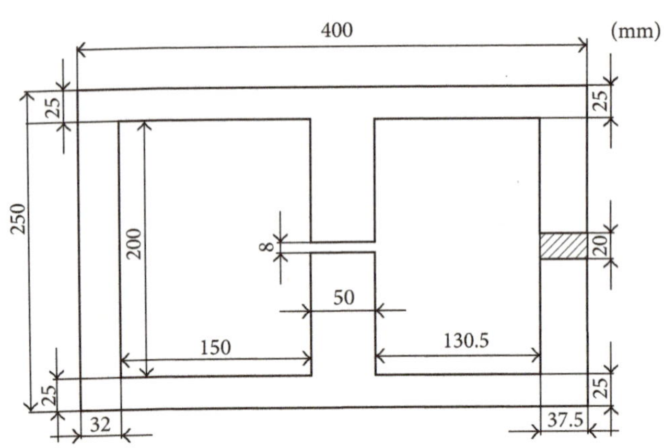

Figure 6: The dimensions of the three legged iron core prototype.

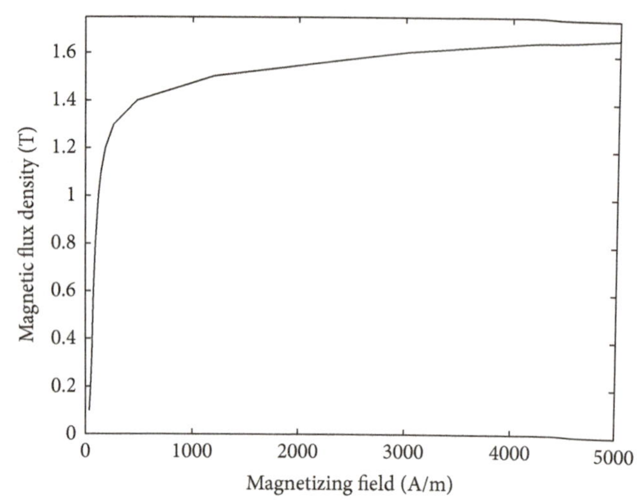

Figure 7: The magnetic properties of the three legged iron core prototype.

excited by the coil having 786 turns, which was estimated to be sufficient with rather low currents. In the middle leg of the core there is the air gap of 8 mm, as the testing area of the magnetocaloric materials. The third leg of the core is excited by the permanent magnet with a remanence of 1.3 T. By changing the ampere-turns of the first leg, the different distributions of the field are investigated in the air gap leg. Of interest was to find the current for which the two fields (ac and dc) are contributing the same. The current of 3 A (close to approximated 2.5 A) was found to be the appropriate one, in the case when having 786 turns.

In Figure 8 the static simulation results for the case when there is the equal contribution of the two fields in the leg with the air gap are shown, having the dc current of 3 A and reaching the averaged field in the air gap $B = 0.28$ T.

4. Experimental Results

The three legged iron core proposed and studied in the previous section has been built. The iron core consists of

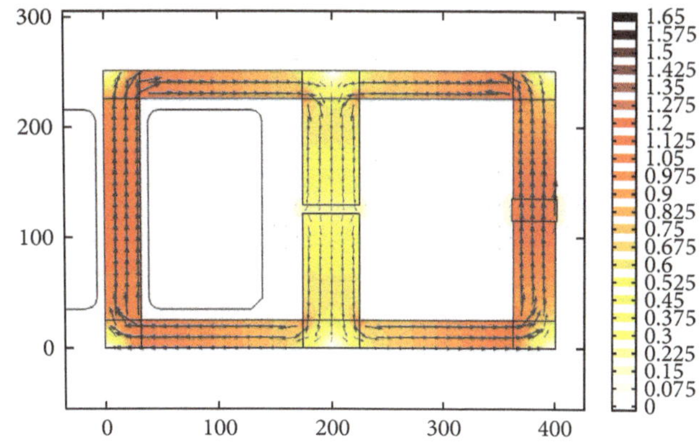

FIGURE 8: Static simulation results of the flux density distribution in the three legged iron core in the case of 3×786 ampere-turns.

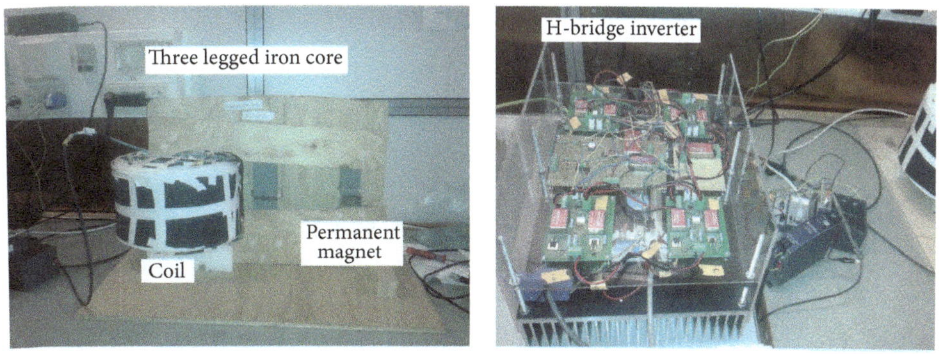

(a) (b)

FIGURE 9: Experimental setup of the iron core (a) and power electronics (b).

0.5 mm thick sheets of silicon steel M350-50A with C6 isolation coating (organic based coating), having the dimensions and magnetic properties according to Figures 6 and 7. The core was laminated to decrease the eddy currents. The sheets were stacked into a support structure and compressed using polyamide threaded rods. The coil having 786 turns is attached to the first leg of the core, with the cable-cross section of 6 mm^2. The air gap of 8 mm is placed in the middle leg of the iron core, that is, the area for the testing of magnetocaloric materials. Neodymium iron boron permanent magnet with dimensions 2 cm × 4 cm × 4 cm (magnet grade N42) and a remanence of 1.3 T is mounted in the third leg. The three legged iron core is shown on Figure 9(a).

The coil in the first leg of the iron core is supplied by H-bridge inverter, as described in Section 2. The inverter with drive circuit is shown in Figure 9(b). The real-time implementation of the algorithm was made by FPGA using a Labview/CompactRio real-time module, which is plugged into a PC. The experiments have been done by feeding a coil with an inductance of $L = 0.36$ H. The total dc voltage provided from the external dc supply 360 V/15 A was 80 V. Switching frequency was set to 1 kHz and fundamental frequency was 5 Hz. No filters were applied in pre- or postprocessing of the results. The results of the experimental verification are shown in Figure 10. In Figure 10(a) the current in the coil $i_L(t)$ is shown, having a maximum of ≈3 A. The current is obviously distorted, due to the magnetic saturation in the part of the core close to the coil leg and due to the effect of the bias introduced by the permanent magnet. The magnetic flux density is shown in the Figure 10(b), having the peak value B very close to the simulated 0.28 T and it is measured in the air gap with a hall effect magnetometer. The peak power consumed from the supply in this case was 184 W.

The peak values from of the coil current and the field in the air gap in Figure 10 show the good agreement with the simulation done in Comsol for a static I-B pairs. The measured field in the air gap is slightly higher than the simulated one, which can be caused by the current being slightly higher and having the averaged field across the air gap in the simulation.

However, for testing the magnetocaloric effect and for the magnetocaloric applications, having a large adiabatic temperature change, ΔT_{ad}, and reaching high entropy change, ΔS, are equally important. One has to optimize simultaneously since the two are not directly coupled. To be really of interest for the magnetocaloric applications, the large ΔT_{ad} and ΔS should be reached for the intermediate field changes, that is, preferably $\Delta B \leq 1.4$ T. Accordingly, most of the studies are focused on the fields $B \approx 1$ T. For this reason, the preliminary testing system presented in this paper needs to be scaled up, to be able to reach the biased alternating field 0-1 T. In terms

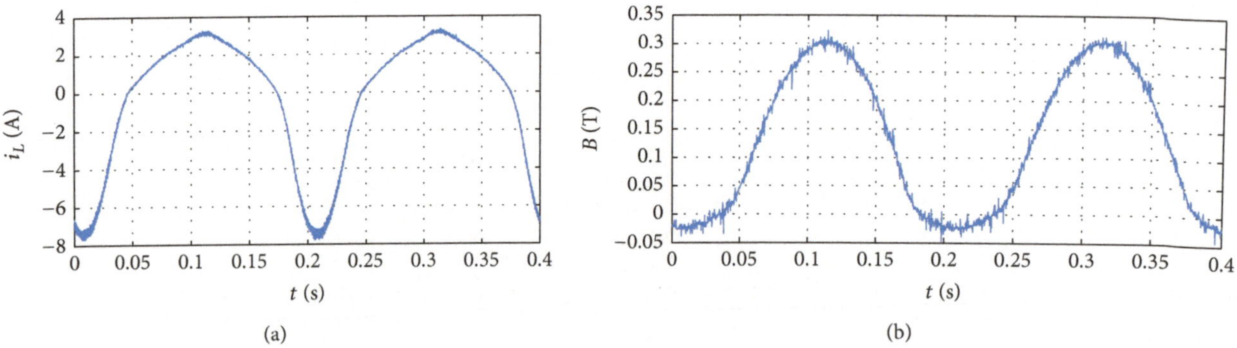

FIGURE 10: Experimental results: (a) load current, (b) measured flux density in the air gap.

of the three legged core dimension limitations, the maximum field in the air gap obtained for this setup was 0.4 T, for a highly distorted current having a maximum ≈30 A. For this measurement the effect of the ac field was dominated, and the resulting field in the air gap is not properly biased. With more robust magnetic core it is possible to reach the desired field in the air gap, without entering in the saturation in the coil leg.

5. Conclusion

A new and improved method of characterizing magnetocaloric materials using a power converter based test platform is presented in this work. It helps to eliminate the limitations of conventional test methods using magnetic machines with moving parts and adds a lot of flexibility in the control of flux density and frequency. In particular, the method is based on combining the rather conventional power electronics with essential control and additional protections, and the novel idea to employ a specific magnetic circuit which can produce biased alternating magnetic field in the air gap, for the testing of the magnetocaloric materials.

The simulation and experimental results are presented to validate the new test platform performance. The proposed prototype has shown the match performance between the simulations done for static I-B pairs and the experiments. Alternating and properly biased magnetic flux density was measured in the air gap, having the maximum very close to the expected value.

The new method explores the complete advantages of power electronic based excitation compared to the machine based one and provides improved overall performance. Since the scaled down prototype showed the expected performance, a new prototype with the improved iron core able to reach the required intermediate fields, that is, $B \approx 1$ T, needs to be built and tested.

Conflict of Interests

The authors declare that there is no conflict of interests regarding the publication of this paper.

Acknowledgments

This study was carried out with funding from the Swedish Research Council, Grant no. 2012-4706. The authors would like to thank the KIC InnoEnergy Ph.D. school for providing all the added value educational activities. Thanks are due to SweGRIDS, EIT KIC InnoEnergy, Swedish Energy Agency, and STandUP for Energy for their support.

References

[1] K. A. Gschneidner Jr. and V. K. Pecharsky, "Thirty years of near room temperature magnetic cooling: where we are today and future prospects," *International Journal of Refrigeration*, vol. 31, no. 6, pp. 945–961, 2008.

[2] H. Yibole, F. Guilliou, L. Zhang, N. H. van Dijk, and E. Brück, "Direct measurement of the magnetocaloric effect in MnFe(P,X)(X = As, Ge, Si) materials," *Journal of Physics D: Applied Physics*, vol. 47, no. 7, Article ID 075002, 2014.

[3] B. Yu, M. Liu, P. W. Egolf, and A. Kitanovski, "A review of magnetic refrigerator and heat pump prototypes built before the year 2010," *International Journal of Refrigeration*, vol. 33, no. 6, pp. 1029–1060, 2010.

[4] V. K. Pecharsky, K. A. Gschneidner Jr., Y. Mudryk, and D. Paudyal, "Making the most of the magnetic and lattice entropy changes," *Journal of Magnetism and Magnetic Materials*, vol. 321, no. 21, pp. 3541–3547, 2009.

[5] O. Gutfleisch, M. A. Willard, E. Brück, C. H. Chen, S. G. Sankar, and J. P. Liu, "Magnetic materials and devices for the 21st century: stronger, lighter, and more energy efficient," *Advanced Materials*, vol. 23, no. 7, pp. 821–842, 2011.

[6] A. Kitanovski and P. W. Egolf, "Innovative ideas for future research on magnetocaloric technologies," *International Journal of Refrigeration*, vol. 33, no. 3, pp. 449–464, 2010.

[7] R. Bjørk, C. R. H. Bahl, A. Smith, and N. Pryds, "Review and comparison of magnet designs for magnetic refrigeration," *International Journal of Refrigeration*, vol. 33, no. 3, pp. 437–448, 2010.

[8] Avago Technologies, "2.5 Amp Gate Drive Optocoupler with Integrated (VCE) Desaturation Detection and Fault Status Feedback," HCPL-316J datasheet, April 2009.

[9] N. Mohan, T. M. Undeland, and W. P. Robbins, *Power Electronics: Converters, Applications, and Design*, John Wiley & Sons, New York, NY, USA, 3rd edition, 2003.

18

I/Q Imbalance Problem in SC-FDMA System with DCT and DFT Basis Functions

F. S. Al-kamali, Hefdhallah Sakran, and N. A. Odhah

Department of Electronics, Faculty of Engineering and Architecture, Ibb University, Ibb, Yemen

Correspondence should be addressed to F. S. Al-kamali; faisalalkamali@yahoo.com

Academic Editor: Ramón M. Rodríguez-Dagnino

In-phase/quadrature-phase (I/Q) imbalance is one of the most serious concerns in the practical implementation of the direct conversion receiver architecture. This paper investigates and analyzes the impacts of the I/Q imbalance on the performance of the single-carrier frequency-division multiple access (SC-FDMA) system with different basis functions by computer simulations only. The suitable phase and amplitude of I and Q branches that can be utilized in SC-FDMA system are studied and determined through simulations for different I/Q imbalance scenarios and different subcarrier mapping schemes. Simulation results show that the phase and amplitude of I and Q branches have significant effects on the bit error rate (BER) performance of the SC-FDMA system.

1. Introduction

Due to high spectral efficiency and robustness against frequency-selective fading, orthogonal frequency-division multiple access (OFDMA) has been chosen as a downlink transmission technique in 3GPP long-term evaluation (LTE) standards [1]. Well-known drawbacks of the OFDMA system are the high peak-to-average power ratio and sensitivity to carrier frequency offset (CFO), which result in severe performance degradation. Due to these drawbacks, the use of OFDMA in the uplink mobile environments was found to be impractical to establish [2]. Single-carrier frequency-division multiple access (SC-FDMA) system has been chosen as an alternative technique in the uplink transmission in 3GPP LTE standards due to its low peak-to-average power ratio [1, 3]. The performance of the SC-FDMA system suffers from the CFO and its sensitivity is lower than that of the OFDMA system [3]. Besides the CFO, there are other radio frequency impairments in direct conversion receiver (DCR) of these systems, such as dc-offset and in-phase and quadrature-phase (I/Q) imbalances, inducing further distortion.

In recent years, DCR has drawn a lot of attention due to its low power consumption and low implementation cost. However, some mismatches in DCR can seriously degrade the system performance [4–7]. The I/Q imbalance is due to the amplitude and phase mismatches between the I and Q branch of the local oscillator whereas the CFO is due to the mismatch of carrier frequency at the transmitter and receiver. In [8–10], It is shown that the I/Q imbalance and CFO can cause a serious intercarrier interference and multiple access interference in SC-FDMA systems. In [8], the issue of CFO in SC-FDMA system is investigated for different basis functions: the discrete Fourier transform (DFT) and the discrete cosine transform (DCT). A new joint equalization and CFO for multiple-input multiple-output SC-FDMA system are proposed and investigated in [9].

The issue of the I/Q imbalance problem in multicarrier systems was extensively investigated in the literature [5–7]. However, to the best of our knowledge, a few papers have only discussed the I/Q imbalance problem in DFT based SC-FDMA (DFT-SC-FDMA) system [10]. Moreover, the issue of the I/Q imbalance problem in DCT based SC-FDMA (DCT-SC-FDMA) system is not studied in the literature. All that motivated us to do the research of this paper.

The main contribution of this work is to study the issue of the I/Q imbalance problem in DFT-SC-FDMA and DCT-SC-FDMA systems. The effects of the I/Q imbalance problem are then compensated using an efficient compensation scheme.

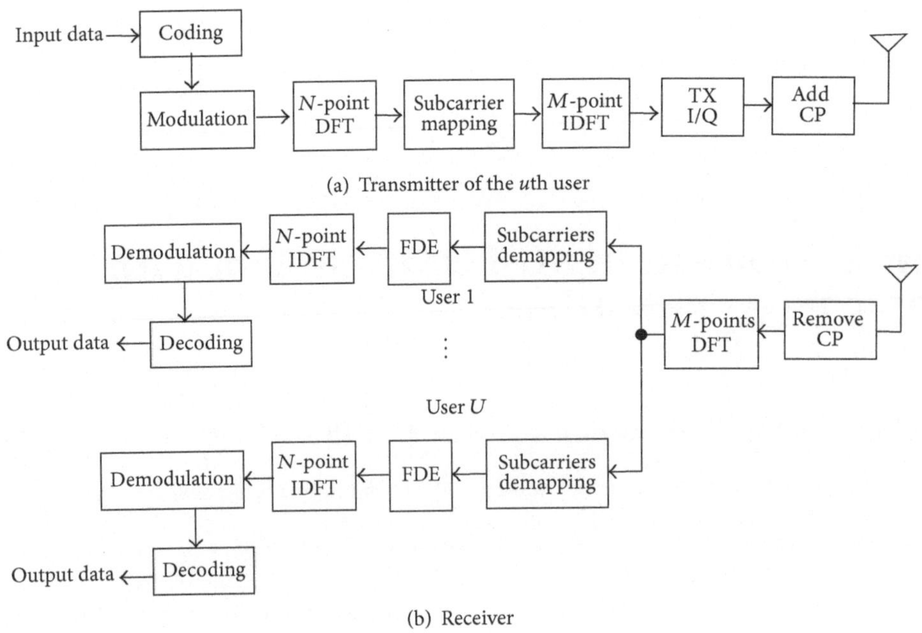

FIGURE 1: Structure of the DFT-SC-FDMA system with the TX I/Q imbalance scenario.

The work in this paper is distinct from the previous research works in this area in that we study the effects of the I/Q imbalance problem using a three-dimensional results. On the other hand, the phase and the amplitude of I and Q branches are taken into consideration simultaneously. We investigate different I/Q imbalance scenarios which are the I/Q imbalance in the transmitter (TX I/Q), the I/Q imbalance in the receiver (RX I/Q), and the I/Q imbalances in the transmitter and receiver (TX-RX I/Q). Simulation results show that I/Q imbalance problem significantly degrades the system performance and its effects on the DCT-SC-FDMA system are greater than that on the DFT-SC-FDMA system. Results also show that the proposed compensation scheme can compensate the effects of the I/Q imbalance problem and improve the system performance.

This paper is organized as follows. In Section 2, we develop the DFT-SC-FDMA system model in the presence of the I/Q imbalance problem. The developed I/Q imbalance compensation scheme is discussed in Section 3. In Section 4, we present the simulation results and the paper is concluded in Section 5.

2. DFT-SC-FDMA System with TX I/Q

In this section, the DFT-SC-FDMA system model in the presence of I/Q imbalance problem is developed. The extension of this model to the DCT-SC-FDMA system is straightforward. More details about the DCT-SC-FDMA system are found in [3].

Figure 1 depicts the block diagram of the DFT-SC-FDMA system with the TX I/Q imbalance scenario. We consider an uplink DFT-SC-FDMA system employing U users. All users have the same number of subcarriers. The input data is encoded and modulated. The modulated symbols are fed to an N-point DFT and then mapped to M subcarriers of assigned subchannels. The signal is then applied to an M-point inverse DFT (IDFT) block.

In SC-FDMA system, there are two subcarrier mapping techniques which are the localized mapping and the interleaved mapping. In localized mapping, the DFT outputs are mapped to a subset of consecutive subcarriers, thereby confining them to only a fraction of the system bandwidth. In interleaved mapping, the DFT outputs of the input data are assigned to subcarriers over the entire bandwidth noncontinuously, resulting in zero amplitude for the remaining subcarriers [1]. In this paper, the interleaved DFT-SC-FDMA is denoted by DFT-IFDMA, the localized DFT-SC-FDMA is denoted by DFT-LFDMA, the interleaved DCT-SC-FDMA is denoted by DCT-IFDMA, and the localized DCT-SC-FDMA is denoted by DCT-LFDMA.

The transmitted signal of the uth user ($u = 1, 2, \ldots, U$) can be formulated as follows:

$$\overline{\mathbf{x}}_u = \mathbf{F}_M^{-1}\mathbf{M}_u\mathbf{F}_N\widetilde{\mathbf{x}}_{\mathbf{u}}, \tag{1}$$

where $\widetilde{\mathbf{x}}_u$ is an $N \times 1$ vector containing the modulated symbols of the uth user. $\mathbf{F}_N$ is an $N \times N$ DFT matrix. $\mathbf{F}_M^{-1}$ is an $M \times M$ IDFT matrix. $\mathbf{M}_u$ is an $M \times N$ matrix describing the subcarriers mapping of the uth user. Also $M = Q \cdot N$, where Q is the maximum number of users that can transmit, simultaneously.

The entries of $\mathbf{M}_u$ for both the localized and the interleaved systems are given by [1, 3]

$$\mathbf{M}_u = \left[\mathbf{0}_{(u-1)N\times N}; \mathbf{I}_N; \mathbf{0}_{(M-uN)\times N}\right],$$

$$\mathbf{M}_u = \left[\mathbf{0}_{(u-1)\times N}; \mathbf{u}_1^T; \mathbf{0}_{(Q-u)\times N}; \ldots; \mathbf{0}_{(u-1)\times N}; \mathbf{u}_N^T; \mathbf{0}_{(Q-u)\times N}\right], \tag{2}$$

where the $\mathbf{I}_N$ and $\mathbf{0}_{Q' \times N}$ matrices denote the $N \times N$ identity matrix and the $Q' \times N$ all-zero matrix, respectively. $\mathbf{u}_l$ ($l = 1, 2, \ldots, N$) denotes the unit column vector, of length N, with all-zero entries except at l. The TX I/Q imbalance problem distorts the ideal transmitted signal as follows:

$$\mathbf{x_u} = \alpha_u \overline{\mathbf{x}}_u + \beta_u \overline{\mathbf{x}}_u^*, \quad (3)$$

where $\overline{\mathbf{x}}_u^*$ is the conjugate of $\overline{\mathbf{x}}_u$. The two complex scalars α_u and β_u are given by [9]

$$\begin{aligned} \alpha_u &= \cos\phi_u + j\varepsilon_u \sin\phi_u, \\ \beta_u &= \varepsilon_u \cos\phi_u - j \sin\phi_u, \end{aligned} \quad (4)$$

where ε_u and ϕ_u denote the phase and amplitude imbalances between I and Q branches of the transmitted signal of the uth user, respectively. When the matching of I and Q balances is ideal, that is, $\varepsilon_u = 1$ and $\phi_u = 0$, then $\alpha_u = 1$ and $\beta_u = 0$. At the end of the transmitter, the cyclic prefix (CP) is appended to $\mathbf{x_u}$. The length of the CP must be longer than that of the channel. The channel after removal of the CP at the receiver can be described by an equivalent $M \times M$ circulant channel matrix $\mathbf{H_u}$, whose first column is given by $\mathbf{h}_u = [h_u(0), h_u(1), \ldots, h_u(L-1), 0_{1\times(M-L)}]$. $h_u(n)$ ($n = 0, 1, 2, \ldots, L-1$) is the multipath channel impulse response between the uth user and the base station. L is the length of the channel impulse response.

At the receiver side, the CP is removed from the received signal and the received signal can be written as follows:

$$\mathbf{r} = \sum_{u=1}^{U} \mathbf{H}_u \mathbf{x}_u + \mathbf{n}, \quad (5)$$

where $\mathbf{x}_u$ is an $M \times 1$ vector representing the block of the transmitted symbols of the uth user. $\mathbf{n}$ is an $M \times 1$ vector describing the additive noise. It contains independent identically distributed zero-mean additive white Gaussian noise. Since the circulant matrix can be diagonalized by DFT matrix [3], the received signal block after DFT is described as follows:

$$\mathbf{R} = \sum_{u=1}^{U} \mathbf{\Lambda}_u \mathbf{F}_M \left(\alpha_u \overline{\mathbf{x}}_u + \beta_u \overline{\mathbf{x}}_u^*\right) + \mathbf{N}, \quad (6)$$

where $\mathbf{\Lambda}_u$ is an $M \times M$ diagonal matrix containing the DFT of the circulant sequence of $\mathbf{H}_u$; that is, $\mathbf{H_u} = \mathbf{F}_M^{-1} \mathbf{\Lambda}_u \mathbf{F}_M$ [3]. $\mathbf{N}$ is the DFT of $\mathbf{n}$.

After that, the demapping, the frequency domain equalization (FDE), and the IDFT operations are performed to provide the estimate of the modulated symbols as follows:

$$\widehat{\overline{\mathbf{x}}}_u = \mathbf{F}_N^{-1} \mathbf{W}_u \mathbf{M}_u^T \mathbf{R}, \quad (7)$$

where $\mathbf{W}_u$ is the $N \times N$ FDE matrix of the uth user. $\mathbf{M}_u^T$ is the $N \times M$ subcarriers demapping matrix of the uth user. It is the transport of the subcarrier mapping matrix.

Finally, the demodulation and the decoding processes are performed.

The RX I/Q imbalance problem distorts the ideal received signal as follows:

$$\overline{\mathbf{r}} = \alpha_u \mathbf{r} + \beta_u \mathbf{r}_u^*, \quad (8)$$

where $\mathbf{r}$ is the received signal after removing the CP.

3. Proposed RX I/Q Imbalance Compensation Scheme

One limiting issue in implementing high-speed wireless systems is the impairment associated with analog processing due to component imperfections. I/Q imbalance is a major source of such impairment and it is one of the main practical obstacles in the implementation of DCR. This highlights the need for compensation scheme for IQ imbalance. There have been many reports in the literature on the compensation of the I/Q imbalance in orthogonal frequency-division multiplexing (OFDM) systems [6, 10–12]. Several compensation methods for I/Q imbalance in OFDM systems have been proposed in [6]. In [10] the effects of the I/Q imbalance problem on the performance of the OFDMA and SC-FDMA systems are investigated and compensated. In [11], assuming that the channel frequency response is smooth, a frequency domain estimation method has been proposed to jointly estimate the I/Q imbalance and channel frequency response. I/Q imbalance estimation and compensation in OFDM systems have been studied in [12].

In this paper, the I/Q imbalance compensation scheme in [11] is developed to compensate the effects of the I/Q imbalance in SC-FDMA system. As we will see in the simulation results, the RX I/Q scenario has greater effects on the system performance than that of the TX I/Q scenario and this scenario must be properly compensated at the receiver. Thus, only the RX I/Q compensation will be discussed in this paper. Define the I/Q parameter

$$\mu = \frac{\beta}{\alpha^*}. \quad (9)$$

Then from (8), we can write the vector $\mu\mathbf{r}$ as follows [11]:

$$\mu\mathbf{r} = \frac{\mathbf{r} - \mu\mathbf{r}^*}{1 - |\mu|^2}. \quad (10)$$

Thus if the I/Q parameter μ is known at the receiver, we can use the above equation to compensate the impact of the RX I/Q scenario.

After that, the signal in (10) is transformed into the frequency domain followed by the demapping, the FDE, the IDFT, the demodulation, and the decoding operations.

4. Simulation Results

In this section, the performance of the DFT-SC-FDMA and the DCT-SC-FDMA systems in the presence of I/Q imbalance problem has been evaluated by simulations. Convolutional code with memory 7 and octal generator

Table 1: Simulation parameters.

Parameter	Description
System bandwidth	5 MHz
modulation	QPSK
Cyclic prefix	20 symbols
M	512
N	128
I/Q imbalance scenarios	TX I/Q, RX I/Q, and TX-RX I/Q
Subcarriers mapping method	DFT-IFDMA, DFT-LFDMA, DCT-IFDMA, and DCT-LFDMA
Channel estimation	Perfect
Equalization	MMSE

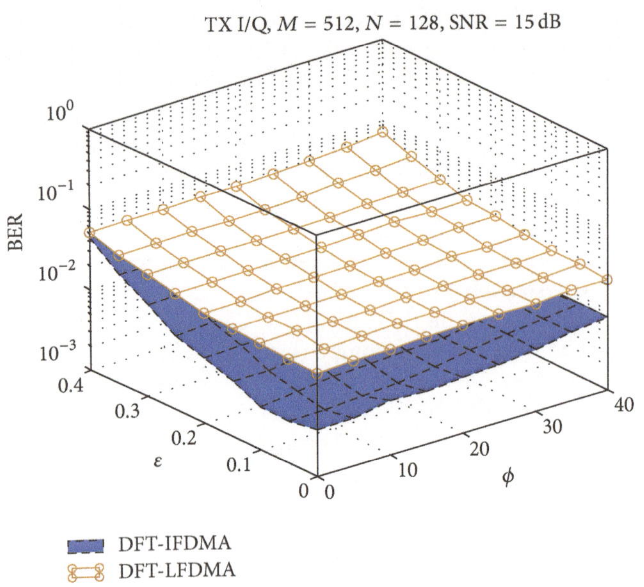

Figure 2: Variation of the BER with both ε and ϕ for the DFT-IFDMA and the DFT-LFDMA systems when the TX I/Q imbalance scenario is used.

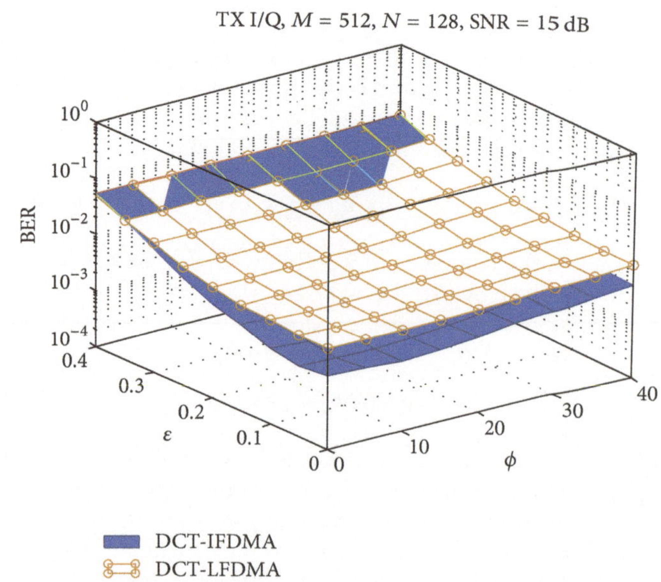

Figure 3: Variation of the BER with both ε and ϕ for the DCT-IFDMA and the DCT-LFDMA systems when the TX I/Q imbalance scenario is used.

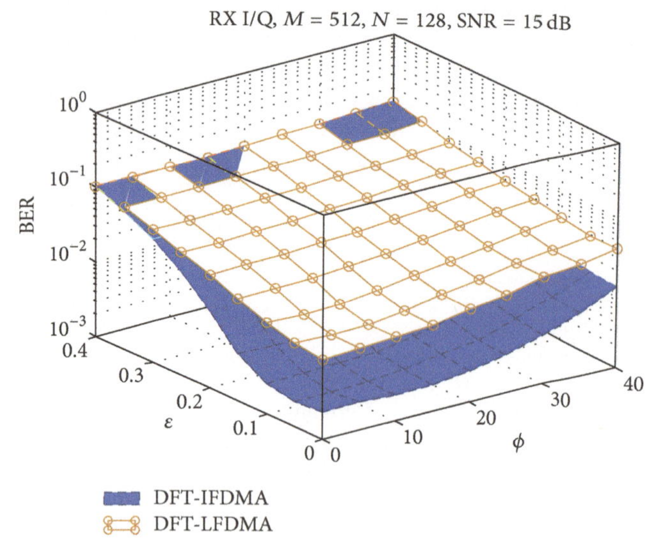

Figure 4: Variation of the BER with both ε and ϕ for the DFT-IFDMA and the DFT-LFDMA systems when the RX I/Q imbalance scenario is used.

polynomials (133, 171) is chosen as channel code. QPSK is used. The channel model used for simulations is vehicular A outdoor channel [13]. The fading was modeled as quasistatic. Simulation parameters are tabulated in Table 1. ε and ϕ are randomly generated for each iteration.

Figures 2 and 3 demonstrate the impact of the TX I/Q imbalance scenario on the performance of the DFT-SCFDMA and DCT-SC-FDMA systems with different subcarriers mapping schemes and a SNR = 15 dB. Figures 2 and 3 show the variation of the BER with both ε and ϕ when the TX I/Q imbalance scenario is considered. On the other hand, ε and ϕ are taken into account simultaneously. It is clearly seen that the performance of the two systems deteriorates at high values of ε and ϕ, especially with the interleaved subcarrier mapping. From these figures, it is clear that for small to moderate values of ε and ϕ, the performances of both systems are insensitive to ε and ϕ.

Figures 4 and 5 show the variation of the BER with both ε and ϕ when the RX I/Q imbalance scenario is considered. It is shown that the performance of the DFT-SC-FDMA and DCT-SC-FDMA systems seriously deteriorates, especially at high values of ε and ϕ. It is also noted that the effects of the RX I/Q imbalance scenario on the DCT-SC-FDMA system are greater than that on the DFT-SC-FDMA system, especially with the interleaved mapping.

From Figures 2 to 5 it is clear that the effects of the RX I/Q imbalance on the performance of the DFT-SC-FDMA and

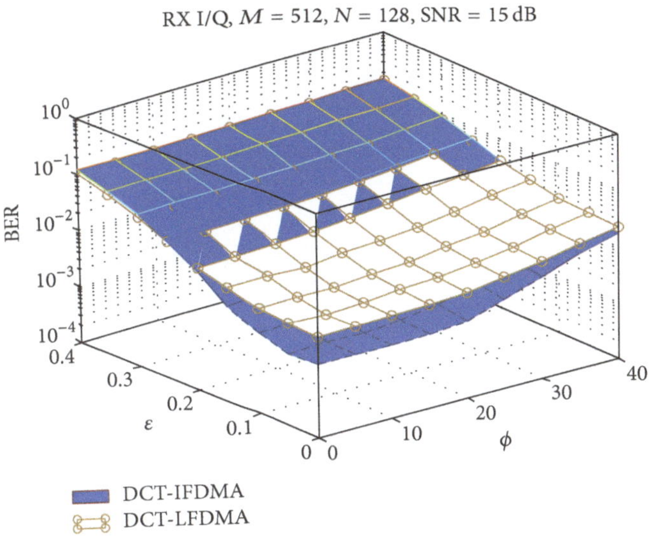

FIGURE 5: Variation of the BER with both ε and ϕ for the DCT-IFDMA and the DCT-LFDMA systems when the RX I/Q imbalance scenario is used.

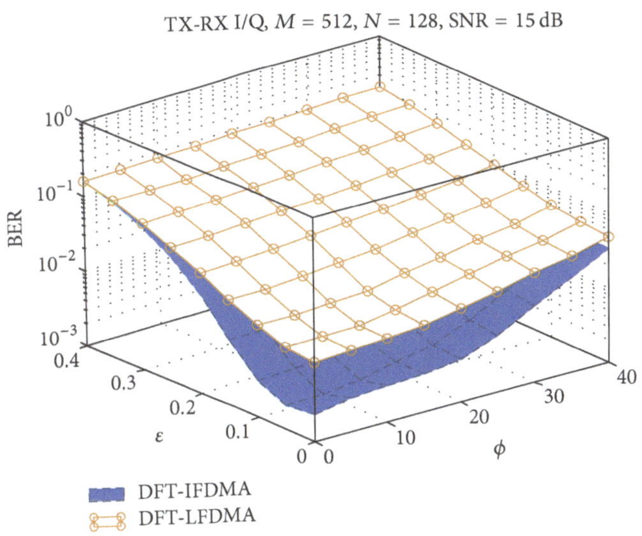

FIGURE 6: Variation of the BER with both ε and ϕ for the DFT-IFDMA and the DFT-LFDMA systems when the TX-RX I/Q imbalance scenario is used.

DCT-SC-FDMA system are greater than that of the TX I/Q imbalance, especially at high ε and ϕ values. On the other hand, the larger the values of ε and ϕ, the higher the BER and the worse the performance. Also, the lower the values of ε and ϕ, the lower the BER and the better the performance. It is also shown that the effects of ε are greater than that of ϕ.

Figures 6 and 7 study the effects of the TX-RX I/Q imbalance scenario on the performance of the DFT-SC-FDMA and DCT-SC-FDMA systems with different subcarriers mapping schemes and a SNR = 15 dB. These figures show that the TX-RX I/Q scenario has the highest effects on the system performance of both systems when compared with that of the TX I/Q and RX I/Q scenarios. Figures 3, 5, and 7 indicate that the performance of the DCT-LFDMA is better than that of the DCT-IFDMA at high ε values, regardless of the ϕ values. This is because that the performance of the interleaved system is more sensitive to the multiple access interference than the performance of the localized system [3].

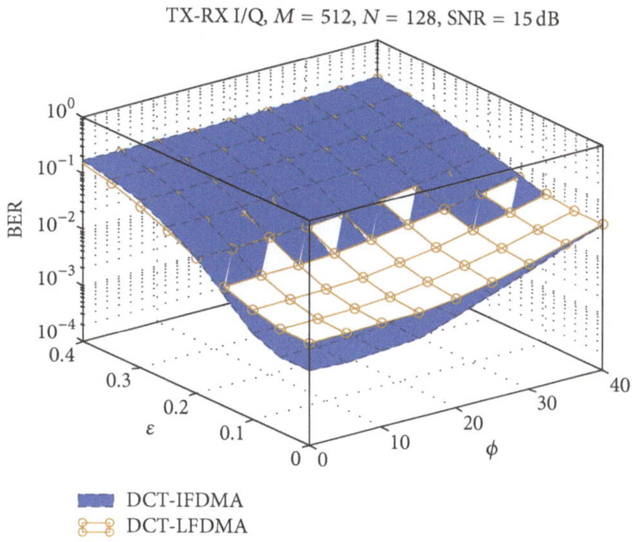

FIGURE 7: Variation of the BER with both ε and ϕ for the DCT-IFDMA and the DCT-LFDMA systems when the TX-RX I/Q imbalance scenario is used.

Figures 8 and 9 show the BER performances versus SNR for the DFT-SC-FDMA and DCT-SC-FDMA systems for different I/Q imbalance scenarios. Each ε_u is a random variable with uniform distribution in $[0, 0.2]$ and each ϕ_u is a random variable with uniform distribution in $[-\pi/8, \pi/8]$. For the TX I/Q scenario, we can clearly see that the BER performance is better than that of the RX I/Q scenario at high SNR. This is because the interference is dominant at high SNR values. However, the performances of these scenarios are nearly the same at low SNR values, when SNR < 5, because the noise is dominant at low SNR region. It is also clear that the performance of the TX-RX scenario is the worst as compared with the other scenarios. Thus, adding an efficient compensation scheme for SC-FDMA systems is very important to compensate the effects of the I/Q imbalance problem.

With a BER of 10^{-3}, the required SNR values for the DCT-IFDMA and DFT-IFDMA systems without TX I/Q imbalance are 16 dB and 16.25 dB, respectively, whereas the required SNR for the DCT-IFDMA and DFT-IFDMA systems with TX I/Q imbalance are 20 dB and 18.75 dB, respectively, which demonstrates a loss of 4 dB and 2.5 dB for the DCT-IFDMA and DFT-IFDMA, respectively. This illustrates that the effects of the TX I/Q imbalance on the DCT-IFDMA system are greater than that on the DFT-IFDMA system.

Figures 8 and 9 also show the performance of the proposed RX I/Q compensation scheme for the DFT-SC-FDMA and DCT-SC-FDMA systems, respectively. It is clearly

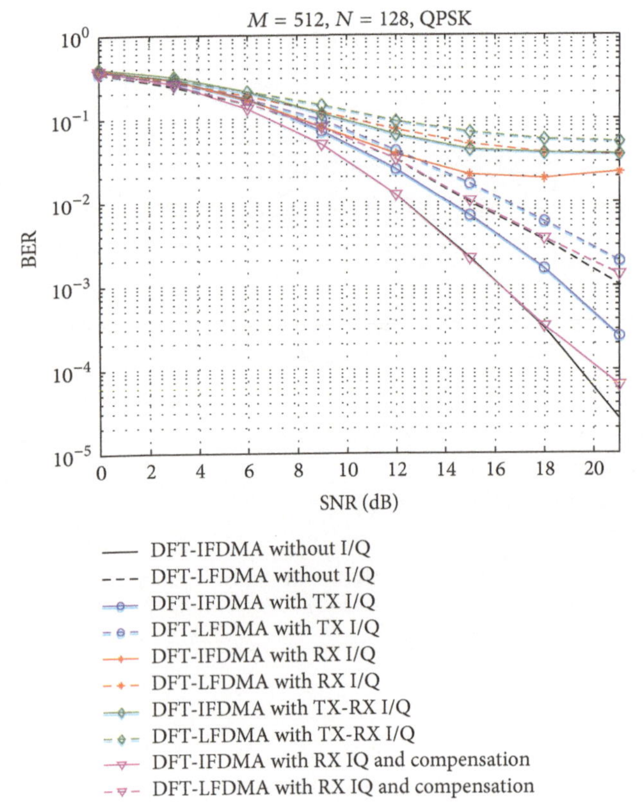

FIGURE 8: BER versus SNR for the DFT-IFDMA and the DFT-LFDMA systems with different I/Q imbalance scenarios and RX I/Q imbalance compensation scheme.

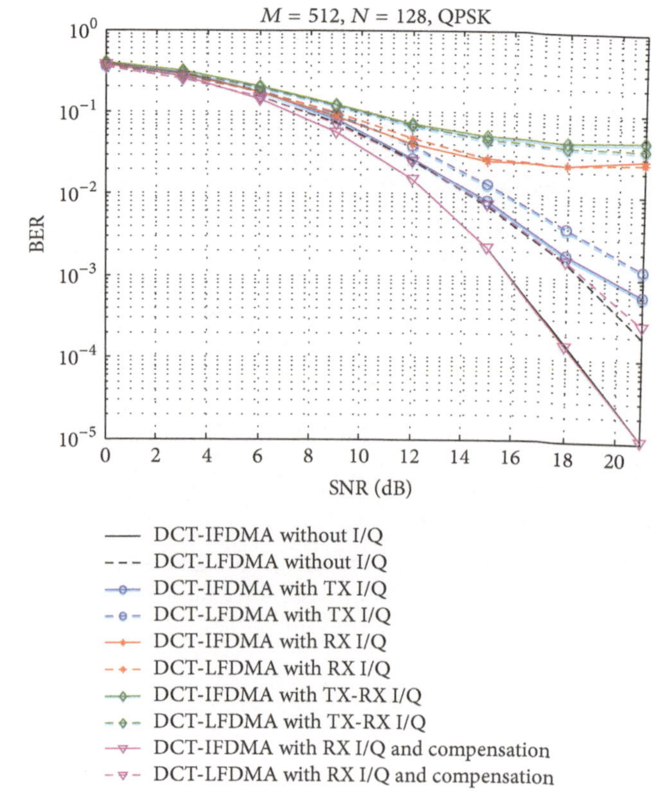

FIGURE 9: BER versus SNR for the DCT-IFDMA and DCT-LFDMA systems with different I/Q imbalance scenarios and RX I/Q imbalance compensation scheme.

seen that the proposed compensation scheme can nearly remove the effects of the RX I/Q imbalance scenario and provide nearly the same performance as that without RX I/Q imbalance.

5. Conclusions

In this paper, the issue of the I/Q imbalance problem in the DFT-SC-FDMA and DCT-SC-FDMA systems was studied and the obtained results might be taken into account in a practical design. It was illustrated that the IQ imbalance problem can cause large degradations in the DFT-SC-FDMA and DCT-SC-FDMA receivers. The RX I/Q scenario has significant effects on the performance degradation compared with that of the TX I/Q scenario and these effects depend on the subcarrier mapping used in the system. It was also noted that the effects of the I/Q imbalance problem on the performance of the DC-SC-FDMA system are greater than that on the performance of the DFT-SC-FDMA system, especially for the interleaved mapping scheme. Simulation results also demonstrated that the difference between the BER values in the region with ε varying from 0 to 0.1 and ϕ varying from 1 to 20° is nearly negligible and the system performance is not sensitive to these parameters in this region. The region of the minimum BER is nearly the same for the DFT-SC-FDMA and DCT-SC-FDMA systems with all I/Q scenarios: the TX I/Q, the RX I/Q, and the TX-RX I/Q.

Also, an efficient RX I/Q compensation scheme for the DFT-SC-FDMA and DCT-SC-FDMA receivers was developed. It was found that the developed scheme can effectively improve the system performance and reduce the sensitivity of the DFT-SC-FDMA and DCT-SC-FDMA receivers to the RX I/Q imbalance problem.

Conflict of Interests

The authors declare that there is no conflict of interests regarding the publication of this paper.

References

[1] H. G. Myung and D. J. Goodman, *Single Carrier FDMA: A New Air Interface for Long Term Evaluation*, John Wiley & Sons, Chichester, UK, 2008.

[2] B. Farhang-Boroujeny, "Filter bank multicarrier modulation: a waveform candidate for 5G and beyond," *Advances in Electrical Engineering*, vol. 2014, Article ID 482805, 25 pages, 2014.

[3] F. E. Abd El-Samie, F. S. Al-kamali, A. Y. Al-Nahari, and M. I. Dessouky, *SC-FDMA for Mobile Communications*, CRC Press, 2013.

[4] B. Razavi, *RF Microelectronics*, Prentice-Hall, Englewood Cliffs, NJ, USA, 1998.

[5] C. L. Liu, "Impacts of I/Q imbalance on QPSK-OFDM-QAM detection," *IEEE Transactions on Consumer Electronics*, vol. 44, no. 3, pp. 984–989, 1998.

[6] A. Tarighat, R. Bagheri, and A. H. Sayed, "Compensation schemes and performance analysis of IQ imbalances in OFDM receivers," *IEEE Transactions on Signal Processing*, vol. 53, no. 8, pp. 3257–3268, 2005.

[7] Y.-H. Chung and S.-M. Phoong, "Joint estimation of I/Q imbalance, CFO and channel response for MIMO OFDM systems," *IEEE Transactions on Communications*, vol. 58, no. 5, pp. 1485–1492, 2010.

[8] F. S. Al-Kamali, M. I. Dessouky, B. M. Sallam, F. Shawki, and F. E. Abd El-Samie, "Equalization and carrier frequency offsets compensation for the SC-FDMA system," *Wireless Personal Communications*, vol. 67, no. 2, pp. 113–138, 2012.

[9] F. S. Al-Kamali, M. I. Dessouky, B. M. Sallam, F. Shawki, W. Al-Hanafy, and F. E. A. El-Samie, "Joint low-complexity equalization and carrier frequency offsets compensation scheme for MIMO SC-FDMA systems," *IEEE Transactions on Wireless Communications*, vol. 11, no. 3, pp. 869–873, 2012.

[10] Y. Yoshida, K. Hayashi, H. Sakai, and W. Bocquet, "Analysis and compensation of transmitter IQ imbalances in OFDMA and SC-FDMA systems," *IEEE Transactions on Signal Processing*, vol. 57, no. 8, pp. 3119–3129, 2009.

[11] J. Tubbax, B. Côme, L. van der Perre et al., "Compensation of IQ imbalance and phase noise in OFDM systems," *IEEE Transactions on Wireless Communications*, vol. 4, no. 3, pp. 872–877, 2005.

[12] J. Tubbax, A. Fort, L. V. Der Perre et al., "Joint Compensation of IQ imbalance and Frequency Offset in OFDM systems," in *Proceedings of the IEEE Global Telecommunications Conference (GLOBECOM '03)*, pp. 2365–2369, December 2003.

[13] 3rd Generation Partnership Project, "Technical specification group radio access network; user equipment (UE) radio transmission and reception (FDD) (Release 7)," Tech. Rep. 3GPP TS 25.101, Section B.2.2, 2007.

19

Modelling and Simulation of Static Excitation System in Synchronous Machine Operation and Investigation of Shaft Voltage

Arun Kumar Datta,[1] Manisha Dubey,[2] and Shailendra Jain[2]

[1] *Central Power Research Institute, Bhopal 462023, India*
[2] *Maulana Azad National Intitute of Technology, Bhopal 462051, India*

Correspondence should be addressed to Arun Kumar Datta; akdbpl@gmail.com

Academic Editor: Ergina Kavallieratou

Static excitation system (SES) has been implemented in a specially designed synchronous machine installed in a testing laboratory. This is a large capacity single machine operated in dual mode (i.e., motor or generator) with the help of static sources. It is well known that bearings of the rotating machines are vulnerable to the effects of the shaft voltages caused by the static sources. Shaft voltage is the prime concern for this special machine too due to SES. To find out the exact cause of the shaft voltage, SES of this machine has been modelled with Power Systems software. Various waveforms drawn from the model are validated through computer simulations and actual laboratory tests. Sources of shaft voltages are also analysed thereafter with the FFT analysis of the rotor voltage and current waveforms.

1. Introduction

There are various means of supplying dc power to the field winding of an electric machine, for example, dc generator, rotating exciter, and static converter. After the invention of semiconductor devices, static converter has become popular. It is also known as static excitation system (SES). SES came into existence in the beginning of the sixties. Till then the alternators were getting dc field power in the rotor from a dc generator, separately or coupled with the same alternator shaft. SES was experimented successfully on a steam turbine alternator in the year 1962 [1]. Later on it was implemented not only on the new alternators but also on the existing machines. Due to ease in retrofitting, the old rotating exciters were also replaced with the SES. Starting with steam power generator the SES was extended up to the captive power plant in paper and pulp industries, pump storage power plant, electric locomotive, and gas power plant [2–6]. SES has also been implemented in high power short circuit alternator [7].

Short circuit alternator is basically a synchronous generator in a high power test laboratory which supplies high value of short circuit currents during test on electrical power equipment. A strong dc field is essential to establish and maintain the alternator terminal voltage during the short circuit test. Old testing laboratories were equipped with a separately excited high power dc generator run by an induction motor. A permanent magnet dc generator feeds the field of high power dc generator. In cascading mode a high power dc is produced which is fed into alternator rotor through slip rings and brushes. Viewing the benefits of SES, short circuit alternators are also started equipped with high power thyristors and fast acting microprocessor controller. One such 1500 MVA short circuit alternator with SES is taken for study in this paper (see Figure 1). Other than SES one more static source that is static frequency converter is connected with the stator terminal of this alternator. With these two static sources the alternator can also be run as a motor [8]. In the beginning machine starts and runs as a motor. During short circuit test the same machine is converted to alternator to deliver the power to the equipment under test.

While rotation, shaft of a high speed large machine is normally charged with certain amount of voltage with respect

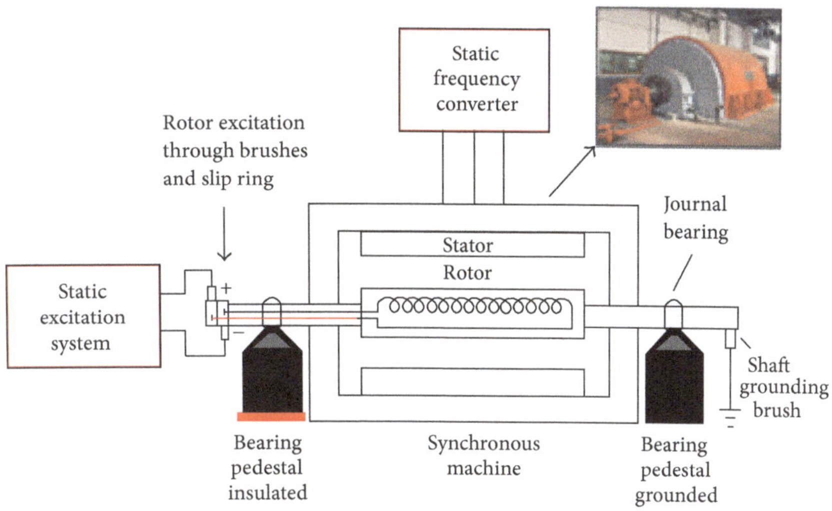

FIGURE 1: Short circuit alternator with the connected system.

to earth. This unwanted voltage is known as shaft voltage. It is a very old and common phenomenon in any electrical machine. Shaft voltage was noticed in the beginning of the 20th century by the engineers while studying different motor failure cases. Shaft voltage in those days was based only on sinusoidal power sources. It was noticed mainly due to unbalance in magnetic fluxes because of various reasons. Application of nonsinusoidal or static sources came into practice many years after the invention of semiconductor devices. With this some new types of shaft voltages appeared [9]. It is proven that shaft voltage under static power source is more predominant due to electrostatic phenomenon rather than pure sine wave operation [10]. Static excitation system has also added a new type of shaft voltage on the basic shaft voltages [11].

Bearing currents are the outcome of presence of shaft voltages. There are three major types of bearing currents: circulating current, dv/dt current, and discharge current [12–14]. Each of these currents flows under different conditions. Other than the mechanical reasons, majority of the premature bearing failures are due to the flow of these bearing currents. Common mode voltage (CMV) [15] is also responsible to flow current through machine bearings to ground. Many authors [16–18] presented different inverters topologies to eliminate CMV. Various filtering techniques (e.g., active, passive, or hybrid) are also proposed to safeguard the motor from the ill effects of CMV [19–21].

Most of the shaft voltage and bearing current problems and solutions related to work were carried out on induction motor [22, 23]. This paper creates a different approach by taking a high power short circuit alternator for the study.

SES has great operation flexibilities compared to the other methods of excitation. It is also cheaper than the old conventional motor-generator set. Fast response characteristic, easy monitoring control, and troubleshooting are some other advantages of SES. In spite of several benefits, generation of shaft voltage is the major concern among the machines with static sources. The shaft voltage problem in the short circuit alternator has already been reported earlier [24–26].

This paper begins with the description of SES. Waveforms at various key points are recorded with the high speed recorder and stored for further analysis. A Simulink model has been prepared with the help of different parameters from converter bridge. The model is run and the results are compared with the actual waveforms recorded from the system. A discussion on the shaft voltage is brought out with these results.

2. SES: Description

SES converts line frequency AC to DC which further energises field winding in the rotor. Excitation level in a machine is needed to vary according to the different modes of the machine. This is only possible with a phase controlled rectifier. Now, to improve the power factor and to reduce the harmonic level of the system, a poly-pulse AC-DC power conversion [27] is required. To maintain this theory a 12-pulse converter configuration is adopted for SES in short circuit alternator. It is formed by a series combination of two 6-pulse thyristor based silicon-controlled rectifier (SCR). Rotor current is very high value in alternator mode. To cope with this current two thyristors are put in parallel in each leg of the bridge. Detailed drawing is depicted in Figure 2.

The circuit begins with a specially designed converter transformer at the bridge input. It is a step down transformer with one primary and two secondary windings. Secondary windings are in star and delta fashion to get a 30° phase shift between them. Rectified six-pulse outputs of each converter bridge are connected in series to form a 12-pulse output. This current is controlled by changing the thyristor gate pulses from the control cubicle. A dc circuit breaker does the switching action between SES and rotor circuit. SES panel is divided into two main parts (see Figure 3). One is power cubicle, consisting of thyristors and snubber circuits. The second part is

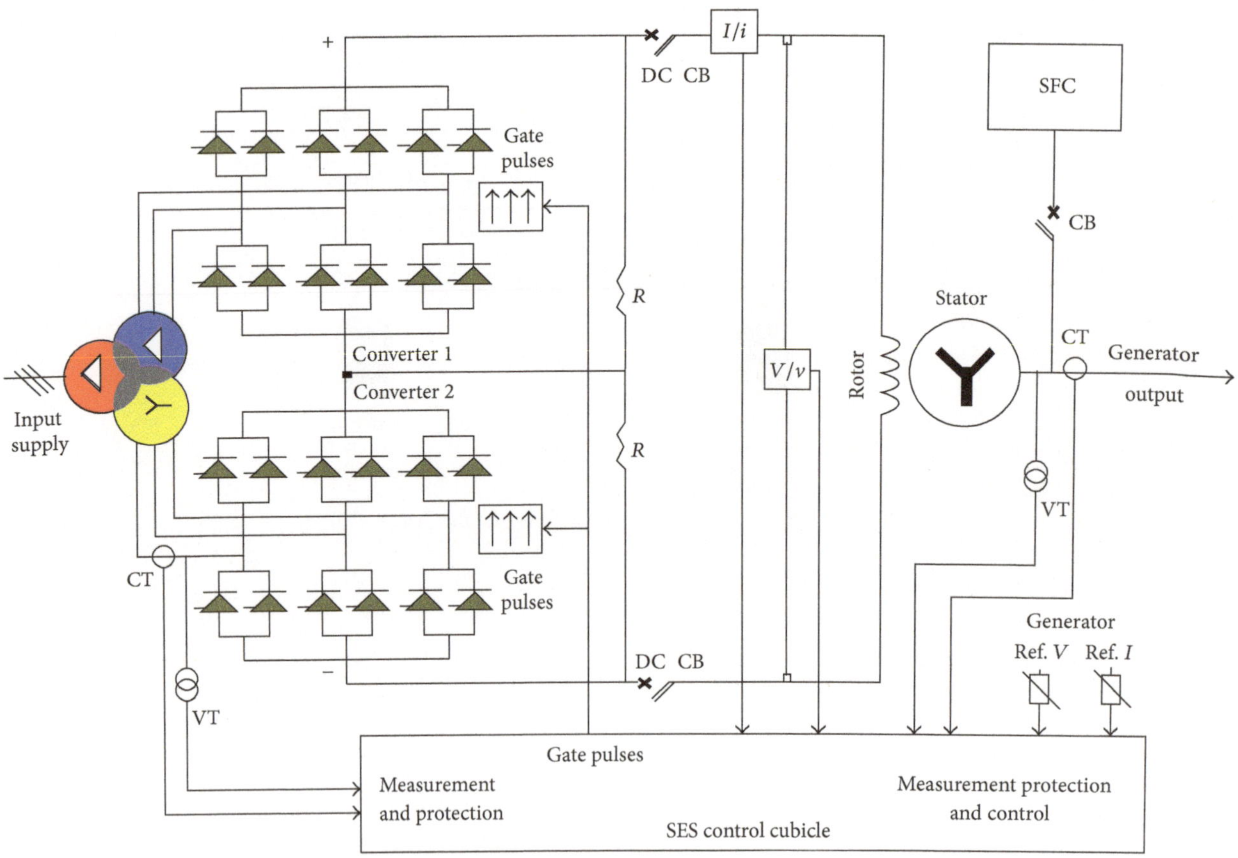

FIGURE 2: Circuit diagram of static excitation system.

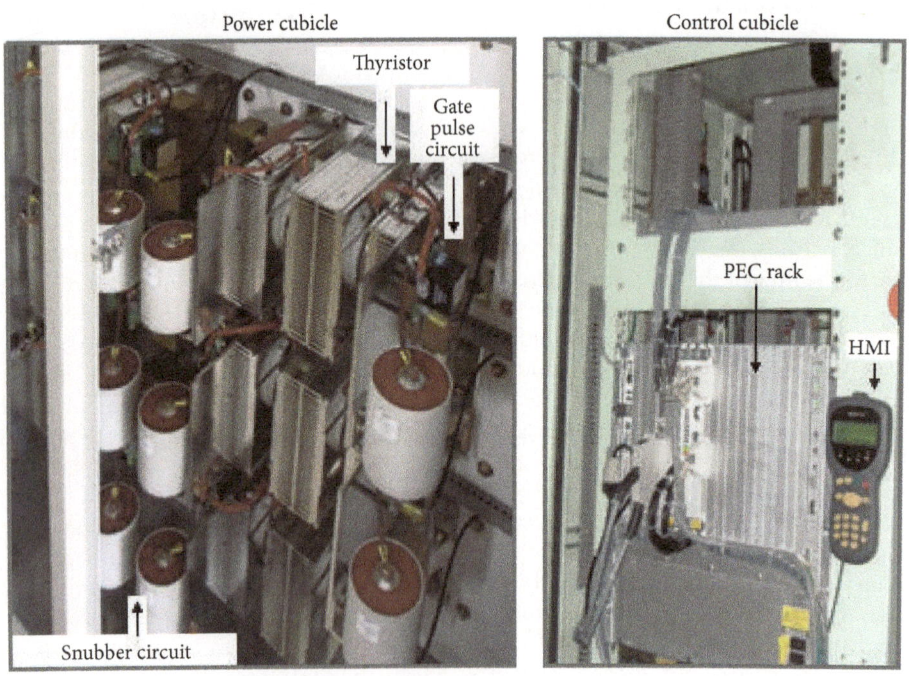

FIGURE 3: Static excitation system for 1500 MVA short circuit alternator.

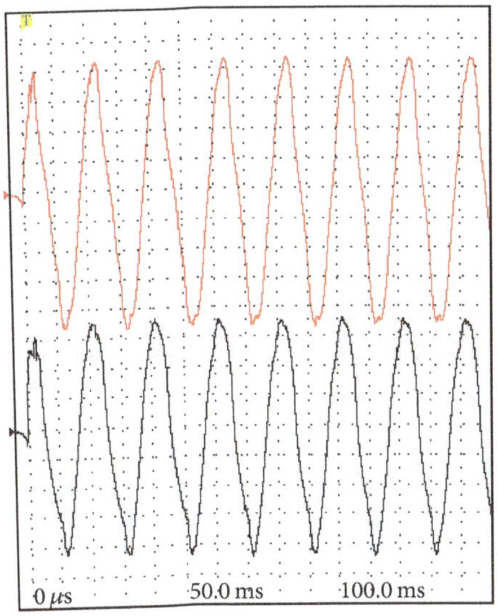

FIGURE 4: Converter input voltages from transformer (top: star winding 950 Vrms, bottom: delta winding 950 Vrms).

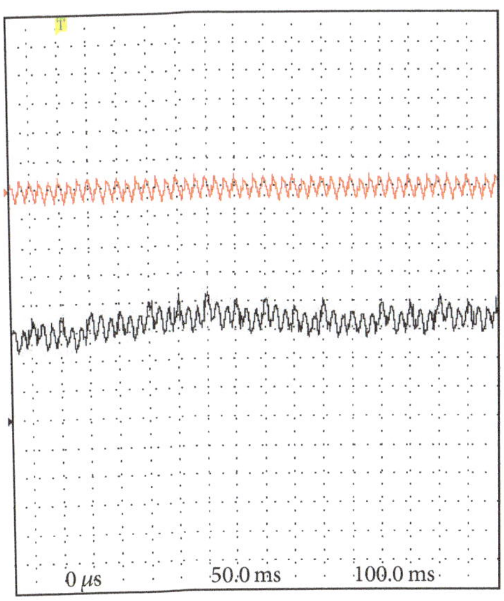

FIGURE 5: Rotor supply for a 3 kV stator output (top: field voltage 30 V, bottom: field current 400 A).

the CPU with other electronic control circuitry and feedback loops. It is named as power electronic controller (PEC) and acts very fast in a μsec range during short circuit test sequence [28]. A human machine interface (HMI) fitted on the controller visualises all faults and abnormality records in the SES.

3. SES: Waveforms

SES waveforms are measured at converter input and output and displayed in Figures 4, 5, and 6. Figure 4 depicts the star and delta input voltages of the converter bridge. A 30° phase shift between the two voltages can be observed (see Figure 4). Converter output applied on the alternator field winding and current flowing through it are shown in Figure 5. This is for a 3 kV alternator output (see Figure 6).

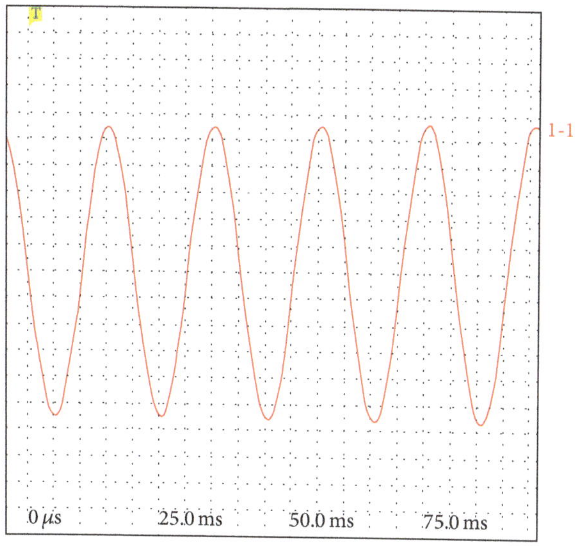

FIGURE 6: Generator output 3 kVrms, 50 Hz.

TABLE 1: SES model parameter values.

Parameter	Value
Input source	33 kV, 3 phases
Transformer nominal power	5 MVA
Transformer primary input	33 kV, 50 Hz
Transformer secondary output star winding	950 V
Transformer secondary output delta winding	950 V
Thyristor bridge	3 arms
Snubber resistance	500 Ω
Snubber capacitance	0.1 μF
Forward resistance	1 mΩ
Forward voltage	0.8 V
Generator field resistance	0.01 Ω
Generator field inductance	20 mH

4. SES: Modelling

Modelling of SES is done to understand its intricacies in a simple manner. It has already been mentioned that SES is a combination of two thyristor bridges in series connection. The same has been modelled in Simulink [29] platform (see Figure 7) with one supply source, input transformer, two thyristor converters, pulse generators, and other circuitries. The model is formed with the variables listed in Table 1. All these parameters are taken from the actual converter.

5. Simulation Results and Discussion on Shaft Voltage

After running simulation the scope window displays different waveforms (see Figure 8). For better representation rotor voltage, rotor current, and reference current are plotted (see Figure 9) through MATLAB script file. The simulated traces

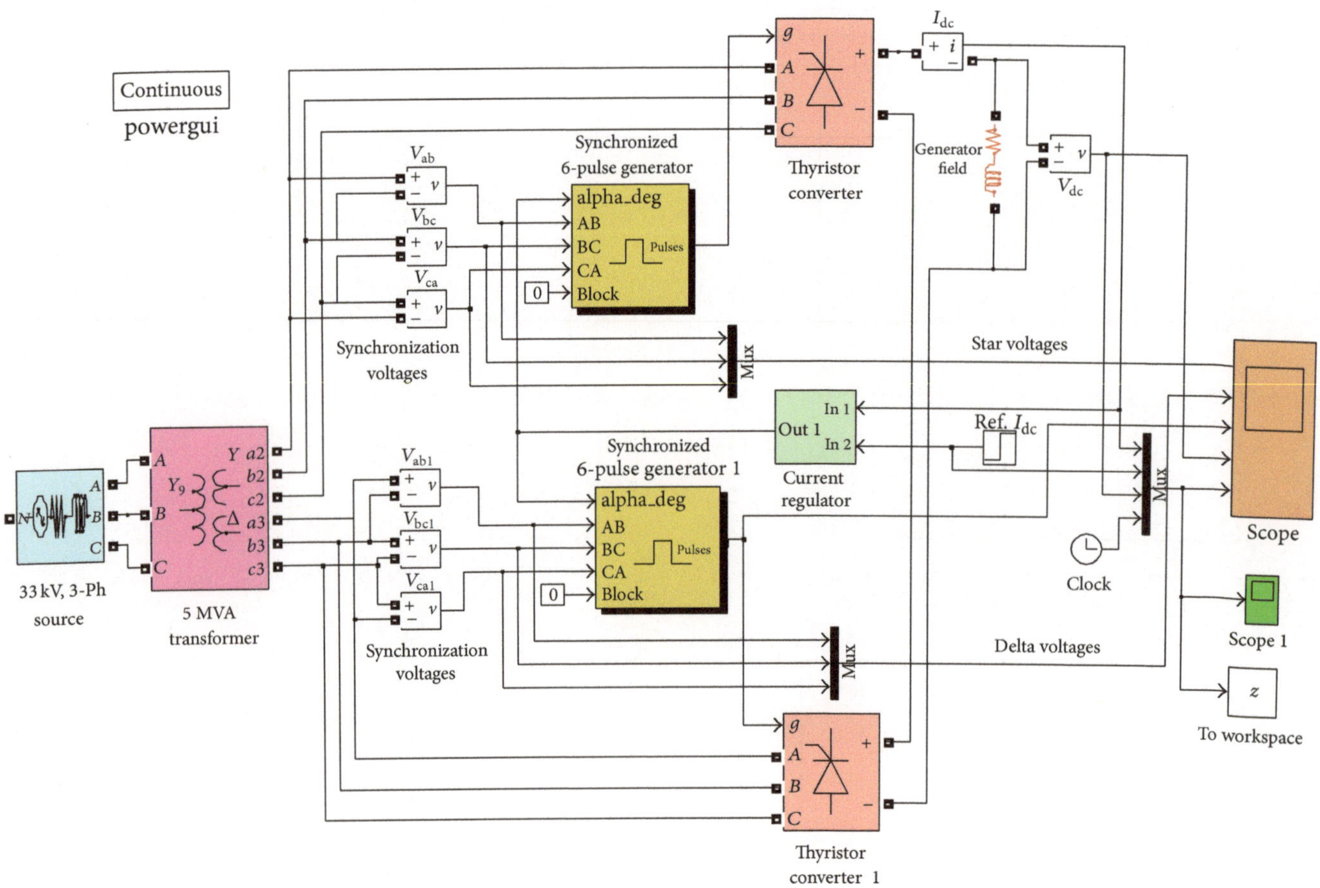

Figure 7: Simulink model of SES.

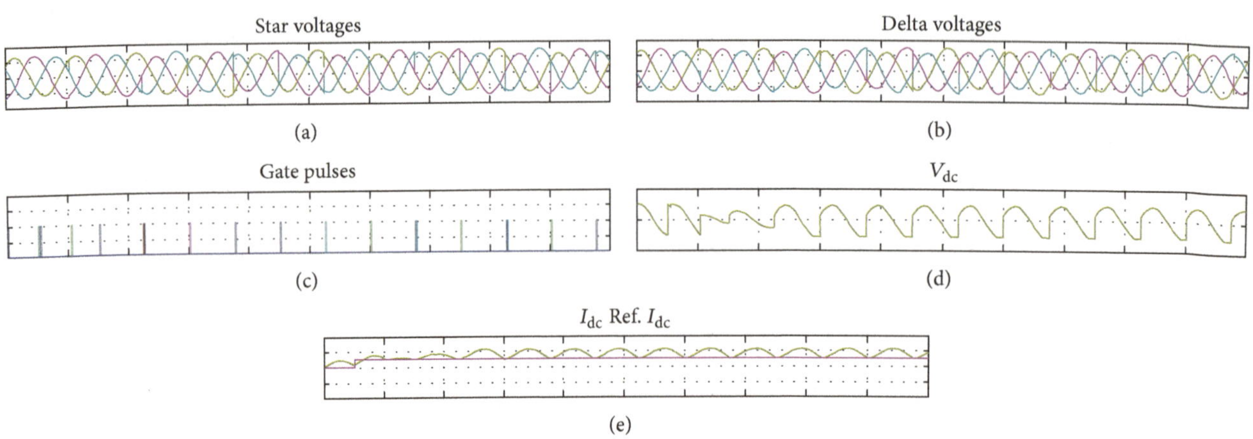

Figure 8: Scope view of simulation run (a) star voltage, (b) delta voltage, (c) gate pulses, (d) dc voltage, and (e) dc current and ref. dc current.

and actual waveforms from the measurements points of SES are compared which show similarities. Rotor voltage and current are pulsating in nature. It contains numbers of low and high order harmonics. This is confirmed by Fourier transform. Figure 10 shows the fast Fourier transform (FFT) analysis for actual rotor voltage whereas Figure 11 is the FFT for simulated waveform. Similarly FFT for rotor current waveforms are depicted in Figures 12 and 13.

Presence of harmonics and other high frequency signals causes the capacitive and inductive couplings in the complete alternator system. Winding insulation, air, and lubricating oil work as dielectric medium for forming the distributed capacitances in the machine. These capacitances (also called parasitic capacitances) form in stator to ground, stator to rotor, stator to shaft, rotor to shaft, bearing to ground, cable to ground, and also supply source to ground. Leakage current

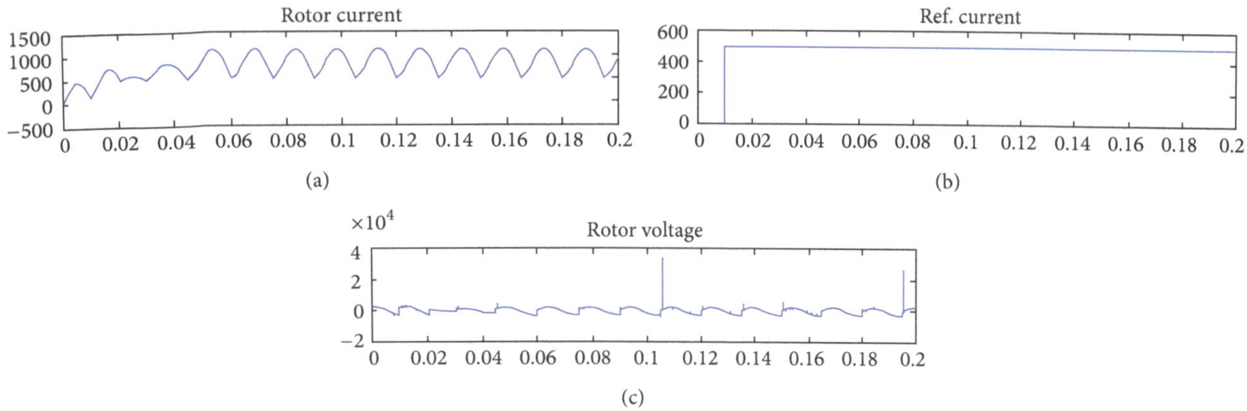

FIGURE 9: SES simulation results (a) rotor current (A), (b) reference current (A), and (c) rotor voltage (V).

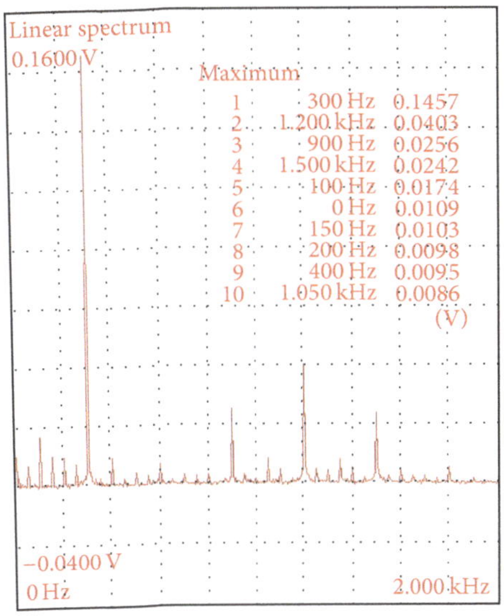

FIGURE 10: FFT for actual rotor voltage.

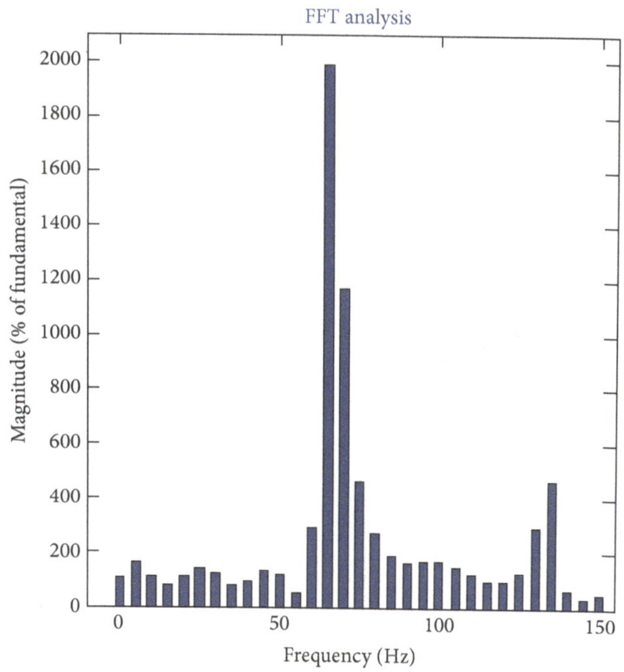

FIGURE 11: FFT for simulated rotor voltage.

($I_c = C \cdot dv/dt$) flows all along the stator and rotor windings due to this parasitic capacitance (C) and high dv/dt. This leakage current generates high frequency flux. Harmonics and other high frequency spikes in the stator and rotor power supply also generate high frequency flux. These cumulative fluxes link with winding, core, frame, shaft, and other metallic parts. In this way the inductive coupling is formed. Capacitive and inductive couplings together give an additive effect for the generation of shaft voltages for this long shaft machine.

As long as the shaft is insulated from ground at one end, then circulating current will not flow. In the presence of voltage, shaft acts like a charged capacitor with lubricating oil as a dielectric medium. Shaft is kept at ground potential with two grounding brushes. But during high speed rotation it is not possible to keep shaft at ground potential all the time. When the contact resistance of the grounding brushes is high (due to poor contact or contamination) the charged shaft discharges creating short high current impulses. Electrostatic discharges (see Figures 14 and 15) also happen between shaft and bearing when dielectric breakdown of lubricating oil film takes place due to close proximity or high voltage. The discharge current pulses contain high amount of energy. This energy is the prime source of bearing erosion and is commonly known as fluting or electric discharge machining (EDM) [30]. This EDM can damage bearing, shaft, and other nearby metal parts. As an example, pitting on a shaft due to shaft discharge is shown in Figure 16.

6. Conclusion

With the prime features, such as quick response, easy monitoring, control, and troubleshooting, SES has made the first choice of the designer for machine field excitation system. The SES discussed in this system is responsible for operating a single machine under different mode. Without any prime

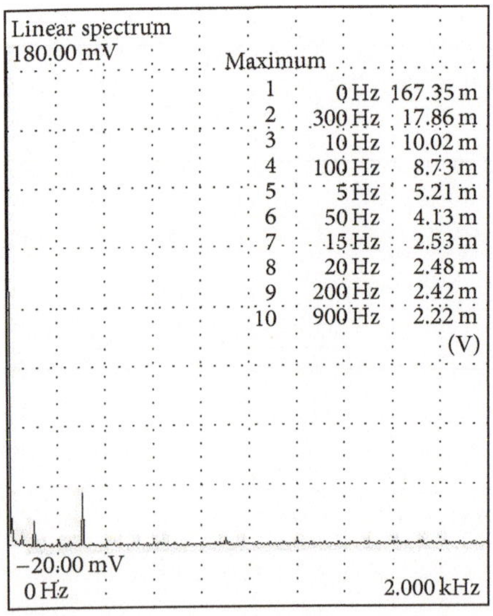

Figure 12: FFT for actual rotor current.

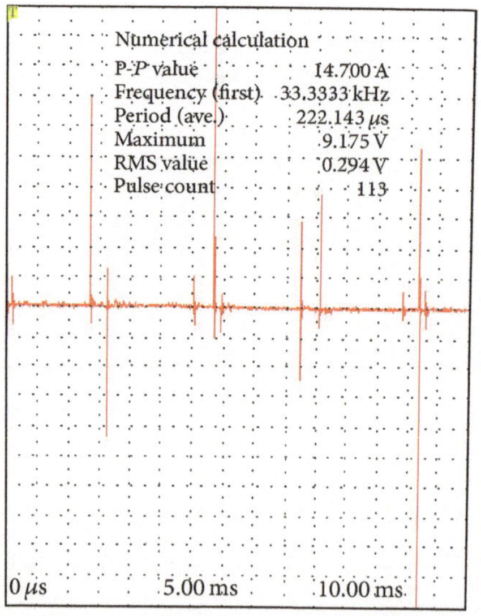

Figure 14: Recorded signals of EDM pulses.

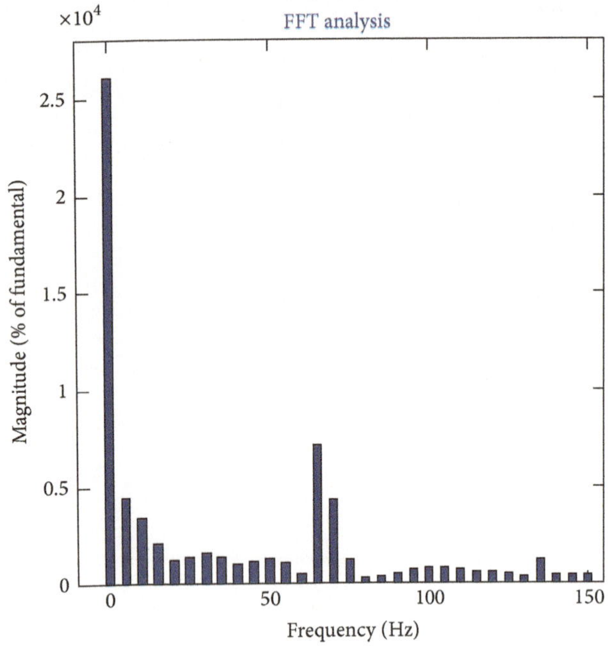

Figure 13: FFT for simulated rotor current.

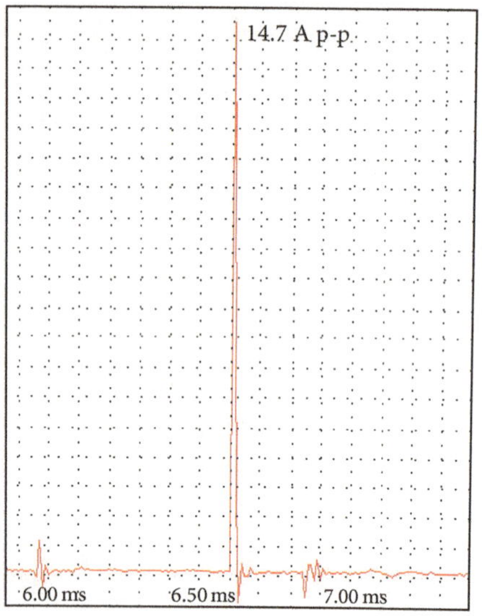

Figure 15: Close look on a single EDM pulse.

mover this large machine can be started softly without much loading of the supply source. The machine can run as a motor or generator. The SES can also be used for braking or reversal operation of the machine. The SES applied in a short circuit alternator is analysed with the recorded waveforms. The same is modelled with the help of Simulink in MATLAB. Simulated waveforms are compared with the real ones found similar in nature.

In spite of several advantages the negative side of this technology is the generation of shaft voltage. Sparking on the grounding brushes is the evidence of presence of shaft voltages. Many reasons are behind this unwanted voltage, for example, capacitive coupling, inductive coupling, and common mode voltage. Thyristors switching in the converter bridge and high level of harmonics in current and voltage cause capacitive and inductive couplings. This was proven by the FFT analysis on current and voltage waveforms. Both the simulated and actual measurements show that the waveforms contain high percentage of harmonics.

Among the various types of bearing currents discharge mode current or EDM is the most harmful causing erosion on the shaft and bearings. Life of machine is reduced by this effect. Large machine requires huge investments on shaft repair and bearing replacement. To arrest the premature bearing failure further study in this field is needed to eliminate the shaft voltage.

FIGURE 16: Pitting on journal shaft due to EDM.

Conflict of Interests

The authors declare that there is no conflict of interests regarding the publication of this paper.

Acknowledgment

Authors are thankful to the management of Central Power Research Institute, India, for supporting this work.

References

[1] L. J. Lane, D. F. Rogers, and P. A. Vance, "Design and tests of a static excitation system for industrial and utility steam turbine-generators," *AIEE*, vol. 80, no. 3, pp. 1077–1085, 1962.

[2] R. C. Schaefer, "Applying static excitation systems," *IEEE Industry Applications Magazine*, vol. 4, no. 6, pp. 41–49, 1998.

[3] Y.-Z. Zhang, "Study of process of starting pumped storage machines by static frequency converter with field current controlled," in *Proceedings of the 2nd International Conference on Signal Processing Systems (ICSPS '10)*, pp. V1224–V1227, July 2010.

[4] R. B. Fisher, "Introduction of static frequency converters on SEPTA's 25-Hz commuter rail system," in *1990 ASME/IEEE Joint Railroad Conference*, pp. 149–155, April 1990.

[5] S.-H. Park, S.-H. Hwang, J.-M. Kim, H.-S. Ryu, and J.-H. Lee, "A starting-up control algorithm of large synchronous generation motor for gas turbosets," in *Proceedings of the IEEE International Symposium on Industrial Electronics (ISIE '08)*, pp. 502–508, Cambridge, UK, June 2008.

[6] H. Taguchi, S. Tamai, Y. Hosokawa, and A. Ando, "APS control method for gas turbine startup by SFC," in *Proceedings of the International Power Electronics Conference*, pp. 264–269, IEEE, 2010.

[7] A. K. Datta, M. Dubey, N. R. Mondal, and B. V. Raghavaiah, "Motor-less operation of short circuit generator—a CPRI perspective," in *Proceedings of the International Conference on Electrical Power and Energy Systems (ICEPES '10)*, pp. 439–445, MANIT, Bhopal, India, August 2010.

[8] A. K. Datta, M. A. Ansari, N. R. Mondal, and B. V. Raghavaiah, "A novel use of power electronics: prime mover-less alternator with static drive & excitation system," *International Journal of Electronics & Communication Technology*, vol. 3, no. 1, pp. 472–475, 2012.

[9] A. K. Datta, M. Dubey, and S. Jain, "Study of shaft voltage & bearing currents in electrical machines," in *Proceedings of the IEEE Students' Conference on Electrical, Electronics and Computer Science (SCEECS '12)*, pp. 1–4, Bhopal, India, March 2012.

[10] J. M. Erdman, R. J. Kerkman, D. W. Schlegel, and G. L. Skibinski, "Effect of PWM inverters on AC motor bearing currents and shaft voltages," *IEEE Transactions on Industry Applications*, vol. 32, no. 2, pp. 250–259, 1996.

[11] C. Ammann, K. Reichert, R. Joho, and Z. Posedel, "Shaft voltages in generators with static excitation systems—problems and solution," *IEEE Transactions on Energy Conversion*, vol. 3, no. 2, pp. 409–419, 1988.

[12] S. Chen and T. A. Lipo, "Circulating type motor bearing current in inverter drives," *IEEE Industry Applications Magazine*, vol. 4, no. 1, pp. 32–38, 1998.

[13] A. Muetze and A. Binder, "Calculation of circulating bearing currents in machines of inverter-based drive systems," *IEEE Transactions on Industrial Electronics*, vol. 54, no. 2, pp. 932–938, 2007.

[14] A. Muetze, "On a new type of inverter-induced bearing current in large drives with one journal bearing," *IEEE Transactions on Industry Applications*, vol. 46, no. 1, pp. 240–248, 2010.

[15] H.-S. Ryu, B. Kim, J. Lee, and I. Lim, "A study of synchronous motor drive using static frequency converter," in *Proceedings of the 12th International Power Electronics and Motion Control Conference (EPE-PEMC '06)*, pp. 1496–1499, IEEE, September 2006.

[16] N. Zhu, D. Xu, B. Wu, N. R. Zargari, M. Kazerani, and F. Liu, "Common-mode voltage reduction methods for current-source converters in medium-voltage drives," *IEEE Transactions on Power Electronics*, vol. 28, no. 2, pp. 995–1006, 2013.

[17] R. S. Kanchan, P. N. Tekwani, and K. Gopakumar, "Three-level inverter scheme with common mode voltage elimination and dc link capacitor voltage balancing for an open-end winding induction motor drive," *IEEE Transactions on Power Electronics*, vol. 21, no. 6, pp. 1676–1683, 2006.

[18] M. R. Baiju, K. K. Mohapatra, R. S. Kanchan, and K. Gopakumar, "A dual two-level inverter scheme with common mode voltage elimination for an induction motor drive," *IEEE Transactions on Power Electronics*, vol. 19, no. 3, pp. 794–805, 2004.

[19] K. K. Yuen, H. S. Chung, and V. S. Cheung, "An active low-loss motor terminal filter for overvoltage suppression and common-mode current reduction," *IEEE Transactions on Power Electronics*, vol. 27, no. 7, pp. 3158–3172, 2012.

[20] Z. Zhao, Y. Zhong, H. Gao, L. Yuan, and T. Lu, "Hybrid selective harmonic elimination PWM for common-mode voltage reduction in three-level neutral-point-clamped inverters for variable

speed induction drives," *IEEE Transactions on Power Electronics*, vol. 27, no. 3, pp. 1152–1158, 2012.

[21] H. Akagi and K. Isozaki, "A hybrid active filter for a three-phase 12-pulse diode rectifier used as the front end of a medium-voltage motor drive," *IEEE Transactions on Power Electronics*, vol. 27, no. 1, pp. 69–77, 2012.

[22] F. Wang, "Motor shaft voltages and bearing currents and their reduction in multilevel medium-voltage PWM voltage-source-inverter drive applications," *IEEE Transactions on Industry Applications*, vol. 36, no. 5, pp. 1336–1341, 2000.

[23] H. Akagi and T. Doumoto, "An approach to eliminating high-frequency shaft voltage and ground leakage current from an inverter-driven motor," *IEEE Transactions on Industry Applications*, vol. 40, no. 4, pp. 1162–1169, 2004.

[24] A. K. Datta, M. Dubey, and S. Jain, "Effect of static power supply in alternator used for short-circuit testing-observation of shaft voltage," *IEEE Transactions on Power Electronics*, 2014.

[25] A. K. Datta, M. Dubey, and S. Jain, "Investigation of bearing currents in dual mode operation of synchronous machine with static excitation system," *Electrical and Electronics Engineering*, vol. 2, no. 4, pp. 45–53, 2013.

[26] A. K. Datta, M. Dubey, and S. Jain, "Study of shaft voltage on short circuit alternator with static frequency converter," *International Journal of Electrical, Electronic Science and Engineering*, vol. 7, no. 11, pp. 692–698, 2013.

[27] T.-J. Liang, J.-F. Chen, C.-L. Chu, and K.-J. Chen, "Analysis of 12 pulse phase control AC/DC converter," in *Proceedings of the 3rd IEEE International Conference on Power Electronics and Drive Systems (PEDS '99)*, pp. 779–783, Hong Kong, July 1999.

[28] A. K. Datta, G. Venkateswarlu, M. A. Ansari, and N. R. Mondal, "Excitation control during short circuit test sequence of 1500 MVA short circuit generator," in *Proceedings of the International Conference on Advances in Computer, Electronics & Electrical Engineering (ICACEEE '12)*, pp. 207–211, Mumbai, India, March 2012.

[29] The MathWorksTM , Inc., USA, Release 2008a (R2008a).

[30] S. Bell, T. J. Cookson, S. A. Cope et al., "Experience with variable-frequency drives and motor bearing reliability," *IEEE Transactions on Industry Applications*, vol. 37, no. 5, pp. 1438–1446, 2001.

20

Effect of DC Link Control Strategies on Multiterminal AC-DC Power Flow

Shagufta Khan and Suman Bhowmick

Department of Electrical Engineering, Delhi Technological University, Delhi 110042, India

Correspondence should be addressed to Shagufta Khan; khan.shagufta7@gmail.com

Academic Editor: George E. Tsekouras

For power-flow solution of power systems incorporating multiterminal DC (MTDC) network(s), five quantities are required to be solved per converter. On the other hand, only three independent equations comprising two basic converter equations and one DC network equation exist per converter. Thus, for solution, two additional equations are required. These two equations are derived from the control specifications adopted for the DC links. Depending on the application, several combinations of valid control specifications are possible. Each combination of a set of valid control specifications is known as a control strategy. The number of control strategies increases with an increase in the number of the DC terminals or converters. It is observed that the power-flow convergence of integrated AC-MTDC power systems is strongly affected by the control strategy adopted for the DC links. This work investigates the mechanism by which different control strategies affect the power-flow convergence pattern of AC-MTDC power systems. To solve the DC variables in the Newton-Raphson (NR) power-flow model, sequential method is considered in this paper. Numerous case studies carried out on a three-terminal DC network incorporated in the IEEE-300 bus test system validate this.

1. Introduction

Multiterminal DC (MTDC) systems have been proven to be more versatile on account of their ability to fully exploit the economic and technical advantages of HVDC technology. Since the success of the first MTDC system "Sardinia-Corsica-Italy Scheme" in 1991, the application of MTDC systems has shown a steady increase. Further, with the rapid progress of renewable energy sources and offshore wind farms, MTDC applications are increasing. MTDC transmits bulk power from several generating stations to several load centres. It is more flexible and economical, being capable of connecting more than two asynchronous AC systems [1–4].

Now, for planning, operation, and control of power systems incorporated with MTDC links, power-flow solution of such systems is required. Power-flow solution of integrated AC-MTDC systems requires five quantities to be solved per converter. These include the DC voltage, the DC current, the control angle, the converter transformer tap ratio, and the converter power factor. On the other hand, only three independent equations comprising two basic converter equations and one DC network equation exist per converter. Thus, for solution, two additional equations are usually required. These two equations are derived from the control specifications adopted for the DC links. Thus, mathematically, the control specifications are used to bridge the gap between the number of independent equations and the number of unknown quantities. Control specifications usually include specified values of converter transformer tap ratio, converter control angle, DC voltage, DC current, or DC power. Depending on the application, several combinations of valid control specifications are possible. Each combination of a set of valid control specifications is known as a control strategy. The number of possible control strategies increases drastically with increase in the number of the DC terminals or converters. Out of a myriad of combinations, only some control strategies are practically adopted in practice.

To carry out power flow in AC-MTDC systems, two different algorithms have generally been reported in the literature. These are known as the unified and the sequential method, respectively. Some excellent research works on the unified and the sequential power-flow method are presented in [5–19]. Unlike the unified method, the sequential method is easier to implement and poses lesser computational burden

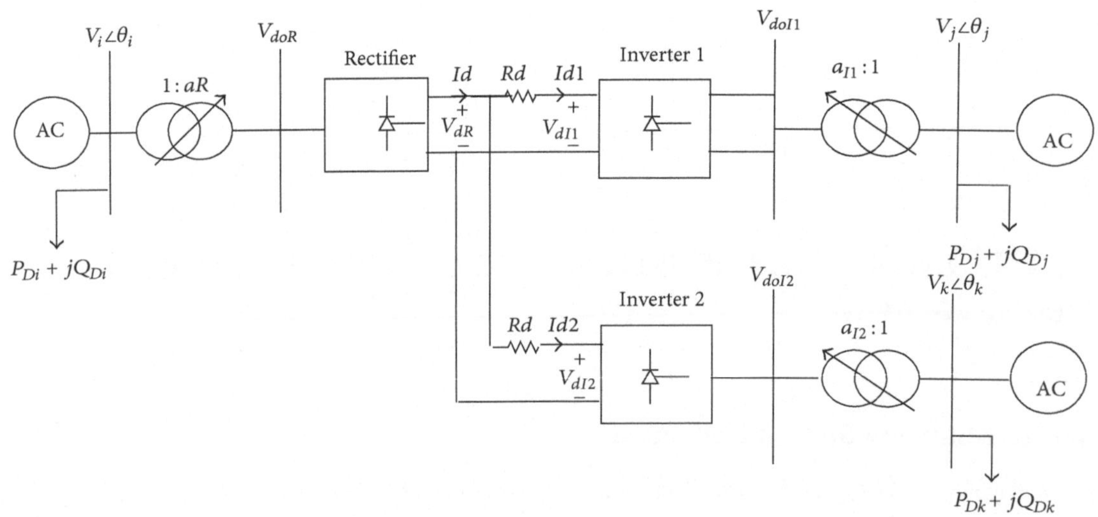

FIGURE 1: Three-terminal HVDC link between buses "*i*," "*j*," and "*k*" of an existing power system network.

due to the smaller size of the Jacobian matrix. Consequently, in this work, only the sequential AC-DC power-flow algorithm has been considered.

In the sequential AC-DC power-flow algorithm, the AC and DC systems are solved separately in each iteration and are coupled by injecting an equivalent amount of real and reactive power at the terminal AC buses. It is observed that each different control strategy affects the sequential power-flow convergence in a uniquely different manner. For standard control strategies, that is, constant DC voltage or current or power, the convergence rate can be improved by solving the DC and AC systems independently. On the other hand, for nonstandard ones like constant tap ratios and constant terminal voltage, the convergence may suffer. The mechanism by which this occurs has not been very clear and has not been exclusively addressed in the literature. This motivated the authors to investigate how different control strategies affect the power-flow convergence. The work is validated by numerous case studies carried out on a three-terminal DC network incorporated in the IEEE 300-bus test system. Although a large number of control strategies are possible with the three-terminal DC network, however, due to a lack of space, only results corresponding to nine different control strategies are reported here.

This paper is arranged as follows: in Section 2, the mathematical modelling of the AC-MTDC system is presented. Section 3 details a few of the control strategies possible for a three-terminal DC network. In Section 4, the AC-MTDC power-flow equations are presented, with the DC link powers acting as an equivalent load on the converter AC buses. Section 5 details the case studies carried out with a three-terminal DC network incorporated in the IEEE 300-bus test system. The conclusions are presented in Section 6.

2. AC-MTDC System Modelling

Figure 1 shows a typical AC power system incorporating a three-terminal DC network. The DC network contains two HVDC links. The first link is connected in the branch "*i-j*" between any two system buses "*i*" and "*j*" of the network while the second one is connected in the branch "*i-k*" between system buses "*i*" and "*k*." The three converters representing one rectifier and two inverters are connected to the AC system at buses "*i*," "*j*," and "*k*," respectively, through their respective converter transformers. The effect of the DC links is accounted for as equivalent amount of real and reactive power injections P_{dR}, Q_{dR}, P_{dI1}, Q_{dI1}, P_{dI2}, and Q_{dI2} at the converters' AC terminal buses "*i*," "*j*," and "*k*," respectively. Although these power injections are not shown in Figure 1, they are included in the analysis by appropriate modifications of the power-flow equations as detailed later in Section 4.

Prior to the selection of variables and formulation of the equations, several basic assumptions are required which are generally accepted in the analysis of steady state DC converter operation [1–4]. These are as follows.

(1) The three AC voltages at the terminal bus bars are balanced and sinusoidal.

(2) The operation of the converters is perfectly balanced.

(3) The direct currents and voltages are smooth.

(4) The converter transformers are lossless and their magnetizing admittances are ignored.

Subsequently, the DC and AC equations are combined together which requires the translation of the converter equations into the per-unit system in order to use them with AC system per-unit equations. The base values for the DC system are defined as the following.

Selection of base values for the DC system:

$$V_{\text{dc base}} = kV_{\text{ac base}}; \quad \text{where } k = \frac{3\sqrt{2}}{\pi}n_b,$$

$$I_{\text{dc base}} = \frac{\sqrt{3}}{k}I_{\text{ac base}},$$

TABLE 1: Some control strategies for a 3-terminal DC link.

Control strategies	Specified quantities: Included in DC load flow	Specified quantities: Excluded in DC load flow	Unknown quantities
1	P_{dI1}, P_{dI2}, V_{dR}	$\alpha_R, \gamma_{I1}, \gamma_{I2}$	$V_{dI1}, V_{dI2}, P_{dR}, I_{dI1}, I_{dI2}, I_{dR}, a_R, a_{I1}, a_{I2}, \cos\phi_R, \cos\phi_{I1}, \cos\phi_{I2}$
2	P_{dI1}, I_{dI2}, V_{dR}	$\alpha_R, \gamma_{I1}, \gamma_{I2}$	$V_{dI1}, V_{dI2}, P_{dI2}, P_{dR}, I_{dI1}, I_{dR}, a_R, a_{I1}, a_{I2}, \cos\phi_R, \cos\phi_{I1}, \cos\phi_{I2}$
3	P_{dI2}, I_{dI1}, V_{dR}	$\alpha_R, \gamma_{I1}, \gamma_{I2}$	$V_{dI1}, V_{dI2}, P_{dI1}, P_{dR}, I_{dI2}, I_{dR}, a_R, a_{I1}, a_{I2}, \cos\phi_R, \cos\phi_{I1}, \cos\phi_{I2}$
4	I_{dI1}, I_{dI2}, V_{dR}	$\alpha_R, \gamma_{I1}, \gamma_{I2}$	$V_{dI1}, V_{dI2}, P_{dI1}, P_{dI2}, P_{dR}, I_{dR}, a_R, a_{I1}, a_{I2}, \cos\phi_R, \cos\phi_{I1}, \cos\phi_{I2}$
5	P_{dI1}, P_{dI2}, V_{dR}	a_R, a_{I1}, a_{I2}	$V_{dI1}, V_{dI2}, P_{dR}, I_{dI1}, I_{dI2}, I_{dR}, a_R, a_{I1}, a_{I2}, \cos\phi_R, \cos\phi_{I1}, \cos\phi_{I2}$
6	I_{dI1}, I_{dI2}, V_{dR}	a_R, a_{I1}, a_{I2}	$V_{dI1}, V_{dI2}, P_{dI1}, P_{dI2}, P_{dR}, I_{dR}, a_R, a_{I1}, a_{I2}, \cos\phi_R, \cos\phi_{I1}, \cos\phi_{I2}$
7	P_{dI1}, I_{dI2}, V_{dR}	a_R, a_{I1}, a_{I2}	$V_{dI1}, V_{dI2}, P_{dI2}, P_{dR}, I_{dI1}, I_{dR}, a_R, a_{I1}, a_{I2}, \cos\phi_R, \cos\phi_{I1}, \cos\phi_{I2}$
8	I_{dI1}, P_{dI2}, V_{dR}	a_R, a_{I1}, a_{I2}	$V_{dI1}, V_{dI2}, P_{dI1}, P_{dR}, I_{dI2}, I_{dR}, a_R, a_{I1}, a_{I2}, \cos\phi_R, \cos\phi_{I1}, \cos\phi_{I2}$
9	$P_{dI1}, P_{dI2}, V_{dI1}$	$\alpha_R, \gamma_{I1}, \gamma_{I2}$	$V_{dIR}, V_{dI2}, P_{dR}, I_{dI1}, I_{dI2}, I_{dR}, a_R, a_{I1}, a_{I2}, \cos\phi_R, \cos\phi_{I1}, \cos\phi_{I2}$

$$
\begin{aligned}
Z_{\text{dc base}} &= k^2 Z_{\text{ac base}}, \\
R_{\text{dc base}} &= \frac{3}{\pi} n_b X_{c\,\text{base}},
\end{aligned}
\tag{1}
$$

which culminate in the per-unit AC-DC system equations as given in the following.

Basic converter equations for a 3-terminal DC network in per-unit system:

$$
\begin{aligned}
V_{dR} &= a_R V_i \cos\alpha_R - X_c I_{dR}, \\
V_{dI1} &= a_{I1} V_j \cos\gamma_{I1} - X_c I_{dI1}, \\
V_{dI2} &= a_{I2} V_k \cos\gamma_{I2} - X_c I_{dI2}, \\
V_{dR} &= a_R V_i \cos\phi_R, \\
V_{dI1} &= a_{I1} V_j \cos\phi_{I1}, \\
V_{dI2} &= a_{I2} V_k \cos\phi_{I2}, \\
I_{dI1} &= \frac{V_{dR} - V_{dI1}}{R_d}, \\
I_{dI2} &= \frac{V_{dR} - V_{dI2}}{R_d}, \\
I_{dR} &= I_{dI1} + I_{dI2}, \\
P_{dR} &= V_{dR} I_{dR}, \\
P_{dI1} &= V_{dI1} * \left[\frac{V_{dR} - V_{dI1}}{R_d}\right], \\
P_{dI2} &= V_{dI2} * \left[\frac{V_{dR} - V_{dI2}}{R_d}\right].
\end{aligned}
\tag{2}
$$

From the above, it can be observed that twelve independent equations exist as against a total of eighteen unknowns. Thus, for a complete solution of the HVDC quantities, six quantities (two per converter) are needed to be specified. These are derived from the control specifications adopted for the DC links. Each combination of a set of valid control specifications is known as a control strategy. Theoretically, the number of possible control strategies increases drastically with increase in the number of the DC terminals or converters. However, only a few of the possible control strategies are practically adopted in practice. This is explained in the next section.

3. MTDC Control Strategies

As already discussed in the previous section, for a three-terminal DC network, a myriad of control strategies are possible. However, due to a lack of space, only nine typical control strategies are considered in this paper. These are detailed in Table 1.

3.1. Control Strategy 1. In this control strategy, the DC voltage and the firing angle are specified for the rectifier. On the other hand, the active powers for both inverters are specified along with their extinction angles.

3.2. Control Strategy 2. In this control strategy, the DC voltage and the firing angle are specified for the rectifier. For inverter 1, the active power is specified while the DC current is specified for inverter 2. In addition, the extinction angles of both inverters are specified.

3.3. Control Strategy 3. In this control strategy, at the rectifier terminal, the DC voltage and the firing angle are specified. The DC current is specified at inverter 1 while the active power is specified at inverter 2. Also, the extinction angles of both inverters are specified.

3.4. Control Strategy 4. In this control strategy, the DC voltage and the firing angle are specified at the rectifier end while the DC currents and the extinction angles are specified for both inverters.

3.5. Control Strategy 5. In this control strategy, the DC voltage is specified for the rectifier while the active powers are specified for both inverters. In addition, the tap settings for all the three converter transformers are specified.

3.6. Control Strategy 6. In this control strategy, the DC voltage is specified for the rectifier while the DC currents are specified for both inverters. The tap settings for all the three converter transformers are also specified.

3.7. Control Strategy 7. In this control strategy, at the rectifier terminal, the DC voltage is specified. The active power is specified at inverter 1 while the DC current is specified at inverter 2. In addition, the tap settings are specified for all the three converter transformers.

3.8. Control Strategy 8. In this control strategy, the DC voltage is specified at the rectifier terminal. For inverter 1, the DC current is specified while for inverter 2 the active power is specified. The tap settings for all the three converter transformers are also specified.

3.9. Control Strategy 9. In this control strategy, the firing angle is specified for the rectifier while the DC voltage is specified for inverter 1. In addition, the values of the active powers and the extinction angles are specified for both inverters.

4. AC-DC Power-Flow Equations

As already discussed in Section 1, for solving the AC-MTDC system, two additional equations per converter are required, which are derived from the control specifications adopted for the system. For the three-terminal DC link considered in this paper, this translates into six specified variables. These six specified variables are divided into two subsets. The first set of specified variables is used only for the DC load flow and corresponds to specifying active power or current at all terminals except at the voltage setting terminal. In this paper, the DC load flow is solved by the Newton-Raphson method.

Subsequently, for solving the AC power flow, the effects of the DC links are included in the power-flow equations by injecting an equivalent amount of real and reactive power at the terminal AC buses connected to the converters. The values of these active and reactive powers are computed using the specified variables and the variables computed from the DC power flow. This results in appropriate modifications of the mismatch equations at the converter terminal AC buses. These are given below.

For any AC bus "i," which is not connected to any DC link, the mismatches in the active and reactive power injections are given, respectively, by

$$\begin{aligned}\Delta P_i &= P_i^{\mathrm{sp}} - \sum_{k=1}^{n} V_i V_k Y_{ik} \cos\left(\theta_i - \theta_k - \phi_{ik}\right), \\ \Delta Q_i &= Q_i^{\mathrm{sp}} - \sum_{k=1}^{n} V_i V_k Y_{ik} \sin\left(\theta_i - \theta_k - \phi_{ik}\right).\end{aligned} \tag{3}$$

The DC network contains two HVDC links. The first link is connected in the branches "i" and "j" between any two buses "i" and "j" of the network while the second one is connected in the branch "i-k" between system buses "i" and "k." The three converters representing one rectifier and two inverters are connected to the AC system at buses "i," "j," and "k," respectively, through their respective converter transformers. The effect of the DC link can be incorporated in the AC power flow as equivalent active and reactive power injections "P_{dR}" and "Q_{dR}" at the rectifier bus "i," along with "P_{dI1}" and "Q_{dI1}" and "P_{dI2}" and "Q_{dI2}" at the inverter buses "j" and "k," respectively. Therefore, the mismatches in the active and reactive power injections can be written as

$$\begin{aligned}\Delta P_i &= P_i^{\mathrm{sp}} - \sum_{p=1}^{n} V_i V_p Y_{ip} \cos\left(\theta_i - \theta_p - \phi_{ip}\right) - P_{dR}, \\ \Delta Q_i &= Q_i^{\mathrm{sp}} - \sum_{p=1}^{n} V_i V_p Y_{ip} \sin\left(\theta_i - \theta_p - \phi_{ip}\right) - Q_{dR}, \\ \Delta P_j &= P_j^{\mathrm{sp}} - \sum_{p=1}^{n} V_j V_p Y_{jp} \cos\left(\theta_j - \theta_p - \phi_{jp}\right) + P_{dI1}, \\ \Delta Q_j &= Q_j^{\mathrm{sp}} - \sum_{p=1}^{n} V_j V_p Y_{jp} \sin\left(\theta_j - \theta_p - \phi_{jp}\right) - Q_{dI1}, \\ \Delta P_k &= P_k^{\mathrm{sp}} - \sum_{p=1}^{n} V_k V_p Y_{kp} \cos\left(\theta_k - \theta_p - \phi_{kp}\right) + P_{dI2}, \\ \Delta Q_k &= Q_k^{\mathrm{sp}} - \sum_{p=1}^{n} V_k V_p Y_{kp} \sin\left(\theta_k - \theta_p - \phi_{kp}\right) - Q_{dI2},\end{aligned} \tag{4}$$

where $P_{dR} = V_{dR} I_{dR}$, $Q_{dR} = P_{dR} \tan\phi_R$, $P_{dI1} = V_{dI1} I_{dI1}$, $Q_{dI1} = P_{dI1} \tan\phi_{I1}$, $P_{dI2} = V_{dI2} I_{dI2}$, and $Q_{dI2} = P_{dI2} \tan\phi_{I2}$. In the above equations, the equivalent active power injections "P_{dR}," "P_{dI1}," and "P_{dI2}" are usually specified or can be very easily computed by manipulation of the specified variables. However, for the equivalent reactive power injections Q_{dR}, Q_{dI1}, and Q_{dI2}, the case is different, depending on the control strategy adopted for the DC links. For control strategies 1, 2, 3, 4, and 9, ϕ_R, ϕ_{I1}, and ϕ_{I2} (and hence Q_{dR}, Q_{dI1}, and Q_{dI2}) can be computed by manipulation of the specified variables. However, for control strategies 5, 6, 7, and 8, ϕ_R, ϕ_{I1}, and ϕ_{I2} (and hence Q_{dR}, Q_{dI1}, and Q_{dI2}) are dependent on both the specified variables and the AC state variables, which are updated in every iteration. This affects the convergence pattern. The steps involved in the computation of the active and reactive power injections for only two typical control strategies 1 and 5 are detailed in Table 7. In control strategy 5 the reactive power injections are dependent on the AC power-flow iterative process while in control strategy 1 they are independent of it. Although the steps involved in the computation of the power injections pertaining to the rest of the control strategies are not shown due to limitations of space, they can be computed in ways similar to control strategies 1 and 5.

It is important to note the conventions of the signs of the equivalent real and reactive power injections representing the DC link. It is assumed that the rectifier consumes both real and reactive power from the AC grid while the inverters supply real power and consume reactive power.

5. Case Studies and Results

Numerous case studies were carried out on a three-terminal DC network comprising one rectifier and two inverters incorporated in the IEEE 300-bus test system [20]. All the three converters are connected to their respective AC buses by tap changing transformers. Although a large number of control strategies are possible with the three-terminal DC network, however, due to limitations of space, only results corresponding to nine different control strategies are reported here. For all the case studies, the commutating reactance and the DC link resistance were chosen as 0.1 p.u. and 0.01 p.u., respectively. The number of bridges "n_b" for all the converters was taken to be equal to 2. To minimize the reactive power requirement at the rectifier and inverter terminals and the overall system losses, the rectifier firing angles and the inverter extinction angles are kept within 5° to 15° and 15° to 22°, respectively, for all the case studies. In a similar manner, the tap ratios of the converter transformers are kept between 0.85 and 1.15. A convergence tolerance of 10^{-12} p.u. was uniformly adopted for all the case studies. In each of the case studies, "NI" refers to the number of iterations taken by the algorithm to converge to the specified tolerance (10^{-12} p.u.). In general, "NI" is representative of the degree of convergence of a power-flow study.

Case Study I. In this case study, two HVDC links are considered. The first link is incorporated between buses "25-26" and the second one between buses "25-232." The converter connected to bus number 25 acts as a rectifier. On the other hand, both converters connected to buses 26 and 232 act as inverters. Control strategy 1 is adopted for the control of the DC links. Corresponding to this strategy, the active power flows on the links 25-26 and 25-232 are set to 0.3 p.u. and 0.1 p.u., respectively. The DC voltage on the rectifier side is set at 1 p.u. The firing angle for the rectifier and the extinction angles for both inverters are set to 5° and 15°, respectively. The power-flow solution is shown in columns 5–8 of Table 2 (corresponding to control strategy 1). The state variables pertaining to the AC and DC systems are denoted as ACSV and DCSV, respectively. From Table 2, it is observed that the power-flow convergence pattern in this case is similar to that of the base case power flow (without HVDC link), converging in seven iterations. This is because the power injections representing the DC links can be computed prior to the AC power flow, as elaborated from the steps given in the first column of Table 7. This is reiterated from the convergence characteristics corresponding to the base case and case study 1, as shown in Figures 2 and 3, respectively. The convergence characteristic plots the error (maximum absolute mismatch in p.u.) against the number of iterations "NI."

Case Study II. In this case study, two HVDC links are connected between buses "213-214" and "213-216." While the converter connected to bus number 213 acts as a rectifier, both converters connected to buses 214 and 216 act as inverters. In this case, control strategy 2 is adopted for the control of the DC links. The rectifier end DC voltage is specified along with the firing angle. While the first inverter acts on constant power mode, the second one acts on the constant current mode. In addition, the extinction angles for both inverters are specified. These values are detailed in columns 1–4 of Table 2 (corresponding to control strategy 2). The power-flow solution is detailed in columns 5–8 of Table 2. The convergence pattern is similar to that of control strategy 1, as the DC link power injections can be computed prior to the AC power flow. The steps involved in the computation of the DC link power injections (Table 7), although similar in nature to control strategy 1, are not shown due to limitations of space.

Case Study III. In this case, two HVDC links are connected between buses "109-146" and "109-147." The converter connected to bus number 109 acts as a rectifier, while both converters connected to buses 146 and 147 act as inverters. For control of the DC links, control strategy 3 is selected. The rectifier end DC voltage is specified along with the firing angle. While the first inverter acts on constant current mode, the second one acts on the constant power mode. In addition, the extinction angles for both inverters are specified. These values are detailed in columns 1–4 of Table 2. The power-flow solution is detailed in columns 5–8 of Table 2 (corresponding to control strategy 3). From Table 2, it is observed that the convergence pattern obtained in this case is again similar to control strategies 1 and 2, which is expected as the power injections can be computed prior to the AC power flow.

Case Study IV. In this case study, two HVDC links are connected between buses "101-102" and "101-104." The rectifier is connected to bus number 101. The inverters are connected to buses 102 and 104, respectively. For control of the two DC links, control strategy 4 is adopted. The rectifier end DC voltage is specified along with the firing angle. Both inverters are operated at constant current and constant extinction angle modes. The values of the specified quantities are detailed in columns 1–4 of Table 3. The power-flow solution is detailed in columns 5–8 of Table 3 (corresponding to control strategy 4). It is observed that as expected the convergence pattern is similar to that obtained with control strategies 1 to 4 (since the DC link power injections are known before the AC power flow).

Case Study V. In this case, two HVDC links are incorporated between buses "86-87" and "86-90." The converter connected to bus number 86 acts as a rectifier while both converters connected to buses 87 and 90 act as inverters. Control strategy 5 is selected for the control of the DC links. In this strategy, the rectifier end DC voltage is specified while both inverters are operated in the constant power mode. In addition, the tap ratios of all the three converter transformers are specified. Columns 1–4 of Table 3 show these specified values. The power-flow solution is shown in columns 5–8 of Table 3 (corresponding to control strategy 5). From the power-flow solution, it is observed that, unlike control strategies 1 to 4, with this control strategy, the convergence pattern is adversely affected. This is because for this control strategy the computation of the reactive power injections (for the DC links) is dependent on the AC power-flow variables

TABLE 2: Study of IEEE-300 bus system with a three-terminal DC network.

HVDC links		HVDC link specification		Power-flow solution			
Rectifier bus	Inverter buses	Control strategy 1 Spec. values		AC terminal buses ACSV		HVDC variables DCSV	
						V_{dI1}	0.998
		P_{dI1} (p.u.)	0.2	V_{25}	0.9632	V_{dI2}	0.999
						P_{dR}	0.3005
						I_{dI1}	0.2004
	26	P_{dI2} (p.u.)	0.1	θ_{25}	-2.122°	I_{dI2}	0.1001
						I_{dR}	0.3005
25						a_R	1.0386
		V_{dR} (p.u.)	1	V_{26}	0.9564	a_{I1}	1.103
						a_{I2}	1.0441
		α_R (deg.)	5	θ_{26}	-3.715°	$\cos\phi_R$	0.9766
						$\cos\phi_{I1}$	0.9377
	232	γ_{I1} (deg.)	15	V_{232}	0.9841	$\cos\phi_{I2}$	0.9563
		γ_{I2} (deg.)	15	θ_{232}	-1.792°	NI	7
		Control strategy 2 Spec. values		AC terminal buses ACSV		HVDC variables DCSV	
						V_{dI1}	0.994
		P_{dI1} (p.u.)	0.6	V_{213}	1.0067	V_{dI2}	0.996
	214					P_{dI2}	0.4036
						P_{dR}	1.0036
		I_{dI2} (p.u.)	0.4	θ_{213}	-21.69°	I_{dI1}	0.6036
						I_{dR}	1.0036
213						a_R	1.0972
		V_{dR} (p.u.)	1	V_{214}	1.0087	a_{I1}	1.0822
						a_{I2}	1.0187
		α_R (deg.)	5	θ_{214}	-21.125°	$\cos\phi_R$	0.9053
						$\cos\phi_{I1}$	0.9106
	216	γ_{I1} (deg.)	15	V_{216}	1.0528	$\cos\phi_{I2}$	0.9286
		γ_{I2} (deg.)	15	θ_{216}	-21.39°	NI	7
		Control strategy 3 Spec. values		AC terminal buses ACSV		HVDC variables DCSV	
						V_{dI1}	0.999
		P_{dI2} (p.u.)	0.1	V_{109}	0.9993	V_{dI2}	0.996
	146					P_{dR}	0.1101
						P_{dI1}	0.0101
		I_{dI1} (p.u.)	0.01	θ_{109}	5.6163°	I_{dI2}	0.1001
						I_{dIR}	0.1101
109						a_R	1.1151
		V_{dR} (p.u.)	1	V_{146}	0.9618	a_{I1}	1.0871
						a_{I2}	1.1439
		α_R (deg.)	5	θ_{146}	-7.1728°	$\cos\phi_R$	0.8974
	147					$\cos\phi_{I1}$	0.9564
		γ_{I1} (deg.)	15	V_{147}	0.9947°	$\cos\phi_{I2}$	0.878
		γ_{I2} (deg.)	15	θ_{147}	-4.9558°	NI	7

Table 3: Study of IEEE-300 bus system with a three-terminal HVDC network.

HVDC link		HVDC link specification		Power-flow solution			
Rectifier bus	Inverter buses	Control strategy 4 Spec. values		AC terminal buses ACSV		HVDC variables DCSV	
						V_{dI1}	0.995
		I_{dI1} (p.u.)	0.5	V_{101}	0.9621	V_{dI2}	0.996
	102					P_{dI1}	0.3984
						P_{dI2}	0.4975
		I_{dI2} (p.u.)	0.4	θ_{101}	−16.34°	P_{dR}	0.9
						I_{dR}	0.9
101						a_R	1.137
		V_{dR} (p.u.)	1	V_{102}	0.997	a_{I1}	1.0851
						a_{I2}	1.0616
		α_R (deg.)	5	θ_{102}	−16.658	$\cos\phi_R$	0.9139
						$\cos\phi_{I1}$	0.9197
	104	γ_{I1} (deg.)	15	V_{104}	1.0103	$\cos\phi_{I2}$	0.9286
		γ_{I2} (deg.)	15	θ_{104}	−18.43°	NI	7
		Control strategy 5 Spec. values		AC terminal buses ACSV		HVDC variables DCSV	
						V_{dI1}	0.997
		P_{dI1} (p.u.)	0.3	V_{86}	1.007	V_{dI2}	0.998
						P_{dR}	0.5
						I_{dI1}	0.3009
	87	P_{dI2} (p.u.)	0.2	θ_{86}	−13.843	I_{dI2}	0.2004
						I_{dR}	0.5021
86						α_R	15.266
		V_{dR} (p.u.)	1	V_{87}	0.996	γ_{I1}	20.363
						γ_{I2}	17.314
		a_R	1.08	θ_{87}	−23.842	$\cos\phi_R$	0.9187
						$\cos\phi_{I1}$	0.91
	90	a_{I1}	1.1	V_{90}	0.9694	$\cos\phi_{I2}$	0.9359
		a_{I2}	1.1	θ_{90}	−23.842	NI	12
		Control strategy 6 Spec. values		AC terminal buses ACSV		HVDC variables DCSV	
						V_{dI1}	0.998
		I_{dI1} (p.u.)	0.2	V_{77}	1.052	V_{dI2}	0.999
	84					P_{dI1}	0.1996
						P_{dI2}	0.0999
		I_{dI2} (p.u.)	0.1	θ_{77}	−6.815°	P_{dR}	0.3
						I_{dR}	0.3
77						α_R	14.212
		V_{dR} (p.u.)	1	V_{84}	1.0272	γ_{I1}	17.647
						γ_{I2}	17.985
		a_R	1.01	θ_{84}	−11.694	$\cos\phi_R$	0.9412
	86					$\cos\phi_{I1}$	0.9342
		a_{I1}	1.04	V_{86}	1.02	$\cos\phi_{I2}$	0.9417
		a_{I2}	1.09	θ_{86}	−12.81°	NI	10

and is updated in every iteration. This is elaborated from the steps involved in the computation of the power injections as shown in the second column of Table 7. From Table 7, it can be observed that the computation of the reactive power injections is dependent on the iterative process. This affects the convergence pattern. This is reiterated by comparing the convergence characteristics for this case, as shown in Figure 4, with those of Figures 2 and 3.

The bus voltage profile for this control strategy is shown in Figure 5. It is observed that the bus voltage profiles undergo very slight changes at only the AC terminal buses connected to the converters.

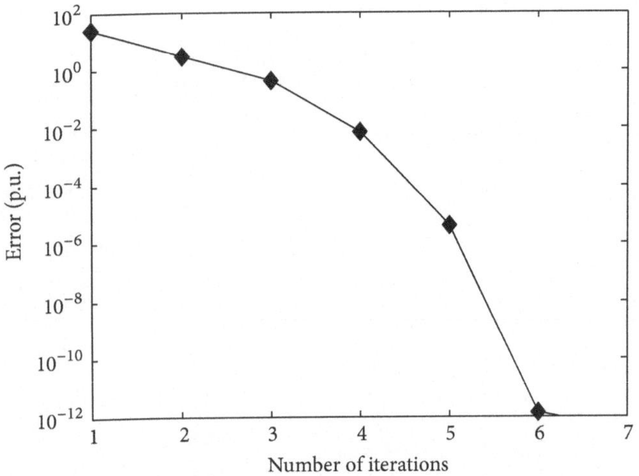

FIGURE 2: Convergence characteristics for the base case power flow in IEEE-300 bus test system.

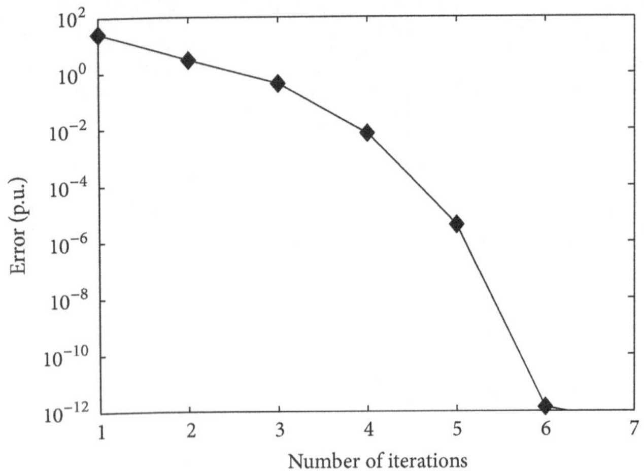

FIGURE 3: Convergence characteristics for case study I of Table 2 with control strategy 1.

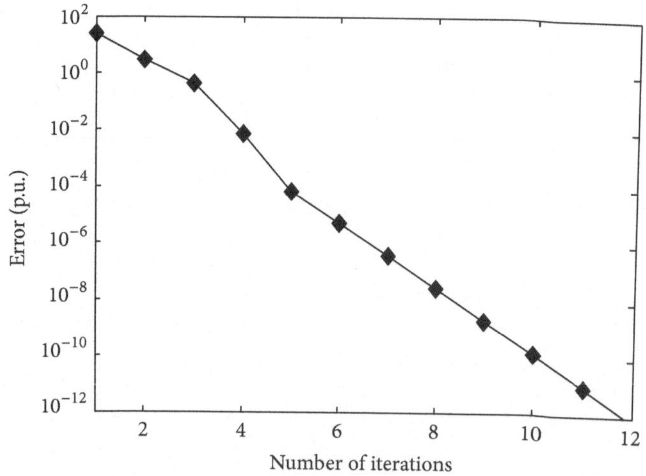

FIGURE 4: Convergence characteristics for case study V of Table 3 with control strategy 5.

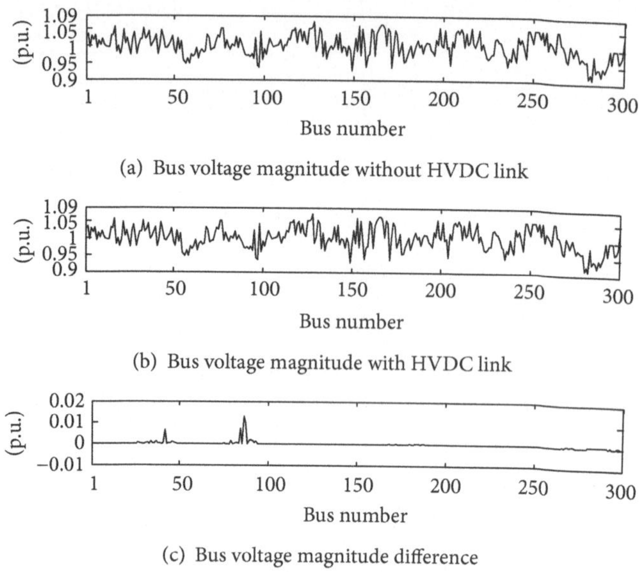

FIGURE 5: Bus voltage profile for case study V with control strategy 5.

It is important to note that due to a limitation of space only the bus voltage profile corresponding to control strategy 5 is shown. This is because among all the case studies carried out the poorest power-flow convergence pattern was observed corresponding to this control strategy.

Case Study VI. In this case, the two HVDC links were incorporated between buses "77-84" and "77-86." While the converter connected to bus number 77 acts as a rectifier, both converters connected to buses 84 and 86 act as inverters. For control of the DC links, control strategy 6 is adopted. In this control strategy, the rectifier end DC voltage is specified. On the other hand, both inverters are operated in the constant current mode. Also, the tap ratios of all the three converter transformers are specified. While columns "1–4" of Table 3 show the values of the specified quantities, columns 5–8 detail the power-flow solution (corresponding to control strategy 6). From the power-flow solution it is observed that in this case the convergence pattern is adversely affected as in case of that of control strategy 5. This is because the converter equivalent reactive power injections are required to be updated in every iteration of the AC power flow. This is also true for the subsequent case studies VII and VIII carried out employing control strategies 7 and 8, respectively. Although the steps involved in the computation of the power injections with control strategies 6 to 8 are similar to those with control strategy 5 (second column, Table 7), they are not shown due to limitations of space.

The convergence characteristics corresponding to this case study are shown in Figure 6. By comparing Figures 5 and 6, the similarity in the convergence characteristics between the case studies with control strategies 4 and 6 can be observed.

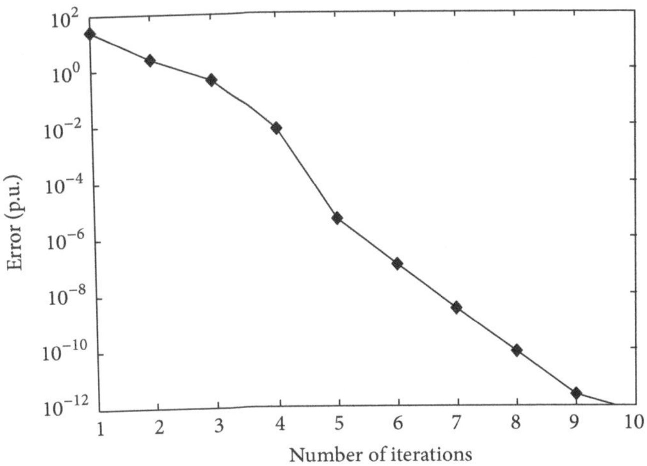

FIGURE 6: Convergence characteristics for case study VI of Table 3 with control strategy 6.

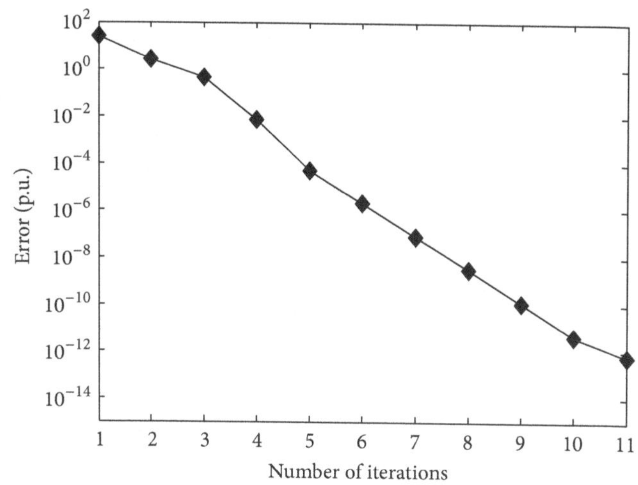

FIGURE 7: Convergence characteristics for case study VII of Table 4 with control strategy 7.

Case Study VII. In this case study, two HVDC links are incorporated between buses "68-173" and "68-174." The converter connected to bus number 68 acts as a rectifier while both converters connected to buses 173 and 174 act as inverters. Control strategy 7 is selected for the control of the DC links. The first inverter is operated in the constant power mode while the second one operates in the constant current mode. In addition, the tap ratios of all the three converter transformers are specified. The specified values are given in columns 1–4 of Table 4. The power-flow solution is shown in columns 5–8 of Table 4 (corresponding to control strategy 7). From the power-flow solution, it is observed that, similar to control strategies 5 and 6, the convergence pattern for this control strategy is adversely affected. This is expected as the reactive power injections are dependent on the iterative process of the AC power flow. This is reiterated from the convergence characteristics for this case study, as shown in Figure 7.

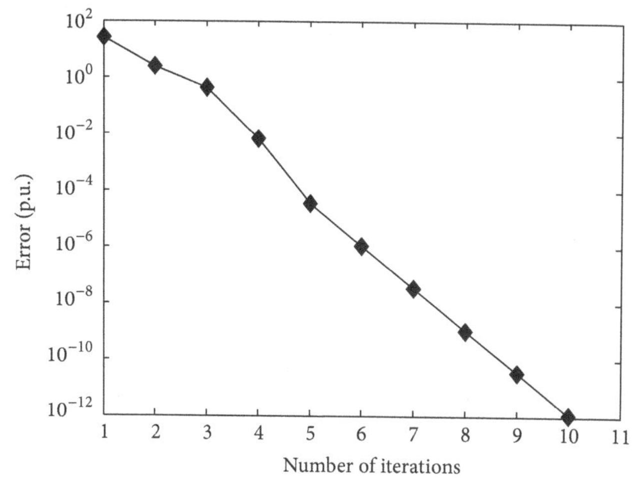

FIGURE 8: Convergence characteristics for case study VIII of Table 4 with control strategy 8.

Case Study VIII. In this case, two HVDC links are incorporated between buses "15-31" and "15-74." The converter connected to bus number 15 acts as a rectifier while both converters connected to buses 31 and 74 act as inverters. Control strategy 8 is selected for the control of the DC links. The rectifier operates in the constant voltage mode. The first and the second inverters are operated in the constant current and constant power modes, respectively. In addition, the tap ratios of all the three converter transformers are specified. Columns 1–4 of Table 4 show these specified values. The power-flow solution is shown in columns 5–8 of Table 4 (corresponding to control strategy 8). From the power-flow solution, it is again observed that, similar to control strategies 5–7, the convergence pattern is adversely affected by this control strategy. The convergence characteristics for this case study, as shown in Figure 8, emphasise this fact.

Case Study IX. In this case, two HVDC links are incorporated between buses "266-270" and "266-271." The converter connected to bus number 266 acts as a rectifier while both converters connected to buses 270 and 271 act as inverters. Control strategy 9 is selected for the control of the DC links. In this strategy, the rectifier is operated in the constant firing angle mode. Both inverters are operated in constant power and constant extinction angle mode. Columns 1–4 of Table 4 show these specified values. The power-flow solution is shown in columns 5–8 of Table 4 (corresponding to control strategy 9). From Table 4, it is observed that, in this case study, the convergence pattern is better than those with control strategies 5–8. This is because the power injections representing the DC link can be computed a priori and are independent of the iterative process of the AC power flow. This is reiterated by the convergence characteristics for this case study, as shown in Figure 9.

The convergence patterns for the control strategies 2, 3, 4, and 9, although similar to control strategy 1, are not shown due to the limitations of space.

TABLE 4: Study of IEEE-300 bus system with three-terminal HVDC network.

HVDC link		HVDC link specification		Power-flow solution			
Rectifier bus	Inverter buses	Control strategy 7		AC terminal buses		HVDC variables	
		Spec. values		ACSV		DCSV	
		P_{dI1} (p.u.)	0.5	V_{68}	1.019	V_{dI1}	0.995
						V_{dI2}	0.9995
	173					P_{dI2}	0.05
						P_{dR}	0.5525
		I_{dI2} (p.u.)	0.1	θ_{68}	−19.200	I_{dI1}	0.5025
68						I_{dR}	0.6025
						α_R	12.373
		V_{dR} (p.u.)	1	V_{173}	1.0363	γ_{I1}	22.275
						γ_{I2}	21.965
		a_R	1.06	θ_{173}	−18.849	$\cos\phi_R$	0.9256
						$\cos\phi_{I1}$	0.8809
	174	a_{I1}	1.09	V_{174}	1.0315	$\cos\phi_{I2}$	0.9228
		a_{I2}	1.05	θ_{174}	−20.83	NI	11
		Control strategy 8		AC terminal buses		HVDC variables	
		Spec. values		ACSV		DCSV	
		I_{dI1} (p.u.)	0.04	V_{15}	1.0095	V_{dI1}	0.9996
						V_{dI2}	0.998
	31					P_{dI1}	0.2299
						P_{dR}	0.4299
		P_{dI2} (p.u.)	0.2	θ_{15}	−9.0928	I_{dI2}	0.2004
15						I_{dR}	0.2404
						α_R	12.733
		V_{dR} (p.u.)	1	V_{31}	1.0101	γ_{I1}	21.787
						γ_{I2}	22.78
		a_R	1.04	θ_{31}	−11.3	$\cos\phi_R$	0.9525
						$\cos\phi_{I1}$	0.9249
	74	a_{I1}	1.07	V_{74}	1.0224	$\cos\phi_{I2}$	0.9038
		a_{I2}	1.08	θ_{74}	−10.272	NI	11
		Control strategy 9		AC terminal buses		HVDC variables	
		Spec. values		ACSV		DCSV	
		P_{dI1} (p.u.)	0.4	V_{25}	0.9965	V_{dR}	0.996
						V_{dI2}	0.997
	270					P_{dR}	0.7025
						I_{dI1}	0.4016
		P_{dI2} (p.u.)	0.3	θ_{25}	−9.2266	I_{dI2}	0.3009
						I_{dR}	0.7025
266						a_R	1.0437
		V_{dI1} (p.u.)	1	V_{26}	0.9966	a_{I1}	1.1118
						a_{I2}	1.1358
		α_R (deg.)	5	θ_{26}	−9.1557	$\cos\phi_R$	0.9576
	271					$\cos\phi_{I1}$	0.9025
		γ_{I1} (deg.)	15	V_{232}	0.9362	$\cos\phi_{I2}$	0.9376
		γ_{I2} (deg.)	15	θ_{232}	−7.6891	NI	6

TABLE 5: Study of power-flow convergence pattern with active power in the second DC link.

HVDC links		HVDC link specification		Power-flow solutions	
Rectifier bus	Inverter buses	Control strategy 5		AC terminal buses	HVDC variables
85	88, 233	$P_{dI1} = 0.5$ (p.u.) $V_{dR} = 1$ (p.u.) $a_R = 1.14$ $a_{I1} = 1.1$ $a_{I2} = 1.12$	P_{dI2} (p.u.) 0.08	$\mathbf{V}_{85} = 0.963\angle - 18.368$ $\mathbf{V}_{88} = 0.99\angle - 19.710$ $\mathbf{V}_{233} = 0.967\angle - 18.23$	$V_{dI1} = 0.995; V_{dI2} = 0.9992; P_{dIR} = 0.5826; I_{dI1} = 0.5025; I_{dI2} = 0.05; I_{dR} = 0.5526; \gamma_{I1} = 16.3001; \gamma_{I2} = 21.7912; \alpha_R = 15.7287; \cos\phi_{I1} = 0.9137; \cos\phi_{I2} = 0.9212; \cos\phi_R = 0.9096$; NI = 14; CT = 93.294982;
			0.1	$\mathbf{V}_{85} = 0.9635\angle - 18.36$ $\mathbf{V}_{88} = 0.99\angle - 19.715$ $\mathbf{V}_{233} = 0.9674\angle - 17.79$	$V_{dI1} = 0.995; V_{dI2} = 0.999; P_{dIR} = 0.6026; I_{dI1} = 0.5025; I_{dI2} = 0.1001; I_{dR} = 0.6026; \gamma_{I1} = 16.3001; \gamma_{I2} = 21.3669; \alpha_R = 15.133; \cos\phi_{I1} = 0.9137; \cos\phi_{I2} = 0.922; \cos\phi_R = 0.9105$; NI = 15; CT = 95.894141;
			0.15	$\mathbf{V}_{85} = 0.9611\angle - 18.314$ $\mathbf{V}_{88} = 0.99\angle - 19.7325$ $\mathbf{V}_{233} = 0.9647\angle - 17.30$	$V_{dI1} = 0.995; V_{dI2} = 0.9985; P_{dIR} = 0.6528; I_{dI1} = 0.5025; I_{dI2} = 0.1502; I_{dR} = 0.6528; \gamma_{I1} = 16.3001; \gamma_{I2} = 20.2677; \alpha_R = 13.5279; \cos\phi_{I1} = 0.9137; \cos\phi_{I2} = 0.9242; \cos\phi_R = 0.9127$; NI = 15; CT = 93.322688;
			0.2	$\mathbf{V}_{85} = 0.9588\angle - 18.268$ $\mathbf{V}_{88} = 0.99\angle - 19.7537$ $\mathbf{V}_{233} = 0.962\angle - 16.815$	$V_{dI1} = 0.995; V_{dI2} = 0.998; P_{dIR} = 0.7029; I_{dI1} = 0.5025; I_{dI2} = 0.2004; I_{dR} = 0.7029; \gamma_{I1} = 16.3001; \gamma_{I2} = 19.1062; \alpha_R = 11.6988; \cos\phi_{I1} = 0.9137; \cos\phi_{I2} = 0.9263; \cos\phi_R = 0.9149$; NI = 16; CT = 101.913689;
			0.25	$\mathbf{V}_{85} = 0.9564\angle - 18.228$ $\mathbf{V}_{88} = 0.99\angle - 19.7786$ $\mathbf{V}_{233} = 0.9593\angle - 16.32$	$V_{dI1} = 0.995; V_{dI2} = 0.9975; P_{dR} = 0.7532; I_{dI1} = 0.5025; I_{dI2} = 0.2506; I_{dR} = 0.7532; \gamma_{I1} = 16.3001; \gamma_{I2} = 17.8704; \alpha_R = 9.522; \cos\phi_{I1} = 0.9137; \cos\phi_{I2} = 0.9284; \cos\phi_R = 0.9171$; NI = 17; CT = 135.807279;
			0.3	$\mathbf{V}_{88} = 0.99\angle - 19.8072$ $\mathbf{V}_{233} = 0.9566\angle - 15.84$ $\mathbf{V}_{88} = 0.99\angle - 19.8072$	$V_{dI1} = 0.995; V_{dI2} = 0.997; P_{dIR} = 0.8034; I_{dI1} = 0.5025; I_{dI2} = 0.3009; I_{dR} = 0.8034; \gamma_{I1} = 16.3001; \gamma_{I2} = 16.544; \alpha_R = 6.602; \cos\phi_{I1} = 0.9137; \cos\phi_{I2} = 0.9305; \cos\phi_R = 0.9194$; NI = 19; CT = 118.803820

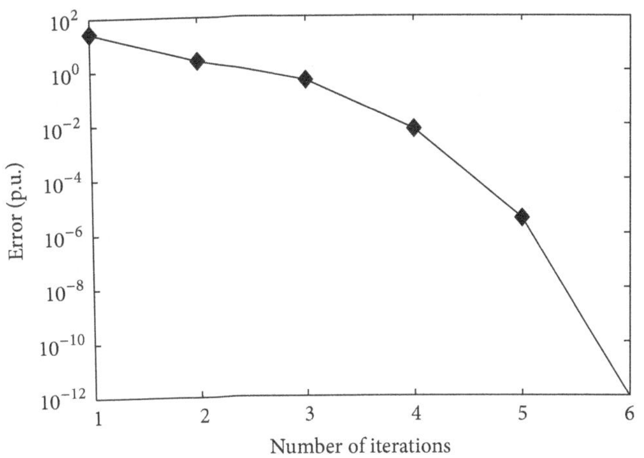

FIGURE 9: Convergence characteristics for case study IX of Table 4 with control strategy 9.

During the numerous studies carried out in the above AC-DC system, it was also observed that, for a particular control strategy adopted, the power-flow convergence pattern is also affected by the values of the controlled variables corresponding to that strategy. Now, the value of the DC link power transfer is one of the most important control variables. Hence, two more case studies were conducted to observe the effect of the variation of the DC link power transfer on the power-flow convergence pattern. Since the poorest convergence pattern is observed with control strategy 5 (Table 3), hence, for both of the case studies, this strategy is adopted. The details of the case studies are given below.

Case Study X. In this case study, the power transfer over the first DC link "P_{dI1}" is maintained to a constant value of 0.5 p.u. On the other hand, the power transfer over the second DC link "P_{dI2}" is gradually increased from 0.08 p.u. to 0.3 p.u. The value of the specified quantities is shown in columns 3-4 of Table 5.

The power-flow solutions are detailed in columns 5-6 of Table 5. The variation of "NI" with active power in the second DC link is shown in Figure 4. From Figure 10, it is observed that "NI" increases as "P_{dI2}" is increased.

Case Study XI. In this case study, the power transfer over the first DC link "P_{dI1}" is gradually increased from 0.1 p.u. to 0.28 p.u. while maintaining the power transfer over the second one (P_{dI2}) to a constant value of 0.2 p.u. The specified quantities and the power-flow solutions are shown in columns 3-4 and 5-6 of Table 6, respectively. The variation of "NI" with active power in the first DC link is shown in Figure 11. From Figure 11, it is again observed that "NI" increases as "P_{dI1}" is increased.

6. Conclusions

For power-flow solution of an integrated AC-DC system, control specifications adopted for the system bridge the

TABLE 6: Study of power-flow convergence pattern with active power in the first DC link.

HVDC links		HVDC link specification		Power-flow solutions	
Rectifier bus	Inverter buses	Control strategy 5		AC terminal buses	HVDC variables
85	88, 233	$P_{dI2} = 0.2$ (p.u.) $V_{dR} = 1$ (p.u.) $a_R = 1.08$ $a_{I1} = 1.1$ $a_{I2} = 1.1$	P_{dI1} (p.u.) 0.1	$\mathbf{V}_{85} = 0.9831\angle - 17.3494$ $\mathbf{V}_{88} = 0.99\angle - 20.2874$ $\mathbf{V}_{233} = 0.9858\angle - 15.9583$	$V_{dI1} = 0.999; V_{dI2} = 0.998; P_{dIR} = 0.3005; I_{dI1} = 0.1001; I_{dI2} = 0.2004; I_{dR} = 0.3005; \gamma_{I1} = 22.0973; \gamma_{I2} = 20.1488; \alpha_R = 14.0408; \cos\phi_{I1} = 0.9174; \cos\phi_{I2} = 0.9203; \cos\phi_R = 0.9418$; NI = 14; CT = 87.437434;
			0.15	$\mathbf{V}_{85} = 0.9808\angle - 17.4693$ $\mathbf{V}_{88} = 0.99\angle - 20.2093$ $\mathbf{V}_{233} = 0.9836\angle - 16.075$	$V_{dI1} = 0.9985; V_{dI2} = 0.998; P_{dIR} = 0.3506; I_{dI1} = 0.1502; I_{dI2} = 0.2004; I_{dR} = 0.3506; \gamma_{I1} = 21.4575; \gamma_{I2} = 19.801; \alpha_R = 12.2691; \cos\phi_{I1} = 0.9169; \cos\phi_{I2} = 0.9224; \cos\phi_R = 0.9441$; NI = 14; CT = 87.093402;
			0.2	$\mathbf{V}_{85} = 0.9785\angle - 17.5935$ $\mathbf{V}_{88} = 0.99\angle - 20.1333$ $\mathbf{V}_{233} = 0.9815\angle - 16.196$	$V_{dI1} = 0.998; V_{dI2} = 0.998; P_{dIR} = 0.4008; I_{dI1} = 0.2004; I_{dI2} = 0.2004; I_{dR} = 0.4008; \gamma_{I1} = 20.7986; \gamma_{I2} = 19.4495; \alpha_R = 10.1946; \cos\phi_{I1} = 0.9164; \cos\phi_{I2} = 0.9244; \cos\phi_R = 0.9463$; NI = 15; CT = 92.089820;
			0.25	$\mathbf{V}_{85} = 0.9762\angle - 17.7218$ $\mathbf{V}_{88} = 0.99\angle - 20.0592$ $\mathbf{V}_{233} = 0.9794\angle - 16.3212$	$V_{dI1} = 0.9975; V_{dI2} = 0.998; P_{dIR} = 0.4510; I_{dI1} = 0.5025; I_{dI2} = 0.2004; I_{dR} = 0.4510; \gamma_{I1} = 20.1179; \gamma_{I2} = 19.0945; \alpha_R = 7.5728; \cos\phi_{I1} = 0.916; \cos\phi_{I2} = 0.9246; \cos\phi_R = 0.9485$; NI = 15; CT = 91.472996;
			0.28	$\mathbf{V}_{85} = 0.9748\angle - 17.8008$ $\mathbf{V}_{88} = 0.99\angle - 20.0158$ $\mathbf{V}_{233} = 0.978\angle - 16.3983$	$V_{dI1} = 0.9972; V_{dI2} = 0.998; P_{dR} = 0.4812; I_{dI1} = 0.2808; I_{dI2} = 0.2004; I_{dR} = 0.4812; \gamma_{I1} = 19.6985; \gamma_{I2} = 18.8797; \alpha_R = 5.4231; \cos\phi_{I1} = 0.9157; \cos\phi_{I2} = 0.9276; \cos\phi_R = 0.9498$; NI = 16; CT = 99.869132

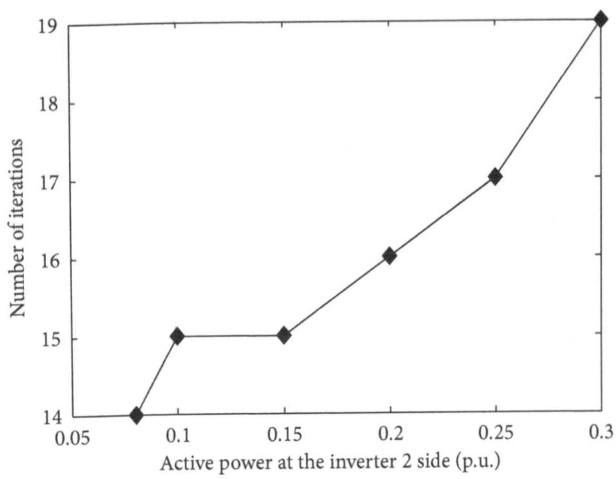

FIGURE 10: Variation of "NI" with active power in second DC link.

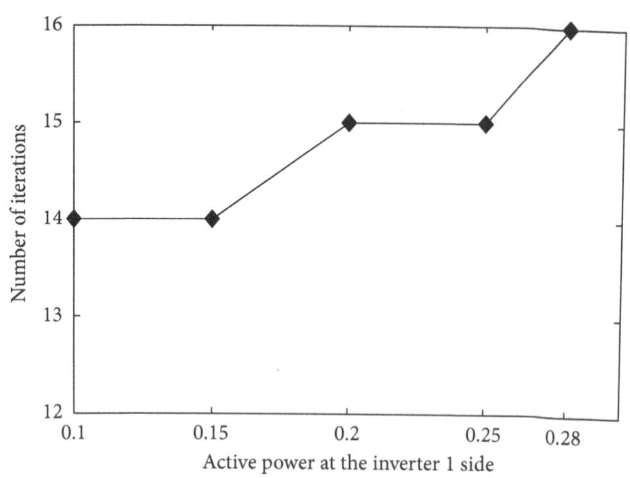

FIGURE 11: Variation of "NI" with active power in first DC link.

gap between the number of independent equations and the number of unknown quantities. Depending on the application, several combinations of valid control specifications are possible. Each combination of a set of valid control specifications comprises a control strategy. It is observed that the power-flow convergence is strongly affected by the control strategy adopted. This work investigates the mechanism by which different control strategies affect the convergence of the iterative process. It is observed that, for a majority of the possible control strategies, the equivalent power injections at the converter terminal buses are independent of the NR iterative loop. However, for others, the equivalent reactive power injections need to be computed in every NR iteration. This affects the convergence of the algorithm. This is validated by multiple case studies with application of nine different control strategies on a three-terminal DC network incorporated in the IEEE 300-bus test system.

From the case studies, it was observed that the power-flow convergence pattern for integrated AC-DC systems depends upon a number of factors. Firstly, it depends on the control strategy adopted for the DC link. Secondly, for a particular control strategy adopted, it depends upon the values of the electrical quantities specified in the control strategy. In addition, it also depends upon the location of the DC link, that is, the system buses between which the link is incorporated.

TABLE 7: Steps to compute active and reactive power injections in control strategies 1 and 5.

(a)

Control strategy 1

Specified quantities	Unknown quantities
$P_{dI1}, P_{dI2}, V_{dR}, \alpha_R, \gamma_{I1}, \gamma_{I2}$	$V_{dI1}, V_{dI2}, P_{dR}, I_{dI1}, I_{dI2}, I_{dR}, a_R, a_{I1}, a_{I2}, \cos\phi_R, \cos\phi_{I1}, \cos\phi_{I2}$

Step 1: compute V_{dI1} and V_{dI2} using DC load flow

Step 2: compute $I_{dI1} = \dfrac{V_{dR} - V_{dI1}}{R_d}$

Step 3: compute $I_{dI2} = \dfrac{V_{dR} - V_{dI2}}{R_d}$

Step 4: compute $I_{dR} = I_{dI1} + I_{dI2}$

Step 5: compute $P_{dR} = V_{dR} I_{dR}$

Step 6: compute $\cos\phi_R = \dfrac{V_{dR}\cos\alpha_R}{V_{dR} + X_c I_{dR}}$

Step 7: compute $Q_{dR} = P_{dR}\tan\phi_R$

Step 8: compute $\cos\phi_{I1} = \dfrac{V_{dI1}\cos\gamma_{I1}}{V_{dI1} + X_c I_{dI1}}$

Step 9: compute $\cos\phi_{I2} = \dfrac{V_{dI2}\cos\gamma_{I2}}{V_{dI2} + X_c I_{dI2}}$

Step 10: compute $Q_{dI1} = P_{dI1}\tan\phi_{I1}$

Step 11: compute $Q_{dI2} = P_{dI2}\tan\phi_{I2}$

Note: P_{dI1} and P_{dI2} are specified. P_{dR}, Q_{dR}, Q_{dI1}, and Q_{dI2} can be computed prior to the AC power flow and hence are independent of the iterative loop.

(b)

Control strategy 5

Specified quantities	Unknown quantities
$P_{dI1}, P_{dI2}, V_{dR}, a_R, a_{I1}, a_{I2}$	$V_{dI1}, V_{dI2}, P_{dR}, I_{dI1}, I_{dI2}, I_{dR}, \alpha_R, \gamma_{I1}, \gamma_{I2}, \cos\phi_R, \cos\phi_{I1}, \cos\phi_{I2}$

Step 1: compute V_{dI1} and V_{dI2} using DC load flow

Step 2: compute $I_{dI1} = \dfrac{V_{dR} - V_{dI1}}{R_d}$

Step 3: compute $I_{dI2} = \dfrac{V_{dR} - V_{dI2}}{R_d}$

Step 4: compute $I_{dR} = I_{dI1} + I_{dI2}$

Step 5: compute $P_{dR} = V_{dR} I_{dR}$

Step 6: compute $\cos\phi_R = \dfrac{V_{dR}}{a_R V_i}$

Note 1: V_i is an AC power-flow variable and is updated in every iteration. Hence, $\cos\phi_R$ changes in every iteration.

Step 7: compute $Q_{dR} = P_{dR}\tan\phi_R$

Step 8: compute $\cos\phi_{I1} = \dfrac{V_{dI1}}{a_{I1} V_j}$

Step 9: compute $\cos\phi_{I2} = \dfrac{V_{dI2}}{a_{I2} V_k}$

Note 2: V_j and V_k are also AC power-flow variables and are updated in every iteration, along with $\cos\phi_{I1}$ and $\cos\phi_{I2}$.

Step 10: compute $Q_{dI1} = P_{dI1}\tan\phi_{I1}$

Step 11: compute $Q_{dI2} = P_{dI2}\tan\phi_{I2}$

Note 3: P_{dI1} and P_{dI2} are specified. P_{dR} can be computed prior to the AC power flow. However, Q_{dR}, Q_{dI1}, and Q_{dI2} depend upon ϕ_R, ϕ_{I1}, and ϕ_{I2}, respectively, and need to be updated in every iteration.

The above research work could be further extended by investigating and comparing the effects of power-flow algorithms other than the sequential method. The effect of power flow based on soft computing techniques on the convergence pattern can be investigated in light of different control methodologies and computation platforms.

Appendices

A. Selection of Base Values for the DC System

Consider

$$V_{\text{ac base}} = V\,(\text{line to line rms voltage}),$$
$$I_{\text{ac base}} = \frac{S_{\text{base}}}{\sqrt{3}V_{\text{ac base}}}, \quad (A.1)$$
$$Z_{\text{ac base}} = \frac{V_{\text{ac base}}}{\sqrt{3}I_{\text{ac base}}}.$$

B. Computation of Active and Reactive Power Injections in Different Control Strategies

As already explained in Section 4, for solving the AC power flow, the equivalent active and reactive power injections at the AC terminal buses connected to the converters need to be computed. The methods involved in the computation of the power injections for different control strategies are all different. However, they can be broadly categorized into two distinct ones. While in the first one the power injections are independent of the AC power flow, in the second one, they are dependent on the iterative process. Control strategies 1, 2, 3, 4, and 9 belong to the former category while control strategies 5, 6, 7, and 8 belong to the latter. The detailed steps involved in the computation of the power injections for control strategies 1 and 5 are shown in Table 7.

From Table 7, it can be observed that the active power injections can be computed prior to the AC power flow. It is also observed that, in control strategy 1, the quantities ϕ_R, ϕ_{I1}, and ϕ_{I2} and hence the reactive power injections Q_{dR}, Q_{dI1}, and Q_{dI2} can be computed prior to the AC power flow by manipulation of the specified variables. On the other hand, in control strategy 5, they are dependent on the AC power-flow variables and thus need to be updated in every iteration.

The steps corresponding to the other control strategies can be detailed in a similar manner and are not shown due to limitations of space.

List of Symbols

S_{base}: Base MVA
$V_{\text{ac base}}$, $I_{\text{ac base}}$, and $Z_{\text{ac base}}$: AC base voltage, AC base current, and AC base impedance, respectively
$V_{\text{dc base}}$, $I_{\text{dc base}}$, and $Z_{\text{dc base}}$: DC base voltage, DC base current, and DC base impedance, respectively
n_b: Number of bridges
X_c: Commutating reactance
V_d, I_d: DC voltage and current, respectively
ϕ_R: Power factor angle at the rectifier end
ϕ_{I1}, ϕ_{I2}: Power factor angles at inverter 1 and inverter 2 ends, respectively
α_R: Firing angle of the rectifier
γ_{I1}, γ_{I2}: Extinction angles of inverter 1 and inverter 2, respectively
V_{dR}: DC voltage at the rectifier end
V_{dI1}, V_{dI2}: DC voltages at inverter 1 and inverter 2 ends, respectively
a_R: Tap ratio of the converter transformer at the rectifier end
a_{I1}, a_{I2}: Tap ratios of the converter transformers at inverter 1 and inverter 2 ends, respectively
R_d: DC link resistance
$V_i\angle\theta_i$: AC bus voltage magnitude (rms) and phase angle at ith bus
$V_j\angle\theta_j$: AC bus voltage magnitude (rms) and phase angle at jth bus
$V_k\angle\theta_k$: AC bus voltage magnitude (rms) and phase angle at kth bus
V_{doR}: No load direct voltage at the rectifier side
V_{doI1}, V_{doI2}: No load direct voltages at inverter 1 and inverter 2 ends, respectively
P_{Di}, Q_{Di}: Active and reactive power demand at bus "i," respectively
P_i, Q_i: Net active and reactive power injection at bus "i," respectively
P_{dR}, Q_{dR}: Equivalent active and reactive power injection of the rectifier, respectively
P_{dI1}, P_{dI2}: Equivalent active power injections of inverter 1 and inverter 2, respectively
Q_{dI1}, Q_{dI2}: Equivalent reactive power injections of inverter 1 and inverter 2, respectively.

Conflict of Interests

The authors declare that there is no conflict of interests regarding the publication of this paper.

References

[1] J. Arrillaga and N. R. Watson, *Computer Modelling of Electrical Power Systems*, John Wiley & Sons, Chichester, UK, 2nd edition, 2001.

[2] P. Kundur, *Power System Stability and Control*, Tata McGraw-Hill Publishing, 2007.

[3] X. F. Wang, Y. H. Song, and M. Irving, *Modern Power Systems Analysis*, Springer, Berlin, Germany, 2008.

[4] K. R. Padiyar, *HVDC Power Transmission Systems*, New Age International Publishers, 2nd edition, 2012.

[5] T. Smed, G. Andersson, G. B. Sheble, and L. L. Grigsby, "A new approach to AC/DC power flow," *IEEE Transactions on Power Systems*, vol. 6, no. 3, pp. 1238–1244, 1991.

[6] J. Reeve, G. Fahmy, and B. Stott, "Versatile load flow method for multi terminal HVDC systems," *IEEE Transactions on Power Apparatus and Systems*, vol. 96, no. 3, pp. 925–933, 1977.

[7] M. E. El-Hawary and S. T. Ibrahim, "A new approach to AC-DC load flow analysis," *Electric Power Systems Research*, vol. 33, no. 3, pp. 193–200, 1995.

[8] K. R. Padiyar and V. Kalyanaraman, "Power flow analysis in MTDC-AC systems—A new approach," *Electric Machines and Power Systems*, vol. 23, no. 1, pp. 37–54, 1995.

[9] Y. K. Fan, D. Niebur, C. O. Nwankpa, H. Kwanty, and R. Fischl, "Multiple power flow solutions of small integrated AC/DC power systems," in *Proceedings of the IEEE International Symposium on Circuits and Systems (ISCAS '00)*, vol. 2, pp. 224–227, Geneva, Switzerland, 2000.

[10] C. Liu, B. Zhang, Y. Hou, F. F. Wu, and Y. Liu, "An improved approach for AC-DC power flow calculation with multi-infeed DC systems," *IEEE Transactions on Power Systems*, vol. 26, no. 2, pp. 862–869, 2011.

[11] C. Liu, A. Bose, and Y. Hou, "Discussion of the solvability of HVDC systems power flow with a sequential method," *Electric Power Systems Research*, vol. 92, pp. 155–161, 2012.

[12] S. Messalti, S. Belkhiat, S. Saadate, and D. Flieller, "A new approach for load flow analysis of integrated AC-DC power systems using sequential modified Gauss-Seidel methods," *European Transactions on Electrical Power*, vol. 22, no. 4, pp. 421–432, 2012.

[13] U. Kiliç and K. Ayan, "Optimizing power flow of AC-DC power systems using artificial bee colony algorithm," *International Journal of Electrical Power & Energy Systems*, vol. 53, no. 1, pp. 592–602, 2013.

[14] Y. Sun, G. Zhang, W. Xu, and J. G. Mayordomo, "A harmonically coupled admittance matrix model for AC/DC converters," *IEEE Transactions on Power Systems*, vol. 22, no. 4, pp. 1574–1582, 2007.

[15] J. Yu, W. Yan, W. Li, C. Y. Chung, and K. P. Wong, "An unfixed piecewise-optimal reactive power-flow model and its algorithm for AC-DC systems," *IEEE Transactions on Power Systems*, vol. 23, no. 1, pp. 170–176, 2008.

[16] R. Wiget and G. Andersson, "DC optimal power flow including HVDC grids," in *Proceedings of the IEEE Electrical Power and Energy Conference (EPEC '13)*, pp. 1–6, August 2013.

[17] D. L. H. Aik and G. Andersson, "Analysis of voltage and power interactions in multi-infeed HVDC systems," *IEEE Transactions on Power Delivery*, vol. 28, no. 2, pp. 816–824, 2013.

[18] U. Kiliç, K. Ayan, and U. Arifoğlu, "Optimizing reactive power flow of HVDC systems using genetic algorithm," *International Journal of Electrical Power and Energy Systems*, vol. 55, pp. 1–12, 2014.

[19] M. J. Carrizosa, F. D. Navas, G. Damm, and F. Lamnabhi-Lagarrigue, "Optimal power flow in multi-terminal HVDC grids with offshore wind farms and storage devices," *International Journal of Electrical Power & Energy Systems*, vol. 65, pp. 291–298, 2015.

[20] http://www.ee.washington.edu/research/pstca/.

21

Row-Based Dual V_{dd} Assignment, for a Level Converter Free CSA Design and Its Near-Threshold Operation

Dipankar Saha, Aanan Chatterjee, Sayan Chatterjee, and C. K. Sarkar

Department of Electronics & Telecommunication Engineering, Jadavpur University, Kolkata, India

Correspondence should be addressed to Dipankar Saha; dipsah_etc@yahoo.co.in

Academic Editor: Changhwan Shin

Subthreshold circuit designs are very much popular for some of the ultra-low power applications, where the minimum energy consumption is the primary concern. But, due to the weak driving current, these circuits generally suffer from huge performance degradation. Therefore, in this paper, we primarily targeted analyzing the performance of a near-threshold circuit (NTC), which retains the excellent energy efficiency of the subthreshold design, while improving the performance to a certain extent. A modified row-based dual V_{dd} 4-operand carry save adder (CSA) design has been reported in the present work using 45 nm technology. Moreover, to find out the effectiveness of the near-threshold operation of the 4-operand CSA design, it has been compared with the other design styles. From the simulation results, obtained for the frequency of 20 MHz, we found that the proposed scheme of CSA design consumes 3.009×10^{-7} Watt of average power (P_{avg}), which is almost 90.9% lesser than that of the conventional CSA design, whereas, looking at the perspective of maximum delay at output, the proposed scheme of CSA design provides a fair 44.37% improvement, compared to that of the subthreshold CSA design.

1. Introduction

Subthreshold digital circuit design is a well-practiced technique, for implementing the highly energy-constrained, ultra-low power applications such as implanted sensors, pacemakers, and mobile peripheral processors [1, 2]. But the primary challenge, that limits its usage only to low performance systems, is the weak driving current. For the subthreshold or near-threshold operation, the MOS transistor is provided with a gate-to-source voltage which is either lower or else nearer to the threshold voltage (v_{th}) of the device. At the same time, the supply voltage (V_{dd}) can be scaled below the v_{th} or else can be set somewhat nearer to the v_{th}. Thus, achieving the minimum power consumption, which leads to a longer battery lifetime, can be possible by using this technique [2]. However, the aforesaid advantage in energy consumption comes at the cost of performance degradation and that is mainly due to the fact that the charging and discharging of the load capacitances of the circuit (with the change in logic function) are actually driven by the weak subthreshold leakage current [3].

Now, it has been observed that a notable improvement in the performance of a CMOS circuit is possible, if we do a little bit of sacrifice in the energy consumption perspective [3]. And, this is the concept which triggers an increasing usage of near-threshold circuits (NTCs). To have the more precise definition, a circuit that operates with a supply voltage which is equal or slightly greater than the v_{th} is called the NTC [4].

On the other hand, assigning the dual V_{dd} scheme to a CMOS circuit can be very effective in reducing both the dynamic and the leakage power [5, 6]. It provides the higher supply voltage (V_{ddH}) to timing critical logic gates, whereas the other noncritical logic gates of the circuit are actually driven by a lower supply voltage (V_{ddL}). Therefore, with this dual V_{dd} technique, it is possible to reduce the overall power consumption, without degrading the performance of the circuit too much [2, 4]. Moreover, the use of the V_{ddH} to speed up the timing critical logic gates and the V_{ddL} to the noncritical logic gates for minimizing the total power of the circuit requires the additional level-shifters which causes extra power consumption as well as area overhead [7]. Now, considering the case of NTC, the key advantage lies in the fact

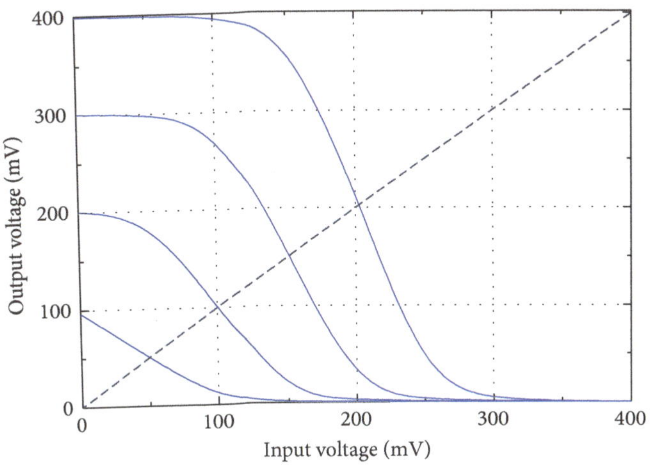

FIGURE 1: Voltage transfer characteristics curve of the CMOS inverter circuit.

that the value of V_{ddH} and V_{ddL} used in the circuit happens to be very close to each other. Thus, such a small difference in two supply voltages can eliminate the requirement of voltage level-shifters [4]. Thereby, properly selecting the subset of the logic gates which needs to be assigned with the V_{ddH}, we can significantly improve the performance of the circuit at an affordable power cost [4].

Though, the assignment of dual V_{dd} can be extremely interesting in case of NTCs, but looking at the physical design implementation part, this approach may cause an extra cost [4]. To reduce this extra cost of routing overhead, we may go for the row-based dual V_{dd} assignment, where the different rows of circuit are prioritized based on their time criticality, and according to that the rows residing in the critical path are driven by the V_{ddH}, while the rest of the rows in the circuit are provided with the V_{ddL} [4]. Now, in this work, to find out the effectiveness of the row-based dual V_{dd} assignment in case of NTCs, the scheme is implemented on an example circuit, which is actually the 4-operand CSA, as described in [8]. The rest of the paper is organized as follows. Section 2 introduces the details of several design issues for the subthreshold circuits. In Section 3, the row-based dual V_{dd} assignment for a 4-operand CSA has been presented, whereas the near-threshold operation of the 4-operand CSA and its performance analysis has been illustrated in Section 4. Section 5 describes the conclusion of this work.

2. Subthreshold Circuit Design Issues

2.1. Modeling the Minimum Energy Point. In case of subthreshold operation ($V_{dd} < v_{th}$), the current that flows through the channel of a transistor is mainly due to diffusion [9]. Now, for the purpose of estimating the minimum energy point of a subthreshold circuit, we can take the help of the current model which serves as the basis for the entire analysis [9]. Assuming that total drain current in subthreshold regime is equal to the subthreshold current (I_{sub}) and considering "n" as the subthreshold slope factor ($n = 1 + C_d/C_{ox}$), V_T as the thermal voltage ($V_T = KT/q$), η as the linearized drain induced barrier lowering (DIBL) coefficient, and S as the subthreshold slope, the I_{sub} can be represented as [3]

$$I_{sub} = I_0 \times 10^{((V_{gs}-v_{th}-\eta V_{ds})/S)} \left(1 - \exp\frac{-V_{ds}}{V_T}\right), \tag{1}$$

where I_0 denotes the drain current at V_{gs} (gate to source voltage) equal to v_{th} (threshold voltage) and the V_{ds} (drain to source voltage) dependence in the "quasisaturation" region has been modeled using the η [9].

Again, for a subthreshold circuit, the gate delay is expressed by the following [3]:

$$t_d = \left[\frac{(K \cdot C_L \cdot V_{dd})}{I_{on}}\right], \tag{2}$$

where K is denoting the delay fitting parameter and C_L is giving the value of the output load capacitance of the gate.

Now, for the $V_{gs} = V_{ds} = V_{dd} \gg V_T$, we can rewrite (2) as [3]

$$t_d = \frac{(K \cdot C_L \cdot V_{dd})}{I_0 \times 10^{(((\eta+1)V_{dd}-v_{th})/S)}}. \tag{3}$$

Thus, the propagation delay of the gate exponentially depends on the V_{dd} as well as the v_{th}.

Next, the total energy consumed per cycle (assuming rail to rail swing, i.e., $V_{gs} = V_{dd}$ for "ON" current) by a single gate can be expressed as [3]

$$E_{total} = E_{dynamic} + E_{leak}, \tag{4}$$

where $E_{dynamic} = (\alpha_{0\to1}) \cdot C_L \cdot V_{dd}^2$ and $E_{leak} = I_{leakage} \times V_{dd} \times t_d$.

$I_{leakage}$ denotes the amount of leakage current, whereas $\alpha_{0\to1}$ gives the low to high activity of the output of the gate [3].

2.2. Optimum Sizing of the Various Logic Gates

2.2.1. Subthreshold Voltage Transfer Characteristics (VTC) of the CMOS Inverter Circuit. For the 45 nm technology node, the SPICE model which is used for the purpose of simulation has the threshold voltage for the NMOS which is set to 0.469 Volt, whereas for the PMOS it is set to −0.418 Volt. Figure 1 shows the voltage transfer characteristics (VTC) curves of an inverter circuit, where the supply voltage is varied from 0 to 0.4 Volt (with an increment of 0.1 Volt), to inspect the behavior of the circuit in the subthreshold region. It is observed that for the ratio of the width of the PMOS (W_p) to the width of the NMOS (W_n) around 4 : 1 there is a sharp transition at the output, whenever the input value crosses the $V_{dd}/2$ level.

2.2.2. Subthreshold XOR Gate Using Transmission Gate Logic. Figure 2(a) shows the conventional transmission gate based 8-transistor XOR that works at ultra-low voltages [9]. Besides, the use of transmission gates in the design helps to balance the number of parallel devices which are operating with the minimum voltage [9].

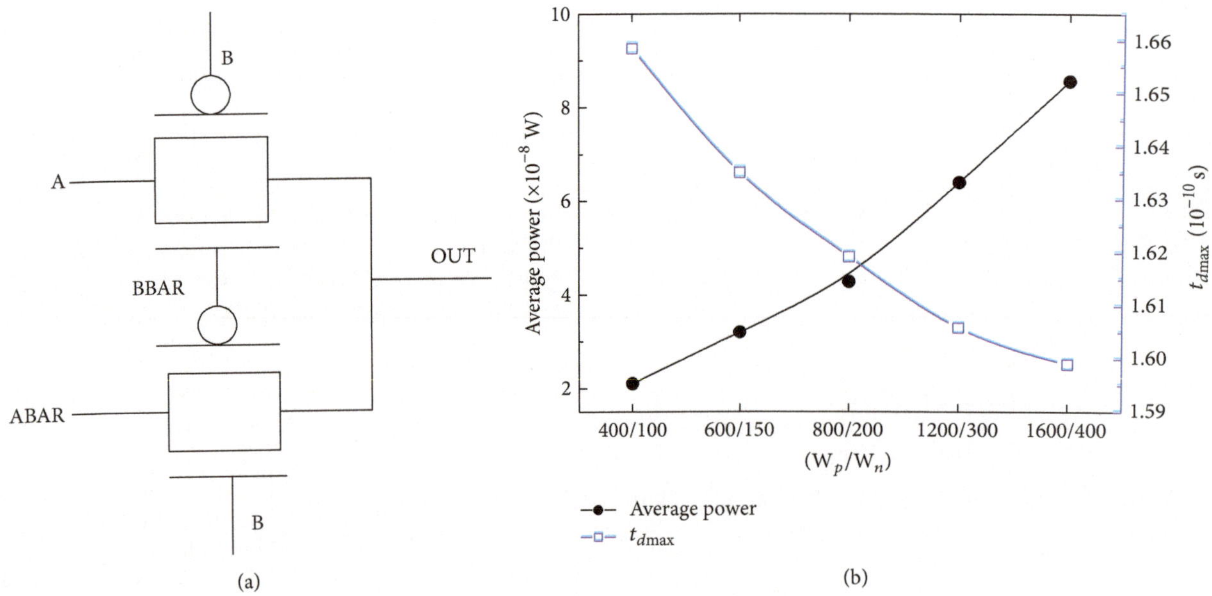

FIGURE 2: (a) 8-transistor transmission gate XOR [9]. (b) Plot of P_{avg} and $t_{d\,\max}$ versus (W_p/W_n).

Figure 2(b) illustrates the plot of the values of the P_{avg} and the maximum delay at output ($t_{d\,\max}$), with the variation of the W_p and W_n sizes of the transistors used in the circuit. It is seen that $W_p/W_n = 800\text{ nm}/200\text{ nm}$ gives the optimum point where the two curves cross each other.

2.2.3. Subthreshold Operation of a Two-Inverter Chain or a Buffer Circuit. Here a two-inverter chain or a buffer circuit is firstly simulated with a single V_{dd} and thereafter with a dual V_{dd} (where V_{dd1} is taken as 0.4 Volt and V_{dd2} is taken as 0.8 Volt). In the first case, where V_{dd} is set to 0.4 Volt and frequency is 200 MHz, we considered the different W_p/W_n values (maintaining the above-mentioned ratio) for the transistors used in the buffer circuit. When the gate length $(L) = 45$ nm, $W_p/W_n = 800\text{ nm}/200\text{ nm}$, we found that the P_{avg} of the circuit is 3.528×10^{-8} Watt and the $t_{d\,\max}$ is 1.378×10^{-10} Second.

Now, in case of the dual V_{dd} assignment for any CMOS circuit, the major problem occurs when a low input swing starts driving a high V_{dd} gate. So, whenever a high voltage gate has to be driven by a low voltage gate, it becomes obvious to use a level converter (LC) [3]. Thus, the LC is supposed to perform the job of shifting the voltage from a lower level to a higher one. However, as the LCs do not implement any logic function, thereby the usage of a large number of LCs in a circuit may ultimately cause in the area as well as energy overhead [7].

To mitigate this issue, the concept of the use of a second threshold voltage for the PMOS transistors in the higher voltage gates (which are actually driven by the lower voltage gates) has been described in [7]. We followed a similar concept here (as shown in Figure 3), except for the fact that, for the purpose of increasing the threshold of those PMOS transistors, we have actually increased their gate lengths [10].

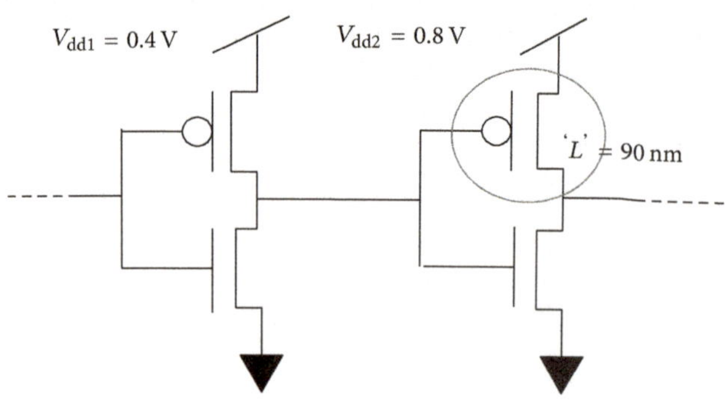

FIGURE 3: Buffer circuit with dual V_{dd}.

The overall performance of this buffer circuit, with a dual V_{dd}, has been described in Table 1.

From Table 1, it can be seen that the best case results are obtained when the gate length of the PMOS transistor in the higher voltage inverter circuit is set to 90 nm.

2.3. Obtaining the $V_{dd,optimum}$ for a Full Adder Circuit. Firstly, the full adder (FA) circuit of Figure 4 has been driven by the single V_{dd} [11, 12] and the inputs having the frequency of 200 MHz. This FA circuit (which has actually got no buffer circuits at its sum and carry outputs) will hereafter be called as FA1 if not otherwise mentioned.

Now, to find out the $V_{\text{dd,optimum}}$ for this FA1, we have varied the V_{dd} from 0.1 Volt to 0.8 Volt (with an increment of 0.1 Volt) and measured the changes in the values of I_{leakage} and $t_{d\,\max}$ (Table 2).

It is observed that, for the region of $V_{\text{dd}} = 0.4$ Volt to 0.6 Volt, the E_{leakage} ($=I_{\text{leakage}} \times V_{\text{dd}} \times t_d$) is minimum. But, considering the aspects of E_{dynamic} (which will increase with

TABLE 1: Performance of the buffer circuit with dual V_{dd}.

V_{dd1} (Volt)	V_{dd2} (Volt)	W_p/W_n	L (nm)	Average power (Watt)	Delay, $t_{d\max}$ (Second)	Output swing (mVolt)
0.4	0.8	800/200	45	3.553×10^{-6}	—	450 to 800
			60	1.132×10^{-6}	1.272×10^{-10}	65 to 800
			75	5.664×10^{-7}	1.429×10^{-10}	18 to 800
			90	3.961×10^{-7}	1.516×10^{-10}	10 to 800

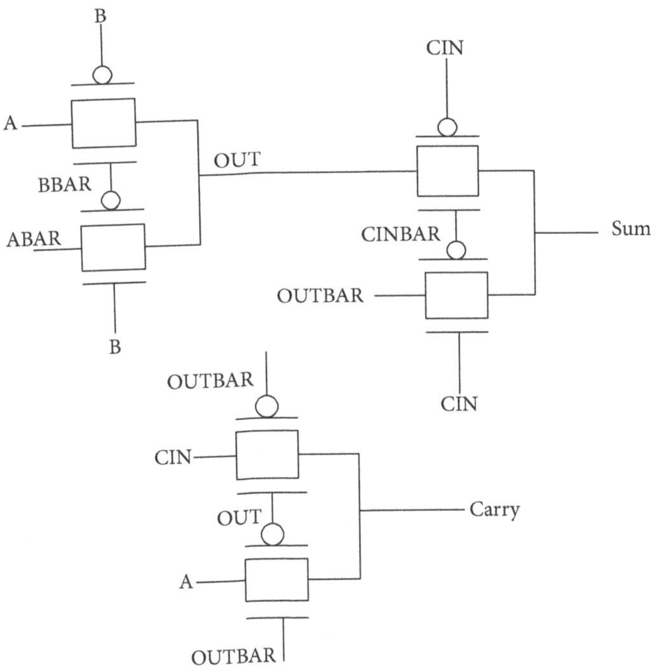

FIGURE 4: Design of the full adder circuit, which is used in FA1 and FA2 blocks [11].

TABLE 2: Variation in $E_{leakage}$ with the change in V_{dd}.

V_{dd} (Volt)	$I_{leakage}$ (Amp.)	Delay, $t_{d\max}$ (Second)	$E_{leakage}$ (Joule)
0.1	0.9×10^{-9}	—	—
0.2	1.8×10^{-9}	1.45×10^{-8}	0.5220×10^{-17}
0.3	3×10^{-9}	3.78×10^{-10}	0.0340×10^{-17}
0.4	5×10^{-9}	10.79×10^{-11}	0.0215×10^{-17}
0.5	7.8×10^{-9}	3.02×10^{-11}	0.0117×10^{-17}
0.6	12×10^{-9}	1.61×10^{-11}	0.0115×10^{-17}
0.7	20×10^{-9}	1.12×10^{-11}	0.0156×10^{-17}
0.8	33×10^{-9}	0.907×10^{-11}	0.0239×10^{-17}

the increase in V_{dd}), we have opted the $V_{dd} = 0.4$ Volt as the $V_{dd,optimum}$ for the FA1 circuit.

In the next, the same full adder circuit of Figure 4 is provided with two buffer circuits at its sum and carry outputs. For those buffers, the first inverter is driven by a supply of V_{dd1}, whereas the second one is driven with the supply voltage which has the value of V_{dd2}. Besides, as mentioned earlier in

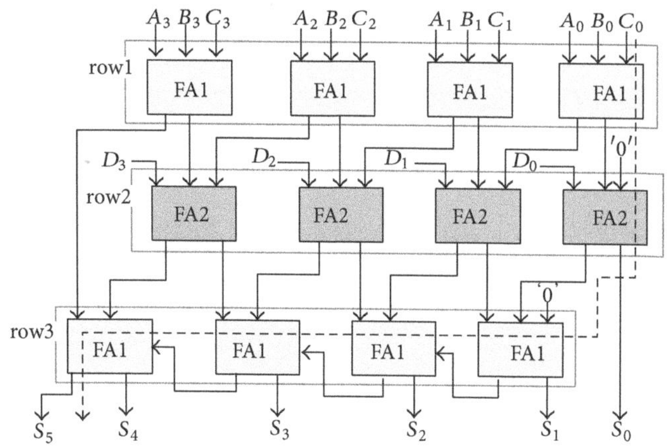

FIGURE 5: Architecture of the 4-operand CSA.

Section 2.2, the length of the PMOS transistor of the second inverter is taken as $L = 90$ nm. Now this FA circuit, which is supplied with the dual V_{dd}, will hereafter be considered as FA2.

Table 3 shows the performance of this FA2 circuit, when the V_{dd1} is set to 0.4 Volt, the frequency is taken as 200 MHz, and the V_{dd2} is varied in between the range of 0.4 Volt to 0.8 Volt.

From Table 3, it can be inferred that the best case result is obtained, considering the power delay product, when the $V_{dd1} = 0.4$ Volt and the $V_{dd2} = 0.5$ Volt.

3. Row-Based Dual V_{dd} Assignment for a 4-Operand CSA

Figure 5 shows a 4-operand CSA, where four 4-bit binary numbers (say, $A_3A_2A_1A_0$, $B_3B_2B_1B_0$, $C_3C_2C_1C_0$, and $D_3D_2D_1D_0$) can be added with an initial carry-in [8]. The upper two rows of the circuit (as shown in Figure 5) form the 4-bit CSA, whereas the third row serves as the carry propagation adder (CPA) [8].

Now, for the purpose of fine tuning the performance, we can opt for the near-threshold operation of this example circuit by selectively using V_{ddH} for the gates which are in the critical path and V_{ddL} for the rest of the gates to reduce the overall power consumption [4]. The dotted line, as shown in Figure 5, is meant for denoting the critical path of the circuit. Moreover, considering the view point of physical design implementation, the approach of row-based dual V_{dd} assignment has been adopted here. For that, the entire circuit

TABLE 3: Performance of the FA2 circuit with dual V_{dd}.

V_{dd1} (Volt)	V_{dd2} (Volt)	Average power (Watt)	Delay, $t_{d\,max}$ (Second)		Average power × Delay (Joule)	
			Sum	Carry	Sum	Carry
0.4	0.8	7.993×10^{-7}	5.237×10^{-10}	4.178×10^{-10}	41.859×10^{-17}	33.395×10^{-17}
0.4	0.7	2.147×10^{-7}	5.472×10^{-10}	4.403×10^{-10}	11.750×10^{-17}	9.454×10^{-17}
0.4	0.6	1.323×10^{-7}	5.703×10^{-10}	4.406×10^{-10}	7.545×10^{-17}	5.829×10^{-17}
0.4	0.5	1.122×10^{-7}	6.084×10^{-10}	4.735×10^{-10}	6.830×10^{-17}	5.315×10^{-17}
0.4	0.4	1.019×10^{-7}	7.415×10^{-10}	6.015×10^{-10}	7.559×10^{-17}	6.132×10^{-17}

TABLE 4: Comparison of different design styles for the 4-operand CSA.

Design style	Supply voltage (Volt)	Average power (Watt)	Delay, $t_{d\,max}$ (Second)
Conventional CSA design	$V_{dd} = 1$	3.307×10^{-6}	2.071×10^{-10}
Proposed scheme of CSA design	$V_{ddL} = 0.4$, $V_{ddH} = 0.5$	3.009×10^{-7}	4.324×10^{-9}
Subthreshold CSA design	$V_{dd} = 0.4$	1.577×10^{-7}	7.774×10^{-9}

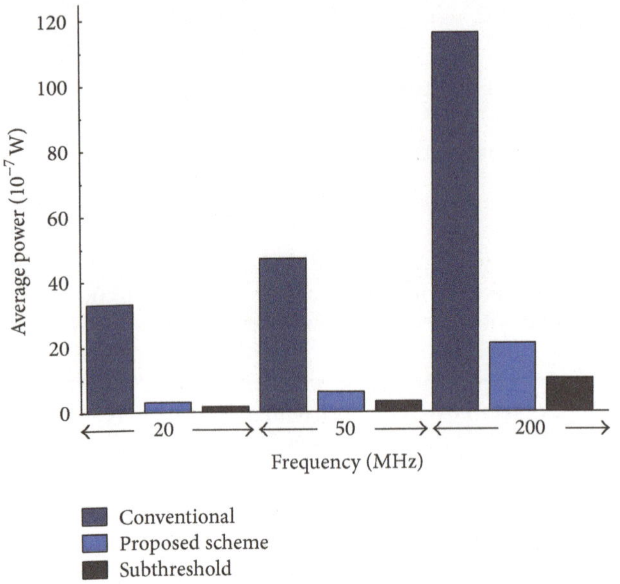

FIGURE 6: Average power consumption values for different frequencies.

is partitioned into three different clusters of row/rows. The first cluster may be formed using the subset of row/rows which is/are not time-critical (hence driven by V_{ddL}), whereas the third cluster can be formed using the subset of row/rows which is/are time-critical (hence driven by V_{ddH}). Now, the row which resides in the second cluster should be studded with the gates which are well-equipped to do the interfacing job in between the row at V_{ddH} and the row at by V_{ddL}.

With this notion, in our modified 4-operand CSA design (as illustrated in Figure 5), row1 is driven by the V_{ddL} (=0.4 Volt), row2 is driven by a dual supply of V_{ddL} (= 0.4 Volt) and V_{ddH} (= 0.5 Volt), and row3 is driven by the V_{ddH} (= 0.5 Volt) only. Furthermore, as the basic building blocks of the CSA design, we have used FA1 blocks for both row1 and row3 and FA2 blocks for the intermediate row2.

4. Near-Threshold Operation of the Proposed Scheme of CSA Design and Its Performance Analysis

When the conventional CSA design of [8] has been simulated with a larger supply voltage (V_{dd} = 1 Volt), for the frequency of 20 MHz, the $t_{d\,max}$ value is obtained as 2.071×10^{-10} Second. But, for the subthreshold operation (V_{dd} = 0.4 Volt) of the same circuit (even though the power consumption reduces drastically), the $t_{d\,max}$ value gets increased to a much higher value of 7.774×10^{-9} Second. Thereby, the application of the subthreshold design is mostly limited to the low performance systems only.

Now, to maintain this excellent energy efficiency of the subthreshold design, while boosting the speed of operation by a significant amount, we can explore the performance of the design for the near-threshold operation [13]. And that is what we have actually done in this work. To evaluate the effectiveness of the near-threshold operation of our modified CSA design, it has been compared with the conventional CSA design as well as the subthreshold CSA design (as shown in Table 4).

While operating for a frequency of 20 MHz, the proposed scheme of CSA design consumes 3.009 $\times 10^{-7}$ Watt of P_{avg}, which is almost 90.9% lesser than that of the conventional CSA design. Again, looking at the delay at output, the proposed scheme of CSA design provides a 44.37% improvement in $t_{d\,max}$, compared to that of the subthreshold CSA design.

Figure 6 illustrates the variation in power consumption values, considering all the three design styles, for different frequencies like 20 MHz, 50 MHz, and 200 MHz.

Following are the key points regarding the performance of the modified 4-operand CSA design presented in this work.

(i) The first one is the flexibility of the choice of any higher supply voltage (as per the requirement) for the gates which are in the critical path. In case where the speed of the subthreshold circuit is important, we can tune it by increasing the V_{ddH} even up to 0.8 Volt.

(ii) From Figure 6, it can be inferred that the proposed scheme works fine not only for an operating frequency of 20 MHz but also for the higher frequency ranges.

(iii) Row-based dual V_{dd} assignment has been incorporated in the proposed scheme of CSA design to facilitate the physical design implementation part.

(iv) But, in the case where performance tuning is not the requirement, rather P_{avg} is the only concern, we may simply go for the subthreshold CSA design where the power consumption is minimum (Table 4).

(v) Lastly, the limits for the upper/lower values of P_{avg} can be obtained through the use of V_{dd} as 1 Volt and V_{dd} as 0.4 Volt, respectively. Simultaneously, at these two V_{dd} limits, we will get the lower/higher amount of $t_{d\,max}$.

The proposed scheme of CSA design fits somewhere in between with a major advantage, that is, the flexibility of the choice of V_{ddH} (which can be used for the purpose of performance tuning).

5. Conclusion

In this work we have mainly focused on the performance analysis of a row-based dual V_{dd} CSA design which operates in the near-threshold voltage regime. For that purpose, we used two supply voltages: V_{ddH} (= 0.5 Volt) and V_{ddL} (= 0.4 Volt). Besides, the entire circuit is partitioned into different clusters of row/rows, and all the logic gates which reside in a particular cluster have been driven by a single supply (may be V_{ddH} or V_{ddL}). Moreover, a fair comparison among the different design styles for a 4-operand CSA has also been presented here. From the results obtained, we can easily infer that the near-threshold operation of the proposed scheme of CSA design can be very much effective in reducing the overall energy consumption, like a subthreshold design. At the same time, it can also be useful in tuning the performance of the circuit so that the maximum delay at output gets reduced.

Conflict of Interests

The authors declare that there is no conflict of interests regarding the publication of this paper.

Acknowledgment

The authors would like to thank SMDP-II Project Lab., IC Design & Fabrication Centre, Jadavpur University, for giving them the opportunity to carry out this work using SPICE Tools.

References

[1] M. Radfar, K. Shah, and J. Singh, "Recent subthreshold design techniques," *Active and Passive Electronic Components*, vol. 2012, Article ID 926753, 11 pages, 2012.

[2] K. Kim and V. D. Agrawal, "Minimum energy CMOS design with dual subthreshold supply and multiple logic-level gates," in *Proceedings of the 12th International Symposium on Quality Electronic Design (ISQED '11)*, pp. 689–694, Santa Clara, Calif, USA, March 2011.

[3] K. Kim and V. D. Agrawal, "True minimum energy design using dual below-threshold supply voltages," in *Proceedings of the 24th IEEE Annual Conference on VLSI Design*, pp. 292–297, January 2011.

[4] M. R. Kakoee and L. Benini, "Fine-grained power and body-bias control for near-threshold deep sub-micron CMOS circuits," *IEEE Journal on Emerging and Selected Topics in Circuits and Systems*, vol. 1, no. 2, pp. 131–140, 2011.

[5] D. E. Lackey, P. S. Zuchowski, T. R. Bednar et al., "Managing power and performance for system-on-chip designs using voltage islands," in *Proceedings of the IEEE/ACM International Conference on Computer Aided Design (ICCAD '02)*, pp. 195–202, 2002.

[6] J. Hu, Y. Shin, N. Dhanwada, and R. Marculescu, "Architecting voltage islands in core-based system-on-a-chip designs," in *Proceedings of the International Symposium on Lower Power Electronics and Design (ISLPED '04)*, pp. 180–185, August 2004.

[7] A. U. Diril, Y. S. Dhillon, A. Chatterjee, and A. D. Singh, "Level-shifter free design of low power dual supply voltage CMOS circuits using dual threshold voltages," in *Proceedings of the 18th International Conference on VLSI Design: Power Aware Design of VLSI Systems*, pp. 159–164, January 2005.

[8] K. S. Yeo and K. Roy, *Low-Voltage, Low-Power VLSI Subsystems*, Tata McGraw Hill, 2009.

[9] A. Wang, B. H. Calhoun, and A. P. Chandrakasan, *Sub-Threshold Design for Ultra Low-Power Systems*, Springer, 2006.

[10] M. Johnson and K. Roy, "Subthreshold leakage control by multiple channel length CMOS (McCMOS)," ECE Technical Report 80, 1997.

[11] S.-M. Kang and Y. Leblebici, *CMOS Digital Integrated Circuit*, Tata McGraw-Hill, 3rd edition, 2003.

[12] S. Basak, D. Saha, S. Mukherjee, S. Chatterjee, and C. K. Sarkar, "Design and analysis of a robust, high speed, energy efficient 18 transistor 1-bit full adder cell, modified with the concept of MVT scheme," in *Proceedings of the 3rd International Symposium on Electronic System Design*, pp. 130–134, December 2012.

[13] B. Zhai, R. G. Dreslinski, D. Blaauw, T. Mudge, and D. Sylvester, "Energy efficient near-threshold Chip Multi-processing," ISLPED, ACM/IEEE, 2007.

22

Musical Rhythms Affect Heart Rate Variability: Algorithm and Models

Hui-Min Wang and Sheng-Chieh Huang

The Department of Electrical Engineering, National Chiao Tung University, Room 720, Engineering Building 5, 1001 University Road, Hsinchu 30010, Taiwan

Correspondence should be addressed to Hui-Min Wang; marcel.wang@gmail.com

Academic Editor: George E. Tsekouras

There were a lot of psychological music experiments and models but there were few psychological rhythm experiments and models. There were a lot of physiological music experiments but there were few physiological music models. There were few physiological rhythm experiments but there was no physiological rhythm model. We proposed a physiological rhythm model to fill this gap. Twenty-two participants, 4 drum loops as stimuli, and electrocardiogram (ECG) were employed in this work. We designed an algorithm to map tempo, complexity, and energy into two heart rate variability (HRV) measures, the standard deviation of normal-to-normal heartbeats (SDNN) and the ratio of low- and high-frequency powers (LF/HF); these two measures form the physiological valence/arousal plane. There were four major findings. Initially, simple and loud rhythms enhanced arousal. Secondly, the removal of fast and loud rhythms decreased arousal. Thirdly, fast rhythms increased valence. Finally, the removal of fast and quiet rhythms increased valence. Our work extended the psychological model to the physiological model and deepened the musical model into the rhythmic model. Moreover, this model could be the rules of automatic music generating systems.

1. Introduction

The relation of music to emotion has been studied for decades and the literature is fruitful [1]. There exist a lot of psychological models between music and emotion [2], but the physiological models between music and emotion are limited [3]. One of the physiological actions, heart rate variability (HRV), which is controlled by the autonomic nervous system (ANS), is tightly connected with emotions [4]. Previously, we had analyzed the relationship between musical rhythms and HRV [5] and built two heuristic models [6, 7]. In this paper, a systematic algorithm is proposed to construct new models.

Musical emotions change with psychophysiological measures and musical features [8], whilst three basic questions are highlighted [9]: how do musical features evoke emotions; how do actions involved in musical emotions progress; and which actions and brain processes are involved in musical emotions. Basically, people feel what music expresses but need not be always; in a simple case, only 61% of 45 participants felt what they perceived [10]. More particularly, a stronger correlation is suggested by a recently developed theory that the aesthetic awe accompanies by being moved (cognitive), emotions (psychological), and thrills (physiological) in the same time [11]. The three associated levels of musical response should be analyzed individually and can be discussed together. In the research about music psychology, sometimes the perceived and felt emotions are examined separately; sometimes the music clips are just labeled artificially; whether they are perceived or felt had not been mentioned yet [12]. The mappings between music space and emotion space [13, 14] employ the following synonyms: music mood detection [15], music emotion measurement [12], characterization [12], recognition [16, 17], classification [12, 18], predicting [19], or modeling [20]; the review articles demonstrate the fruits of experts' interests [16–20].

In a valuable model, four elementary properties are goodness-of-fit [21] (generalization [22]), simplicity [21, 22], predictability [21, 22], and the relation to existing theories [22]. Beyond cognitive reactions, the prediction of emotional reactions is also important in music psychology [22, 23]. In addition to an existing review [24], a survey of the models [25–45] for emotional responses is given in this paper. The acoustic features, emotion spaces, and methods to generate these models are listed in Table 1 with four major observations. Initially, regression analysis [30, 31, 36, 38–40, 42, 43, 45] is widely employed, while some soft computing methods (fuzzy [41]; support vector machine [34, 37]; neural network [35]; K-nearest neighborhood [27]; Gaussian mixture model [25]) are also used. Secondly, beyond the enviable valence and arousal, tension [29, 32, 36, 40, 44] is another common dimension too. Thirdly, the felt emotion [28, 35] is infrequently examined compared to the perceived emotion. Finally, most works employ a wide range of acoustic features, while the interest in a single feature moves from pitch [40, 44] to timbre [28] and rhythm [26]. For the emotion space, Russell [13] proposed the valence/arousal space and Thayer [14] reduced it to four labels; the following synonyms are commonly used: valence as pleasantness [38]; arousal as activity [36, 45]; tension [29, 32, 36, 40, 44] as interest [43], expectancy [40], strength [38], potency [38], and resonance [27].

Whether perceived or felt emotion, the three commonly used methods, including valence/arousal dimensions, lists of basic emotions, and diverse emotion inventories, are not complete in the musical emotion study [46]. To have a more comprehensive overview [4], the underlying mechanisms of the central nervous system (CNS) [46, 47] and the ANS [48–50] should be also deliberated. In the most essential sense, the acoustic properties can be perceived by the nervous system and evoke psychophysiological responses. For example, fast and loud voices usually cause emotional arousal and increased respiratory and heart rates through the auditory and limbic systems [46, 47]. Some related physiological reactions coincided with the emotion expressed in music [48] and rated valence/arousal level [49]. Therefore a valence/arousal model [50] of musical emotion had been built based on the associated physiological measures, including electromyogram, electrocardiogram, skin conductivity, and respiration changes. Extended linear discriminant analysis (pLDA) was employed in the classification of musical emotions.

Since emotion correlates with physiological responses [48–50], too, the models of musical emotions were developed by integrating the acoustic features and physiological measures [51–53]. Despite combining with physiological measures, these models [51–53] tend to be similar to the psychological models mentioned above [25–45] in spite of the same perceived or felt emotion space. In addition, using acoustic features to model [3] the physiological responses seems more radical if the underlying mechanisms [46, 47] are considered. The model [3] employs 11 musical characteristics and the conclusion is that rhythmic features are the major factors of the physiological responses (respiration, skin conductance, and heart rate) to music. It also points out a limitation that some acoustic features would correlate with each other. This makes it difficult to discriminate their relative contributions to the detected relationships.

The rest of this paper is organized as follows. In Section 2, the rhythmic features for modeling and the HRV features for the physiological emotion space are introduced. In Section 3, the experiment and developed algorithm are presented. In Section 4, the models of the relationships between the rhythmic and the HRV features are illustrated in figures, and the equations and tables of statistics are provided. In Section 5, how tempo, complexity, and energy work in the psychological experiments, psychological models, physiological experiments, and physiological models are reviewed and compared. Finally, in Section 6, we summarize the contributions made in this study and suggest the directions for the further research.

2. Preliminary

2.1. Why Are Physiological Models Necessary? To study the relation of musical features to emotion, most psychological models focus on the perceived emotion, as Table 1 demonstrated. However, the perceived emotion need not be equal to the felt emotion [10, 54], and the felt emotion may not have the related physiological responses neither [55]. If the music clips are selected by the participants themselves, the perceived emotion has no statistical significance with the felt emotion in the ranks. Otherwise the differences are statistically significant [54]. Moreover, some patterns of the physiological responses appear while there is no related self-report [55]. Since the perceived emotion, felt emotion, and physiological actions play different roles, a physiological model is necessary beyond the psychological models.

2.2. Valence/Arousal Model. Although four dimensions are necessary for emotion spaces [56], it is difficult to realize a four-dimensional model. Three dimensions are also employed in the psychological models [27, 36, 38, 43]. Some work [57] announced that three dimensions could be reduced into two, without significant loss of goodness-of-fit; it also shows that the dimensional model is better than the discrete (categorical) model in resolution. Thus Russell's two-dimensional valence/arousal model [13] states its merit, within the ebb and flow of relevant woks, in the discipline of psychology.

2.3. A Physiological Valence/Arousal Model. SDNN and LF/HF, two measures of HRV [58], are employed as two dimensions in our physiological valence/arousal model. SDNN is the standard deviation of normal-to-normal heartbeat intervals in time domain and LF/HF is the ratio of low- and high-frequency powers after the fast Fourier transform (FFT); SDNN presents the variation of the circulatory system and LF/HF presents the balance of the sympathetic and parasympathetic nervous systems [59]. HRV is highly correlated with emotion [60–64] and some evidence [65–70] reveals that SDNN is a good indicator of valence in the physiological perspective. In general cases, the normal subjects' SDNNs are higher than the depression subjects' [65–67]. For the normal subjects' case, the SDNN levels of subjects with positive mood are higher than the negative mood [68]. For the depression subjects' case, the SDNN levels of low-depression patients' are higher than the high-depression patients' [69, 70]. All of these findings reach a statistical significance level. Hence SDNN could be a proper indicator as the physiological valence. In addition, increased LF/HF values denote that the sympathetic nerve activity tended to be

TABLE 1: Psychological models of musical features.

Reference	Year	Acoustic features: Number	Acoustic features: Catalog	Acoustic features: Type A: rhythm; B: pitch; C: timbre; D: other	Emotion space		Method
[25]	2012	70	A(1)B(17)C(52)	A: dynamic (1); B: tonal (17); C: spectral (11), timbre (41).	Russell	Perceived	GMM
[26]	2012	11	A(11)	A: rhythm (11).	Thayer	Perceived	C5
[27]	2012	20	A(10)B(6)C(4)	A: rhythm (7), intensity (3); B: pitch and tonality (4), harmony (2); C: timbre (4).	VA-Resonance	Perceived	KNN
[28]	2012	60	C(60)	C: timbre (60).	Thayer	Felt	C4
[29]	2012	9	A(6)B(3)	A: dynamics (1), onset frequency (1), tempo (1), meter (1), rhythmic regularity (1), syncopation (1); B: harmony (1), pitch height (1), melodic expectation (1).	T	Perceived	MPW
[30]	2012	16	A(6)B(10)	A: dynamic (1), crescendo (1), density (1), speed (1), tempo (1), articulation (1); B: direction (1), surprise (1), tendency (1), mode (1), dissonance (1), harmonic surprise (1), harmonic tempo (1), high pitch (1), low pitch (1), closure (1).	D11	Perceived	Regression
[31]	2011	46	A(16)B(10)C(10)D(10)	A: temporal (6), rhythmic (10); B: melody/harmony (10); C: spectral (10); D: lyrics (10).	Russell	Perceived	Regression
[32]	2011	4	A(2)B(2)	A:tempo (1), dynamics (1); B: register (1), contour (1).	T	Perceived	ANOVA
[33]	2011	32	C(32)	C: spectral contrast and MFCCs (20), echo nest timbre (12).	Russell	Perceived	CRF
[34]	2010	292	A(87)B(22)C(18)D(165)	A: rhythm (87); B: chords (22); C: spectral (18); D: lyrics (12), metadata (153).	Thayer	Perceived	SVM
[35]	2009	13	A(3)B(6)C(4)	A: dynamics (2), tempo (1); B: mean pitch (2), pitch variation (3), texture (1); C: timbre (4).	Russell	Felt	NN
[36]	2009	18	A(7)B(8)C(3)	A: dynamics (1), rhythm (3), articulation (3); B: harmony (5), register (3); C: timbre (3).	VAT	Perceived	Regression
[37]	2008	29	B(M)C(M)D(F)	B: pitch (many); C: timbre (many); D: other (few).	Russell	Perceived	SVM
[38]	2008	9	A(4)B(3)D(2)	A: temporal surface (2), dynamics (2); B: register (2), tonality (1); D: other (2).	VA-Strength	Perceived	Regression
[39]	2008	15	A(2)B(7)C(6)	A: loudness (1), volume (1); B: tonal dissonance (2), pure tonal (1), complex tonal (1), multiplicity (1), tonality (1), chord (1); C: spectral centroid (1), sharpness (2), timbral width (1), spectral dissonance (2).	Thayer	Perceived	Regression
[40]	2007	1	B(1)	B: pitch (1).	T	Perceived	Regression
[41]	2006	15	A(7)B(8)	A: loudness (1), duration (6); B: pitch (8).	Thayer	Perceived	Fuzzy
[42]	2006	18	A(3)B(11)C(4)	A: dynamics (2), tempo (1); B: mean pitch (2), pitch variation (3), harmony (5), texture (1); C: timbre (4).	Russell	Perceived	Regression
[43]	2007	8	A(3)B(4)C(1)	A: tempo (1), loudness (1), articulation (1); B: roughness (1), melody (1), ambitus (1), register (1); C: brightness (1).	VA-Interest	Perceived	Regression

TABLE 1: Continued.

Reference	Year	Number	Catalog	Acoustic features Type A: rhythm; B: pitch; C: timbre; D: other	Emotion space		Method
[44]	2005	4	B(4)	B: pitch (stability, proximity, direction, mobility).	T	Perceived	Formula
[45]	2004	6	A(3)B(3)	A: loudness (1), spectral centroid (1), tempo (1); B: melodic (1), contour (1), texture (1).	Russell	Perceived	Regression

Emotion space: V: valence; A: arousal; T: tension; D11: 11 dimensions.
Method: GMM: Gaussian mixture model; C5: classifiers 5; KNN: K-nearest neighborhood; C4: classifiers 4; MPW: moving perceptual window; ANOVA: analysis of variance; CRF: conditional random field; SVM: support vector machine; NN: neural network.

strong; thus LF/HF could be used as a physiological indicator of arousal levels [59].

2.4. Musical Features Employed in the Model. The three most important features of music are rhythm, pitch, and timbre [71–76]. Although pitch [77] and timbre [78] have been employed to recognize emotion in music, the role of rhythm seems more rudimentary. In fact, the rhythmic features [26] had been used in modeling the emotional responses and acquired a reasonable performance.

An all-encompassing representation of the rhythm may be the temporal organization of sound, tightly connected with meter, which refers to the periodic composition in music [23]. Musical rhythm may have its roots in the motor rhythms controlling heart rate, breathing, and locomotion [79], dominated by brain stem the ancient structure [47]; fast, loud sounds produce an increased activation of the brain [47] and some evidence suggests that the musical rhythm can regulate emotion and motivational states [80].

Rhythm is less tractable than pitch and timbre in localizing its specific neural substances [81]. However, many cultures emphasize the rhythm, with two others playing a less crucial role [82]. To study the neural basis of rhythm, brain imaging, psychophysical, and computational approaches have been employed [83]. Beat and meter induction are the fundamental elements of cognitive mechanisms [84], while the representations of metric structure and neural markers of expectation of the beat have been found in both electroencephalography (EEG) and magnetoencephalography (MEG) [85]. Beat perception is innate; newborn infants expect downbeats (onsets of rhythmic cycles), even unmarked by stress or the spectral features [86]. Infants also engage in the rhythmic movement to rhythmically regular sounds and the faster movement tempo is associated with the faster auditory tempo [87]. Although the beat perception is innate, the ability to detect rhythmic changes is more developed in adults than infants [88] and the trained musicians than the untrained individuals [89].

Of all rhythmic features, experts suggest that tempo, complexity (regularity), and energy (intensity, strength, dynamic, loudness, and volume) [15, 27, 90, 91] are the most significant.

3. Methods

3.1. Experiment

3.1.1. Participants. There were 22 healthy subjects, 15 males and 7 females, engaged in the experiment. The average age was 23. None of them had been professionally trained in music.

3.1.2. Musical Stimuli. There were four drum loops in this study (L1 to L4). The parameters (tempo, complexity, and energy) are listed in Table 2 and the rankings are illustrated in Figure 1.

3.1.3. Apparatus. The ECG signal was captured by a 3-channel portable device (MSI E3-80, FDA 510(k) K071085) at 500 Hz sampling rate from the chest surface of the body. Only the channel-1 data was taken to be analyzed.

3.1.4. Procedures. All experiments were carried out in moderate temperature, humidity, and light with subjects sitting and wearing headphones (eyes closed) in a quiet room. Each subject accepted 4 rounds of experiment in different days. Each round took 15 minutes, separated as 5 stages, as Figure 2 illustrated. Stage 1 had 5 minutes to let the subjects calm down. Stage 2 had 3 minutes of rest, as the baseline for the responses of the drum loops. Stage 3 had 2 minutes, as the baseline for the responses after the drum loops. Stage 4 had 3 minutes of stimulus of some drum loop. Stage 5 had 2 minutes of rest. The ECG signal from stage 2 to stage 5 was recorded and separated as epoch 1 (E1) to epoch 4 (E4). Comparison 1 (C1) was the difference of E1 and E3 and comparison 2 (C2) was the difference of E2 and E4.

3.2. Signal Processing

3.2.1. Musical Features. Musical rhythm is expressed by the successive notes that record the relating temporal information. Tempo (T_p) could be decided as the unit, beats per minute (bpm) [92].

TABLE 2: Rhythmic features, model factors, and HRV data collection of four drum loops (L1~L4).

	Definition	L1		L2		L3		L4	
Tp	Tempo	89.60		105.30		139.50		176.50	
Cp	Perceptual complexity	2.73		2.27		1.14		3.86	
Eg	Energy, $\sum(k^2)/(10^{13})$	1.99		1.02		3.83		1.56	
Valence (C1)	−1/Tp	−4.18		−2.23		0.50		2.26	
Valence (C2)	Tp/Eg	−2.33		4.48		−3.34		5.64	
Arousal (C1)	Eg/Cp	−0.02		−0.05		0.27		−0.05	
Arousal (C2)	−Tp × Eg	−0.15		−0.10		−0.42		−0.23	
SDNN (C1)	Standard deviation of all RR intervals of C1	−5.01	±6.95	−0.81	±10.57	−0.10	±11.63	2.27	±12.06
SDNN (C2)	Standard deviation of all RR intervals of C2	−2.66	±7.31	5.17	±9.66	−3.14	±15.52	5.08	±10.29
LF/HF (C1)	Ratio of low frequency and high frequency of C1	−0.04	±1.67	−0.02	±0.81	0.28	±0.89	−0.06	±0.66
LF/HF (C2)	Ratio of low frequency and high frequency of C2	−0.10	±0.76	−0.10	±0.68	−0.41	±0.59	−0.29	±1.05

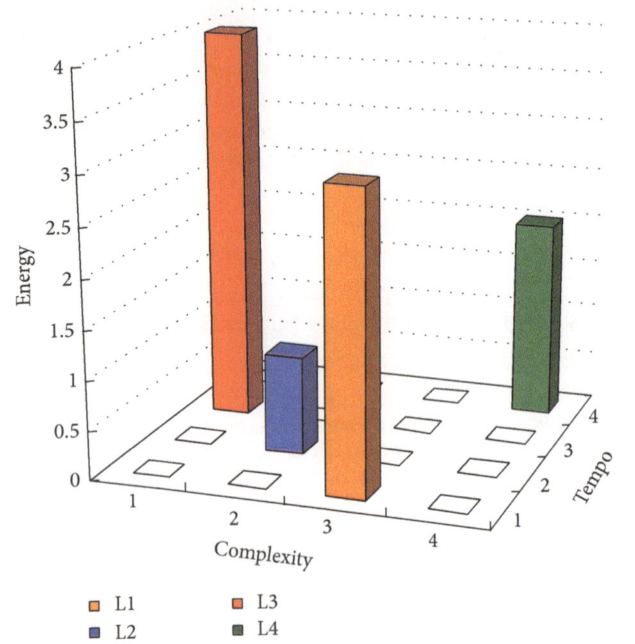

FIGURE 1: Four drum loop patterns (L1~L4) employed in this study with ranked tempo (L4 > L3 > L2 > L1), complexity (L4 > L1 > L2 > L3), and energy (L3 > L1 > L4 > L2).

The other characteristic, perceptual complexity (C_p), was obtained by asking the human subjects to judge the complexity of the rhythms they had listened to. It was assessed by the subjects using a subjective rating of 1 to 4 on a Likert Scale (4 being the most complex) [93].

The energy parameter (E_g) was defined as Σk^2, the summation of square of k, where k is the amplitude of the signal [94, 95].

3.2.2. HRV Features. To acquire the measures of HRV features, QRS detection was the first step [96–98], where R denotes the peak in a heartbeat signal. After the abnormal beats were rejected [99], the mean of R-R intervals

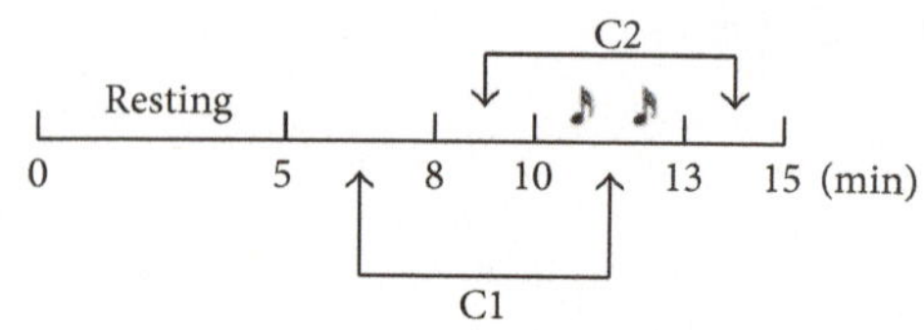

FIGURE 2: The procedure of each experiment: the epoch of 5 min to 8 min is the baseline of the physiological responses during music listening and the epoch of 8 min to 10 min is the baseline of the physiological responses after music listening.

(MRR), standard deviation of normal-to-normal R-R intervals (SDNN), and root of mean of sum of square of differences of adjacent R-R intervals (RMSSD) were measured in time domain [59]. After interpolation [100] (prepared for FFT) and detrending [101] (to filter the respiratory signal), the FFT [102] was applied to calculate the low- (LF) and high-frequency (HF) powers and their ratio (LF/HF). The results of SDNN and LF/HF are listed in Table 2. Four groups of data, SDNN C1, SDNN C2, LF/HF C1, and LF/HF C2, were observed in our analysis.

3.3. Algorithm. For modeling the responses of some HRV measure and the related musical rhythms, our algorithm included 3 steps. Initially, all possible combinations of the rhythmic features were explored. Secondly, the values of the combinations were linearly transformed. Finally, the coefficients of the linear transformations were calculated such that the Euclidian metric between the results of step 2 and the related HRV responses is the minimum.

The stimuli are 4 drum loops (rhythm), named as R_1 to R_4 here:

$$R_{ri}, \quad ri \in \{1, 2, 3, 4\}. \tag{1}$$

Each rhythm relates to some HRV measure H_{ri}:

$$H_{ri}, \quad ri \in \{1, 2, 3, 4\}. \tag{2}$$

For each rhythm, there are three rhythmic features, T_p, C_p, and E_g, named as F_1 to F_3:

$$F_{fi}, \quad fi \in \{1,2,3\}. \tag{3}$$

Since a musical stimulus contains multiple combinations and interactions of various features, it is difficult to realize which feature is contributing to the perceived emotion [45], or physiological responses. Our solution is considering all possible combinations: the influence of each feature F_{fi} is linear (order 1), of no effect (order 0), or inverse (order −1). Although the higher orders are plausible, order one is still the most suitable to construct a simplified model for understanding the relation of the musical rhythms to their related HRV measures:

$$E_{ri} = \prod_{fi=1}^{3} \left(R_{ri}.F_{fi}\right)^{fi.p}, \quad fi.p \in \{-1,0,1\}. \tag{4}$$

Linear transformation is necessary because the units of rhythmic and HRV features are not uniform. All we need are the related correspondences. Consider

$$T_{ri} = x\left(E_{ri}\right) + y, \quad x \in R, \ y \in R. \tag{5}$$

Thus for some subject's HRV responses H_1 to H_4 (e.g., SDNN) to rhythms R_1 to R_4, some combination of rhythmic features can model the relation if the metric between T_{ri} and H_{ri} is the minimum. Euclidean metric [27, 103] is employed in this work:

$$D = \sqrt{\sum_{ri=1}^{4} \left(T_{ri} - H_{ri}\right)^2}. \tag{6}$$

After squaring each side of (6), we can acquire the coefficients of the linear transformation to get the minimum metric if the partial derivatives of D^2 (with respect to x and y) are both zero:

$$D^2 = \sum_{ri=1}^{4} \left(T_{ri} - H_{ri}\right)^2. \tag{7}$$

3.4. Statistics

3.4.1. Outliers. The judgment and removal of outliers are fundamental in the processing of experimental data [104]. In our study, the experimental data was separated into three groups by the minimum metric mentioned in Section 3.3 and the group with larger metric was eliminated.

The basic idea is to rank a group of data by the algorithm we proposed and make the data with larger metric obsolete. There were four groups of data, SDNN (C1), SDNN (C2), LF/HF (C1), and LF/HF (C2). Each group had 22 records of the 22 participants. Each record had four subrecords of drum loops L1 to L4. First, we used the algorithm in each group and ranked the 22 records as 1 to 22 by their minimum metric. After summing the ranks of the four groups of these 22 participants, these participants were partitioned into three levels: small, medium, and large metric (7, 7, and 8 participants). We defined the predictable class with the small and medium metrics and the nonpredictable class with the large metric. The nonpredictable class was excluded from our experimental data. After removing the nonpredictable class, the rhythmic features and the four averages of these four groups of predictable class were modeled by our algorithm again.

3.4.2. ANOVA and t-Test. Repeated measurements have four major advantages: obtaining the individual patterns of change, less subjects, serving as subjects' own controls, and reliability; the disadvantages are the complication by the dependence among repeated observations and less control of the circumstances [105]. In our experiment, the repeated measurements were employed.

Two basic methods of ANOVA are one-way between-groups and one-way within-groups. A more complicated method is two-way factorial ANOVA with one between-groups and one within-groups factor [106]. The repeated-measures ANOVA can be considered as a special case of two-way ANOVA [107]. The formula of repeated measures ANOVA with one within-groups factor (W) and one between-groups factor (B) is listed as [106]

$$y \sim B * W + \text{Error}\left(\frac{\text{Subject}}{W}\right). \tag{8}$$

For each HRV measure, there are two ANOVA tables. W is the participant. If the epoch is fixed, B is the rhythm. If the rhythm is fixed, B is the epoch. Then the Tukey honest significant difference test was used to acquire pair-by-pair comparisons [108] for each between-groups pair.

Finally, the pairwise t-tests [109, 110] were applied for C1 and C2 to realize whether the HRV responses are influenced by some particular rhythm.

4. Results

Table 2 collected the musical features, factors of models, and HRV data in this study. Valence C1 and arousal C1 reveal how rhythmic features influenced the HRV responses while listening to music, and valence C2 and arousal C2 reveal how rhythmic features influenced the HRV responses after listening to music. The values of modified combinations of the rhythmic and HRV features were illustrated in Figures 3(a) and 3(b). Furthermore, (9) to (12) demonstrated the relationships.

Fast tempo enhanced SDNN; that is, people prefer faster tempi:

$$\text{SDNN (C1)} \propto -\frac{1}{T_p}. \tag{9}$$

High intensity and low complexity enhanced LF/HF, the physiological arousal:

$$\frac{\text{LF}}{\text{HF}}\text{(C1)} \propto \frac{E_g}{C_p}. \tag{10}$$

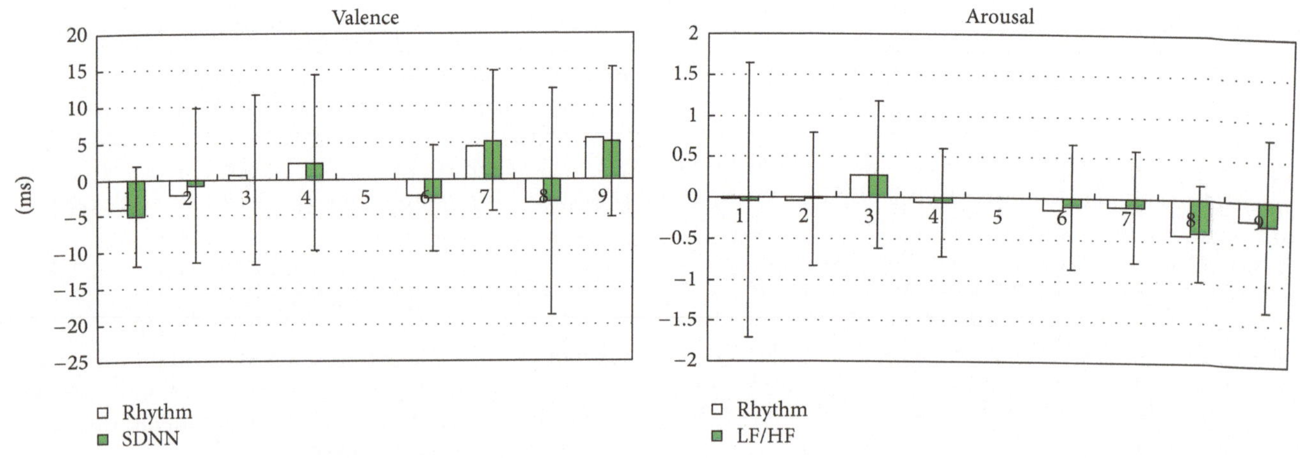

(a) (b)

FIGURE 3: (a) The gray bars represent the averages of SDNN and the white bars represent the approached values of the model. Items 1 to 4 represent drum loops 1 to 4 in C1 and items 6 to 9 represent drum loops 1 to 4 in C2; (b) gray bars represent the averages of LF/HF and white bars represent the approached values of the model. Items 1 to 4 represent drum loops 1 to 4 in C1 and items 6 to 9 represent drum loops 1 to 4 in C2 (note: related numeric data is presented in Table 2).

People prefer faster tempi. However, the sound should not be too loud if we hope the effect can remain after the music stops:

$$\mathrm{SDNN}\,(\mathrm{C2}) \propto \frac{T_p}{E_g}. \tag{11}$$

If fast and loud music stops, people feel relaxed:

$$\frac{\mathrm{LF}}{\mathrm{HF}}\,(\mathrm{C2}) \propto -T_p \times E_g. \tag{12}$$

Table 3 is the statistical results. The SDNN (C2) of the four rhythms had a significance value 0.01. The LF/HF of L3 had a significance value 0.03 while and after listening to music.

5. Discussion

5.1. Advantages of the Proposed Model. There are four major advantages in this work. Initially, the psychological music models are many [2, 25–45], but the physiological music models are rare [3]. Secondly, the psychological music models [2, 25–45] and physiological musical experiments [1, 111] are many, but studies of the psychological rhythm model [26] and the physiological rhythmic experiment [112] are rare. Thirdly, almost all studies concern the responses during the music, but discussions about the responses after the music are limited [113]. Finally, there are many music studies about HRV [1, 111], but there is no model. Our work is the first one.

5.2. Comparison of the Models. There are six levels of musical model illustrated in Figure 4: (a) the model from the results of the psychological experiments, (b) the model of the perceived emotion, (c) the model of the felt emotion, (d) the model from the physiological results, (e) the proposed physiological valence/arousal model while listening to the musical rhythms, and (f) the proposed physiological valence/arousal model after listening to the musical rhythms.

The six levels of model need not reach a consensus. There are four reasons. Initially, the perceived emotion is not always the same as the felt emotion [10, 54]. These two emotions are closer if the music clips are chosen by the participants themselves instead of others. Secondly, some measured physiological responses might not have been reported by self-report [55]. Thirdly, if the music clips are simplified to the form of rhythm or tempo, the responses are changed [112]. Finally, the responses while and after listening to music need not be the same obviously.

Figure 4(a) shows that T_p and E_g dominate arousal and C_p dominates valence in a survey of psychological experiments (pp. 383–392) [1]. The effects of T_p and E_g on arousal are clear. As for valence, the regular and varied rhythms derive positive emotions and the irregular rhythms derive negative emotions (pp. 383–392) [1]. Hence valence is negatively correlated to C_p.

Figure 4(b) shows that T_p [26, 31, 45] and E_g [26, 27, 31, 45] dominate arousal and T_p [27] and C_p [27] dominate valence in the perceived models. The results of arousal for the perceived models are the same as the psychological experiments. But T_p and C_p are viewed as the valence-based features. Whether they are positively correlated to valence or not has not been highlighted in the paper [27].

Figure 4(c) shows that T_p [35] and E_g [35] dominate arousal and T_p [35] dominates valence in the felt model. The results of felt and perceived models are similar. But the valence-based feature is limited at T_p in the felt model [35].

Figure 4(d) shows that T_p [55, 113, 114] and E_g [114] dominate arousal and T_p [55] dominates valence in the physiological experiments. Both psychological studies (Figures 4(a), 4(b), and 4(c)) and physiological studies (Figure 4(d)) have the same results about arousal. But the faster tempi are found to decrease SDNN, the physiological valence [55].

TABLE 3: Statistical P values for epochs and rhythms.

(a) SDNN

Epoch	Pairwise L1-L2	L2-L3	L3-L4	L1–L3	L2–L4	L1–L4	L1~L4			
E1	1.00	1.00	1.00	1.00	1.00	1.00	0.57			
E2	0.40	0.40	0.63	1.00	1.00	1.00	0.15			
E3	1.00	1.00	1.00	0.96	1.00	1.00	0.47			
E4	1.00	1.00	1.00	1.00	1.00	1.00	0.77			
C1	0.64	1.00	1.00	1.00	1.00	0.55	0.26			
C2	0.14	0.14	0.14	1.00	1.00	0.14	0.01	**		
Rhythm	**Pairwise E1-E2**	**E2-E3**	**E3-E4**	**E1–E3**	**E2–E4**	**E1–E4**	**E1~E4**		**C1-C2**	
L1	0.75	0.75	0.94	0.09	0.75	0.07	0.02	*	0.34	
L2	0.02	0.19	1.00	1.00	0.27	1.00	0.07	.	0.12	
L3	1.00	1.00	1.00	1.00	1.00	1.00	0.60		0.41	
L4	1.00	1.00	1.00	1.00	0.53	1.00	0.40		0.40	

(b) LF/HF

Epoch	Pairwise L1-L2	L2-L3	L3-L4	L1–L3	L2–L4	L1–L4	L1~L4			
E1	1.00	0.09	0.36	0.97	1.00	1.00	0.41			
E2	1.00	1.00	1.00	1.00	1.00	1.00	0.77			
E3	1.00	1.00	1.00	1.00	1.00	1.00	0.78			
E4	1.00	1.00	1.00	1.00	1.00	1.00	0.44			
C1	1.00	1.00	1.00	1.00	1.00	1.00	0.80			
C2	1.00	0.66	1.00	1.00	1.00	1.00	0.54			
Rhythm	**Pairwise E1-E2**	**E2-E3**	**E3-E4**	**E1–E3**	**E2–E4**	**E1–E4**	**E1~E4**		**C1-C2**	
L1	1.00	1.00	1.00	1.00	1.00	1.00	0.87		0.90	
L2	0.27	0.74	0.07	1.00	1.00	0.27	0.09	.	0.55	
L3	0.42	1.00	0.45	0.82	0.14	1.00	0.12		0.03	*
L4	1.00	1.00	0.92	1.00	1.00	0.92	0.23		0.50	

The values of HRV measures in C1 are the comparisons (differences) of values of HRV measures in epochs E1 and E3.
The values of HRV measures in C2 are the comparisons (differences) of values of HRV measures in epochs E2 and E4.
L1-L2: drum loops 1 and 2; L1~L4: drum loops 1 to 4.
Significant codes: 0 "***" 0.001 "**" 0.01 "*" 0.05 "." 0.1 " " 1.

Figure 4(e) shows that E_g and C_p dominate arousal and T_p dominates valence. E_g is directly proportional to arousal and C_p is inversely proportional to arousal. As for valence, faster tempi increase SDNN, the physiological valence. This relation is also illustrated in the left part of Figure 3(a).

Figure 4(f) shows that T_p and E_g dominate both arousal and valence after the stimuli. The arousal is negatively correlated to $T_p \times E_g$. Hence if a fast and loud rhythm is removed, the subjects will feel relaxed. Fast tempi increase valence in Figure 4(e), the responses during music. This effect remains after the music stops if the intensity of the music is low.

In the valence perspective, T_p and C_p are considered as two major rhythmic parameters [27]. As for T_p, there exist opposite comments. T_p is positively correlated with valence in the felt models [35] but T_p is negatively correlated with physiological valence since the faster tempo decreased SDNN [55]. The proper conclusion is that the role of T_p is dependent on the context: both slow and fast tempi can derive different emotions (pp. 383–392) [1]. For example, both happy and angry music have fast tempi. As for C_p, its contribution is not defined in the perceived model [27]. Although C_p is negatively correlated with valence in the psychological experiments (pp. 383–392) [1], the same definition has opposite results: the firm rhythm derives positive valence (pp. 383–392) [1] and the firm rhythm derives negative valence [31]. Instead of music, if only the rhythm is considered, higher SDNN is observed with faster tempi in our work, as Figure 4(e) illustrated; and E_g affects the responses of physiological valence after the music stops, as Figure 4(f) illustrated.

In the arousal perspective, T_p and E_g are two of the most dominant features. Our findings also show that if a fast, loud sound (drum loop 3) was removed, LF/HF would decrease, as illustrated in the right part of Figure 3(b). Although T_p

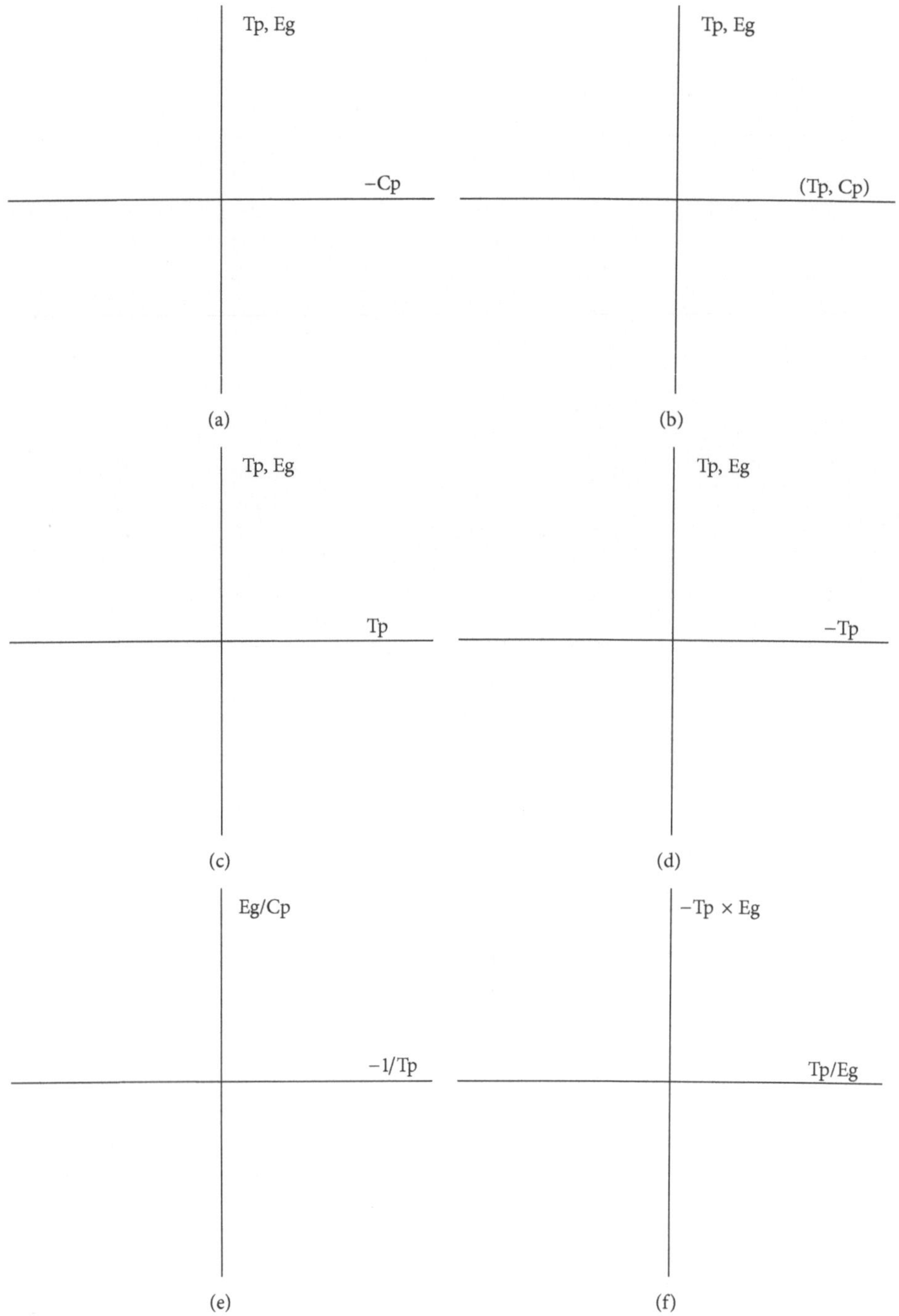

FIGURE 4: Six levels of the relationships between the rhythmic features and the valence/arousal model; the horizontal axis is valence and the vertical axis is arousal. (a) Model from the results of the psychological experiments, (b) model of the perceived emotion, (c) model of the felt emotion, (d) model from the results of the physiological experiments, (e) proposed physiological valence/arousal model while listening to the musical rhythm, and (f) proposed physiological valence/arousal model after listening to the musical rhythm.

is usually considered as the most important factor in all features (pp. 383–392) [1], another research indicates E_g is more important than T_p in arousal [45]; our findings support this opinion, as illustrated in Figure 4(e). C_p is considered as a valence parameter in Figures 4(a) and 4(b). But our results show that simple rhythms enhanced the physiological arousal.

5.3. Statistical Results. Table 3 collects all P values and there are two significant results: the ANOVA of SDNN (C2) for L1~L4 and the t-test of LF/HF (C1, C2) for L3. Although the significant value is more than 0.05, the left part of Figure 3(a) still shows that SDNN, the physiological valence, is positively correlated with the tempi of the drum loops. The right part of Figure 3(a) shows that SDNN is positively correlated with

T_p/E_g ($P < 0.01$). Both L2 and L4 have faster tempi and smaller energies. The left part of Figure 3(b) shows that LF/HF, the physiological arousal, is positively correlated with E_g/C_p. L3, a fast drum loop with a very simple complexity, enhances arousal in the largest degree. The right part of Figure 3(b) shows that after the music stops, the fast and loud rhythms make the participants relaxed. Although there is no statistical significance in Figure 3(b), the t-test result of L3, while and after listening to the music clip, reaches statistical significance ($P < 0.05$).

5.4. Limitations

5.4.1. The Three Rhythmic Parameters. Of the three elementary rhythmic features, tempo is the most definite [92]. T_p is positively correlated with SDNN, the physiological valence (Figure 4(e)). In the other way, T_p is negatively correlated with SDNN [55]. To integrate these two opposite findings, the psychological perspectives (Figures 4(a), 4(b), and 4(c)) are more suitable to explain this dilemma. That is, T_p is a feature of arousal; both positive and negative valence can be measured; it depends on the context (pp. 383–392) [1].

To measure the complexity of a music or rhythm clip, mathematical quantities are better than adjectives. For example, the firm rhythms can derive both positive (pp. 383–392) [1] and negative [31] valence. We can acquire a number from Likert Scale [93]; however, it is not proper as an engineering reference. Of the mathematical studies about rhythmic complexity [115–117], oddity [117] is the closest method for measuring the perceptual complexity. There is a more fundamental issue; two rhythms with different responses may have the same complexity measure.

The last parameter, the total energy of a waveform, is defined as Σk^2, the summation of square of k, where k is the amplitude of the signal [94, 95]. But the intensity within the waveform may not be fixed: large or small variations suggest fear or happiness; the rapid or few changes in intensity may be associated with pleading or sadness (pp. 383–392) [1]. If the waveform is partitioned into small segments [118], the energy feature will be not only an arousal, but also a valence parameter (pp. 383–392) [1]. And the analytic resolution of the musical emotion will increase.

5.4.2. Nonlinearity of the Responses. The linear and inverse functions are simple, intuitive, and useful in a general overview. However, some studies assume that the liking (or valence) of music is an inverted-U curve, dependent on arousal (tempo and energy) [119–121] and complexity [119, 120]. The inverted-U phenomenon may disappear with the professional musical expertise [122]. There is also a plausible transform of the inverted-U curve: although 120 bpm may be a preferred tempo for common people [123], the twin-peak curve was found in both a popular music database and physiological experiments [124]. Finally, the more complex curves are possible [3].

5.5. The Modified Model. T_p and E_g correlate with arousal, as Figures 4(a)–4(d) and Figure 4(f) illustrated, but E_g dominates the arousal [45], as Figure 4(e) showed. To integrate these models, let E'_g be the average energy of all beats. Thus E_g is the product of T_p and E'_g, and (9)–(12) can be modified as (13)–(16). This provides us with an advanced aspect.

The valence of musical responses keeps the same:

$$\text{SDNN (C1)} \propto -\frac{1}{T_p}. \tag{13}$$

Both tempo and energy are arousal parameters now. Moreover, our model demonstrates that the complexity of rhythm is also an arousal parameter:

$$\frac{\text{LF}}{\text{HF}}\text{ (C1)} \propto T_p \times \frac{E'_g}{C_p}. \tag{14}$$

The valence after the musical stimuli is influenced by the energy:

$$\text{SDNN (C2)} \propto \frac{1}{E'_g}. \tag{15}$$

Finally, both tempo and energy influence the arousal after musical stimuli, but tempo dominates the effect:

$$\frac{\text{LF}}{\text{HF}}\text{ (C2)} \propto -T_p^2 \times E'_g. \tag{16}$$

To integrate all psychological and physiological music theories, we summarize briefly that tempo, energy, and complexity are both valence and arousal parameters. The inverted-U curve [119–121] is suggested for valence. Considering the valence after music, people prefer the music of low intensity. For the responses after the musical stimuli, the arousal correlates negatively with tempo and energy whereas tempo is the dominant parameter.

5.6. Criteria for Good Models. There are four criteria for a valuable model [22]. They are generalization (goodness-of-fit [21]) and predictive power, simplicity, and its relation to existing theories. In plain words, a model should be able to explain experimental data used in the model and not used in the model for verification. The parameters should be small relative to the amount of the data. The model should be able to unify two formerly unrelated theories. It will help to understand the underlying phenomenon instead of an input-output routine.

Our proposed model satisfies generalization, simplicity, and the relation to other theories. In our study, the selected model has the least metric (error) among all possible models. It has only three parameters (T_p, C_p, and E_g). And it fills the gap among the psychological and physiological experiments and the models of music and rhythm. The proposed model does not have enough predictive power. It was built for the subjects whose responses are able to be modeled; the nonpredictable class, one-third of all subjects, was excluded. There was no test data in this study. However, the model derived from the experimental data can be integrated well with the literature in this field, as Section 5.5 mentioned.

6. Conclusion and Future Work

A novel experiment, a novel algorithm, and a novel model have been proposed in this study. The experiment explored how rhythm, the fundamental feature of music, influenced the ANS and HRV. And the relationships between musical features and physiological features were derived from the algorithm. Moreover, the model demonstrated how the rhythmic features were mapped to the physiological valence/arousal plane.

The equations in our model could be the rules for music generating systems [125–127]. The model could also be fine-tuned if the rhythmic oddity [117] (for complexity), small segments [118] (for energy), and various curves [3, 119–121, 124] are considered.

Conflict of Interests

The authors declare that there is no conflict of interests regarding the publication of this paper.

Acknowledgments

The authors thank Mr. Shih-Hsiang Lin heartily for his support in both program design and data analysis. The authors also thank Professor Shao-Yi Chien and Dr. Ming-Yie Jan for related discussion.

References

[1] P. N. Juslin and J. A. Sloboda, Eds., *Handbook of Music and Emotion: Theory, Research, Applications*, Oxford University Press, 2010.

[2] Y. H. Yang and H. H. Chen, *Music Emotion Recognition*, CRC Press, 2011.

[3] P. Gomez and B. Danuser, "Relationships between musical structure and psychophysiological measures of emotion," *Emotion*, vol. 7, no. 2, pp. 377–387, 2007.

[4] G. Cervellin and G. Lippi, "From music-beat to heart-beat: a journey in the complex interactions between music, brain and heart," *European Journal of Internal Medicine*, vol. 22, no. 4, pp. 371–374, 2011.

[5] S.-H. Lin, Y.-C. Huang, C.-Y. Chien, L.-C. Chou, S.-C. Huang, and M.-Y. Jan, "A study of the relationship between two musical rhythm characteristics and heart rate variability (HRV)," in *Proceedings of the IEEE International Conference on BioMedical Engineering and Informatics*, pp. 344–347, 2008.

[6] H.-M. Wang, S.-H. Lin, Y.-C. Huang et al., "A computational model of the relationship between musical rhythm and heart rhythm," in *Proceeding of the IEEE International Symposium on Circuits and Systems (ISCAS '09)*, pp. 3102–3105, Taipei, Taiwan, May 2009.

[7] H.-M. Wang, Y.-C. Lee, B. S. Yen, C.-Y. Wang, S.-C. Huang, and K.-T. Tang, "A physiological valence/arousal model from musical rhythm to heart rhythm," in *Proceedings of the IEEE International Symposium of Circuits and Systems (ISCAS '11)*, pp. 1013–1016, Rio de Janeiro, Brazil, May 2011.

[8] C. L. Krumhansl, "Music: a link between cognition and emotion," *Current Directions in Psychological Science*, vol. 11, no. 2, pp. 45–50, 2002.

[9] M. Pearce and M. Rohrmeier, "Music cognition and the cognitive sciences," *Topics in Cognitive Science*, vol. 4, no. 4, pp. 468–484, 2012.

[10] P. Evans and E. Schubert, "Relationships between expressed and felt emotions in music," *Musicae Scientiae*, vol. 12, no. 1, pp. 75–99, 2008.

[11] V. J. Koneĉni, "Does music induce emotion? A theoretical and methodological analysis," *Psychology of Aesthetics, Creativity, and the Arts*, vol. 2, no. 2, pp. 115–129, 2008.

[12] M. Zentner, D. Grandjean, and K. R. Scherer, "Emotions evoked by the sound of music: characterization, classification, and measurement," *Emotion*, vol. 8, no. 4, pp. 494–521, 2008.

[13] J. A. Russell, "A circumplex model of affect," *Journal of Personality and Social Psychology*, vol. 39, no. 6, pp. 1161–1178, 1980.

[14] R. E. Thayer, *The Biopsychology of Mood and Arousal*, Oxford University Press on Demand, 1989.

[15] D. Liu, L. Lu, and H.-J. Zhang, "Automatic mood detection from acoustic music data," in *Proceedings of the International Symposium on Music Information Retrieval*, pp. 81–87, 2003.

[16] Y. E. Kim, E. M. Schmidt, R. Migneco et al., "Music emotion recognition: a state of the art review," in *Proceedings of the International Symposium on Music Information Retrieval*, pp. 255–266, 2010.

[17] Y.-H. Yang and H. H. Chen, "Machine recognition of music emotion: a review," *ACM Transactions on Intelligent Systems and Technology*, vol. 3, no. 3, article 40, 2012.

[18] K. Trohidis, G. Tsoumakas, G. Kalliris, and I. Vlahavas, "Multi-label classification of music into emotions," in *Proceedings of the International Symposium on Music Information Retrieval*, pp. 325–330, September 2008.

[19] M. A. Rohrmeier and S. Koelsch, "Predictive information processing in music cognition. A critical review," *International Journal of Psychophysiology*, vol. 83, no. 2, pp. 164–175, 2012.

[20] G. Widmer and W. Goebl, "Computational models of expressive music performance: the state of the art," *Journal of New Music Research*, vol. 33, pp. 203–216, 2004.

[21] H. Honing, "Computational modeling of music cognition: a case study on model selection," *Music Perception*, vol. 23, no. 5, pp. 365–376, 2006.

[22] H. Purwins, P. Herrera, M. Grachten, A. Hazan, R. Marxer, and X. Serra, "Computational models of music perception and cognition I: the perceptual and cognitive processing chain," *Physics of Life Reviews*, vol. 5, no. 3, pp. 151–168, 2008.

[23] H. Purwins, M. Grachten, P. Herrera, A. Hazan, R. Marxer, and X. Serra, "Computational models of music perception and cognition II: domain-specific music processing," *Physics of Life Reviews*, vol. 5, no. 3, pp. 169–182, 2008.

[24] T. Eerola, "Modeling listeners' emotional response to music," *Topics in Cognitive Science*, vol. 4, no. 4, pp. 607–624, 2012.

[25] J.-C. Wang, Y.-H. Yang, H.-M. Wang, and S.-K. Jeng, "The acoustic emotion gaussians model for emotion-based music annotation and retrieval," in *Proceedings of the 20th ACM International Conference on Multimedia (MM '12)*, pp. 89–98, November 2012.

[26] J. Cu, R. Cabredo, R. Legaspi, and M. T. Suarez, "On modelling emotional responses to rhythm features," in *Proceedings of the Trends in Artificial Intelligence (PRICAI '12)*, pp. 857–860, Springer, 2012.

[27] J. J. Deng and C. Leung, "Music emotion retrieval based on acoustic features," in *Advances in Electric and Electronics*, vol. 155 of *Lecture Notes in Electrical Engineering*, pp. 169–177, 2012.

[28] H. G. Avisado, J. V. Cocjin, J. A. Gaverza, R. Cabredo, J. Cu, and M. Suarez, "Analysis of music timbre features for the construction of user-specific affect model," in *Theory and Practice of Computation*, pp. 28–35, Springer, Berlin, Germany, 2012.

[29] M. M. Farbood, "A parametric, temporal model of musical tension," *Music Perception*, vol. 29, no. 4, pp. 387–428, 2012.

[30] J. Albrecht, "A model of perceived musical affect accurately predicts self-report ratings," in *Proceedings of the International Conference on Music Perception*, pp. 35–43, 2012.

[31] Y.-H. Yang and H. H. Chen, "Prediction of the distribution of perceived music emotions using discrete samples," *IEEE Transactions on Audio, Speech, and Language Processing*, vol. 19, pp. 2184–2196, 2011.

[32] R. Y. Granot and Z. Eitan, "Musical tension and the interaction of dynamic auditory parameters," *Music Perception*, vol. 28, no. 3, pp. 219–246, 2011.

[33] E. M. Schmidt and Y. E. Kim, "Modeling musical emotion dynamics with conditional random fields," in *Proceedings of the 12th International Society for Music Information Retrieval Conference (ISMIR '11)*, pp. 777–782, October 2011.

[34] B. Schuller, J. Dorfner, and G. Rigoll, "Determination of non-prototypical valence and arousal in popular music: features and performances," *EURASIP Journal on Audio, Speech, and Music Processing*, vol. 2010, Article ID 735854, 2010.

[35] E. Coutinho and A. Cangelosi, "The use of spatio-temporal connectionist models in psychological studies of musical emotions," *Music Perception*, vol. 27, no. 1, pp. 1–15, 2009.

[36] T. Eerola, O. Lartillot, and P. Toiviainen, "Prediction of multidimensional emotional ratings in music from audio using multivariate regression models," in *Proceedings of the International Symposium on Music Information Retrieval*, pp. 621–626, 2009.

[37] T.-L. Wu and S.-K. Jeng, "Probabilistic estimation of a novel music emotion model," in *Advances in Multimedia Modeling*, pp. 487–497, Springer, 2008.

[38] G. Luck, P. Toiviainen, J. Erkkilä et al., "Modelling the relationships between emotional responses to, and musical content of, music therapy improvisations," *Psychology of Music*, vol. 36, no. 1, pp. 25–45, 2008.

[39] Y.-H. Yang, Y.-C. Lin, Y.-F. Su, and H. H. Chen, "A regression approach to music emotion recognition," *IEEE Transactions on Audio, Speech and Language Processing*, vol. 16, no. 2, pp. 448–457, 2008.

[40] F. Lerdahl and C. L. Krumhansl, "Modeling tonal tension," *Music Perception*, vol. 24, no. 4, pp. 329–366, 2007.

[41] Y.-H. Yang, C.-C. Liu, and H. H. Chen, "Music emotion classification: a fuzzy approach," in *Proceeding of the Annual ACM International Conference on Multimedia (MM '06)*, pp. 81–84, New York, NY, USA, October 2006.

[42] M. D. Korhonen, D. A. Clausi, and M. E. Jernigan, "Modeling emotional content of music using system identification," *IEEE Transactions on Systems, Man, and Cybernetics B: Cybernetics*, vol. 36, no. 3, pp. 588–599, 2006.

[43] M. Leman, V. Vermeulen, L. de Voogdt, D. Moelants, and M. Lesaffre, "Prediction of musical affect using a combination of acoustic structural cues," *Journal of New Music Research*, vol. 34, no. 1, pp. 39–67, 2005.

[44] E. H. Margulis, "A model of melodic expectation," *Music Perception*, vol. 22, no. 4, pp. 663–714, 2005.

[45] E. Schubert, "Modeling perceived emotion with continuous musical features," *Music Perception*, vol. 21, pp. 561–585, 2004.

[46] K. R. Scherer, "Which emotions can be induced by music? What are the underlying mechanisms? And how can we measure them?" *Journal of New Music Research*, vol. 33, pp. 239–251, 2004.

[47] P. N. Juslin and D. Västfjäll, "Emotional responses to music: the need to consider underlying mechanisms," *Behavioral and Brain Sciences*, vol. 31, no. 5, pp. 559–621, 2008.

[48] L.-O. Lundqvist, F. Carlsson, P. Hilmersson, and P. N. Juslin, "Emotional responses to music: experience, expression, and physiology," *Psychology of Music*, vol. 37, no. 1, pp. 61–90, 2009.

[49] N. S. Rickard, "Intense emotional responses to music: a test of the physiological arousal hypothesis," *Psychology of Music*, vol. 32, no. 4, pp. 371–388, 2004.

[50] J. Kim and E. André, "Emotion recognition based on physiological changes in music listening," *IEEE Transactions on Pattern Analysis and Machine Intelligence*, vol. 30, no. 12, pp. 2067–2083, 2008.

[51] E. Coutinho and A. Cangelosi, *Computational and Psycho-Physiological Investigations of Musical Emotions*, University of Plymouth, 2008.

[52] E. Coutinho and A. Cangelosi, "Musical emotions: predicting second-by-second subjective feelings of emotion from low-level psychoacoustic features and physiological measurements," *Emotion*, vol. 11, no. 4, pp. 921–937, 2011.

[53] K. Trochidis, D. Sears, D. L. Tran, and S. McAdams, "Psychophysiological measures of emotional response to Romantic orchestral music and their musical and acoustic correlates," in *Proceedings of the International Symposium on Computer Music Modelling and Retrieval*, pp. 45–52, 2012.

[54] V. N. Salimpoor, M. Benovoy, G. Longo, J. R. Cooperstock, and R. J. Zatorre, "The rewarding aspects of music listening are related to degree of emotional arousal," *PLoS ONE*, vol. 4, no. 10, Article ID e7487, 2009.

[55] M. D. van der Zwaag, J. H. D. M. Westerink, and E. L. van den Broek, "Emotional and psychophysiological responses to tempo, mode, and percussiveness," *Musicae Scientiae*, vol. 15, no. 2, pp. 250–269, 2011.

[56] J. R. J. Fontaine, K. R. Scherer, E. B. Roesch, and P. C. Ellsworth, "The world of emotions is not two-dimensional," *Psychological Science*, vol. 18, no. 12, pp. 1050–1057, 2007.

[57] T. Eerola and J. K. Vuoskoski, "A comparison of the discrete and dimensional models of emotion in music," *Psychology of Music*, vol. 39, no. 1, pp. 18–49, 2011.

[58] U. R. Acharya, K. P. Joseph, N. Kannathal, C. M. Lim, and J. S. Suri, "Heart rate variability: a review," *Medical and Biological Engineering and Computing*, vol. 44, no. 12, pp. 1031–1051, 2006.

[59] "Heart rate variability: standards of measurement, physiological interpretation and clinical use. Task Force of the European Society of Cardiology and the North American Society of Pacing and Electrophysiology," *Circulation*, vol. 93, no. 5, pp. 1043–1065, 1996.

[60] R. D. Lane, K. McRae, E. M. Reiman, K. Chen, G. L. Ahern, and J. F. Thayer, "Neural correlates of heart rate variability during emotion," *NeuroImage*, vol. 44, no. 1, pp. 213–222, 2009.

[61] P. Rainville, A. Bechara, N. Naqvi, and A. R. Damasio, "Basic emotions are associated with distinct patterns of cardiorespiratory activity," *International Journal of Psychophysiology*, vol. 61, no. 1, pp. 5–18, 2006.

[62] B. M. Appelhans and L. J. Luecken, "Heart rate variability as an index of regulated emotional responding," *Review of General Psychology*, vol. 10, no. 3, pp. 229–240, 2006.

[63] G. H. E. Gendolla and J. Krüsken, "Mood state, task demand, and effort-related cardiovascular response," *Cognition and Emotion*, vol. 16, no. 5, pp. 577–603, 2002.

[64] R. McCraty, M. Atkinson, W. A. Tiller, G. Rein, and A. D. Watkins, "The effects of emotions on short-term power spectrum analysis of heart rate variability," *The American Journal of Cardiology*, vol. 76, no. 14, pp. 1089–1093, 1995.

[65] A. H. Kemp, D. S. Quintana, and M. A. Gray, "Is heart rate variability reduced in depression without cardiovascular disease?" *Biological Psychiatry*, vol. 69, no. 4, pp. e3–e4, 2011.

[66] C. M. M. Licht, E. J. C. De Geus, F. G. Zitman, W. J. G. Hoogendijk, R. Van Dyck, and B. W. J. H. Penninx, "Association between major depressive disorder and heart rate variability in the Netherlands study of depression and anxiety (NESDA)," *Archives of General Psychiatry*, vol. 65, no. 12, pp. 1358–1367, 2008.

[67] R. M. Carney, R. D. Saunders, K. E. Freedland, P. Stein, M. W. Rich, and A. S. Jaffe, "Association of depression with reduced heart rate variability in coronary artery disease," *The American Journal of Cardiology*, vol. 76, no. 8, pp. 562–564, 1995.

[68] F. Geisler, N. Vennewald, T. Kubiak, and H. Weber, "The impact of heart rate variability on subjective well-being is mediated by emotion regulation," *Personality and Individual Differences*, vol. 49, no. 7, pp. 723–728, 2010.

[69] S. Koelsch, A. Remppis, D. Sammler et al., "A cardiac signature of emotionality," *European Journal of Neuroscience*, vol. 26, no. 11, pp. 3328–3338, 2007.

[70] R. Krittayaphong, W. E. Cascio, K. C. Light et al., "Heart rate variability in patients with coronary artery disease: differences in patients with higher and lower depression scores," *Psychosomatic Medicine*, vol. 59, no. 3, pp. 231–235, 1997.

[71] M. Müller, D. P. W. Ellis, A. Klapuri, and G. Richard, "Signal processing for music analysis," *IEEE Journal on Selected Topics in Signal Processing*, vol. 5, no. 6, pp. 1088–1110, 2011.

[72] K. Jensen, "Multiple scale music segmentation using rhythm, timbre, and harmony," *Eurasip Journal on Advances in Signal Processing*, vol. 2007, Article ID 73205, 2007.

[73] N. Scaringella, G. Zoia, and D. Mlynek, "Automatic genre classification of music content: a survey," *IEEE Signal Processing Magazine*, vol. 23, no. 2, pp. 133–141, 2006.

[74] J.-J. Aucouturier and F. Pachet, "Representing musical genre: a state of the art," *Journal of New Music Research*, vol. 32, pp. 83–93, 2003.

[75] T. Li and M. Ogihara, "Detecting emotion in music," in *Proceedings of the International Symposium on Music Information Retrieval*, pp. 239–240, 2003.

[76] G. Tzanetakis and P. Cook, "Musical genre classification of audio signals," *IEEE Transactions on Speech and Audio Processing*, vol. 10, no. 5, pp. 293–302, 2002.

[77] L. Jaquet, B. Danuser, and P. Gomez, "Music and felt emotions: how systematic pitch level variations affect the experience of pleasantness and arousal," *Psychology of Music*, 2012.

[78] J. C. Hailstone, R. Omar, S. M. D. Henley, C. Frost, M. G. Kenward, and J. D. Warren, "It's not what you play, it's how you play it: timbre affects perception of emotion in music," *The Quarterly Journal of Experimental Psychology*, vol. 62, no. 11, pp. 2141–2155, 2009.

[79] L. Trainor, "Science & music: the neural roots of music," *Nature*, vol. 453, no. 7195, pp. 598–599, 2008.

[80] J. Bispham, "Rhythm in Music: what is it? Who has it? And why?" *Music Perception*, vol. 24, no. 2, pp. 125–134, 2006.

[81] J. A. Grahn, "Neuroscientific investigations of musical rhythm: recent advances and future challenges," *Contemporary Music Review*, vol. 28, no. 3, pp. 251–277, 2009.

[82] K. Overy and R. Turner, "The rhythmic brain," *Cortex*, vol. 45, no. 1, pp. 1–3, 2009.

[83] J. S. Snyder, E. W. Large, and V. Penhune, "Rhythms in the brain: basic science and clinical perspectives," *Annals of the New York Academy of Sciences*, vol. 1169, pp. 13–14, 2009.

[84] H. Honing, "Without it no music: beat induction as a fundamental musical trait," *Annals of the New York Academy of Sciences*, vol. 1252, no. 1, pp. 85–91, 2012.

[85] J. A. Grahn, "Neural mechanisms of rhythm perception: current findings and future perspectives," *Topics in Cognitive Science*, vol. 4, no. 4, pp. 585–606, 2012.

[86] I. Winkler, G. P. Háden, O. Ladinig, I. Sziller, and H. Honing, "Newborn infants detect the beat in music," *Proceedings of the National Academy of Sciences*, vol. 106, pp. 2468–2471, 2009.

[87] M. Zentner and T. Eerola, "Rhythmic engagement with music in infancy," *Proceedings of the National Academy of Sciences of the United States of America*, vol. 107, no. 13, pp. 5768–5773, 2010.

[88] S. E. Trehub and E. E. Hannon, "Conventional rhythms enhance infants' and adults' perception of musical patterns," *Cortex*, vol. 45, no. 1, pp. 110–118, 2009.

[89] E. Geiser, E. Ziegler, L. Jancke, and M. Meyer, "Early electrophysiological correlates of meter and rhythm processing in music perception," *Cortex*, vol. 45, no. 1, pp. 93–102, 2009.

[90] C.-H. Yeh, H.-H. Lin, and H.-T. Chang, "An efficient emotion detection scheme for popular music," in *Proceedings of the IEEE International Symposium on Circuits and Systems (ISCAS '09)*, pp. 1799–1802, Taipei, Taiwan, May 2009.

[91] L. Lu, D. Liu, and H. J. Zhang, "Automatic mood detection and tracking of music audio signals," *IEEE Transactions on Audio, Speech and Language Processing*, vol. 14, no. 1, pp. 5–18, 2006.

[92] S. Dixon, "Automatic extraction of tempo and beat from expressive performances," *Journal of New Music Research*, vol. 30, pp. 39–58, 2001.

[93] R. Likert, "A technique for the measurement of attitudes," *Archives of Psychology*, vol. 22, no. 140, pp. 1–55, 1932.

[94] W. A. Sethares, R. D. Morris, and J. C. Sethares, "Beat tracking of musical performances using low-level audio features," *IEEE Transactions on Speech and Audio Processing*, vol. 13, no. 2, pp. 275–285, 2005.

[95] J. P. Bello, C. Duxbury, M. Davies, and M. Sandler, "On the use of phase and energy for musical onset detection in the complex domain," *IEEE Signal Processing Letters*, vol. 11, no. 6, pp. 553–556, 2004.

[96] W. Wei, C. Zhang, and W. Lin, "A QRS wave detection algorithm based on complex wavelet transform," *Applied Mechanics and Materials*, vol. 239-240, pp. 1284–1288, 2013.

[97] B. U. Köhler, C. Hennig, and R. Orglmeister, "The principles of software QRS detection," *IEEE Engineering in Medicine and Biology Magazine*, vol. 21, no. 1, pp. 42–57, 2002.

[98] J. Pan and W. J. Tompkins, "A real-time QRS detection algorithm," *IEEE Transactions on Biomedical Engineering*, vol. 32, no. 3, pp. 230–236, 1985.

[99] J. Mateo and P. Laguna, "Analysis of heart rate variability in the presence of ectopic beats using the heart timing signal," *IEEE Transactions on Biomedical Engineering*, vol. 50, no. 3, pp. 334–343, 2003.

[100] S. McKinley and M. Levine, "Cubic spline interpolation," http://online.redwoods.edu/instruct/darnold/laproj/fall98/skymeg/proj.pdf.

[101] M. P. Tarvainen, P. O. Ranta-aho, and P. A. Karjalainen, "An advanced detrending method with application to HRV analysis," *IEEE Transactions on Biomedical Engineering*, vol. 49, no. 2, pp. 172–175, 2002.

[102] P. Welch, "The use of fast Fourier transform for the estimation of power spectra: a method based on time averaging over short, modified periodograms," *IEEE Transactions on Audio and Electroacoustics*, vol. 15, pp. 70–73, 1967.

[103] T. Warren Liao, "Clustering of time series data: a survey," *Pattern Recognition*, vol. 38, no. 11, pp. 1857–1874, 2005.

[104] V. Barnett and T. Lewis, *Outliers in Statistical Data*, vol. 1, John Wiley & Sons, 1984.

[105] C. S. Davis, *Statistical Methods for the Analysis of Repeated Measurements*, Springer, New York, NY, USA, 2002.

[106] R. Kabacoff, *R in Action*, Manning Publications, 2011.

[107] L. Pace, *Beginning R: An Introduction to Statistical Programming*, Apress, 2012.

[108] J. Albert and M. Rizzo, *R by Example*, Springer, 2012.

[109] M. Gardener, *Beginning R: The Statistical Programming Language*, Wrox, 2012.

[110] J. Adler and J. Beyer, *R in a Nutshell*, O'Reilly, Frankfurt, Germany, 2010.

[111] S. D. Kreibig, "Autonomic nervous system activity in emotion: a review," *Biological Psychology*, vol. 84, no. 3, pp. 394–421, 2010.

[112] S. Khalfa, M. Roy, P. Rainville, S. Dalla Bella, and I. Peretz, "Role of tempo entrainment in psychophysiological differentiation of happy and sad music?" *International Journal of Psychophysiology*, vol. 68, no. 1, pp. 17–26, 2008.

[113] L. Bernardi, C. Porta, and P. Sleight, "Cardiovascular, cerebrovascular, and respiratory changes induced by different types of music in musicians and non-musicians: the importance of silence," *Heart*, vol. 92, no. 4, pp. 445–452, 2006.

[114] M. Iwanaga, A. Kobayashi, and C. Kawasaki, "Heart rate variability with repetitive exposure to music," *Biological Psychology*, vol. 70, no. 1, pp. 61–66, 2005.

[115] E. Thul and G. T. Toussaint, "Analysis of musical rhythm complexity measures in a cultural context," in *Proceedings of the C^3S^2E Conference (C3S2E '08)*, pp. 7–9, May 2008.

[116] E. Thul and G. T. Toussaint, "On the relation between rhythm complexity measures and human rhythmic performance," in *Proceeding of the C^3S^2E Conference (C3S2E '08)*, pp. 199–204, May 2008.

[117] E. Thul and G. T. Toussaint, "Rhythm complexity measures: a comparison of mathematical models of human perception and performance," in *Proceedings of the International Symposium on Music Information Retrieval*, pp. 14–18, 2008.

[118] Z. Xiao, E. Dellandrea, W. Dou, and L. Chen, "What is the best segment duration for music mood analysis ?" in *Proceedings of the International Workshop on Content-Based Multimedia Indexing (CBMI '08)*, pp. 17–24, London, UK, June 2008.

[119] A. Lamont and R. Webb, "Short- and long-term musical preferences: what makes a favourite piece of music?" *Psychology of Music*, vol. 38, no. 2, pp. 222–241, 2010.

[120] A. C. North and D. J. Hargreaves, "Liking, arousal potential, and the emotions expressed by music," *Scandinavian Journal of Psychology*, vol. 38, no. 1, pp. 45–53, 1997.

[121] S. Abdallah and M. Plumbley, "Information dynamics: patterns of expectation and surprise in the perception of music," *Connection Science*, vol. 21, no. 2-3, pp. 89–117, 2009.

[122] M. G. Orr and S. Ohlsson, "Relationship between complexity and liking as a function of expertise," *Music Perception*, vol. 22, no. 4, pp. 583–611, 2005.

[123] D. Moelants, "Preferred tempo reconsidered," in *Proceedings of the International Conference on Music Perception and Cognition*, pp. 580–583, 2002.

[124] Y.-S. Shih, W.-C. Zhang, Y.-C. Lee et al., "Twin-peak effect in both cardiac response and tempo of popular music," in *Proceedings of the International Conference of the IEEE Engineering in Medicine and Biology Society*, pp. 1705–1708, 2011.

[125] A. P. Oliveira and A. Cardoso, "A musical system for emotional expression," *Knowledge-Based Systems*, vol. 23, no. 8, pp. 901–913, 2010.

[126] I. Wallis, T. Ingalls, and E. Campana, "Computer-generating emotional music: the design of an affective music algorithm," in *Proceedings of the 11th International Conference on Digital Audio Effects (DAFx '08)*, pp. 7–12, September 2008.

[127] S. R. Livingstone, R. Mühlberger, A. R. Brown, and A. Loch, "Controlling musical emotionality: an affective computational architecture for influencing musical emotions," *Digital Creativity*, vol. 18, no. 1, pp. 43–53, 2007.

23

A Photoplethysmography Melanin Evaluation System by Modified Boltzmann Transport Equation (BTE)

Sheng-Chieh Huang, Hui-Min Wang, and Shi-Han Luo

Department of Electrical Engineering, National Chiao Tung University, Room 720, Engineering Building 5, 1001 University Road, Hsinchu 30010, Taiwan

Correspondence should be addressed to Sheng-Chieh Huang; schuang@cn.nctu.edu.tw

Academic Editor: José R. C. Piqueira

With the advance in cosmetic medical technology in recent years, more and more people get cosmetic medical treatments, especially skin whitening treatments. Nevertheless, people usually assess the effect of skin whitening products by vision, which is subjective and will be different from each person. To acquire the value of melanin concentration objectively, people need to go to cosmetic medical clinics. This will cause inconvenience to people. This paper develops a novel evaluation platform based on optical assessment methods, which employ different absorption and scattering properties to different wavelengths of light in human tissue to obtain melanin concentration. Moreover, this paper proposes a new method that compensates the interaction between epidermis and dermis to acquire the melanin concentration more accurately. The novel platform designed in this paper is smaller and consumes lower-power and smaller when comparing to other conventional devices in market.

1. Introduction

The number of medical devices used in the global market is expected to be much more in the future [1], while noninvasive treatments and skin care products are more accepted to the mass population due to their safety and efficiency and then result in the fast growing of the global cosmetic market [2]. Among various techniques, photoplethysmography (PPG) is a noninvasive electrooptic, low-cost, and simple technique that can be analyzed to provide the information of physiological status and skin properties, such as blood oxygenation, melanin, blood volume, heart rate, and other features of a human [3–6]. To evaluate the melanin concentration and other tissue properties, most bibliographies are usually based on Boltzmann transfer equation (BTE) [7–12], Beer-Lambert law simplified from the BTE [13–18], or empirical equations derived from diffuse reflectance model and Boltzmann transfer equation [19–22]. Nevertheless, the light scattered from the dermis (where hemoglobin mainly exists) will interact with the tissue in the epidermis and the main absorber is melanin. This will cause the nonlinearity and nonindependence of contribution from each concentration. To overcome this problem, this study proposes a new method that compensates the nonlinearity of chromophores to calculate the melanin concentration.

Skin consists of three layers: the epidermis, the dermis, and the subcutis; most of the melanin exists in the epidermis [23]. There are five layers in the epidermis, including basal layer, spiny layer, granular layer, clear layer, and horny layer [24–27]. Melanin is generated from tyrosine in melanocytes. It has two forms; one is eumelanin, which contributes a brown-black color and is more general; the other one is pheomelanin, which contributes a yellow-red color [28–30].

The optical properties of tissue include multiple scattering, absorption, optical pathways in skin, and the penetration depth of light. The electromagnetic spectrum plays a significant role in interacting with biological systems [31]. Initially, cells, fibers, and chromophores are specific relevance to the discussion about the propagation of light in skin, although skin is composed of various kinds of cellular level elements

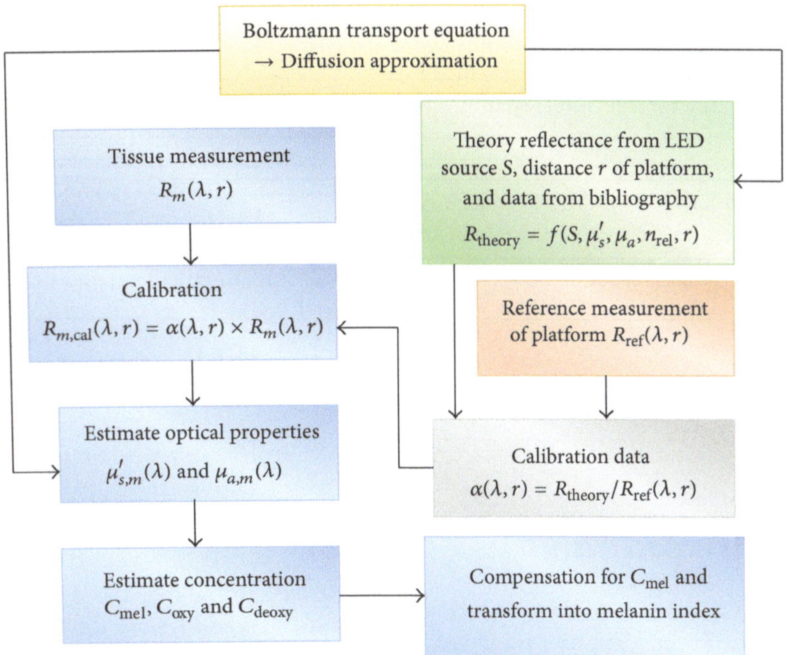

FIGURE 1: Algorithm procedure to obtain tissue properties.

[32, 33]. Mie scattering is used to represent the scattering property when the size of the constituent is approaching to the wavelength of the incident such as mitochondria, nuclei, and collagen fibers [31, 34]. Secondly, with a low absorption coefficient of the element, light can propagate longer distance without large attenuation [35]. Absorption in skin over the visible and near-infrared spectrum is mainly caused by many kinds of chromophores: melanin, hemoglobin, water, bilirubin, and beta-carotene [13, 21, 36–40]. The absorption spectra of hemoglobin have been mentioned in many studies [37, 38, 41, 42]. Bilirubin and beta-carotene give the yellowish-olive color of an individual's skin due to their absorption band [40, 43–45]. Thirdly, for optical pathways in skin, literature assumes that the centers of absorption and scattering are uniformly spread across skin layers, and anisotropy induced by preferential alignments of collagen fibers is insignificant [32, 37, 46, 47]. Finally, the penetration depth of light depends on the wavelength of light [32, 48, 49].

PPG has been studied and widely used in medicine because it is a simple, low-cost, and noninvasive optical technique that can be used to detect microcirculation of tissue [6, 50, 51]. Appropriate filtering and amplification can be used to obtain nonpulsatile part and pulsatile part of PPG signal [50] for the following pulse wave analysis. Finally, according to the principles of near-infrared optical tomography, the design approach of the systems is discussed [13, 15, 48, 52–56]. The main three methods are continuous-wave (CW), time-domain (TD), and frequency-domain (FD). CW uses sources with constant intensity to interrogate the tissue. The configuration of the system is the easiest and the cost is also the lowest.

The rest of this paper is organized as follows. In Section 2, we introduce the algorithm for obtaining tissue properties and the method for compensating the chromophore. In Section 3, a novel evaluation platform, Physiskin, is proposed and the algorithm introduced was realized in this platform. In Section 4, the experiments are illustrated and discussed. In Section 5, we conclude this work and point out the future work.

2. Algorithm

These steps are illustrated in Figure 1 and will be explained in each subsection. Initially, the Boltzmann transport equation is introduced to model the light propagation in the turbid media from Sections 2.1 to 2.4. Secondly, this study describes the method to calculate the tissue properties in Section 2.5. Thirdly, after obtaining the necessary parameter and concerning the constant contribution of water, the tissue properties of a subject are calculated as well as the concentration of each chromophore from a linear equation in Section 2.6. Finally, the Empirical Mode Decomposition method is employed to compensate the nonlinear relationship between the epidermis and the dermis for the melanin concentration measured in Section 2.7.

2.1. Modeling Light Propagation in Scattering Media. In this study, we use the transport theory, also known as radiative transfer theory, from the previous works to describe the light propagation in turbid media. Boltzmann transport equation is applied to describe the scattering and absorbing events of light propagation in the media [8, 9, 57–60]. However, in most practical cases, this equation cannot be solved precisely. It is necessary to consider an approximate approach. Diffusion approximation, which is validated in the strongly scattering media, is one of the simplified approaches. This approach is valid in most biological tissues due to the very large coefficient in scattering than absorbing [59]. The diffusion approximation is performed in the semi-infinite geometry to obtain the appropriate solution for the interrogation of biological tissues [59, 60].

2.2. Boltzmann Transport Equation (BTE). The Boltzmann transport equation can be used to properly describe the movement of particles in scattering and absorbing media [11, 59, 60]. The propagation of electromagnetic waves in scattering media can be described with the Boltzmann transport equation as follows:

$$\frac{1}{c}\frac{\partial L(\mathbf{r},\boldsymbol{\Omega},t)}{\partial t}+\nabla\cdot L(\mathbf{r},\boldsymbol{\Omega},t)\boldsymbol{\Omega} = -(\sigma+\beta)L(\mathbf{r},\boldsymbol{\Omega},t) + \sigma\iint_{4\pi} L(\mathbf{r},\boldsymbol{\Omega}',t)\,f(\boldsymbol{\Omega}'\cdot\boldsymbol{\Omega})\,d\boldsymbol{\Omega}' + Q(\mathbf{r},\boldsymbol{\Omega},t), \tag{1}$$

where the radiance L is in units and is the vector pointing to the interesting direction. The linear scattering coefficient and the absorption coefficient represent the inverses of the mean free paths for scattering and absorption, respectively. Moreover, $f(\boldsymbol{\Omega}'\cdot\boldsymbol{\Omega})$, the normalized differential scattering cross section, satisfies

$$\iint_{4\pi} f(\boldsymbol{\Omega}'\cdot\boldsymbol{\Omega})\,d\boldsymbol{\Omega}' = 1. \tag{2}$$

The function $f(\boldsymbol{\Omega}'\cdot\boldsymbol{\Omega})$ in (1) and (2) is the normalized probability function of scattering a photon that propagates from direction $\boldsymbol{\Omega}'$ into direction $\boldsymbol{\Omega}$. The source term $Q(\mathbf{r},\boldsymbol{\Omega},t)$ is the power injected into a unit solid angle centered on $\boldsymbol{\Omega}$ in a unit volume at $\mathbf{r}$. The parameter c represents the speed of photons in the medium which is equal to $c = (3\times10^8\,\text{m/s})/n$, where n is the refractive index of the medium, which typically equals 1.40 for biological tissue. As shown in (1), the BTE has both integral and differential terms, and its solution needs initial and boundary conditions to obtain $L(\mathbf{r},\boldsymbol{\Omega},t)$. This makes the equation sophisticated and hard to be solved. Therefore, the diffusion approximation is used to simplify the BTE to obtain the solution that can be analyzed easier.

2.3. Diffusion Approximation in an Infinite Medium. The diffusion approximation hypothesizes that the radiance can be presented as an isotropic fluence rate plus a directional flux when scattering is much stronger than absorption in the media [57, 59, 60]. The equation of the radiance in (1) will be expressed in the following:

$$L(\mathbf{r},\boldsymbol{\Omega},t) = \frac{1}{4\pi}\phi(\mathbf{r},t)+\frac{3}{4\pi}\mathbf{j}(\mathbf{r},t)\cdot\boldsymbol{\Omega}, \tag{3}$$

where

$$\phi(\mathbf{r},t) = \iint_{4\pi} L(\mathbf{r},\boldsymbol{\Omega},t)\,d\boldsymbol{\Omega}$$
$$\mathbf{j}(\mathbf{r},t) = \iint_{4\pi} L(\mathbf{r},\boldsymbol{\Omega},t)\,\boldsymbol{\Omega}\,d\boldsymbol{\Omega}. \tag{4}$$

We assume the form of the source term is like the radiance in (3):

$$Q(\mathbf{r},\boldsymbol{\Omega},t) = \frac{1}{4\pi}q_0(\mathbf{r},t)+\frac{3}{4\pi}\mathbf{q}_1(\mathbf{r},t)\cdot\boldsymbol{\Omega}. \tag{5}$$

With the assumption of diffusion approximation, we can take (3) and (5) into (1) to obtain a closed set of two equations (6) and (7), where D is the diffusion coefficient and g is defined as the average of the cosine of the scattering angle and is the effective transport scattering coefficient which is $\sigma' = (1-g)\sigma$:

$$\frac{1}{c}\frac{\partial\phi(\mathbf{r},t)}{\partial t}+\nabla\cdot\mathbf{J}(\mathbf{r},t)+\beta\phi(\mathbf{r},t) = q_0(\mathbf{r},t) \tag{6}$$

$$\frac{1}{c}\frac{\partial\mathbf{j}(\mathbf{r},t)}{\partial t} = -\frac{1}{3}\nabla\phi(\mathbf{r},t)-\frac{1}{3D}\mathbf{j}(\mathbf{r},t)+\mathbf{q}_1(\mathbf{r},t) \tag{7}$$

$$D = \frac{1}{3[(1-g)\sigma+\beta]} = \frac{1}{3(\sigma'+\beta)}. \tag{8}$$

If we assume that the source term is isotropic meaning that $\mathbf{q}_1 = 0$ and ignore the time derivative of $\mathbf{J}(\mathbf{r},t)$, which means that the time between photon collisions with scattering particles is much shorter than the variation of $\mathbf{j}(\mathbf{r},t)$ with time in the medium, with this assumption, take (7) into (6), and we rewrite (6) in the form of the photon-diffusion equation as

$$\frac{1}{c}\frac{\partial\phi(\mathbf{r},t)}{\partial t}+\nabla^2D\phi(\mathbf{r},t)+\beta\phi(\mathbf{r},t) = q_0(\mathbf{r},t). \tag{9}$$

With utilizing differential equation skills and assuming that intensity of source is fixed, defined as S, the photon-diffusion equation can be solved. The solution of fluence rate $\phi(\mathbf{r},t)$ in the macroscopically homogeneous and infinite medium is shown in the following:

$$\phi(r,t) = \frac{S}{4\pi rD}\exp\left[-r\left(\frac{\beta}{D}\right)^{1/2}\right]. \tag{10}$$

2.4. Light Propagation in a Semi-Infinite Medium. It is important that these optical techniques for clinical, diagnostic, and therapeutic applications are usually noninvasive, which means the light source and the detector must be mounted on the surface of the tissue. This kind of measurement is called semi-infinite geometry [60]. Nevertheless, the infinite medium geometry is that the source and detector are both immersed in the medium, which cause the solutions of the diffusion approximation to be violated near the surface. Hence, the boundary of tissue is unavoidable and the diffusion approximation must be concerned with the boundary condition to avoid enormous errors, such as 50% or more in the measurement of optical properties.

In the semi-infinite geometry, when the boundary condition is utilized in Fresnel reflections at the surface, the violation of diffusion approximation is partially recovered. There are two boundary conditions which can be employed to interpret the Fresnel reflections that bring about the refractive index mismatch at the tissue-air interface. One is the extrapolated boundary condition and the other one is the partial-current boundary condition. The extrapolated boundary and the partial-current configurations are depicted.

Sound physical principles are the basis for both the partial-current and the extrapolated boundary approaches. On the one hand, the partial-current boundary condition is logically corresponding with the limitations of the diffusion approximation. On the other hand, the extrapolated boundary condition is based on the mathematical solution to the transport equation in general situation. Each has its own pros and cons. However, the simulations and experiment measurements in both boundary conditions show that they have similar results to the infinite medium results. Moreover, with some simplifications, the configuration of the partial-current boundary condition resembles the configuration of extrapolated boundary condition. Hence, the solution of the partial-current boundary condition can be connected with the solution of extrapolated boundary condition. We obtain the solution of fluence rate, shown in (11), with considering the boundary condition. The geometry of the extrapolated-boundary condition to interrogate the tissue is as follows:

$$\phi(\mathbf{r},t) = \frac{S}{4\pi D}\left\{\frac{\exp\left[-r_0(\beta/D)^{1/2}\right]}{r_0} - \frac{\exp\left[-r_{0b}(\beta/D)^{1/2}\right]}{r_{0b}}\right\}. \tag{11}$$

The variable is the transport mean free path for the first collision for a single photon. As mentioned before, in the strong scattering, the effective transport scattering coefficient σ' is much larger than the absorption coefficient β. Therefore, we simplify $l_{tr} = (\sigma' + \beta)^{-1} \approx (\sigma')^{-1}$. Other variables in the equation of fluence rate are

$$\begin{aligned} l_s &= \frac{1+R_{\text{eff}}}{1-R_{\text{eff}}}2D \approx \frac{1+R_{\text{eff}}}{1-R_{\text{eff}}}\frac{2}{3}l_{tr} \\ r_0 &= \left[(l_{tr})^2 + \rho^2\right]^{1/2} = \left[\left(\frac{1}{\sigma'+\beta}\right)^2 + \rho^2\right]^{1/2} \\ &\approx \left[(\sigma')^{-2} + \rho^2\right]^{1/2} \\ r_{0b} &= \left[(2l_s + l_{tr})^2 + \rho^2\right]^{1/2} \\ &\approx \left[\left(\frac{4}{3\sigma'}\frac{1+R_{\text{eff}}}{1-R_{\text{eff}}} + \frac{1}{\sigma'}\right)^2 + \rho^2\right]^{1/2}, \end{aligned} \tag{12}$$

where ρ is the horizontal distance between the source and the detector and R_{eff} is the effective reflection coefficient which is typically equal to 0.493 in the biological tissue.

2.5. Reflectance Calibration Model. The optical properties vary from each part of human tissue. Only using simulation to confirm the measurement is not appropriate. Hence, this study uses the optical properties from other studies as reference data [7, 32, 61]. This study chooses the cheek, similar to the forehead, as the measuring part due to the signal-to-noise ratio, practical using, and clear wave [62]. We use the optical properties from bibliographies, the LED source intensity, and source-detector distance of platform to calculate the theoretical reflection intensity from (11). The theoretical reflection intensity depending on these parameters can be defined as

$$R_{\text{theory}} = f\left(S, \mu_s', \mu_a, n_{\text{rel}}, r\right), \tag{13}$$

where S is the source intensity of LED, r is the source-detector distance of platform, μ_s' is scattering coefficient in tissue, μ_a is absorption coefficient in tissue, and n_{rel} is relative refractive index from bibliographies.

In (13), the spectrometer (Ocean optics USB4000) is used to measure the LED source intensity of each wavelength. Due to the difference between the reflection intensity measured from the system in this study and the theoretical reflection intensity calculated from the reference, we fit the reflection difference by

$$\alpha(\lambda, r) = \frac{R_{\text{theory}}}{R_{\text{ref}}(\lambda, r)}, \tag{14}$$

where $\alpha(\lambda, r)$ is called the calibration coefficient. It is the specific wavelength of light. r is the source-detector distance.

To calculate α, we use the platform in this study to measure the reflection intensity, $R_{\text{ref}}(\lambda, r)$, of a yellow race human's cheek as the criterion for the calibration coefficient. With obtaining the theoretical reflectance calculated by (13) and $R_{\text{ref}}(\lambda, r)$ measured by the system, the calibration coefficient $\alpha(\lambda, r)$ can be obtained. After obtaining the calibration coefficient $\alpha(\lambda, r)$, all the data $R_m(\lambda, r)$ acquired from the platform in the future will be calibrated as

$$R_{m,\text{cal}}(\lambda, r) = \alpha(\lambda, r) \times R_m(\lambda, r). \tag{15}$$

With (15), the tissue properties can be calculated.

2.6. Chromophores Concentration. We assume that the chromophores giving the absorption coefficient are principally melanin, oxyhemoglobin, deoxyhemoglobin, and water in the spectral band window which this study adopts. The residual three unknown chromophores' concentration can be derived from (16). This equation has been shown in many studies for the determination of human tissue chromophores concentration in vivo [7, 11, 14, 61, 63–66]:

$$\mu_a(\lambda) = \sum_i \varepsilon_i(\lambda) C_i, \tag{16}$$

where $\varepsilon_i(\lambda)$ is the extinction coefficient of a chromophore at a specific wavelength and C_i is the concentration of a specific chromophore.

In (16), the chromophores are all independent of each other, meaning that the chromophores will not have interaction with each other. Hence, the absorption coefficient can be the summation from each chromophore with its extinction coefficient. Furthermore, it is obviously observed that if we want to derive three chromophores' concentration, we need three different wavelengths which can produce three different absorption coefficients. Based on the absorption spectra, this study chooses 640, 700, and 940 nm of LED wavelength to minimize the influence of the unwanted chromophores and interrogate the tissue. The 640 nm is mainly for melanin, and 700 nm and 940 nm are for the oxyhemoglobin, which is at the left side of the isobestic point (800 nm), and deoxyhemoglobin, which is at the right side of the isobestic point. Therefore, the concentration of each chromophore can be extracted from the following:

$$\begin{aligned}\mu_{a,640\,\mathrm{nm}} &= \varepsilon_{\mathrm{HbO_2},640\,\mathrm{nm}} C_{\mathrm{oxy}} + \varepsilon_{\mathrm{Hb},640\,\mathrm{nm}} C_{\mathrm{deoxy}} \\ &\quad + \varepsilon_{\mathrm{mel},640\,\mathrm{nm}} C_{\mathrm{mel}} \\ \mu_{a,700\,\mathrm{nm}} &= \varepsilon_{\mathrm{HbO_2},700\,\mathrm{nm}} C_{\mathrm{oxy}} + \varepsilon_{\mathrm{Hb},700\,\mathrm{nm}} C_{\mathrm{deoxy}} \\ &\quad + \varepsilon_{\mathrm{mel},700\,\mathrm{nm}} C_{\mathrm{mel}} \\ \mu_{a,940\,\mathrm{nm}} &= \varepsilon_{\mathrm{HbO_2},940\,\mathrm{nm}} C_{\mathrm{oxy}} + \varepsilon_{\mathrm{Hb},940\,\mathrm{nm}} C_{\mathrm{deoxy}} \\ &\quad + \varepsilon_{\mathrm{mel},940\,\mathrm{nm}} C_{\mathrm{mel}}.\end{aligned} \tag{17}$$

The experiment for each subject takes about 25 seconds, and the first three seconds will be deleted due to the transient time caused by the low-pass filter designed in the software. Then, take the physiological signal filtered and calibrated to acquire the tissue properties in time series. With these tissue properties, we can obtain the mean melanin concentration and dynamic hemoglobin concentration. Furthermore, the dynamic oxygen saturation in tissue can be calculated as

$$S_tO_2 = \frac{C_{\mathrm{HbO_2}}}{C_{\mathrm{HbO_2}} + C_{\mathrm{Hb}}}. \tag{18}$$

2.7. Compensation and Calibration of Melanin Concentration. The concentration of each chromophore calculated from (16) is based on the independence of each other. From the configuration of optical path way [32], the light incident on the skin surface is propagated into the epidermis and dermis under scattering and absorption. Some portion of the scattered light coming back from the dermis to the epidermis may interact with the melanin in the epidermis again and then finally be emitted to the skin surface that can be detected by the sensor. Therefore, this propagation course causes the nonlinearity of (16). The absorbance effects in the epidermis and dermis are not easily separated [67].

However, the concentration of oxyhemoglobin and deoxyhemoglobin should not have large variation at the same part of skin interrogated between individuals [68, 69]. From our experiment results, we have found out that the hemoglobin concentration we acquired does not have large difference between individuals. Furthermore, we find out a critical feature that can be used to compensate the melanin concentration.

We assume that more heartbeat power obtained from the physiological signal gives more blood information. The reason is that when less light is absorbed by the melanin in the epidermis, more light can propagate into the dermis and then the constituents of the physiological signal received contain more pulsatile information. Consequently, the heartbeat power can be used to compensate the melanin concentration.

Nevertheless, the physiological signal that the system received has too much noise and signal unwanted. Although high frequency noise can be filtered out by using low-pass filter, the artifact or some noise that comes from high frequency band with the intermodulation still cannot be removed. Thus, using Fast-Fourier transform (FFT) directly to acquire the heartbeat power may be wrong due to the complexity mentioned before. To obtain the signal without using the filter, we introduce a method called Empirical Mode Decomposition (EMD) [3] that can obtain the signal without high frequency noise and artifact and can acquire the correct heartbeat power.

EMD has been proposed as a method that can analyze time series data. It has extensive applications to extract noisy nonlinear and nonstationary signals. EMD decomposes a signal into a finite set of AM-FM constituents. These constituents are called Intrinsic Mode Functions (IMFs), which is a counterpart to the common harmonic function. An IMF is not limited to be a narrow band signal but can be modulated in both amplitude and frequency.

To implement the needed decomposition of functions, we first specify all the local maxima and the local minima in the signal $x(t)$. Then, a cubic line is used to link all the local maxima to produce the upper envelope. The lower envelope is also produced by linking all the local minima. Both envelopes should include all the data between them. Define m_1 as the mean of the envelope. The difference between the data and m_1 is defined as h_1, a typical IMF,

$$x(t) - m_1 = h_1. \tag{19}$$

After extracting the first IMF, we specify $r_1 = x(t) - h_1$, which is called the residual. We continue the procedures to specify all local maxima and the local minima in the signal r_1 and then obtain the second IMF, called h_2. By following the steps mentioned above, we obtain a set of IMFs and a residual. The criterion specified to stop sifting is that when there are less than two extreme values in defining the envelope. Finally, we can express the data $x(t)$ in a series by

$$x(t) = \sum_{k=1}^{K} h_k + r(t), \quad k = 1, 2, \ldots, K. \tag{20}$$

In all IMFs, we search the IMF in our data analysis that has the most heartbeat power by using FFT. The mean melanin calculated by (17) is then compensated with the power of the found IMF calculated by FFT and the compensation equation can be expressed as

$$C'_{\mathrm{mel(measure)}} = \left(\frac{IMF_{\mathrm{HB\_power,ref}}}{IMF_{\mathrm{HB\_power,measure}}}\right)^2 \times C_{\mathrm{mel(measure)}}. \tag{21}$$

The square of IMF used to compensate the melanin concentration in (21) is to increase the effect of heartbeat power. With more heartbeat power measured of the individual, this individual is supposed to have less melanin concentration. The block diagram of Empirical Mode Decomposition (EMD) is illustrated in Figure 2. From the processes mentioned above, the average melanin concentration can be calculated and compensated.

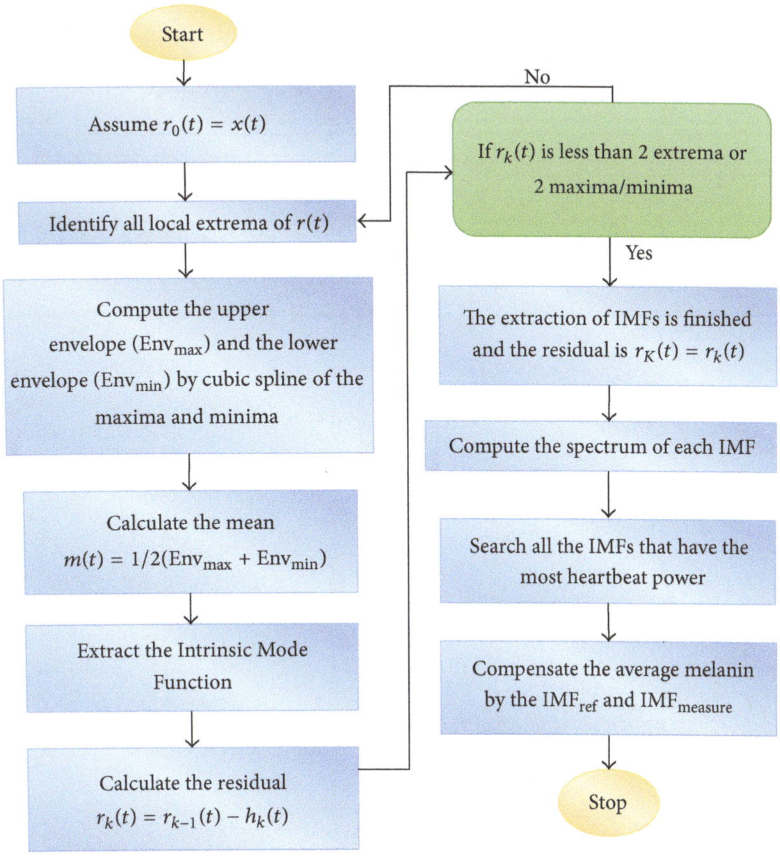

FIGURE 2: Block diagram of EMD for the compensation of melanin.

After obtaining the modified melanin concentration, this study transforms the melanin concentration into melanin index for easier differentiation with the same form like [70, 71]. The equation of transformation from melanin concentration to melanin index is shown in (22). Three states of the transformation are employed to fit the melanin index from 0 to 999 because the reference subject is not the lightest or the darkest person. Moreover, the Mexameter MX18 is employed to calibrate the melanin index acquired from the Physiskin and four subjects are investigated as the reference. The equation from MX18 to calibrate the melanin index is in (23) and will be rounded to integer. The melanin index of the Mexameter ranges from 0 to 999:

$$\begin{gathered}\text{if } (C_{\text{mel,sub}} < 0.1 \times C_{\text{mel,ref}}) \\ \text{MI} = \left(\frac{112.5 \times C_{\text{mel,sub}}}{C_{\text{mel,ref}}}\right) \times 10 \\ \text{else if } (C_{\text{mel,sub}} < C_{\text{mel,ref}}, C_{\text{mel,sub}} \geq 0.1 \times C_{\text{mel,ref}}) \\ \text{MI} = 100 + \left(\frac{125 \times C_{\text{mel,sub}}}{C_{\text{mel,ref}}}\right) \\ \text{else} \\ \text{MI} = 225 + \left(\frac{18.75 \times C_{\text{mel,sub}}}{C_{\text{mel,ref}}}\right)\end{gathered} \tag{22}$$

$$\text{MI} = \text{round}\ (0.9929 \times \text{MI} - 96.467), \tag{23}$$

where $C_{\text{mel,sub}}$ is the average melanin concentration of the subject, $C_{\text{mel,ref}}$ is the average melanin concentration of the reference, and MI is melanin index.

With melanin index, people can compare their melanin pigmentation with each other or make sure that the skin whitening products are surely effective to them. All of these algorithms are first realized on Matlab 2009b and implemented as an application program for demonstration in microprocessor. If the data is uploaded on the personal computer with 32-bit operating system and i7 CPU@2.67 GHz, it takes about 30 seconds to show all final results that consist of average hemoglobin, heart rate, and melanin index.

3. Implementation

Due to large attenuation of light propagating in the medium, about 100 dB, and the noise that interferes with the signal, an appropriate design of the system is quite important to have good performance in subsequent analysis [72]. The signal first passes a transimpedance amplifier (TIA), band-pass filters (BPF), a multiplexer (MUX), and a constant gain amplifier (AMP). Later, the signal is converted to digital signal by an 8-bit analog-to-digital converter (ADC) inside the microprocessor and then sent to the computer for data analysis. The schematic of circuit blocks is shown in Figure 3. The power supply used is 5 V and the source voltage, which is 3.3 V and is not shown in Figure 3, is produced in the system

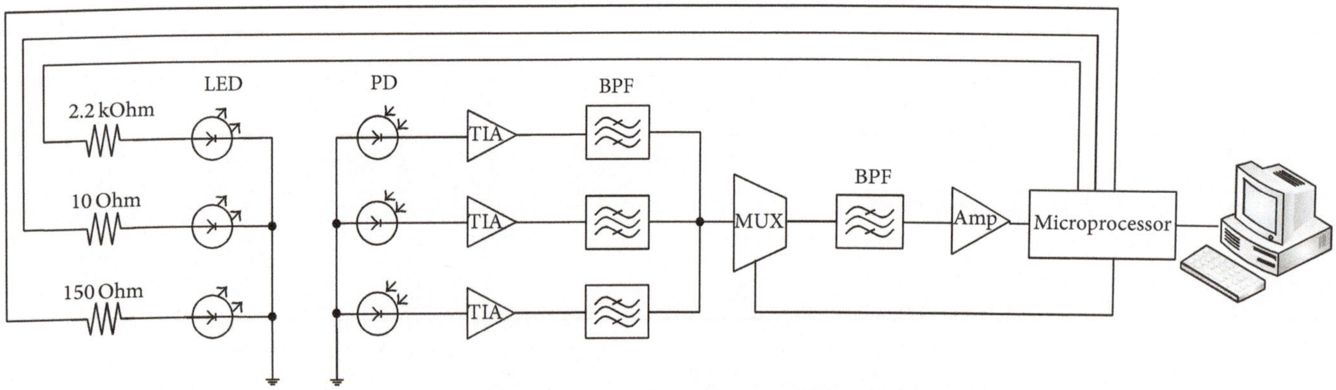

FIGURE 3: Schematic of each block in the apparatus.

(a) (b)

FIGURE 4: (a) Front of the PCB. (b) Back of the PCB.

by a low-dropout (LDO) regulator. To implement a high gain and good SNR system, the gain for each stage is arranged and the band of the filter is selected properly.

In this study, the circuit of the system is implemented on the PCB. Figures 4(a) and 4(b) show the front and the back of the PCB. The LEDs and PDs are all on the front. They were turned on and arranged in the same form as in the bread board. Other components and ICs are arranged on the back of PCB to avoid the uncomfortable condition during the experiment.

4. Evaluation

4.1. Comparison of Devices Performance. To show the performance of the device implemented in this study, the device is compared with some commercially conventional devices. Although there are still many devices that can interrogate the skin and obtain the tissue properties, this study only takes devices that have similar system characters like Physiskin, such as being portable and easy to use, into account. Table 1 illustrates the comparison between devices [73–75]. The platform, Physiskin, designed in this study has ultra-low power as compared with other devices.

4.2. Experiment of Correlation between Physiskin and Mexameter MX18 for Melanin Concentration. Twenty subjects, 15 males and 5 females, aged from 20 to 29 years are measured in this experiment. Each subject's left cheek is measured by Mexameter MX18 and the Physiskin on the same part.

TABLE 1: Comparison between conventional devices and device in this study.

	Skintel melanin reader	MU99	Soft plus	Physiskin (this study)
Wavelength (nm)	640 700 910	568 660 870	660 875	640 700 940
Power consumption	—	3.6 W	60 W	0.2 W
Index range	0–99	0–999	0–100	0–999
Measurement time	1 sec	1 sec	—	1 sec
Measure hemoglobin	✗	✓	✗	✓
Measure melanin	✓	✓	✓	✓

Melanin indices of the two devices are recorded in Table 2. Moreover, the correlation between the MX18 and Physiskin is about 0.987 and the average error is 4.85, only about 0.5% in the range of melanin index. It shows that the Physiskin can match with the MX18.

5. Conclusion

Based on the diffusion approximation simplified from BTE, this study designs a system that can evaluate the tissue to acquire melanin concentration of an individual. However,

TABLE 2: Experiment result of melanin index from MX18 and Physiskin.

Subject	MX18	Physiskin	Diff.
1	144	146	2
2	165	167	2
3	165	169	4
4	167	169	2
5	173	173	0
6	167	176	9
7	176	176	0
8	178	176	2
9	176	182	6
10	188	182	6
11	177	184	7
12	183	186	3
13	189	186	3
14	187	191	4
15	188	194	6
16	191	197	6
17	232	219	13
18	239	230	9
19	273	286	13
20	300	300	0
Avg.	192.9	194.45	4.85

the equations used to obtain the concentration of each chromophore are based on the noninteraction between epidermis and dermis. This assumption is not enough because some light propagating through the dermis will scatter and go back to the epidermis and then will be received by the detector. Therefore, a new compensation method based on EMD is proposed in this paper and shows a good operation on compensating the melanin concentration. In addition to the PCB implementation, the IC solution of the system will be implemented to improve the performance of the system.

Conflict of Interests

The authors declare that there is no conflict of interests regarding the publication of this paper.

References

[1] P. Boyer, B. I. Morshed, and T. Mussivand, "Medical device market in China," *Artificial Organs*, vol. 39, no. 6, pp. 520–525, 2015.

[2] International Society of Aesthetic Plastic Surgery, *ISAPS Internation Survey on Aesthetic/Cosmetic Procedures Performed in 2011*, International Society of Aesthetic Plastic Surgery, 2011.

[3] B. Prathyusha, T. S. Rao, and D. Asha, "Extraction of respiratory rate from PPG signals using PCA and EMD," *International Journal of Research in Engineering and Technology*, vol. 1, no. 2, pp. 164–184, 2012.

[4] J. L. Reuss and D. Siker, "The pulse in reflectance pulse oximetry: modeling and experimental studies," *Journal of Clinical Monitoring and Computing*, vol. 18, no. 4, pp. 289–299, 2004.

[5] E. Kviesis-Kipge, J. Zaharans, O. Rubenis, and A. Grabovskis, "A photoplethysmography device for multipurpose blood circulatory system assessment," in *SPIE/OSA European Conference on Biomedical Optics*, pp. 1–7, May 2011.

[6] J. Allen, "Photoplethysmography and its application in clinical physiological measurement," *Physiological Measurement*, vol. 28, article R1, 2007.

[7] J. B. Fishkin, O. Coquoz, E. R. Anderson, M. Brenner, and B. J. Tromberg, "Frequency-domain photon migration measurements of normal and malignant tissue optical properties in a human subject," *Applied Optics*, vol. 36, no. 1, pp. 10–20, 1997.

[8] Z. Sun, Y. Huang, and E. M. Sevick-Muraca, "Precise analysis of frequency domain photon migration measurement for characterization of concentrated colloidal suspensions," *Review of Scientific Instruments*, vol. 73, no. 2, pp. 383–393, 2002.

[9] T. H. Pham, O. Coquoz, J. B. Fishkin, E. Anderson, and B. J. Tromberg, "Broad bandwidth frequency domain instrument for quantitative tissue optical spectroscopy," *Review of Scientific Instruments*, vol. 71, no. 6, pp. 2500–2513, 2000.

[10] F. Bevilacqua, A. J. Berger, A. E. Cerussi, D. Jakubowski, and B. J. Tromberg, "Broadband absorption spectroscopy in turbid media by combined frequency-domain and steady-state methods," *Applied Optics*, vol. 39, no. 34, pp. 6498–6507, 2000.

[11] S. Fantini and M. A. Franceschini, "Frequency-domain techniques for tissue spectroscopy and imaging," in *Handbook of Optical Biomedical Diagnostics*, V. V. Tuchin, Ed., chapter 7, pp. 1–52, The Society of Photo-Optical Instrumentation Engineers, Bellingham, Wash, USA, 1st edition, 2002.

[12] E. Gratton, S. Fantini, M. A. Franceschini, G. Gratton, and M. Fabiani, "Measurements of scattering and absorption changes in muscle and brain," *Philosophical Transactions of the Royal Society B: Biological Sciences*, vol. 352, no. 1354, pp. 727–735, 1997.

[13] P. Rolfe, "In vivo near-infrared spectroscopy," *Annual Review of Biomedical Engineering*, vol. 2, no. 7, pp. 715–754, 2000.

[14] A. Bozkurt, A. Rosen, H. Rosen, and B. Onaral, "A portable near infrared spectroscopy system for bedside monitoring of newborn brain," *BioMedical Engineering Online*, vol. 4, article 29, 11 pages, 2005.

[15] S. Fantini, M. A. Franceschini, J. B. Fishkin, B. Barbieri, and E. Gratton, "Quantitative determination of the absorption spectra of chromophores in strongly scattering media: a light-emitting-diode based technique," *Applied Optics*, vol. 33, no. 22, pp. 5204–5213, 1994.

[16] J. B. Fishkin, P. T. C. So, A. E. Cerussi, S. Fantini, M. A. Franceschini, and E. Gratton, "Frequency-domain method for measuring spectral properties in multiple-scattering media: methemoglobin absorption spectrum in a tissuelike phantom," *Applied Optics*, vol. 34, no. 7, pp. 1143–1155, 1995.

[17] B. J. Tromberg, N. Shah, R. Lanning et al., "Non-invasive *in vivo* characterization of breast tumors using photon migration spectroscopy," *Neoplasia*, vol. 2, no. 1-2, pp. 26–40, 2000.

[18] W. Karlen, S. Raman, J. M. Ansermino, and G. A. Dumont, "Multiparameter respiratory rate estimation from the photoplethysmogram," *IEEE Transactions on Biomedical Engineering*, vol. 60, no. 7, pp. 1946–1953, 2013.

[19] G. Zonios and A. Dimou, "Modeling diffuse reflectance from semi-infinite turbid media: application to the study of skin optical properties," *Optics Express*, vol. 14, no. 19, pp. 8661–8674, 2006.

[20] D. Yudovsky and L. Pilon, "Rapid and accurate estimation of blood saturation, melanin content, and epidermis thickness

from spectral diffuse reflectance," *Applied Optics*, vol. 49, no. 10, pp. 1707–1719, 2010.

[21] G. Zonios, A. Dimou, I. Bassukas, D. Galaris, A. Tsolakidis, and E. Kaxiras, "Melanin absorption spectroscopy: new method for noninvasive skin investigation and melanoma detection," *Journal of Biomedical Optics*, vol. 13, no. 1, Article ID 014017, 2008.

[22] R. Marchesini, A. Bono, and M. Carrara, "In vivo characterization of melanin in melanocytic lesions: spectroscopic study on 1671 pigmented skin lesions," *Journal of Biomedical Optics*, vol. 14, no. 1, Article ID 014027, 8 pages, 2009.

[23] D. J. Gawkrodger, *Dermatology: An Illustrated Colour Text*, Churchill Livingstone, London, UK, 3rd edition, 2002.

[24] Weedon and David, *Weedon's Skin Pathology Essentials*, Churchill Livingstone, London, UK, 1st edition, 2011.

[25] J. G. Marks and J. J. Miller, *Lookingbill and Mark's Principles of Dermatology*, Elsevier, 4th edition, 2006.

[26] P. M. Elias, "The skin barrier as an innate immune element," *Seminars in Immunopathology*, vol. 29, no. 1, pp. 3–14, 2007.

[27] E. Proksch, J. M. Brandner, and J.-M. Jensen, "The skin: an indispensable barrier," *Experimental Dermatology*, vol. 17, no. 12, pp. 1063–1072, 2008.

[28] M. Brenner and V. J. Hearing, "The protective role of melanin against UV damage in human skin," *Photochemistry and Photobiology*, vol. 84, no. 3, pp. 539–549, 2008.

[29] P. Meredith and J. Riesz, "Radiative relaxation quantum yields for synthetic eumelanin," *Photochemistry and Photobiology*, vol. 79, no. 2, pp. 211–216, 2004.

[30] S. Jacques, Optical Absorption of Melanin, 1998, http://omlc.ogi.edu/spectra/melanin/index.html.

[31] T. Vo-Dinh, *Biomedical Photonics Handbook*, CRC Press, Boca Raton, Fla, USA, 2003.

[32] T. Igarashi, K. Nishino, and S. K. Nayer, *The Appearance of Human Skin*, Department of Computer Science, Columbia University, New York, NY, USA, 2005.

[33] A. Dunn, *Light scattering properties of cells [Ph.D. thesis]*, University of Texas at Austin, 1997.

[34] S. L. Jacques, "Origins of tissue optical properties in the UVA, visible, and NIR regions," *Trends in Optics and Photonics: Advances in Optical Imaging and Photon Migration*, vol. 2, pp. 364–371, 1996.

[35] G. B. Altshuler and I. Yaroslavsky, *Absorption Characteristic of Tissue as a Basis for the Optimal Wavelength Choice in Photodermatology*, Palomar Medical Technologies, Burlington, Mass, USA, 2004.

[36] Columbia University Press, "Absorption spectrum," in *The Columbia Encyclopedia*, Columbia University Press, 6th edition, 2008.

[37] R. R. Anderson and J. A. Parrish, "The optics of human skin," *Journal of Investigative Dermatology*, vol. 77, no. 1, pp. 13–19, 1981.

[38] A. R. Young, "Chromophores in human skin," *Physics in Medicine and Biology*, vol. 42, no. 5, pp. 789–802, 1997.

[39] A. Krishnaswamy and G. V. G. Baranoski, "A biophysically-based spectral model of light interaction with human skin," *Computer Graphics Forum*, vol. 23, no. 3, pp. 331–340, 2004.

[40] H. Nugroho, A. F. M. Hani, R. Jolivot, and F. Marzani, "Melanin type and concentration determination using inverse model," in *Proceedings of the 3rd National Postgraduate Conference (NPC '11)*, pp. 1–7, Kuala Lumpur, Malaysia, September 2011.

[41] S. Prahl, *Optical Absorption of Hemoglobin*, Oregon Medical Laser Center, Portland, Ore, USA, 1998, http://omlc.ogi.edu/spectra/hemoglobin/.

[42] W. G. Zijlstra, A. Buursma, and W. P. Meeuwsen-van der Roest, "Absorption spectra of human fetal and adult oxyhemoglobin, de-oxyhemoglobin, carboxyhemoglobin, and methemoglobin," *Clinical Chemistry*, vol. 37, no. 9, pp. 1633–1638, 1991.

[43] S. Prahl, *Optical Absorption of Water*, 1998, http://omlc.ogi.edu/spectra/water/index.html.

[44] S. Prahi, *Bilirubin*, 2012, http://omlc.ogi.edu/spectra/PhotochemCAD/html/119.html.

[45] S. Prahl, *Beta-Carotene*, 2012, http://omlc.ogi.edu/spectra/PhotochemCAD/html/041.html.

[46] P. H. Andersen, "Reflectance spectroscopic analysis of selected experimental dermatological models with emphasis on cutaneous vascular reactions," *Skin Research and Technology*, vol. 3, no. 1, pp. 5–58, 1997.

[47] S. Nickell, M. Hermann, M. Essenpreis, T. J. Farrell, U. Krämer, and M. S. Patterson, "Anisotropy of light propagation in human skin," *Physics in Medicine and Biology*, vol. 45, no. 10, pp. 2873–2886, 2000.

[48] B. J. Tromberg, N. Shah, R. Lanning et al., "Non-invasive in vivo characterization of breast tumors using photon migration spectroscopy," *Neoplasia*, vol. 2, no. 1-2, pp. 26–40, 2000.

[49] M. C. Moreira, R. N. do Prado, and A. Campos, "Application of high-brightness LEDs in tissue human and their therapeutic interaction," in *Proceedings of the IEEE Industry Applications Society Annual Meeting (IAS '09)*, pp. 1–6, IEEE, Houston, Tex, USA, October 2009.

[50] R. Dresher, *Wearable forehead pulse oximetry: minimization of motion and pressure artifacts [M.S. thesis]*, Biomedical Engineering, The Faculty of the Worcester Polytechnic Institute, 2006.

[51] J. Kraitl and H. Ewald, "Optical non-invasive methods for characterization of the human health status," in *Proceedings of the 21st International Conference on Sensing Technology*, Palmerston North, New Zealand, 2005.

[52] S. Ma and B. Fang, "Optical image reconstruction using frequency-domain simulations: optimization of detector positions," in *Proceedings of International Conference on Modelling, Identification and Control (ICMIC '12)*, pp. 68–73, Wuhan, China, June 2012.

[53] A. G. Orlova, I. V. Turchin, V. I. Plehanov et al., "Frequency-domain diffuse optical tomography with single source-detector pair for breast cancer detection," *Laser Physics Letters*, vol. 5, no. 4, pp. 321–327, 2008.

[54] D. M. Hueber, M. A. Franceschini, H. Y. Ma et al., "Non-invasive and quantitative near-infrared haemoglobin spectrometry in the piglet brain during hypoxic stress, using a frequency-domain multidistance instrument," *Physics in Medicine and Biology*, vol. 46, no. 1, pp. 41–62, 2001.

[55] K. S. No and P. H. Chou, "Mini-FDPM and heterodyne mini-FDPM: handheld non-invasive breast cancer detectors based on frequency-domain photon migration," *IEEE Transactions on Circuits and Systems I: Regular Papers*, vol. 52, no. 12, pp. 2672–2685, 2005.

[56] C. Mansouri and N. H. Kashou, "New window on optical brain imaging medical development, simulations and applications," in *Selected Topics on Optical Fiber Technology*, M. Yasin, S. W. Harun, and H. Arof, Eds., chapter 11, pp. 228–271, InTech, Rijeka, Croatia, 2012.

[57] K. M. Case and P. F. Zweifel, *Linear Transport Theory*, Addison-Wesley, Reading, Mass, USA, 1967.

[58] J. B. Fishkin and E. Gratton, "Propagation of photon-density waves in strongly scattering media containing an absorbing semi-infinite plane bounded by a straight edge," *Journal of the Optical Society of America A: Optics and Image Science, and Vision*, vol. 10, no. 1, pp. 127–140, 1993.

[59] S. Fantini, M. A. Franceschini, and E. Gratton, "Semi-infinite-geometry boundary problem for light migration in highly scattering media: a frequency-domain study in the diffusion approximation," *Journal of the Optical Society of America B*, vol. 11, no. 10, p. 2128, 1994.

[60] R. C. Haskell, L. O. Svaasand, T.-T. Tsay, T.-C. Feng, M. S. McAdams, and B. J. Tromberg, "Boundary conditions for the diffusion equation in radiative transfer," *Journal of the Optical Society of America A*, vol. 11, no. 10, pp. 2727–2741, 1994.

[61] R. M. P. Doornbos, R. Lang, M. C. Aalders, F. W. Cross, and H. J. C. M. Sterenborg, "The determination of in vivo human tissue optical properties and absolute chromophore concentrations using spatially resolved steady-state diffuse reflectance spectroscopy," *Physics in Medicine and Biology*, vol. 44, no. 4, pp. 967–981, 1999.

[62] C. J. Pujary, *Investigation of photodetector optimization in reducing power consumption by a noninvasive pulse oximeter sensor [M.S. thesis]*, Faculty of the Worcester Polytechnic Institute, 2004.

[63] Y. Yang, H. Liu, X. Li, and B. Chance, "Low-cost frequency-domain photon migration instrument for tissue spectroscopy, oximetry, and imaging," *Optical Engineering*, vol. 36, no. 5, pp. 1562–1569, 1997.

[64] S. Fantini, M. A. Franceschini-Fantini, J. S. Maier, S. A. Walker, B. Barbieri, and E. Gratton, "Frequency-domain multichannel optical detector for noninvasive tissue spectroscopy and oximetry," *Optical Engineering*, vol. 34, no. 1, pp. 32–42, 1995.

[65] B. J. Tromberg, O. Coquoz, J. B. Fishkin et al., "Non-invasive measurements of breast tissue optical properties using frequency-domain photon migration," *Philosophical Transactions of the Royal Society B: Biological Sciences*, vol. 352, no. 1354, pp. 661–668, 1997.

[66] T. H. Pham, R. Hornung, H. P. Ha et al., "Noninvasive monitoring of hemodynamic stress using quantitative near-infrared frequency-domain photon migration spectroscopy," *Journal of Biomedical Optics*, vol. 7, no. 1, pp. 34–44, 2002.

[67] I. Nishidate, Y. Aizu, and H. Mishina, "Estimation of melanin and hemoglobin in skin tissue using multiple regression analysis aided by Monte Carlo simulation," *Journal of Biomedical Optics*, vol. 9, no. 4, pp. 700–710, 2004.

[68] M. G. Sowa, A. Matas, B. J. Schattka, and H. H. Mantsch, "Spectroscopic assessment of cutaneous hemodynamics in the presence of high epidermal melanin concentration," *Clinica Chimica Acta*, vol. 317, no. 1-2, pp. 203–212, 2002.

[69] P. Välisuo, I. Kaartinen, V. Tuchin, and J. Alander, "New closed-form approximation for skin chromophore mapping," *Journal of Biomedical Optics*, vol. 16, no. 4, Article ID 046012, 2011.

[70] G. Zonios, J. Bykowski, and N. Kollias, "Skin melanin, hemoglobin, and light scattering properties can be quantitatively assessed in vivo using diffuse reflectance spectroscopy," *Journal of Investigative Dermatology*, vol. 117, no. 6, pp. 1452–1457, 2001.

[71] S. Doherty, R. Cohen, F. Whitney, O. Bradley, and U. Seckel, *New Methods for Guidance of Light-Based Treatments Using Objective Melanin Measurements*, Palomar Medical Technologies, 2011.

[72] T. Tu, Y. Chen, J. Zhang, X. Intes, and B. Chance, "Analysis on performance and optimization of frequency-domain near-infrared instruments," *Journal of Biomedical Optics*, vol. 7, no. 4, pp. 643–649, 2002.

[73] Cynosure, Skintel Melanin Reader, http://www.cynosure.com/product/skintel-melanin-reader/.

[74] J. F. Limited, Skin Pigmentation Analyzer MU 99, 2007, http://www.joinfree.com.hk/sa/SA_010101_en.html.

[75] Callegari, *Soft Plus*, 2014, http://pdf.medicalexpo.com/pdf/callegari/soft-plus/67937-70784.html.

Performance of Various Metaheuristic Techniques for Economic Dispatch Problem with Valve Point Loading Effects and Multiple Fueling Options

Ijaz Ahmed,[1,2] Ali Rauf Rao,[1] Alaf Shah,[1] Engr Alamzeb,[1] and Junaid Ali Khan[1]

[1] *Faculty of Electrical Engineering, Hamdard University, Islamabad Campus, Islamabad 44000, Pakistan*
[2] *Department of Electrical Engineering, Bahria University, Islamabad Campus, Islamabad 44000, Pakistan*

Correspondence should be addressed to Ijaz Ahmed; ijazhere2000@gmail.com

Academic Editor: Mamun B. Ibne Reaz

The paper presents the metaheuristic approaches based on pattern search and simulated annealing hybridized with sequential quadratic programming, a powerful nonlinear adaptive technique for estimation of the finest combination of generated power in a given system at lowest operating cost while sustaining the operating condition of system efficiently. The fuel cost is minimized by satisfying the nonlinear operating conditions of thermal units mainly based on generation capacity constraints, generator ramp limit, power balance constraints, and valve point loading effect and by keeping in view the prohibited operating zones, respectively. About the optimization, a comparative study is made for pattern search (PS) and simulated annealing (SA), as a viable global search technique and sequential quadratic programming, an efficient local optimizer and their hybrid versions. The applicability, stability, and reliability of the designed approaches are validated through comprehensive statistical analysis based on Monte Carlo simulations.

1. Introduction

The performance and designing of electric power system serve two purposes; one is to maintain the economy and the other is to include reliability of the system [1]. Economic dispatch problem (EDP) is one of the significant optimization problems in industrial independent power system and acquires increasing importance as the operating cost and electric energy demand is increasing globally around the planet [2, 3]. The procedure involves dividing the power demand of generating system for online thermal units in such manner that their operating cost is optimal while satisfying the power demand to the user and constraint adequately [3, 4]. Since the operating cost of generating unit is inflated an optimal EDP can lead to prominent cost saving with reducing the atmospheric emission of thermal power plant [4]. In the energy crises age, a number of engineers and scientists are taking interest in the optimized solution of this industrial problem [5] by keeping in view the constraints like generator ramp limit, power balance, and valve point loading effect (VPLE) and prohibited operating zones [6, 7]. However the facts of EDP with multiple fueling option (MFO) and MFO combined with VPLE conditions is not well addressed simultaneously, which are the motivations for taking this study with metaheuristic computing approach.

Traditional numerical models and programming strategies have been developed, like lambda iteration method, quadratic programming, Lagrangian relaxation algorithm, base point participation factor, gradient method linear programming, reduced gradient method, piecewise linear cost function, Newton method, and dynamic programming [8, 9], but all these EDP tools are flexible and applicable if the function between operating cost and fuel or heat are piecewise linear and increasing categorically [10]. Modern approaches use artificial neural network models which show dominance on classical techniques [11], while its parameters can overwrite during the learning in complex constraints based EDP problems. In order to overcome this issue the metaheuristic approaches based on tabu search, Hopfield, hill climbing, evolutionary computation, swarm intelligence, and ant/bee colony optimizations can outperform [12, 13].

In the recent development on optimization technique for EDP problems, multiperiod hydrothermal economic dispatch is employed with an efficient local search based on interior point method [14] that has an issue of getting stuck in the local minima. The solution of unit commitment problem is addressed by a hybrid scheme based on modified subgradient method and simulating annealing algorithm [15]; this combination of derivative based and derivative free method is good but with the constraints of high computational budget involved in its computations. The predictor corrector primal dual interior point algorithm is well presented in [16] in terms of computational cost for solving EDP problem; however the mean level of accuracy is not well matured. Parametric approach involved in parameter setting free harmony search and parallel adaptive algorithm schemes are established for economic dispatch and environmental power dispatch problems, respectively [17, 18]. With all this recent development, this paper presents the comprehensive analysis in terms of minimum fuel cost, level of accuracy in the optimizer, statistical analysis on computational budget needed, and convergence rate as well, on complex dynamical version of EDP problem involving MFO and VPLE conditions. The presented work is compared with [19–21], in which EDP problem has been exploited by evolutionary strategy, standard particle swarm optimization (PSO), Gaussian PSO, and chaotic PSO, respectively.

The remaining of this paper is organized as follows: Section 2 describes the formulation of an EDP along with the constraints. The proposed methodology along with the procedural hybrid step is revealed in Section 3. Section 4 discusses the computational procedure and analyzes the results when applied to the scenarios of EDPs with 13, 15, and 20 thermal units. The last section of the paper outlines the conclusions drawn from the simulation and results along with some directions for the future research.

2. Economic Dispatch Problem Formulation

The objective of EDP is to find the finest combination of generated power in a given system at lowest operating cost while sustaining the operating condition of system efficiently by satisfying the nonlinear operating conditions formulated as follow.

2.1. Fuel Cost Minimization. Assume we have a system where n thermal unit committed generating real power P_{gi} and total fuel cost is F_{ci} (P_{gi}); usually the fuel cost curve of committed units is assumed to be quadratic function of the real power output of generators. Consider

$$F_{TC}\big|_{\min} = \sum_{i=1}^{NG} F_{ci}\left(P_{gi}\right) = \sum_{i=1}^{NG}\left(A_i + B_i P_{gi} + C_i P_{gi}^2\right), \quad (1)$$

where NG is the number of committed generators and P_{gi} is real output power of generator in megawatt (MW) and can be represented as a real vector $P_{gi} = [p_{g1} + p_{g2} + p_{g3} + \cdots + p_{gNG}]$, while F_{TC} and F_{Ci} are the cost of all committed generators and cost of the ith generator in terms of rupees/hour (R/h), respectively. A_i, B_i, and C_i are the fuel-cost coefficient of ith generator.

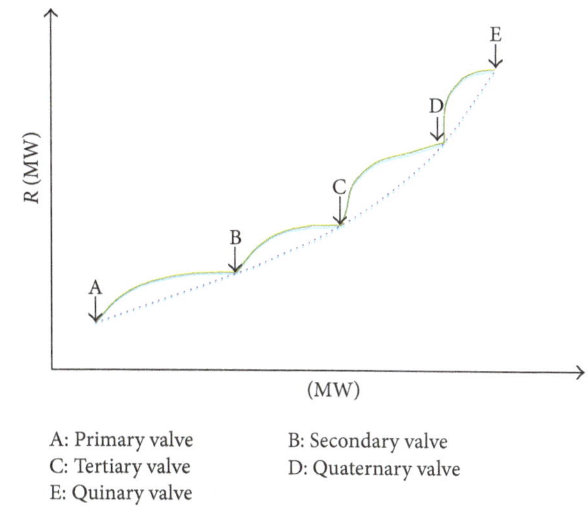

Figure 1: Cost function with primary, secondary, tertiary, quaternary, and quinary valves.

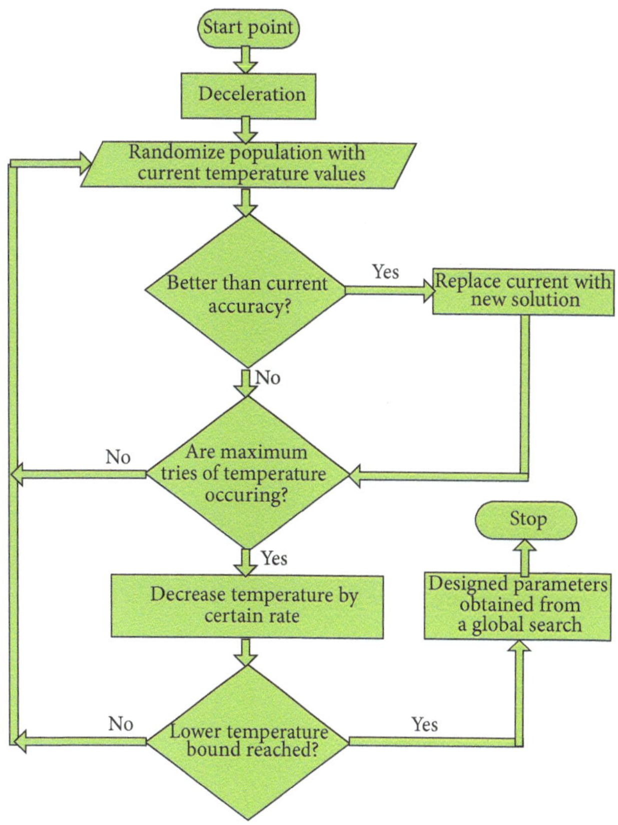

Figure 2: Flow diagram of metal of annealing algorithm.

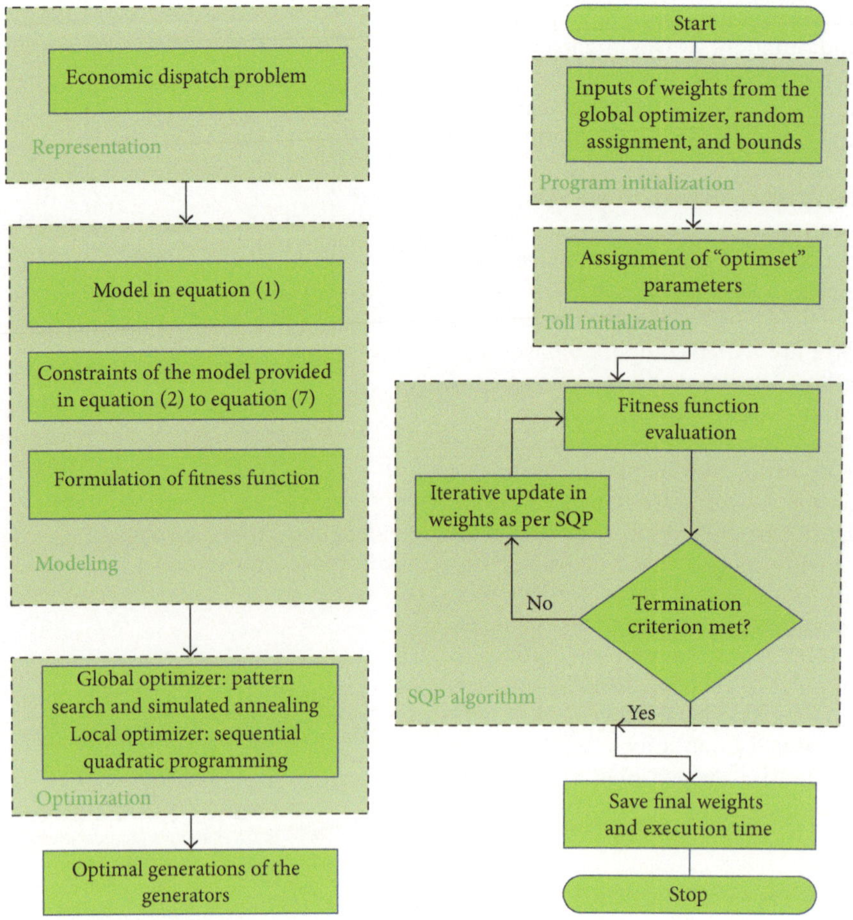

(a) (b)

FIGURE 3: Process of proposed solution (a) overall procedure and (b) optimization of fitness function using hybrid computation.

2.2. Inequality Constraint

2.2.1. Generation Capacity Constraint. For the stable operation of each thermal unit the power generated, that is, P_{gi}, must be restricted by minimum power that must be generated and maximum power that a unit can deliver as follows:

$$P_{gi}(\text{min}) \leq P_{gi} \leq P_{gi}(\text{max}), \quad i = 1, 2, 3, \ldots, \text{NG}. \tag{2}$$

2.2.2. Generator Ramp Limit. With reference of [4] considering the ramp rate limit of generator, generation capacity constraint equation (2) can be modified as given below:

$$\begin{aligned} &\text{Max}\left(P_{gi}(\text{min}), P_{gi}^{0} - \text{DR}_i\right) \\ &\quad \leq P_{gi} \leq \text{Min}\left(P_{gi}(\text{max}), P_{gi}^{0} + \text{UR}_i\right), \\ &\qquad i = 1, 2, 3, \ldots, \text{NG}, \end{aligned} \tag{3}$$

where P_{gi}^{0}, UR_i, and DR_i are the previous operating point, upper ramp limit, and down ramp limit of the ith generator, respectively.

2.2.3. Power Balance Constraint. Usually all the power plants are not working on same place they are spread geographically and also they must have some transmission power losses. The total power generated by committed thermal unit must be equal to the power demand of electric network plus the losses associated in power transmission [22]:

$$\sum_{i=1}^{\text{NG}} P_{gi} = P_D + P_{NL}, \tag{4a}$$

$$P_{NL} = \sum_{i=1}^{\text{NG}} \sum_{j=1}^{\text{NG}} P_i B_{ij} P_j, \quad p_i = p_{gi} - p_D, \tag{4b}$$

$$\begin{aligned} \beta_{ij} &= \frac{R_{ij}\cos\left(\theta_i - \theta_j\right)}{\left|V_i\right|\left|V_j\right|\cos\theta_i\cos\theta_j}, \\ \theta_j &= \delta_i - \theta_i, \quad \theta_i = \tan\left(\frac{Q_i}{P_i}\right), \end{aligned} \tag{5}$$

where P_D is the power demand, NG is number of generators, and P_{NL} is the transmission power loss collective equation that is also known as coordination equation. The most

TABLE 1: Parameter setting/values using PS and SA.

PS		SA	
Parameters	Settings	Parameters	Setting
Poll method	MADS positive basis 2N	Start point size	15
Poll order	Random	Population size	300
Mesh size	(1, 15)	Annealing function	Boltzmann
Expansion factor	2.5	Reannealing interval	150
Contraction factor	0.7	Temperature update	Exponential
Penalty factor	150	Initial temperature	100
Initial range	[0-1]	Bound	Problem specific
Maximum function Evaluations	30000	Maximum function evaluation	45000
Maximum iterations	1500	Maximum iterations	1500
Others	Default	Others	Default

TABLE 2: Parameter setting/values using SQP method.

Sequential quadratic programming	
Parameters	Settings
Start point	Final weights of PS or SA
Bounds	Specific lower and upper limit of generators
X-tolerance	10^{-12}
Function tolerance	10^{-12}
Hessian	BFGS
Scaling	Objective and constraints
Maximum iterations	1000
Maximum function evaluations	1500
Others	Default

approximate and important method for transmission network losses as a function of generator power is β coefficient method based on fact that, under normal operating conditions, the transmission loss is quadratic in the injected bus real power. Equation (4b) represents the general form of the loss formula using β coefficient method. Here NB, P_iP_j, Q_iQ_j, and Z_{ij} are the number of buses, the real power, reactive power of ith and jth buses, and elements of impedance matrix, respectively. However in this study β coefficient method will be used so β_{ij} are the loss constant of coefficient β.

2.3. Valve-Point Loading Effect (VPLE). For a power plant having NG committed units have multivalve opening sequence to meet the fluctuating loads demand. VPLE introduce ripples in the input-output characteristics of generating units resultantly making the heat curve rate nonconvex with increase in unit loading. Heat rate curve decreases between the opening points of two valves as shown in Figure 1 [23].

VPLE introduces ripples in the heat rate curve in multi valve opening turbine as shown in Figure 1, and for a perfect modeling of EDP a filter in the form of sinusoidal function must be added in objective function in this paper as given below:

$$F_{ci}\left(P_{gi}\right) = A_i + B_iP_{gi} + C_iP_{gi}^2 + \left|e_i \times \sin\left(f_i \times \left(P_{gi}^{\min} - P_{gi}\right)\right)\right|, \quad (6)$$

where e_i and f_i are the coefficient of ith unit that represents valve-point effect.

2.4. Prohibited Operating Zones. According to [3] practical generating units contain forbidden operating regions also known as "prohibited operating zones" due to machinery errors such as boiler pumps motors faults which ultimately leads input-output curve nonlinear and this can be modeled as follows:

$$P_{gi} \in \begin{cases} P_{gi}^{\min} \le P_{gi} \le P_{gi1}^{NL} \\ P_{gik-1}^{U} \le P_{gi} \le P_{gik}^{NL} \\ P_{gizI}^{U} \le P_{gi} \le P_{gi}^{\max}, \end{cases} \quad (7)$$

where P_i is the number of prohibited zones in the ith generator curve, k is the index of prohibited zone, and P_{gik}^{NL} and P_{gik}^{NL} are lower and upper limit of kth prohibited zone of the ith generator.

2.5. EDP with Multiple-Fueling Option (MFO). Another significant consideration involved in designing of EDP optimal cost is unit multiple fueling. A thermal unit has multiple cost rate curves because unit is fueled by different sources and needs to operate the unit on lower curve which result in producing hybrid cost function (HCF) and HCF contains information about unit operation or type of fuel used in generating operation. Practically multiple fuel sources are supplied to the thermal units and a single thermal unit can be represented by multiple quadratic cost functions; each cost function represents the type of fuel used and each unit must

TABLE 3: Fitness evaluation values and their respective computational time for proposed schemes.

Scenarios	PS		SA		SQP		PS-SQP		SA-SQP	
	F_{val}	τ	F_{val}	τ	F_{val}	τ	F_{val}	τ	F_{val}	τ
Case 1	$2.01E-13$	0.29975	$1.25E-15$	26.9359	$5.24E-19$	0.09749	$8.95E-21$	0.22732	$8.40E-19$	0.05562
Case 2	$1.84E-11$	1.68357	$1.20E-16$	29.8246	$1.22E-11$	2.98696	$2.90E-13$	1.10707	$1.00E-19$	0.24536
Case 3	$2.01E-13$	0.29975	$1.25E-15$	26.9359	$5.24E-19$	0.09749	$8.95E-21$	0.22732	$8.40E-19$	0.05562

TABLE 4: Reliability based on 100 independent runs for fifteen units system with $P_D = 2630$ MW.

Method	Min_{cost}	$\text{Mean}_{\text{cost}}$	STD_{cost}	Max_{cost}	Mean_{CPU} time
Evolutionary strategy [19]	32568.54	32620.00	—	32710.00	0.48
PSO [20]	33858.00	33039.00	—	3331.000	26.59
Genetic algorithm [20]	33113.00	33228.00	—	3337.000	49.31
Gaussian PSO [21]	32508.12	35122.79	1918.62	38044.42	0.69
Chaotic PSO [21]	32775.68	35340.70	1942.84	38104.54	0.59
Proposed					
PS	33154.10	34908.12	1712.34	36.001.14	0.640587
SA	33124.24	35012.54	1612.35	36312.12	22.78067
SQP	32567.45	33467.14	1712.12	35123.90	0.418991
PS-SQP	32005.00	32145.23	1735.12	32568.28	1.059578
SA-SQP	32000.14	32112.12	1659.38	32412.02	23.199661

pin point the economic fuel to burn and fuel cost function formulation is as follows [23]:

$$F_{ci}\left(P_{gi}\right) = \begin{cases} A_{i1} + B_{i1}P_{gi} + C_{i1}P_{gi}^2, & \text{fuel } 1, P_{gi}^{\min} \le P_{gi} \le P_{gi1} \\ A_{i2} + B_{i2}P_{gi} + C_{i2}P_{gi}^2, & \text{fuel } 2, P_{gi} \le P_{gi} \le P_{gi2} \\ \vdots & \vdots \\ A_{ik} + B_{ik}P_{gi} + C_{ik}P_{gi}^2, & \text{fuel } k, P_{gik-1} \le P_{gi} \le P_{gi}^{\max}, \end{cases} \tag{8}$$

where A_{ik}, B_{ik}, and C_{ik} are the cost coefficient of generator ith using k fuels.

2.6. EDP Considering Both VPLE and MFO. To obtain the realistic and transparent EDP solution in this paper both MFO and VPLE model are considered. Mathematical modeling for this consideration is combination of both cost functions given in (6) and (8). Consider

$$F_{ci}\left(P_{gi}\right) = \begin{cases} A_{i1} + B_{i1}P_{gi} + C_{i1}P_{gi}^2 + \left|e_{i1} \times \sin\left(f_{i1} \times \left(P_{gi1}^{\min} - P_{gi1}\right)\right)\right|, & \text{fuel } 1, P_{gi}^{\min} \le P_{gi} \le P_{gi1} \\ A_{i2} + B_{i2}P_{gi} + C_{i2}P_{gi}^2 + \left|e_{i2} \times \sin\left(f_{i2} \times \left(P_{gi2}^{\min} - P_{gi2}\right)\right)\right|, & \text{fuel } 2, P_{gi} \le P_{gi} \le P_{gi2} \\ \vdots & \vdots \\ A_{ik} + B_{ik}P_{gi} + C_{ik}P_{gi}^2 + \left|e_{ik} \times \sin\left(f_{ik} \times \left(P_{gik}^{\min} - P_{gik}\right)\right)\right|, & \text{fuel } k, P_{gik-1} \le P_{gi} \le P_{gi}^{\max}. \end{cases} \tag{9}$$

3. Proposed Methodology

The brief introduction of optimizers used for optimization of EDP is narrated along with the necessary detail of hybrid algorithm in the form of procedural steps.

3.1. Simulated Annealing. Simulated annealing is heuristic as well as probabilistic method proposed by Scott Kirkpatrick, C. Daniel Gelatt, and Mario P. Vecchi in 1983 for localizing better approximation to find global minimum of given cost function that may be surrounded with several local minima in a large search space [24]. SA is more efficient as compared with hill climbing and is a good choice for engineers and researchers to handle complex nonlinear systems especially problems like EDP due to its coding leniency to get guarantee finest optimal solutions. The method works on the principle of randomly initializing of finite solution space of specific size N on the basis of scattered neighborhood points that mimic from the principle of thermal annealing of solids [25]. The decreasing of temperature at a specific rate in each iteration resembles the optimization of the merit function of EDP subject to varying the heat values as adaptive parameters of a specific problem. The candidate solution is updated based on the values of the fitness in terms of mean square error of the previous possible solution along with the index term; on this basis the iterative process matures towards

TABLE 5: Reliability based on 100 independent runs for twenty units system with $P_D = 2500$ MW.

Method	Min_{cost}	$\text{Mean}_{\text{cost}}$	STD_{cost}	Max_{cost}	Mean_{CPU} time (s)
Lambda iteration method [38]	62456.6391	—	—	—	0.033
Hopfield neural networks [38]	62456.6341	—	—	—	0.006
Gaussian PSO [21]	60526.15	61123.36	330.89	62757.19	0.50
Chaotic PSO [21]	60803.51	61223.95	507.56	62982.19	0.37
Proposed					
PS	59829.15	61256.21	1342.56	63125.56	1.054175
SA	59803.45	61002.34	1423.89	62340.16	32.31666
SQP	57231.43	59012.45	1712.35	60123.23	0.294808
PS-SQP	58123.98	58563.65	976.140	58901.26	1.348983
SA-SQP	57120.21	58011.78	658.341	58764.38	33.643791

TABLE 6: Data for thirteen thermal units of generating unit capacity and coefficient [37].

Units	$P_i^{\min}$ (MW)	$P_i^{\max}$ (MW)	a (\$/MW^2)	b (\$/MW^2)	c (\$/MW^2)
1	0	680	0.00028	8.10	550
2	0	360	0.00056	8.10	309
3	0	360	0.00056	8.10	307
4	60	180	0.00028	7.74	240
5	60	180	0.00324	7.74	240
6	60	180	0.00324	7.74	240
7	60	180	0.00324	7.74	240
8	60	180	0.00324	7.74	240
9	60	180	0.00324	7.74	240
10	40	120	0.00284	8.60	126
11	40	120	0.00284	8.60	126
12	55	120	0.00284	8.60	126
13	55	120	0.00284	8.60	126

the region of convergence of the fitness function. SA belongs to the family of global search techniques so it has various start points in the start of the algorithm to avoid local minimum. Usually the number of iterations of SA is very large while the computational budget is quite reasonable for most of the test bench problem. Although, it has been applied in various areas of engineering, some key applications are found in crystallographic refinement, neural computing [26], optimal reconfiguration of distribution networks, wireless communication, phase balancing, van der pol oscillators involved in mechanical engineering [27], and combinatorial optimization problem [28]. The generic flow diagram used for the simulated annealing algorithm is provided in Figure 2.

3.2. Pattern Search. Pattern search (PS) is numerical optimization method proposed by Hooke and Jeeves [29] for statistical problems and main advantages of PS are that it is efficient and computational structure is simple to understand and also it does not required gradient of cost function so it can be used in functions that are not constant or not differentiable and method is also known as direct-search, derivative-free approaches. Pattern search algorithm searches for a sequence of points to reach an optimal solution by initializing a set of points called mesh. The mesh formation procedure is involved by adding a pattern vector to the initial point and if PS acquires a point in mesh that refines the objective function then the new point becomes the current point at the next step of the algorithm [30].

Usually the practical nonlinear optimization problem of power system lies outside the reach of the standard optimization techniques. The practical EDP problems can be optimized by PS which is computationally efficient and easy to implement as compared to other heuristic techniques [31]. PS initializes by computing sequence of point that may or may not be reached optimally and established a set point called mesh around these given points. This current point could be a starting point provided by user or extracted from previous step of algorithm. Current points are then added to a pattern vector which is scalar multiple of a set of vectors for mesh formation. If an objective function is improved by point in mesh then the new point becomes the current point at the next iteration [32].

Pattern search is efficient optimization techniques applied in areas of parallel computing, pattern character recognition, graphic user interference thermal control systems, nonlinear and optimal control theory [33], and so forth.

3.3. Sequential Quadratic Programming. The SQP algorithm is one of the nonlinear programming techniques used extensively by the research community for constraints optimization problems. The strength of the scheme is well-established in terms of efficiency, accuracy, and percentage of successful solutions over a large number of test problems. Thum and Schilling [34] have provided excellently a brief introduction and description of SQP algorithms. As the physics of SQP is based on local search techniques so it is more probable to get stuck in the local minimum so some of the iterations during Monte Carlo simulation can show divergence or least accuracy. However the rate of convergence and level of accuracy for successful runs is dramatically high for SQP algorithms. The detailed history, mathematical formulation, importance, and application related with SQP methods can be seen in [35, 36]. The SQP methods have been applied on broad disciplines of engineering and applied science from

TABLE 7: (a) Data for fifteen thermal units of generating unit capacity and coefficient [19–21]. (b) Data for fifteen thermal units of generating units of ramp-rate limits and prohibited zones of generating units.

(a)

Units	P_i^{min} (MW)	P_i^{max} (MW)	a (\$/MW^2)	b (\$/MW^2)	c (\$/MW^2)
1	150	455	0.000299	10.1	671
2	150	455	0.000183	10.2	574
3	20	130	0.001126	8.8	374
4	20	130	0.001126	8.8	374
5	150	470	0.000205	10.4	461
6	135	460	0.000301	10.1	630
7	135	465	0.000364	9.8	548
8	60	300	0.000338	11.2	227
9	25	162	0.000807	11.2	173
10	25	160	0.001203	10.7	175
11	20	80	0.003586	10.2	186
12	20	80	0.005513	9.9	230
13	25	85	0.000371	13.1	225
14	15	55	0.001929	12.1	309
15	15	55	0.004447	12.4	323

(b)

Units	P_i^0 (MW)	UR$_i$ (MW/h)	DR$_i$ (MW/h)	Prohibited zones Zone 1	Zone 2	Zone 3
1	400	80	120			
2	300	80	120	[185 255]	[305 335]	[420 450]
3	105	130	130			
4	100	130	130			
5	90	80	120	[180 200]	[305 335]	[390 420]
6	400	80	120	[230 255]	[365 395]	[430 455]
7	350	80	120			
8	95	65	100			
9	105	60	100			
10	110	60	100			
11	60	80	80			
12	40	80	80	[30 40]	[55 65]	
13	30	80	80			
14	20	55	55			
15	20	55	55			

the day of their birth to today; for instance, few recently published articles addressing SQP technique are referred to for interested readers in [37, 38].

Now the intention is to formulate the overall procedure and the optimization of the unknown designed parameters of the units generation along with the fuel cost; the process of the proposed scheme is provided in Figure 3. The modeling is performed based on relations provided in (1) and relations of constraints, respectively. The optimization is performed using global versions of simulated annealing and pattern search hybrid with sequential quadratic programming, an efficient local search technique. In Figure 3(b) the designed parameters obtained from global optimizer are provided as a start point to SQP for further tuning of the design. The algorithm used for the optimization of EDP in the form of the logical steps based on hybrid computational technique is as follows.

Step 1 (population generation). A random and well scattered population of size M is generated with agents of length exactly equal to the number of power generating units N.

In general the ith agent of the candidate solution is represented by $\mathbf{c}_i = [c_{i1}, c_{i2}, \dots, c_{iN}]$, $c_{ij} \in \Re : L_b \le c_{ij} \le U_b$ $\forall i = 1, 2, 3, \dots, N$, $j = 1, 2, 3, \dots, M$, where L_b and U_b are the lower and upper possible limits to generate power.

Step 2 (initialization). Initialize and assign the general and specific values for parameter setting involved in pattern search and simulated annealing algorithm as provided in Table 1.

TABLE 8: Data for twenty thermal generating units capacity and coefficients (20 units) [19–21].

Units	P_i^{min} (MW)	P_i^{max} (MW)	a (\$/MW^2)	b (\$/MW^2)	c (\$/MW^2)
1	150	600	0.00068	18.19	1000
2	50	200	0.00071	19.26	970
3	50	200	0.00650	19.80	600
4	50	200	0.00500	19.10	700
5	50	160	0.00738	18.10	420
6	20	100	0.00612	19.26	360
7	25	125	0.00790	17.14	490
8	50	150	0.00813	18.92	660
9	50	200	0.00522	18.27	765
10	30	150	0.00573	18.92	770
11	100	300	0.00480	16.69	800
12	150	500	0.00310	16.76	970
13	40	160	0.00850	17.36	900
14	20	130	0.00511	18.70	700
15	25	185	0.00398	18.70	450
16	20	80	0.07120	14.26	370
17	30	85	0.00890	19.14	480
18	30	120	0.00713	18.92	680
19	40	120	0.00622	18.47	700
20	30	100	0.00773	19.79	850

Step 3 (fitness evaluation). Find the fitness of each agent using the fitness evaluation function ξ as given below in the form of mean square error (MSE):

$$\xi = \left(\frac{\sum_{j=1}^{N} \left(\left| C_{op}^{j} - \widehat{C}_{op}^{j} \right|^2 \right)}{N} \right), \tag{10}$$

where $\widehat{C}_{op}^{i}$ is the approximated operating cost for jth generator.

Step 4 (termination criteria). The algorithm is terminated on the basis of the following criteria.

(a) The predefined fitness value is achieved.

(b) The total number of generations is executed.

(c) Predefined value of the function tolerance is achieved.

(d) The desired function evaluations are performed.

If any of the above criteria are met then go to Step 6 otherwise simulate Step 5.

Step 5 (reproduction). New population is generated using the operations evolved in PS and SA as per the setting given in Table 1 and then go to Step 3.

Step 6 (refinement via SQP method). One of the best results obtained by PS and SA is provided to SP as a start point for further refinement, and the parameter setting and values used for SQP algorithm are provided in Table 2.

Step 7 (storage). Store the global best of this cycle and repeat the procedure from Steps 3 to 7, to get sufficient numbers of independent runs to have sufficient data for performing analysis.

Step 8 (statistical analysis). The statistical analysis is made on the basis of four fundamental performance modes like best, worst, mean, and standard deviation and skewness.

4. Simulation and Results

In this section, a scenario based simulation is performed based on three case studies composed of thirteen, fifteen, and twenty generating units, respectively, for EDP problem in which the objective functions were ramp rate limit, prohibited operating zones in the power system operation; MFO and VPLE and transmission losses are also employed to demonstrate efficiency of proposed schemes using MATLAB built in subroutines of the optimizers like PS, SA, SQP, PS-SQP, and SA-SQP, respectively, whose parameters setting and values are provided in Tables 1 and 2. The reference data sheet for thirteen thermal units [37] with load demand of 3080 MW is provided in Appendix A. The data set used for fifteen thermal units [19–21] along with generating unit capacity, coefficients, ramp rate limits, and prohibited zones is provided in Appendix B while the load demand expected to be determined is 2630 MW. However, the data sheet for twenty thermal units of generating unit capacity and coefficients [38] are provided in Appendix C, where system supplies a total load demand of 2500 MW.

One of the best designed parameters is in fact the power generating values of generators and is presented in Figure 4 at semilog scale in order to widen the small effects in the power values. The lowest cost of the power generation can be obtained by taking the generating values from the hybrid schemes based on PS-SQP and SA-SQP, respectively, from Figures 4(a), 4(b), and 4(c) in fitness evaluation function provided in (10) along with the subsequent relations given in (1) and expressions of EDP constraints. The x-axis in Figure 4 defines the number of thermal units taken in the study and y-axis demonstrates the solvers PS, SA, SQP, PS-SQP, and SA-SQP, respectively, while z-axis provides the optimized values of the power for thermal units.

The optimized final values along with computational time for PS, SA, SQP, PS-SQP, and SA-SQP are provided in Table 3. The time analysis carried out for this paper is on Intel(R) Core (TM) i5-3210M CPU @ 2.50 GHz, 2.5 GHz, 4.0 GB of RAM, and running with MATLAB version 2012a. It is quite evident from the table that level of accuracy is dominated in case of hybrid schemes as compared to global search and local search techniques separately. However the computational budget of hybrid approach is a little bit more as compared to other presented schemes but this effect can be over shed due to the high accuracy attained by hybrid schemes. The local search scheme seems to be outperformed in terms of both accuracy and computational cost; however one cannot rely on one of these best values of SQP as the local search techniques are more probable to get stuck in the local optima due to bad initial guess. Another observation is that as the number of

Table 9: Reliability based on 100 independent runs for fifteen units system with $P_D = 3080$ MW [37].

Method	Minimum cost	Mean cost	STD of cost	Maximum cost	Mean CPU time
Proposed					
PS	17524.78	17659.34	1543.72	17860.12	0.394373
SA	17598.19	17623.78	1872.83	17765.23	14.15776
SQP	17653.11	17723.14	1712.34	189023.1	0.148936
PS-SQP	17201.98	17543.76	1612.35	17678.91	0.543309
SA-SQP	17491.17	17502.12	1709.12	17233.24	14.306696

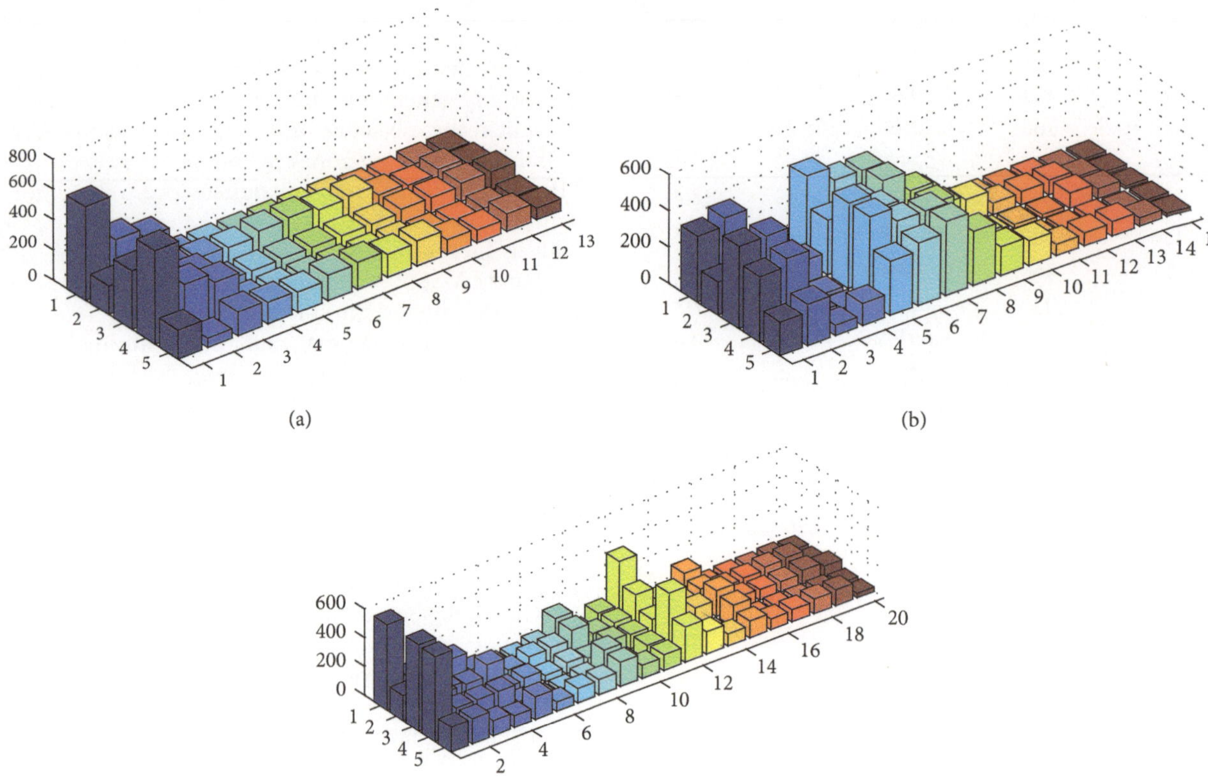

Figure 4: Analysis of the optimizers for 100 independent runs for (a) thirteen generating units, (b) fifteen generating units, and (c) twenty generating units.

generating units increases the level of accuracy reduces while computational time increases due to increase in the EDP complexity.

The reliability of the stochastic algorithms is being tested by a comprehensive statistical analysis in terms of minimum of cost (Min_{cost}), maximum value of cost (Max_{cost}), mean cost value ($\text{Mean}_{\text{cost}}$), and standard deviation of the cost (STD_{cost}). The analysis is not only performed on cost value but also performed on computational time needed for the optimization in term of mean CPU (Mean_{CPU}) time. The results of the Monte Carlo simulations for stochastic solvers are narrated in Tables 4 and 5 for fifteen and twenty units, respectively, while for thirteen units the same analysis is provided in Appendix D.

It is quite evident from Tables 4 and 5 that the results of the proposed schemes are in good agreement in terms of reliability, effectiveness, economical cost, and computational budget required in terms of time complexity; however the hybrid approaches PS-SQP and SA-SQP show more supremacy than that of PS, SA, and SQP. The computational budget of hybrid schemes seems to computationally heavy of order of just few seconds; however this aspect seems to be minimal at the cost of level of accuracy achieved.

As the proposed scheme is based on heuristic computation one of the best results cannot define the effectiveness and reliability of the schemes; therefore in order to see the effectiveness of given stochastic methods the results are obtained for hundred independent runs rather on a single successful run. The fitness given in (10) obtained on all independent iterations is plotted in Figure 5 for three variations of generating units.

The percentage of convergence is also calculated by fixing a certain level of the threshold ($\xi \leq 10^{-10}$) in the level of fitness obtained by all proposed algorithms and the results

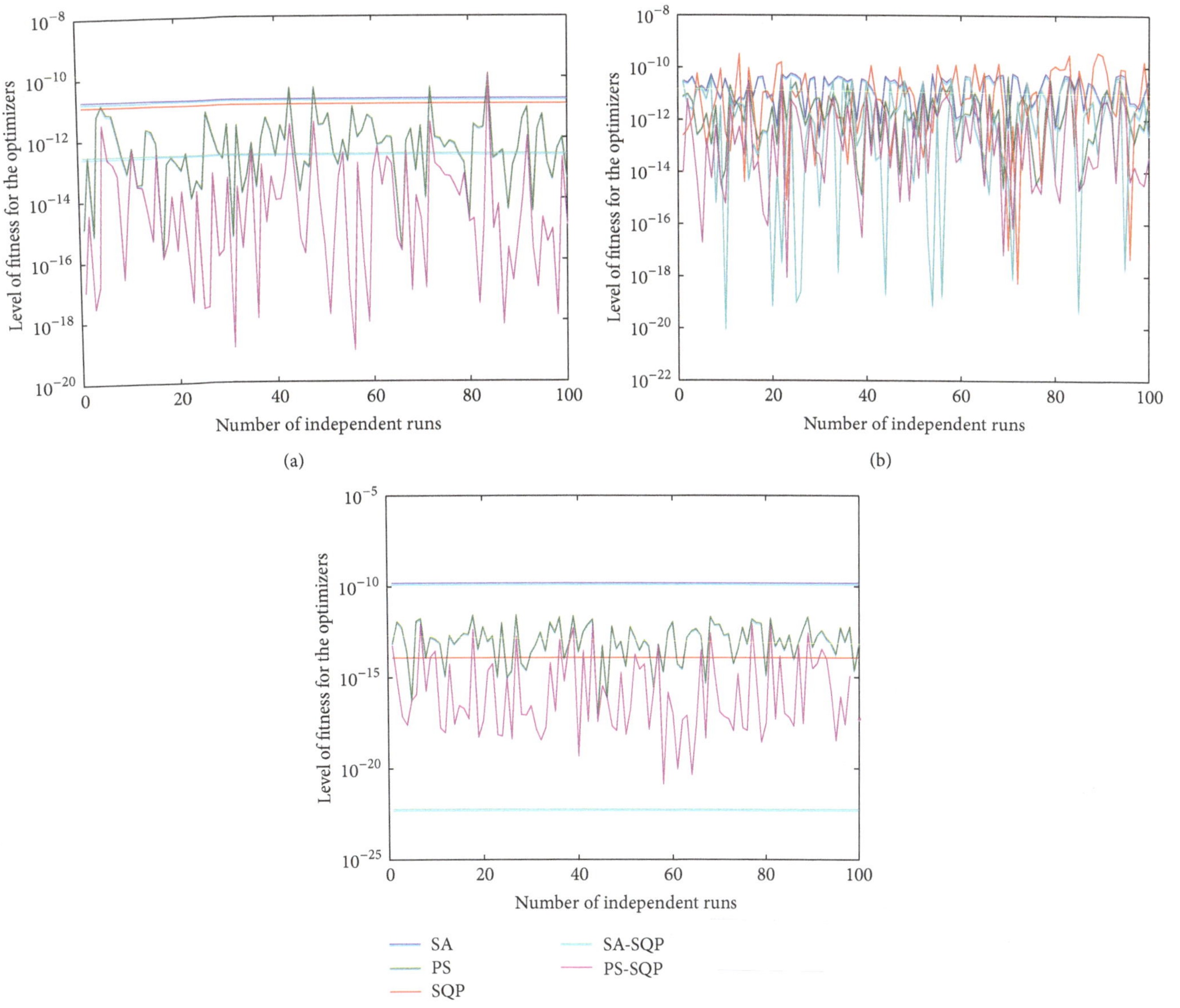

FIGURE 5: Process of proposed solution (a) overall procedure and (b) optimization of fitness function using hybrid computation.

are plotted in Figure 6 at four hundred independent runs of the given algorithms. It is observed from Figure 6(a) that all the optimizers provide 100% convergence for EDP with thirteen thermal units while for the case of fifteen units the convergence for SA is 97% and 100% for rest of the algorithms. Scenario three that is relatively stiff does not show convergence for PS on a threshold of ($\xi \leq 10^{-10}$), while the other optimizers like SA, SQP, PS-SQP, and SA-SQP are 100% convergent as evident from Figure 6(c).

5. Conclusions

On the basis of the results obtained by various optimizers in the last sections, the following conclusion can be drawn.

The proposed scheme based on SA-SQP and PS-SQP hybrid algorithms can provide reliable, effective, and well optimized cost of EDP for the scenarios of thirteen, fifteen, and twenty thermal units for a demand of 3080 M, 2630 MW, and 2500 MW, respectively.

The comparative study between the solvers is carried out in terms of absolute cost, computational complexity, and fitness value achieved by the PS, SA, SQP, PS-SP, and SA-SP algorithms. The hybrid algorithms are found to be better than that of global and local search techniques applied independently for all variants of ESPs.

The results of statistical analysis determined for 100 independent runs of the design approaches show that the values in terms of mean, standard deviation, and minimum of the total cost error associated with PS-SQP and SA-SQP are considerably more dominated than SA, PS, and SQP techniques. The value of the mean execution time for finding the optimal weight with proposed schemes shows

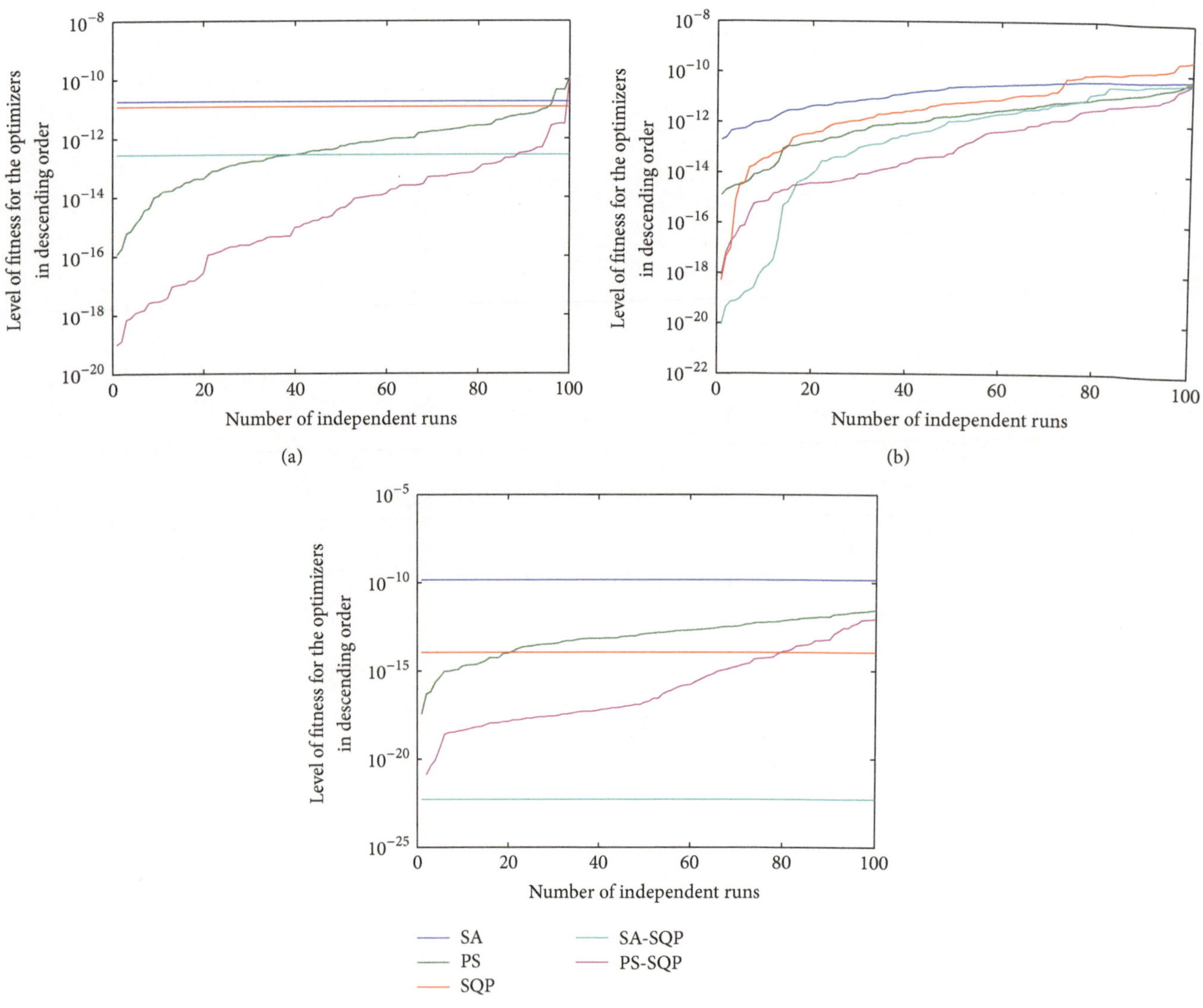

Figure 6: Process of proposed solution (a) overall procedure and (b) optimization of fitness function using hybrid computation.

that the hybrid approaches take relatively more time than the other three techniques but this exception can be overshadow due to its utmost accuracy and consistently convergent performance.

At a fitness threshold of $\varepsilon \leq 10^{-10}$, the algorithms based on hybrid schemes on average provide 100% convergence while the convergences of SA and PS are observed to be in a good range of acceptance, although in same situations PS remains consistent due to the same creation of mesh in the polling and searching domain.

The main advantages of given schemes lie in their ease in implementation, simplicity in the concept, minimal computational complexity, less hardware implementation burden, and suitability for real-time applications. In future, one may look for biological inspired computational method to analyze synthesis and optimize the other dynamical constraint based EDPs problems.

Appendices

A. Generating Units Capacity and Coefficients (13 Units)

See Table 6.

B. (a) Generating Units Capacity and Coefficients (15 Units) (b) Ramp-Rate Limits and Prohibited Zones (15 Units)

See Table 7.

C. Generating Units Capacity and Coefficients (20 Units)

See Table 8.

TABLE 10: Optimal solution of the thirteen, fifteen, and twenty thermal units, respectively, using solvers involving PS, SA, SQP, PS-SQP, and SA-SQP.

Solver	$P1$	$P2$	$P3$	$P4$	$P5$	$P6$	$P7$	$P8$	$P9$	$P10$	$P11$	$P12$	$P13$	$P14$	$P15$	$P16$	$P17$	$P18$	$P19$	$P20$
PS	616.84	320.36	256.93	60.00	60.00	60.00	60.00	60.00	60.00	40.00	40.00	55.00	55.00							
SA	167.57	124.92	102.42	126.36	177.93	134.57	114.18	176.51	159.49	103.92	106.64	105.45	116.08							
SQP	389.86	213.26	213.10	77.79	77.79	77.79	77.79	77.79	77.79	101.77	101.77	120.00	120.00							
PS-SQP	615.89	319.52	256.55	60.20	60.20	60.20	60.20	60.20	60.20	40.26	40.26	55.22	55.22							
SA-SQP	53.56	155.96	83.33	117.24	152.98	178.22	179.87	143.53	179.96	111.76	117.59	116.14	117.38							
PS	362.30	406.00	49.38	20.00	470.00	391.00	391.00	188.00	153.00	25.00	20.00	20.00	25.00	15.00	15.00					
SA	288.19	282.66	103.84	118.81	289.05	298.57	325.19	232.56	149.19	139.45	66.08	75.58	79.88	40.72	44.98					
SQP	455.00	455.00	20.00	20.00	470.00	383.22	135.00	300.00	54.64	25.00	20.00	20.00	85.00	55.00	15.00					
PS-SQP	362.30	406.00	49.38	20.00	470.00	391.00	391.00	188.00	153.00	25.00	20.00	20.00	25.00	15.00	15.00					
SA-SQP	348.05	206.00	129.58	129.29	286.20	214.17	387.15	199.47	155.91	150.20	79.58	63.07	83.53	53.47	51.26					
PS	592.55	178.00	178.00	178.00	50.00	20.00	25.00	50.00	178.00	30.00	100.00	406.00	40.00	20.00	153.00	21.00	30.00	30.00	40.00	30.00
SA	198.51	57.90	79.62	126.63	100.81	66.24	93.72	131.98	199.39	131.02	174.43	260.56	99.35	71.17	90.09	71.07	82.59	101.27	107.38	96.09
SQP	600.00	199.98	56.60	50.12	65.91	82.33	43.48	74.14	50.05	111.65	100.00	150.00	61.65	115.64	179.77	79.99	73.71	75.72	99.82	72.93
PS-SQP	592.39	177.98	177.97	177.97	50.02	20.04	25.03	50.03	177.98	30.03	100.01	405.90	40.03	20.03	152.99	21.03	30.03	30.03	40.03	30.03
SA-SQP	200.92	98.28	181.61	85.93	137.73	76.53	76.42	113.69	94.98	88.16	243.88	216.69	57.41	125.65	81.96	78.04	56.07	111.10	107.88	97.10

D. Reliability Comparison Based upon 100 Independent Runs for 15 Units with $P_D = 3080$ MW

See Table 9.

E. Optimal Solution for 13, 15, and 20 Units Using Solvers Involving PS, SA, SQP, PS-SQP, and SA-SQP

See Table 10.

Conflict of Interests

The authors declare that there is no conflict of interests regarding the publication of this paper.

References

[1] P.-H. Chen and H.-C. Chang, "Large-scale economic dispatch by genetic algorithm," *IEEE Transactions on Power Systems*, vol. 10, no. 4, pp. 1919–1926, 1995.

[2] R. K. Swaina, N. C. Sahub, and P. K. Hotac, "Gravitational search algorithm for optimal economic dispatch," *Procedia Technology*, vol. 6, pp. 411–419, 2012, Proceedings of the 2nd International Conference on Communication, Computing & Security (ICCCS '12).

[3] R. Poli, "Analysis of the publications on the applications of particle swarm optimisation," *Journal of Artificial Evolution and Applications*, vol. 2008, Article ID 685175, 10 pages, 2008.

[4] Z.-L. Gaing, "Particle swarm optimization to solving the economic dispatch considering the generator constraints," *IEEE Transactions on Power Systems*, vol. 18, no. 3, pp. 1187–1195, 2003.

[5] D. He, F. Wang, and Z. Mao, "A hybrid genetic algorithm approach based on differential evolution for economic dispatch with valve-point effect," *International Journal of Electrical Power & Energy Systems*, vol. 30, no. 1, pp. 31–38, 2008.

[6] B. Naama, H. Bouzeboudja, and A. Allali, "Solving the economic dispatch problem by using Tabu Search algorithm," in *Proceedings of the TerraGreen International Conference on Advancements in Renewable Energy and Clean Environment*, pp. 694–701, February 2013.

[7] M. A. Abido, "Multiobjective evolutionary algorithms for electric power dispatch problem," *IEEE Transactions on Evolutionary Computation*, vol. 10, no. 3, pp. 315–329, 2006.

[8] F. S. Wen and C. S. Chang, "Probabilistic approach for fault-section estimation in power systems based on a refined genetic algorithm," *IEE Proceedings—Generation, Transmission and Distribution*, vol. 144, no. 2, pp. 160–168, 1997.

[9] L. dos Santos Coelho and V. C. Mariani, "Particle swarm approach based on quantum mechanics and harmonic oscillator potential well for economic load dispatch with valve-point effects," *Energy Conversion and Management*, vol. 49, no. 11, pp. 3080–3085, 2008.

[10] J. S. Al-Sumait, J. K. Sykulski, and A. K. Al-Othman, "Solution of different types of economic load dispatch problems using a pattern search method," *Electric Power Components and Systems*, vol. 36, no. 3, pp. 250–265, 2008.

[11] L. D. S. Coelho and V. C. Mariani, "An improved harmony search algorithm for power economic load dispatch," *Energy Conversion and Management*, vol. 50, no. 10, pp. 2522–2526, 2009.

[12] T. Yalcinoz, H. Altun, and M. Uzam, "Economic dispatch solution using a genetic algorithm based on arithmetic crossover," in *Proceedings of the IEEE Porto Power Tech Conference (PPT '01)*, Porto, Portugal, September 2001.

[13] W.-M. Lin, F.-S. Cheng, and M.-T. Tsay, "An improved tabu search for economic dispatch with multiple minima," *IEEE Transactions on Power Systems*, vol. 17, no. 1, pp. 108–112, 2002.

[14] D. T. Phan and A. Koc, "Optimization approaches to security-constrained unit commitment and economic dispatch with uncertainty analysis," in *Optimization and Security Challenges in Smart Power Grids*, Energy Systems, pp. 1–37, Springer, Berlin, Germany, 2013.

[15] Ü. B. Filik and M. Kurban, "Solving unit commitment problem using modified subgradient method combined with simulated annealing algorithm," *Mathematical Problems in Engineering*, vol. 2010, Article ID 295645, 15 pages, 2010.

[16] A. R. Balbo, M. A. da Silva Souza, E. C. Baptista, and L. Nepomuceno, "Predictor-corrector primal-dual interior point method for solving economic dispatch problems: a postoptimization analysis," *Mathematical Problems in Engineering*, vol. 2012, Article ID 376546, 26 pages, 2012.

[17] J. Li, D. Niu, and Y. Wu, "A parallel adaptive particle swarm optimization algorithm for economic/environmental power dispatch," *Mathematical Problems in Engineering*, vol. 2012, Article ID 271831, 14 pages, 2012.

[18] D. Barbagallo, E. di Nitto, D. J. Dubois, and R. Mirandola, "A bio-inspired algorithm for energy optimization in a self-organizing data center," in *Self-Organizing Architectures*, vol. 6090 of *Lecture Notes in Computer Science*, pp. 127–151, 2010.

[19] A. Pereira-Neto, C. Unsihuay, and O. R. Saavedra, "Efficient evolutionary strategy optimization procedure to solve the non-convex economic dispatch problem with generator constraints," *IEEE Proceeding—Generation, Transmission and Distribution*, vol. 152, no. 5, pp. 653–660, 2005.

[20] T. A. A. Victoire and A. E. Jeyakumar, "Discussion of "particle swarm optimization to solving the economic dispatch considering the generator constraints"," *IEEE Transactions on Power Systems*, vol. 19, no. 4, pp. 2121–2122, 2004.

[21] L. D. S. Coelho and C.-S. Lee, "Solving economic load dispatch problems in power systems using chaotic and Gaussian particle swarm optimization approaches," *International Journal of Electrical Power and Energy Systems*, vol. 30, no. 5, pp. 297–307, 2008.

[22] D. P. Kothari and J. S. Dhilon, *Power System Optimization*, PHI Learning, New Delhi, India, 2010.

[23] L. D. S. Coelho, R. C. Souza, and V. C. Mariani, "Improved differential evolution approach based on cultural algorithm and diversity measure applied to solve economic load dispatch problems," *Mathematics and Computers in Simulation*, vol. 79, no. 10, pp. 3136–3147, 2009.

[24] E. H. Aarts, J. H. Korst, and P. J. van Laarhoven, "A quantitative analysis of the simulated annealing algorithm: a case study for the traveling salesman problem," *Journal of Statistical Physics*, vol. 50, no. 1-2, pp. 187–206, 1988.

[25] S. Kirkpatrick, C. D. Gelatt, and M. P. Vecchi, "Optimization by simulated annealing," *Science*, vol. 220, no. 4598, pp. 671–680, 1983.

[26] V. Černý, "Thermodynamical approach to the traveling salesman problem: an efficient simulation algorithm," *Journal of Optimization Theory and Applications*, vol. 45, no. 1, pp. 41–51, 1985.

[27] J. A. Khan, M. A. Z. Raja, and I. M. Qureshi, "Novel approach for a van der Pol oscillator in the continuous time domain," *Chinese Physics Letters*, vol. 28, no. 11, Article ID 110205, 2011.

[28] H.-D. Chiang, J.-C. Wang, O. Cockings, and H.-D. Shin, "Optimal capacitor placements in distribution systems. I: a new formulation and the overall problem," *IEEE Transactions on Power Delivery*, vol. 5, no. 2, pp. 634–642, 1990.

[29] R. Hooke and R. Jeeves, "Direct search solution of numerical and statistical problems," *Journal of Association for Computing Machinery*, vol. 8, pp. 212–229, 1961.

[30] E. D. Dolan, R. M. Lewis, and V. Torczon, "On the local convergence of pattern search," *SIAM Journal on Optimization*, vol. 14, no. 2, pp. 567–583, 2003.

[31] M. A. Raja, J. A. Khan, and I. M. Qureshi, "A new stochastic approach for solution of Riccati differential equation of fractional order," *Annals of Mathematics and Artificial Intelligence*, vol. 60, no. 3-4, pp. 229–250, 2010.

[32] T. R. Ramanan, M. Iqbal, and K. Umarali, "A particle swarm optimization approach for permutation flow shop scheduling problem," *International Journal for Simulation and Multidisciplinary Design Optimization*, vol. 5, article A20, 6 pages, 2014.

[33] A. M. Elaiw, A. M. Shehata, and M. A. Alghamdi, "A model predictive control approach to combined heat and power dynamic economic dispatch problem," *Arabian Journal for Science and Engineering*, vol. 39, no. 10, pp. 7117–7125, 2014.

[34] S. Thum and R. Schilling, "Optimization of hydraulic machinery bladings by multilevel CFD techniques," *International Journal of Rotating Machinery*, vol. 2, pp. 161–167, 2005.

[35] Z. A. Vale, C. Ramos, P. Faria et al., "Comparison between deterministic and meta-heuristic methods applied to ancillary services dispatch," in *Trends in Applied Intelligent Systems*, vol. 6096 of *Lecture Notes in Computer Science*, pp. 731–741, Springer, Berlin, Germany, 2010.

[36] A. Ş. Uyar, B. Türkay, and A. Keleş, "A novel differential evolution application to short-term electrical power generation scheduling," *International Journal of Electrical Power and Energy Systems*, vol. 33, no. 6, pp. 1236–1242, 2011.

[37] K. P. Wong and Y. W. Wong, "Genetic and genetic/simulated-annealing approaches to economic dispatch," *IEE Proceedings: Generation, Transmission and Distribution*, vol. 141, no. 5, pp. 507–513, 1994.

[38] C.-T. Su and C.-T. Lin, "New approach with a Hopfield modeling framework to economic dispatch," *IEEE Transactions on Power Systems*, vol. 15, no. 2, pp. 541–545, 2000.

Permissions

All chapters in this book were first published in AEE, by Hindawi Publishing Corporation; hereby published with permission under the Creative Commons Attribution License or equivalent. Every chapter published in this book has been scrutinized by our experts. Their significance has been extensively debated. The topics covered herein carry significant findings which will fuel the growth of the discipline. They may even be implemented as practical applications or may be referred to as a beginning point for another development.

The contributors of this book come from diverse backgrounds, making this book a truly international effort. This book will bring forth new frontiers with its revolutionizing research information and detailed analysis of the nascent developments around the world.

We would like to thank all the contributing authors for lending their expertise to make the book truly unique. They have played a crucial role in the development of this book. Without their invaluable contributions this book wouldn't have been possible. They have made vital efforts to compile up to date information on the varied aspects of this subject to make this book a valuable addition to the collection of many professionals and students.

This book was conceptualized with the vision of imparting up-to-date information and advanced data in this field. To ensure the same, a matchless editorial board was set up. Every individual on the board went through rigorous rounds of assessment to prove their worth. After which they invested a large part of their time researching and compiling the most relevant data for our readers.

The editorial board has been involved in producing this book since its inception. They have spent rigorous hours researching and exploring the diverse topics which have resulted in the successful publishing of this book. They have passed on their knowledge of decades through this book. To expedite this challenging task, the publisher supported the team at every step. A small team of assistant editors was also appointed to further simplify the editing procedure and attain best results for the readers.

Apart from the editorial board, the designing team has also invested a significant amount of their time in understanding the subject and creating the most relevant covers. They scrutinized every image to scout for the most suitable representation of the subject and create an appropriate cover for the book.

The publishing team has been an ardent support to the editorial, designing and production team. Their endless efforts to recruit the best for this project, has resulted in the accomplishment of this book. They are a veteran in the field of academics and their pool of knowledge is as vast as their experience in printing. Their expertise and guidance has proved useful at every step. Their uncompromising quality standards have made this book an exceptional effort. Their encouragement from time to time has been an inspiration for everyone.

The publisher and the editorial board hope that this book will prove to be a valuable piece of knowledge for researchers, students, practitioners and scholars across the globe.

List of Contributors

Manel Allous
Department of Electrical Engineering, LACS Laboratory, National Engineering School of Tunis,Université de Tunis El Manar, P.O. Box 37, Le Belvedere, 1002 Tunis, Tunisia

Nadia Zanzouri
Department of Electrical Engineering, LACS Laboratory, National Engineering School of Tunis,Université de Tunis El Manar, P.O. Box 37, Le Belvedere, 1002 Tunis, Tunisia
Preparatory Engineering Institute of Tunis 2, Université de Tunis, Rue Jawaher Lel Nahrou-Monfleury, 1089 Tunis, Tunisia

Mehdi Teimouri
Faculty of New Sciences and Technologies, University of Tehran, Tehran 14399, Iran

Ahmadreza Hedayat
Newracom, Irvine, CA 92618, USA

Vladimir Filaretov, Konstantin Gorshkov, and Sergey Kurganov
Department of Electrical Engineering, Ulyanovsk State Technical University, 32 Severny Venets Street, Ulyanovsk 432027, Russia

Leonid P. Yaroslavsky
Department of Physical Electronics, School of Electrical Engineering, Tel Aviv University, 69978 Tel Aviv, Israel

Arman Sharifi
Department of Electrical Engineering, Islamic Azad University, Kermanshah Branch, Kermanshah 6718997551, Iran

Ignazio Bertuglia, Giuseppe Consentino and Michele Laudani
Power Transistors Division, STMicroelectronics, Catania, Italy

Abdelghani Dendouga and Slimane Oussalah
Division Microelectronique et Nanotechnologie, Centre de Développement des Technologies Avancées (CDTA), Citéé 20 Août 1956, 16000 Algiers, Algeria

Damien Thienpont
OMEGA Micro, Microelectronics Design Center, Polytechnic School, LLR Aile 4, 91128 Palaiseau Cedex, France

Abdenour Lounis
Laboratoire de l'Accélérateur Linéaire (LAL), Centre Scientifique d'Orsay, Université de Paris Sud XI, Bat 200, BP 34,91898 Orsay Cedex, France

Luke Jebaraj and Kumar Sriram
Department of Electrical and Electronics Engineering, St.Anne's College of Engineering and Technology, Panruti, Tamil Nadu 607 110, India

Charles Christober Asir Rajan
Department of Electrical and Electronics Engineering, Pondicherry Engineering College, Puducherry 605014, India

Behrouz Farhang-Boroujeny
ECE Department, University of Utah, Salt Lake City, UT 84112, USA

Hongze Li, Liuyang Cui and Sen Guo
School of Economics and Management, North China Electric Power University, Beijing 102206, China

Vinay Kumar Jadoun, Nikhil Gupta, K. R. Niazi and Anil Swarnkar
Malaviya National Institute of Technology, Jaipur 302017, India

Uthman Baroudi
Computer Engineering Department, King Fahd University of Petroleum and Minerals, Dhahran 31261, Saudi Arabia

Amin-ud-din Qureshi
Systems Engineering, King Fahd University of Petroleum and Minerals, Dhahran 31261, Saudi Arabia

Samir Mekid
Mechanical Engineering Department, King Fahd University of Petroleum and Minerals, Dhahran 31261, Saudi Arabia

Jitendra Mohan and Bhartendu Chaturvedi
Department of Electronics and Communication Engineering, Jaypee Institute of Information Technology, Noida 201304, India

Sudhanshu Maheshwari
Department of Electronics Engineering, Z. H. College of Engineering and Technology, Aligarh Muslim University, Aligarh 202002, India

Jakub Osuský and Stanislav Števo
Institute of Control and Industrial Informatics, Faculty of Electrical Engineering and Information Technology
Slovak University of Technology, Ilkovičova 3, 812 19 Bratislava, Slovakia

Nagendra Singh
Department of Electrical Engineering, Mewar University, Chittorgarh, Rajasthan, India

Yogendra Kumar
Department of Electrical Engineering, Maulana Azad National Institute of Technology, Bhopal, India

Andreas F.Molisch, Giuseppe Caire, Dilip Bethanabhotla and Mingyue Ji
Department of Electrical Engineering, University of Southern California, Los Angeles, CA 90089, USA

David Ott
Intel Corporate Research, 5000WChandler Boulevard, Chandler, AZ 85226, USA

Jeffrey R. Foerster
Intel Corporate Research, 5000WChandler Boulevard, Chandler, AZ 85226, USA
Intel Corporate Research, 2111 NE 25th Avenue, Hillsboro, OR 97124, USA

Deepak Elamalayil Soman, Jelena Loncarski, Lisa Gerdin, Petter Eklund, Sandra Eriksson and Mats Leijon
Department of Engineering Sciences, Ångström Laboratory, Uppsala University, Lägerhyddsvägen 1, P.O. Box 534, 751 21 Uppsala, Sweden

F. S. Al-kamali, Hefdhallah Sakran and N. A. Odhah
Department of Electronics, Faculty of Engineering and Architecture, Ibb University, Ibb, Yemen

Arun Kumar Datta
Central Power Research Institute, Bhopal 462023, India

Manisha Dubey and Shailendra Jain
Maulana Azad National Intitute of Technology, Bhopal 462051, India

Shagufta Khan and Suman Bhowmick
Department of Electrical Engineering, Delhi Technological University, Delhi 110042, India

Dipankar Saha, Aanan Chatterjee, Sayan Chatterjee and C. K. Sarkar
Department of Electronics & Telecommunication Engineering, Jadavpur University, Kolkata, India

Hui-MinWang and Sheng-Chieh Huang
The Department of Electrical Engineering, National Chiao Tung University, Room 720, Engineering Building 5, 1001 University Road, Hsinchu 30010, Taiwan

Sheng-Chieh Huang, Hui-MinWang and Shi-Han Luo
Department of Electrical Engineering, National Chiao Tung University, Room 720, Engineering Building 5,1001 University Road, Hsinchu 30010, Taiwan

Ijaz Ahmed
Faculty of Electrical Engineering, Hamdard University, Islamabad Campus, Islamabad 44000, Pakistan
Department of Electrical Engineering, Bahria University, Islamabad Campus, Islamabad 44000, Pakistan

Ali Rauf Rao, Alaf Shah, Engr Alamzeb and Junaid Ali Khan
Faculty of Electrical Engineering, Hamdard University, Islamabad Campus, Islamabad 44000, Pakistan